Memristive Computing

Yongbin Yu · Xiangxiang Wang · Xiao Feng ·
Jiarun Shen · Nyima Tashi · Pinaki Mazumder

Memristive Computing

Science Press
Beijing

Springer

Yongbin Yu
School of Information and Software Engineering
University of Electronic Science and Technology of China
Chengdu, Sichuan, China

Xiangxiang Wang
School of Information and Software Engineering
University of Electronic Science and Technology of China
Chengdu, Sichuan, China

Xiao Feng
School of Information and Software Engineering
University of Electronic Science and Technology of China
Chengdu, Sichuan, China

Jiarun Shen
School of Information and Software Engineering
University of Electronic Science and Technology of China
Chengdu, Sichuan, China

Nyima Tashi
School of Information Science and Technology
Tibet University
Lhasa, Xizang, China

Pinaki Mazumder
Department of Electrical Engineering and Computer Science
University of Michigan–Ann Arbor
Ann Arbor, MI, USA

ISBN 978-981-96-0331-2 ISBN 978-981-96-0332-9 (eBook)
https://doi.org/10.1007/978-981-96-0332-9

Jointly published with Science Press
The print edition is not for sale in China mainland. Customers from China mainland please order the print book from: Science Press.

This Springer imprint is published by the registered company Springer Nature Singapore Pte Ltd.
The registered company address is: 152 Beach Road, #21-01/04 Gateway East, Singapore 189721, Singapore

Foreword

The memristor, as the fourth fundamental circuit element, has emerged as a cornerstone of next-generation computing since its theoretical inception and subsequent physical realization. Its intrinsic ability to retain historical states while enabling energy-efficient, brain-inspired computations positions it as a transformative force in nonlinear circuits and systems, secure communication, signal processing, and artificial intelligence (AI). *Memristive Computing* masterfully navigates the expansive landscape of memristor-centered in-memory computing research—from foundational principles to cutting-edge applications—offering readers a comprehensive synthesis of theory, design, and innovation. This monograph arrives at a pivotal moment, as the global scientific community strives to transcend the limitations of traditional computing paradigms and embrace memristor-based neuromorphic architectures that emulate the efficiency and adaptability of biological systems.

The authors' 17-year systematic exploration, beginning with the memristor's fundamental characteristics and emulators, provides essential groundwork for researchers. Subsequent chapters delve into memristive family, chaotic circuits, en/decryption systems, and adaptive filters, showcasing the memristor's versatility in addressing real-world challenges such as secure data transmission and dynamic signal processing. Particularly compelling are the discussions on memristor-based neural networks and swarm intelligence, which illuminate pathways toward scalable, energy-efficient AI systems capable of rivaling biological cognition. The rigorous dynamic analysis studies memristive neural networks through mathematical models to understand stability and learning processes, which are essential for reliable AI systems. Memristor-based swarm intelligence conceptualizes memristor networks as agents in a swarm, where collective behavior emerges from individual interactions. This could optimize distributed computing or solve complex problems efficiently.

What distinguishes this work is its interdisciplinary vision. By integrating insights from electronic engineering, computer science, software engineering, mathematical science, and neuroscience, the authors foster a holistic understanding of memristive computing—a critical step toward cultivating the next generation of scientists and engineers equipped with cross-disciplinary computational thinking based on the fourth fundamental circuit element. This aligns seamlessly with China's strategic priorities in advancing AI and neuromorphic technologies, as outlined in national initiatives such as the “New Generation Artificial Intelligence Development Plan”.

To graduate students and seasoned professionals alike, this monograph serves as both a deep reference and a catalyst for innovation. It not only demystifies complex computational intelligence phenomena but also inspires novel applications in sustainable computing, large language models, full self-driving, and robotics. The authors' academic rigor, coupled with their forward-looking perspectives, ensures that this monograph will remain relevant as memristive technology evolves.

February 2025

You He
Academician
Chinese Academy of Engineering
Beijing, China

Preface

From May 25 to 27, 2008, as an organizer, I participated in the 6th IEEE International Conference on Communications, Circuits, and Systems (ICCCAS) held in Xiamen, China. During the opening ceremony, we invited Prof. Leon Chua from the Department of Electrical Engineering and Computer Sciences at the University of California, Berkeley, to deliver a keynote speech on the "Memristor". This marked the first time I had the privilege of listening to the "Father of the Memristor" explain his groundbreaking theory in person. Earlier that year, in April 2008, Hewlett-Packard (HP) Lab announced that they had experimentally confirmed the existence of the memristor and successfully fabricated a functional physical model. Their findings were swiftly published as a brief communication in the May 1, 2008, issue of Nature. Notably, Prof. Leon Chua had first theorized the memristor in 1971, proposing it as the fourth fundamental circuit element—joining the resistor, capacitor, and inductor—to complete the framework of electronic circuit theory. In June 2008, shortly after earning my Ph.D., I joined the School of Computer Science and Engineering at the University of Electronic Science and Technology of China (UESTC), officially embarking on my journey into memristor-based research. This series of events—spanning the conference, HP's experimental breakthrough, and Chua's seminal lecture—profoundly shaped my academic trajectory and deepened my commitment to advancing memristive computing.

A pivotal moment occurred on April 19, 2019, when I attended a lecture by Prof. Leon Chua titled "Demystifying the Five Enigmas of Nonvolatile Memristors" and presented our team's progress in memristor research during a 3-hour in-person discussion. Later that year, from November 11 to 14, 2019, I participated in the Chua Memristor Research Institute International Conference: Theory, Devices, and Applications, where I engaged directly with Prof. Chua to address critical challenges in memristor studies. With the invaluable mentorship of Prof. Leon Chua—the "Father of the Memristor"—our team has achieved significant breakthroughs in memristor research. Throughout the publication process of this monograph, we received extensive and invaluable guidance via email from the father of the memristor. His profound insight, meticulous critique, and visionary perspective significantly shaped the theoretical framework and technical rigor of this work. While confidentiality precludes explicit attribution, we express our deepest gratitude for his selfless intellectual generosity and unwavering commitment to advancing memristive computing.

From 2013 to 2014, supported by the China Scholarship Council, I conducted research on memristor circuits as a visiting scholar at the University of Michigan, Ann Arbor, under the guidance of Prof. Pinaki Mazumder, who is listed as a coauthor on the cover of the monograph. Our collaborative work culminated in contributions to the monograph of Neuromorphic Circuits for Nanoscale Devices (River Publishers, 2015). Later, during my 2016–2017 visit to the University of California, Santa Barbara, I focused on memristor-based neural networks, laying the groundwork for ultra-low-power computing architectures. The two research exchanges in the United States played a pivotal role in advancing the global perspective and technical depth of our memristor studies, fostering cross-border collaborations that significantly accelerated innovation. Moreover, international collaborations transformed our research from localized explorations into globally recognized contributions, ensuring that memristive computing remains at the forefront of neuromorphic engineering.

During the period of 2008–2025, my research group had at first focused on charge/flux-controlled memristor circuit and its chaotic circuit for text/image en/decryption, and then extensively worked on filter circuits for signal processing by using memristive family. Subsequently, we developed memristive agent and in-memory computing of swarm intelligence including dominant ant colony optimization, fish swarm algorithm, and genetic algorithm. Furthermore, we discussed the dynamic analysis of memristive neural network. Finally, we developed memristor-based neural network such as cellular neural network, convolutional neural network, VGG, and LSTM by using memristive neurons. Throughout the course of this research, we extend our profound gratitude to Academician Nyima Tashi listed as a coauthor on the cover of the monograph for his invaluable guidance and mentorship. All the works were supported by the following projects:

1. UESTC Projects: Modeling and Dynamical Properties of Memristor Circuits (2010), and Research on Novel Memristive Transistors (2014).
2. National Key Laboratory Fund: Simulation and Fabrication of Memristors (2013).
3. National Natural Science Foundation of China (NSFC): Memristor-Based Neural Networks for Ultra-Low-Power Circuits (2016), Memristor Network-Based Biocomputing and Swarm Intelligence Optimization (2022), Research on Spatiotemporal Dynamic Heterogeneous Modeling and Impulsive Control Synchronization of Complex Memristive Neural Networks (2024).
4. Sichuan Provincial Key R&D Program in AI: Dynamics of Memristive Multilayer Neural Networks and Applications in Pigmentary Skin Diseases (2019), Swarm Intelligence-Optimized Memristive Neural Networks (2023).
5. National Science and Technology Major Project on New-Generation AI: Research and Application of Automatic Recognition Technology for Tibetan Language and Characters (2022).

Each chapter of this monograph features critical "Comments" that serve dual purposes:

(1) **Retrospective Synthesis**: They systematically consolidate memristor research subfields, key challenges, methodological approaches, and technological advancements, while addressing unresolved controversies and synthesizing foundational theories.
(2) **Prospective Guidance**: They offer forward-looking perspectives to inspire future investigations, identifying emerging challenges and predicting breakthrough opportunities (e.g., memristor as an integrated storage and computing device, in-memory computing).

These comments are designed to provide novel perspectives tailored to diverse audiences, as detailed below:

(1) Researchers can gain clarity on interdisciplinary connections and theoretical gaps, and navigate the evolving technology through curated "research roadmaps".
(2) Engineers can explore the societal and industrial implications of neuromorphic technologies, and uncover actionable strategies for integrating memristors into real-world applications (e.g., edge computing, edge AI).

By bridging historical rigor and futuristic vision, these comments transform the monograph into a dynamic knowledge ecosystem, empowering reader and stakeholder to advance memristor science with both depth and foresight.

I extend the deepest gratitude to all members of our research team whose dedication, creativity, and collaborative spirit have been the backbone of this work. In my laboratory, the contributors are Wenshu Zhang, Yongtuo Li, Lefei Men, Qingqing Hu, Qian Tang, Quanxin Deng, Biao Li, Bo Dang, Chen Zhou, Chenyu Yang, Yancheng Wang, Daijin Yang, Fei Lei, Huan Zhang, Huihui Ma, Jiehong Mo, Man Cheng, Yuanjingyang Zhong, Yifan Li, Ekong Favour, Francisco Fernández Díez, Kwabena Adu, Mingxiao Wang, Shouyi Liang, Xiang Lu, Nijing Yang, Zhengfei Kang, Rongquan Zhang, and Ju Jin. Equally indispensable is the wisdom and mentorship provided by our esteemed professors. Specifically, I would like to thank Prof. Shouming Zhong, Prof. Jingye Cai, Prof. Lin Wang, Prof. Liyong Ren, Prof. Xiaorong Pu, Prof. Xuliang Zhang, and Prof. Yuanzhe Yao of UESTC, Prof. Xiaofeng Liao of Chongqing University, Kaibo Shi of Chengdu University, and Yijing Liu of Sichuan University. Each contribution, big or small, has woven itself into the fabric of this monograph.

Chengdu, China
February 2025

Yongbin Yu

Contents

Chapter 1
Memristor and Its Emulator

Memristors are typically classified into two categories: charge-controlled and flux-controlled memristors [1]. Charge-controlled memristors are devices that achieve information storage based on changes in the resistivity of electrode-electrolyte contact. They are composed of liquid and solid electrolytes. When a voltage is applied, the concentration and resistance of the electrode-electrolyte change, thereby altering the impedance of the circuit and achieving information storage. Flux-controlled memristors, on the other hand, are devices that achieve information storage based on the magnetoresistance effect. They are composed of magnetic materials and metals. When a magnetic field is applied, the magnetoresistance of the magnetic material changes, thereby altering the impedance of the circuit and realizing information storage. Therefore, charge-controlled memristors and flux-controlled memristors differ in their storage principles and structures.

Memristor emulators, as crucial tools for realizing memristive behavior, play a vital role in studying memristor circuit characteristics and verifying related theoretical analyses. These emulators not only simulate the nonlinear properties of memristors but also enable exploration of their performance in various circuit configurations without relying on physical devices. This chapter introduces the latest research work related to memristor models and their emulators in detail. These studies have not only deepened the understanding of memristor characteristics but also provided important support for their deployment in practical applications, driving memristor technology toward a more mature and practical direction.

1.1 Memristor

In 1971, the notion of memristor (a contraction for memory resister) was postulated by Prof. Leon Chua, which is regarded as the fourth fundamental two-terminal circuit element (shown in Fig. 1.1) [2]. In 1976, the concept of memristor was extended to

Y. Yu et al., *Memristive Computing*,
https://doi.org/10.1007/978-981-96-0332-9_1

Fig. 1.1 Four axiomatically defined circuit elements

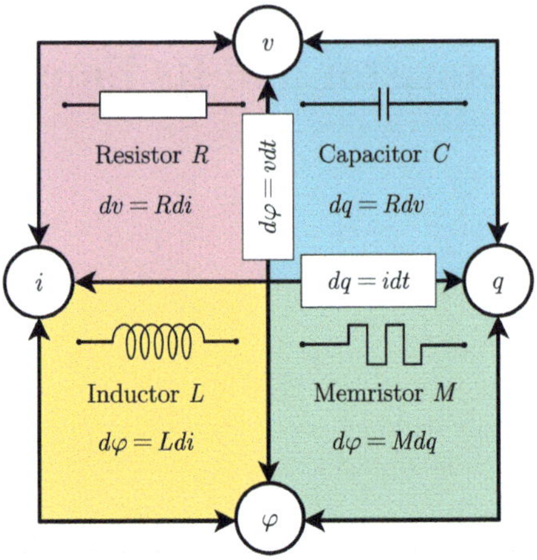

a memristive system [3]. In 2008, R. Stanley Williams and his team at Hewlett-Packard (HP) Labs achieved a significant breakthrough in memristor technology by successfully fabricating an impurity-drift, charge-controlled memristor. This major discovery ignited the boom of research on memristors.

The mathematical models of Memristor can be defined by two circuit variables of charge q and flux φ. When flux is satisfied with $\varphi = \hat{\varphi}(q)$, the memristor described by $M(q) \triangleq d\hat{\varphi}(q)/dq$, is charge-controlled. At the same time, if charge is a scalar function such that $q = \hat{q}(\varphi)$, the memristor, expressed by $W(\varphi) \triangleq d\hat{q}(\varphi)/d\varphi$, is flux-controlled. According to the two constitutive relations, memristor is classified as memristance $M(q)$ and memconductance $W(\varphi)$, whose mathematical models have been presented with their special behaviors and properties. Moreover, a current (i)-controlled and a voltage (v)-controlled memristive system is extended to make classification such that $i = dq/dt$ and $v = d\varphi/dt$. Given sinusoidal excitation, all kinds of memristors exhibit a distinctive "fingerprint" characterized by a pinched hysteresis loop confined to the first and the third quadrants of the $v - i$ plane.

Stimulated by HP's fabrication of memristors, a boom in the research was set off, and one of the outstanding research is the Chua corsage memristor, which will be introduced in this chapter.

Discovered and theorized over half a century ago as the fourth fundamental circuit element [2, 4–6], the memristor was first realized on May 1, 2008, when its existence was highlighted in *Nature*:

THE MEMRISTOR
"Missing circuit element found"

On its cover, there are two typical characteristics for memristor reported in *Nature*.

(1) Nanoscale. The nanoelectronics, physics, chemistry, and nanotechnology were utilized by a small group of ingenious researchers from HP Labs in Palo Alto, California, under the leadership of Dr. R. Stanley Williams, to produce the world's first functioning nanoscale memristor [7]. Dr. Williams' fascinating and inspiring article [8] reveals that his innovative research group experimented with numerous exotic materials and novel approaches over several years before successfully building a functioning nanoscale memristor by sandwiching a thin film of titanium dioxide between two platinum electrodes [8].
(2) Nonvolatile. Dr. Williams also explained in layman's terms that a memristor is a two-terminal device whose resistance depends on the magnitude and polarity of the applied voltage and the duration for which the voltage is applied. When the voltage is turned off, the memristor retains its most recent resistance until the next time it is turned on, regardless of whether that occurs a day or a year later. This attribute [6], referred to as the "remembrance of things past", does not require a power supply, making such memristors nonvolatile.

In light of the above-mentioned properties, the memristor is intended to exploit as memory and synapse.

(1) Memory. This inherent property has been used to clarify many ambiguous and confusing issues [9–12] and is widely exploited in semiconductor memories for smartphones, laptops, PCs, and other applications, such as the Internet of Things and in-memory computing. Recently, the nonvolatile property has also been employed to resolve five long-standing enigmas, as detailed in various studies [13–16].
(2) Synapse. In a biophysical and neuromorphic view, *nonvolatile* memristors are natural models for ***synapses***. However, because nonvolatile memristors are *locally passive two-terminal devices* [13–15], they cannot generate the *all-or-none* signals, also known as action potentials, that are essential for life. It has been proven that [13], with the exception of primitive life forms that can survive without ion channels, most living systems are equipped with ion-channel memristors, which have an edge of chaos kernel—a prime example of the *Chua riddle* [13–18]. According to the first law of thermodynamics, which states there is no free lunch, these life-supporting memristors are volatile. In electronic circuit theory terms, all ***volatile memristors*** must be powered by a DC supply, such as a battery.

Currently, there are some problems with *volatile* memristors, which are listed as follows:

(1) Fabrication problem. Unlike conventional electronic components such as resistors, inductors, capacitors, and diodes, volatile memristors are not commercially

available. They must be designed and fabricated from exotic materials using advanced nanofabrication techniques. For instance, the locally active volatile VO_2 memristor reported in [19] utilizes vanadium dioxide as the raw material and requires electron beam lithography for its fabrication, which can only be performed by well-trained, highly specialized technicians. Even more challenging is the fabrication of a locally active third-order volatile memristor [20], which involves inducing niobium dioxide to enter a Mott transition switching regime—a counterintuitive discovery that earned Sir Nevill Mott the 1977 Nobel Prize in Physics.

(2) Neuromorphic problem. Indeed, building a volatile memristor and extending it to a biomimetic neuron capable of emulating a wide variety of well-known neuromorphic phenomena—such as all-or-none firing, phasic spiking, periodic spiking, bursting with a prescribed number of spikes per burst, burst number adaptation, inhibition-induced spiking, integrator-induced firing, resonator-induced firing, and chaos—is currently an expensive, time-consuming, and nontrivial task.

(3) Dynamical complexity. Not only does one need to be well-versed in advanced nanoscale device fabrication technology but also must have the credentials and resources to access a high-tech nanodevice fabrication facility. And this is just the first hurdle! Uncovering the dynamical complexity of a volatile memristor, essential for its extension to a biomimetic neuron, requires measuring—not simulating via software—the entire DC $V - I$ curve [9, 11, 13] of the actual device. This task has been impossible—until now. All published DC $V - I$ curves of volatile memristors have been incomplete, containing gaps [19, 20] where the most relevant bifurcation dynamics were hidden.

1.2 A Charge-Controlled Memristor

1.2.1 Introduction

Memristor controlled by charge and flux is characterized by a $\varphi - q$ and a $q - \varphi$ nonlinear curves respectively, and a memristive system controlled by current and voltage is characterized by a $v - i$ and $i - v$ nonlinear curves respectively. With a piecewise-linear $\varphi - q$ characteristic curve, charge-controlled and flux-controlled memristors were studied by Makoto Itoh and Leon Chua [21]. With a cubic $q - \varphi$ nonlinear curve, a flux-controlled memristor was proposed by Muthuswamy and Leon Chua [22]. Furthermore, a current-controlled memristive system was presented in 2010 [23], and a non-ideal active voltage-controlled memristive system was studied in 2016 [24]. These memristor models are shown in Table 1.1. Motivated by the fundamental difference in the way memristor model defines, we focus our attention on a novel charge-controlled memristor.

Table 1.1 Memristor models

	Type	Mathematical model	Physical circuit
Charge	Controlled memristor	$\varphi(q) = bq + 0.5(a - b)(\lvert q + 1\rvert - \lvert q - 1\rvert)$ [21]	No
Flux-		$q(\varphi) = \alpha\varphi + \beta\varphi^3$ [22]	Yes
Current	Controlled memristive system	$v_M = \beta\left(x^2 - 1\right) i_M$, $\dot{x} = i_M - \alpha x - i_M x$ [23]	Yes
Voltage-		$i = \left(-G_a + G_b v_0^2\right) v$, $\dot{v}_0 = -\frac{v}{R_1 C_0} - \frac{v_0}{R_2 C_0}$ [24]	Yes

Due to nonlinear feature, memristors are widely used in chaotic circuit [25–30]. Memristors with the memory property are utilized to design nonvolatile nanomemories where binary bits are stored as resistances [31, 32], and other memristors are proposed to work as synaptic weights and built to emulate the human brain in artificial neural networks [33, 34]. However, commercially available memristors have not been fabricated because of the cost and difficulties in the research of nanoscale devices.

In this section, a charge-controlled memristor model defining a quadratic relation between φ and q, and its physical circuit realization are studied. The test results from physical circuit demonstrate the fingerprint of memristor behavior.

1.2.2 Memristor Model

The constitutive relation of a charge-controlled memristor [2] is defined as:

$$\varphi = \hat{\varphi}(q), \tag{1.1}$$

where $\varphi(q)$ is continuous and piecewise-differentiable functions with bounded slopes. The ideal charge-controlled memristor [2] is defined as:

$$\begin{aligned} v_M &= M(q) i_M \\ \dot{q} &= i_M \end{aligned}, \tag{1.2}$$

where $M(q)$ is the memristance and v_M is equal to $d\varphi/dt$. It has been proved that (1.1) and (1.2) are equivalent representations by Chua [35]. Previous memristor research tended to choose cubic nonlinear polynomials about φ and q for the memristor model. In contrast we proposed, we proposed a quadratic polynomial given by:

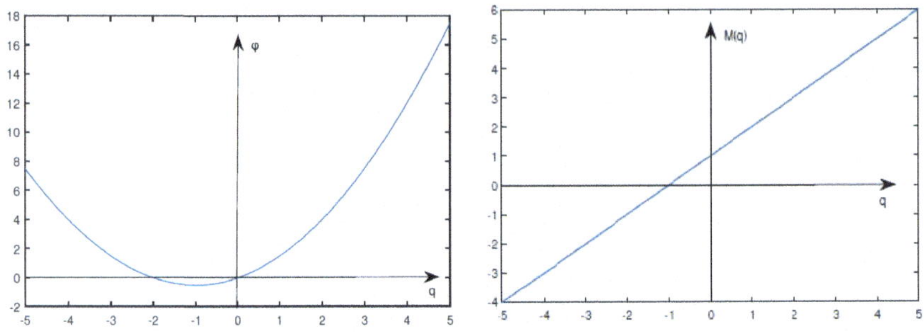

(a) $\varphi - q$ curve (b) $M(q) - q$ curve

Fig. 1.2 Graphical illustration of the $\varphi - q$ curve and $M(q) - q$ curve

$$\varphi = q + \frac{1}{2}q^2. \tag{1.3}$$

The intuition is that smooth nonlinearities should be easier to implement. Indeed, we can also define our polynomial as $q = \varphi + 1/2\varphi^2$, which represents a flux-controlled memristor. Here, without loss of generality, we concentrate on the charge-controlled memristor defined in (1.3). In order to study the relation between memristance $M(q)$ and q, we differentiate both sides of (1.3) with respect to q, as given below:

$$M(q) = \frac{d\varphi}{dq} = 1 + q. \tag{1.4}$$

The $\varphi - q$ curve and $M(q) - q$ curve are plotted in MATLAB and shown in Figs. 1.2 and 1.3, respectively.

The foremost characteristic property of the memristor which distinguishes it from the other basic circuit elements is its pinched hysteresis loop, the Lissajous figure of $(i(t), v(t))$ in the $v - i$ plane. To demonstrate its memristive characteristics, defined

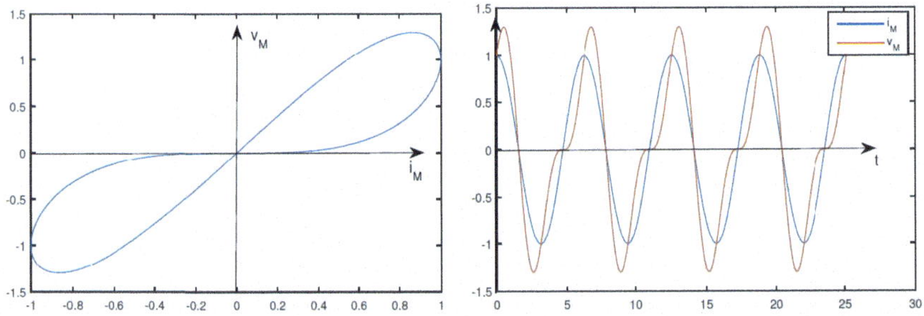

(a) Pinched hysteresis loop (b) Periodic waveforms associated with $i(t)$ and $v(t)$ plotted with $A = 1$ and $\omega = 1$

Fig. 1.3 Lissajous figure and Periodic waveforms

by:

$$\begin{cases} i(t) = A\cos\omega t, & t \geq 0 \\ i(t) = 0 & t < 0 \end{cases}. \tag{1.5}$$

Across the memristor. In order to obtain the corresponding voltage respond $v(t)$ from the constitutive relation, we must use corresponding i and q to represent the voltage response $v(t)$. Differentiating (1.4) with respect to time t, obtain:

$$v_M = \frac{d\varphi}{dt} = qi + i = i(q+1). \tag{1.6}$$

Integrating both sides of (1.5) with respect to t (Assuming the initial charge $q_0 = q(0) = 0$), obtain:

$$q(t) = \int_0^t A\cos\omega\tau d\tau = \frac{A}{\omega}\sin\omega t, \quad t \geq 0. \tag{1.7}$$

Substituting (1.5) and (1.7) into (1.6), we obtain the corresponding voltage v_M as follows:

$$v_M = A\cos\omega t\left(\frac{A}{\omega}\sin\omega t + 1\right) = \frac{A^2}{\omega}\sin\omega t\cos\omega t + A\cos\omega t. \tag{1.8}$$

Plotting the Lissajous figure of $(i(t), v(t))$ in the $v - i$ plane, via (1.8) and (1.5), we obtain the pinched hysteresis loop shown in Fig. 1.3a for $A = 1$ and $\omega = 1$. From the figure, it can be seen that the hysteresis loop is always pinched at the original $(v, i) = (0, 0)$, which describes the characteristic of the memristor.

1.2.3 Implementation Circuit

Since our memristor is a special case of a much more general memristive system, we rewrite our memristor equation in another form:

$$v_M = (x+1)i_M \tag{1.9}$$

$$\dot{x} = i_M, \tag{1.10}$$

where x denotes the state of the system, which is equal to q in this system.

The schematic of the emulator circuit of our memristor is shown in Fig. 1.4. The proposed circuit emulates a charge-controlled memristor element. It consists of seven resistors, one capacitor, two amplifiers (AD712JN), and one multiplier (AD633JN), and they are all commercially available. The whole memristor emulator (enclosed in the box) is powered by 15 V DC power supplies.

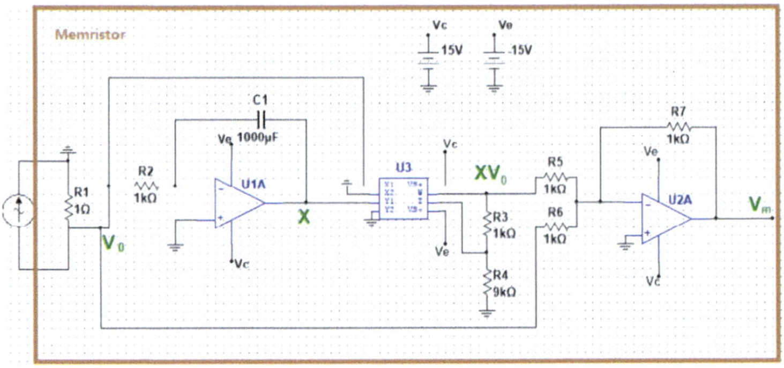

Fig. 1.4 Schematic of the physical circuit

The concept behind realizing a memristor is to first sense the current flowing through the circuit. Therefore, we use a sensing resister R_1. In this case, $R_1 = 1\,\Omega$. Hence,

$$V_0 = -I_m R_1. \tag{1.11}$$

Op-amp U1A realizes the differential (1.10) for the state variable x of memristor. From the schematic, we obtain the equation for the output of U1A:

$$\frac{V_0}{R_2} = -C_1 \frac{dx}{dt}. \tag{1.12}$$

Substituting (1.11) into (1.12), we get

$$-I_m \frac{R_1}{R_2} = -C_1 \frac{dx}{dt}. \tag{1.13}$$

Since we have chosen $R_1 = 1\,\Omega$, $R_2 = 1\,\mathrm{k}\Omega$, $C_1 = 1000\,\mu\mathrm{F}$, x in (1.13) can be simplified as:

$$\dot{x} = I_m. \tag{1.14}$$

Hence, (1.10) has been realized.

Op-amp U2A and multiplier U3 are utilized for the implementation of (1.9). According to the datasheets of the multipliers [Analog, 2010] and the connections shown in the schematic, we can get the output of multiplier U3 is XV_0. Op-amp U2A is an inverting summing amplifier. The output V_m is given by:

$$V_m = -\frac{R_7}{R_5} X V_0 - \frac{R_7}{R_6} V_0. \tag{1.15}$$

Substituting (1.11) into (1.15), we can obtain the following:

$$V_m = I_m R_1 \left(\frac{R_7}{R_5} X + \frac{R_7}{R_6} \right). \tag{1.16}$$

Since we have chosen $R_5 = R_6 = R_7 = R = 1\,\text{k}\Omega$, $R_1 = 1\,\Omega$, V_m in (1.16) can be simplified as:

$$V_m = (X + 1) I_m. \tag{1.17}$$

Hence, (1.9) has been realized.

1.2.4 Result from the Physical Circuit

In order to plot the pinched hysteresis loop, we use the test circuit shown in Fig. 1.5. The current source $I_m(t)$ is configured to be approximately 100 mA amplitude sin-wave. The signals are captured by an oscilloscope which displays the physical circuit at 3 kHz and 30 kHz, respectively. The pinched hysteresis loop, shown in Fig. 1.6a, is obtained by measuring V_0 versus V_m since V_0 is R_1 time of input current.

From Fig. 1.6, we can see that the pinched hysteresis loops show high accuracy with the mathematical analysis of our memristor. Furthermore, compared with Fig. 1.6a, the pinched hysteresis loop at 30 kHz shrinks and tends to a straight line, which is called frequency dependency [31].

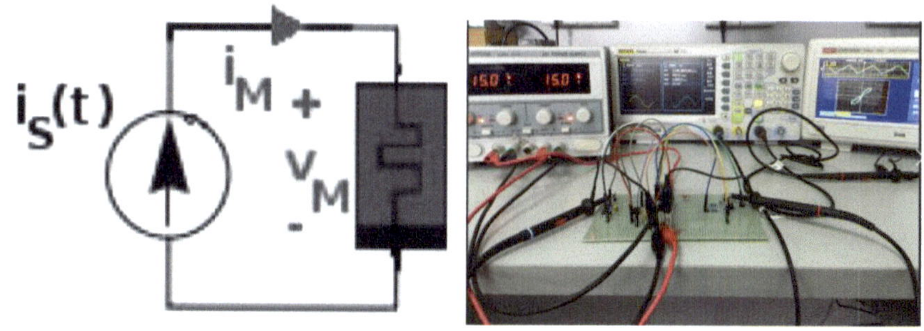

(a) Test circuitmodel. (b) Physical test circuit.

Fig. 1.5 Test circuit model and Physical test circuit

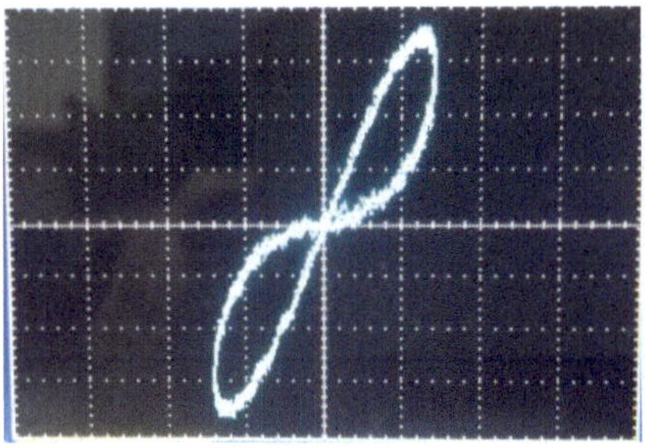

(a) Pinched hysteresis loop, 3kHz.

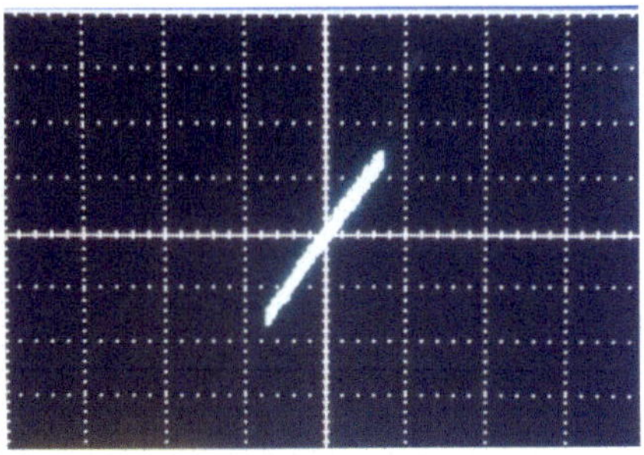

(b) Pincehd hysteresis loop, 30kHz.

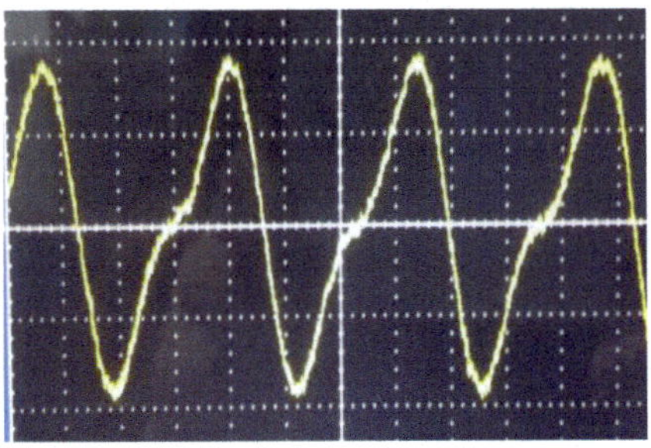

(c) Waveform of output of the memristor V_m.

Fig. 1.6 Results from the physical circuit

1.2.5 Conclusion

In this section, a novel charge-controlled memristor is introduced, inspired by existing mathematical models. A new quadratic relationship between φ and q is defined, and the characteristics of the memristor are illustrated by plotting the $v - i$ Lissajous figure under sinusoidal current excitation, which demonstrates pinched hysteresis. A memristor emulator is presented, and a physical circuit is fabricated. In this circuit, a sinusoidal current is applied, and the voltage response v_M is measured. The resulting Lissajous figure, which aligns perfectly with theoretical analysis, is displayed on an oscilloscope in $x - y$ mode. Due to the various properties of memristors, this circuit realization indicates potential applications such as nonvolatile memory and neural networks. Future research will explore applying the proposed memristor as a synapse in artificial neural networks.

1.3 A Flux-Controlled Memristor

1.3.1 Introduction

Flux-controlled memristors represent a significant category of memristive devices, characterized by a single-valued functional relationship between charge and magnetic flux. Modeling and circuit implementation of flux-controlled memristors not only contribute to a deeper understanding of memristor operating mechanisms but also lay the foundation for developing novel memristor-based circuits and systems. However, existing flux-controlled memristor models often employ simplistic polynomial forms, which struggle to comprehensively describe the nonlinear characteristics of memristors. Furthermore, the design of memristor equivalent circuits faces limitations such as single-ended grounding, which restricts their flexibility in practical applications.

To address these challenges, this section introduces a new third-order flux-controlled memristor model and designs its corresponding equivalent circuit to better describe and implement memristor behavior. By incorporating an additional quadratic term in the charge-flux relationship, the proposed model offers a more general representation compared to existing models. Simultaneously, this study presents a "floating" two-port equivalent circuit design that overcomes the limitations of grounded memristor models, enhancing its adaptability in various circuit configurations. Through theoretical analysis, computer simulations, and physical circuit implementations, this section explores the model's behavior under different input conditions, demonstrating its ability to capture key memristive characteristics and providing new insights and methods for memristor modeling and circuit design.

1.3.2 Modeling of a Novel Third-Order Flux-Controlled Memristor

As mentioned earlier, the charge q in the flux-controlled memristor is a single-valued function of the magnetic flux ϕ. In this section, the function $q(\phi)$ is expanded by the Taylor series, and its expression is as follows:

$$q(\phi) = \sum_{n=0}^{\infty} a_n \phi^n. \tag{1.18}$$

In order to simplify the problem, this section only expands the polynomial to the third order:

$$q = \alpha\phi + \beta\phi^2 + \gamma\phi^3, \tag{1.19}$$

where α, β, and γ are the coefficients of the subterms, respectively, and their values are constants. Compared with the third-order flux-controlled memristor model used in [36–39], the flux-controlled memristor model proposed in this section has an additional $\beta\phi^2$ component in describing the $q-\phi$ relationship, which makes it more general. When $\beta = 0$, the memristor model is consistent with the memristor model proposed in [36–39]. Combining $W(\phi) = \frac{dq(\phi)}{d\phi}$ leads to the corresponding memconductance $W(\phi)$:

$$W(\phi) = \alpha + 2\beta\phi + 3\gamma\phi^2. \tag{1.20}$$

The $q-\phi$ curves and $W-\phi$ curves of the new third-order flux-controlled memristor and the third-order flux-controlled memristor without the quadratic component of ϕ are shown in Fig. 1.7, where $\alpha = 1\text{S}$, $\beta = 1\text{S/Wb}$, and $\gamma = 1\text{S/Wb}^2$.

The following equation can be derived from two fundamental relations: the relation $v(t) = \frac{d\phi(t)}{dt}$ connecting voltage v, flux ϕ, and time t; and the relation $i(t) = W(\phi)v(t)$ linking current i, voltage v, and memductance W.

$$\begin{cases} i(t) = (\alpha + 2\beta\phi + 3\gamma\phi^2)v(t) \\ \phi(t) = \phi_0 + \int_0^t v(\tau)d\tau \end{cases}, \tag{1.21}$$

where ϕ_0 is the initial amount of flux. It can be seen that the output of the flux-controlled memristor is not only related to the input at this moment but also related to the previous input, which reflects the "memory" characteristic of the memristor. In addition, in order to make the volt-ampere characteristic curve of the flux-controlled

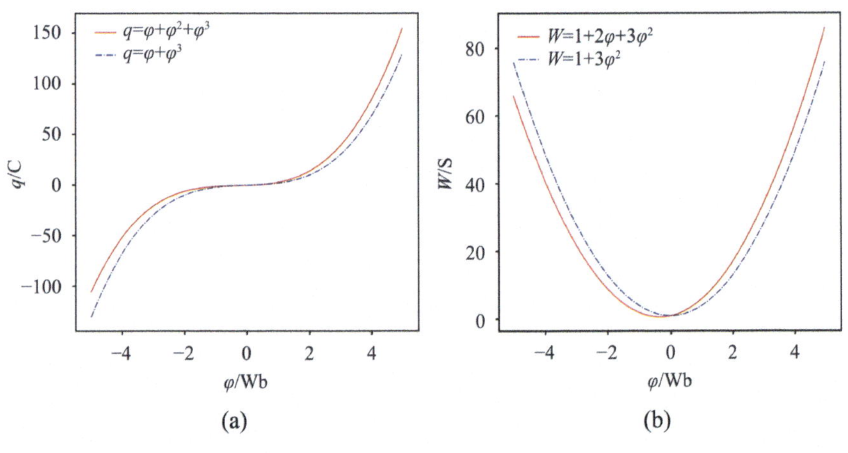

Fig. 1.7 $q-\phi$ curve and $W-\phi$ curve. **a** $q-\phi$ curve. **b** $W-\phi$ curve

memristor always in the first and third quadrants [40, 41], the value of the memconductance is always greater than 0, it is required that α, β, and γ are related to each other:

$$\begin{cases} 4\beta^2 - 4\alpha 3\gamma < 0 \\ 3\gamma > 0 \end{cases}. \tag{1.22}$$

To verify the accuracy of the model, the output voltage signal of a voltage source is set as:

$$\begin{cases} v(t) = A\cos(\omega t) & t \geq 0 \\ v(t) = 0 & t < 0 \end{cases}. \tag{1.23}$$

The expression of current $i(t)$ can be obtained from (1.21) and (1.23), as shown below:

$$i(t) = \left(\alpha A + \frac{3\gamma A^2}{4\omega^2}\right)\cos(\omega t) + \beta\frac{A^2}{\omega}\sin(2\omega t) - \frac{3\gamma A^3}{4\omega^2}\cos(3\omega t). \tag{1.24}$$

Given the input voltage source, the selected amplitude A, frequency ω, and the values of parameters α, β, and γ are shown in Table 1.2.

According to (1.24), the second-end voltage of the third-order flux-controlled memristor and the corresponding current waveforms and volt-ampere characteristic curves can be obtained by simulation in MATLAB software, as shown in Fig. 1.8. From the volt-ampere characteristic curve of the third-order flux-controlled memristor in Fig. 1.8b, it can be seen that the hysteresis curve intersects at the origin and is located in the first and third quadrants, and the shape is oblique "8", indicating that the third-order flux-controlled memristor model accords with the characteristics of the memristor.

Table 1.2 Parameters of the third-order flux-controlled memristor model

Parameter/Unit	Value
A/V	1
$\omega/\mathrm{rad}\cdot\mathrm{s}^{-1}$	π
α/S	1
$2\beta/\mathrm{S}\cdot\mathrm{Wb}^{-1}$	1
$3\gamma/\mathrm{S}\cdot\mathrm{Wb}^{-1}$	1

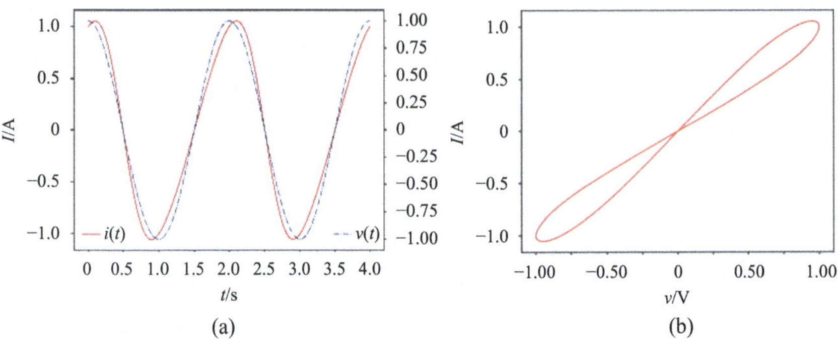

Fig. 1.8 Simulation result of the third-order flux-controlled memristor. (a) Volt-ampere waveform in time domain. (b) Volt-ampere characteristic curve

1.3.3 Equivalent Circuit of Third-Order Flux-Controlled Memristor

1.3.3.1 Principle of "Floating" Two-Port Equivalent Circuit

In the modeling process of the memristor in the previous section, the various operations of the voltage are relative to the zero potential. If the equivalent circuit of the memristor is built directly according to (1.21), then one end of the equivalent circuit must be grounded. To overcome the limitation of one end grounding of memristor equivalent circuit and make it become "floating" two-port, several kinds of floating ground memristor equivalent circuits are proposed in [42–45], but the circuit structures are more complex.

To simplify the circuit, the voltage relative to the zero potential is converted into the "potential difference" [38] between the two ports, and the two-port "floating" expression of the third-order flux-controlled memristor is obtained as follows:

$$\begin{cases} i(t) = (\alpha + 2\beta\phi + 3\gamma\phi^2)(v_1 - v_2) \\ \phi(t) = \phi_0 + \int_0^t [v_1(\tau) - v_2(\tau)]d\tau \end{cases}, \tag{1.25}$$

where v_1 and v_2 represent the voltages across the equivalent circuit of the memristor, respectively.

1.3.3.2 Realization of Third-Order Flux-Controlled Memristor Circuit

The potential difference between the two ports of the memristor equivalent circuit can be calculated by using a subtractor composed of operational amplifiers. Next, the calculated potential difference can be added, multiplied, and integrated. The specific circuit diagram implementation is shown in Fig. 1.9.

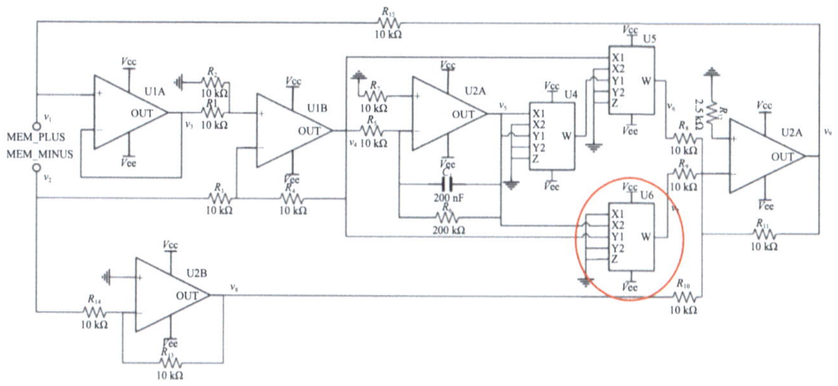

Fig. 1.9 Equivalent circuit diagram of the third-order flux-controlled memristor

The equivalent circuit of the memristor consists of 15 resistors, 1 capacitor, 5 AD712 operational amplifiers, and 3 AD633 multipliers. The whole equivalent circuit is driven by 15V DC power supply. Since the new third-order flux-controlled memristor model presented in this section has an extra $\beta\phi^2$ component, a multiplier U6 is used in this equivalent circuit to compute this component.

The operational amplifier U1A forms a voltage follower, which has the characteristics of high input impedance and low output impedance. The input current rarely enters the operational amplifier, which acts as an isolation from the previous circuit. The output voltage v_3 is equal to the input voltage v_1. The operational amplifier U1B and the resistor $R1$ ~4 form a subtractor. Since the resistance of $R_1 \sim R_4$ are equal, the relationship between the output voltage v_4 and the input voltage v_2 and v_3 is as follows:

$$v_4 = v_3 - v_2. \tag{1.26}$$

The operational amplifier U2A forms an integrator with the resistor $R_5 \sim R_7$ and the capacitor C_1. Since the actual operational amplifier is not ideal, the function of resistor R_7 is to reduce the error. The function of resistance R_6 is to prevent integral saturation or interception caused by integral drift. Also, to minimize the integration error, resistors $R_6 >> R_5$. The relationship between output voltage v_5 and input voltage v_4 is as follows:

$$v_5 = \frac{-1}{R_5 C_1} \int_0^t v_4(\tau) dt = \frac{-1}{R_5 C_1} \phi(t). \tag{1.27}$$

U4~U6 are AD633 multipliers. According to its data manual, the relationship between the input and output is as follows:

$$W = \frac{(X_1 - X_2)(Y_1 - Y_2)}{10V} + Z. \tag{1.28}$$

The expressions of voltage v_6 and voltage v_7 can be calculated directly according to (1.28), as shown below:

$$v_6 = \frac{v_4 v_5^2}{100}, \tag{1.29}$$

$$v_7 = -\frac{v_4 v_5}{10}. \tag{1.30}$$

Operational amplifier U2B forms an inverse amplifier with resistors R_{13} and R_{14}. Since the resistance $R_{13} = R_{14}$, the relationship between output voltage v_8 and input voltage v_2 is as follows:

$$v_8 = -v_2. \tag{1.31}$$

The operational amplifier U3A and the resistor $R_8 \sim R_{12}$ form an inverse adder. Similarly, the function of the resistor R_{12} is to reduce the calculation error. Since the values of the resistance $R_8 \sim R_{11}$ are all $10\,\text{k}\Omega$, the relationship between the output voltage v_9 and the input voltage v_6, v_7, and v_8 is given by:

$$v_9 = -(v_6 + v_7 + v_8). \tag{1.32}$$

Finally, the current flowing through the resistor R_{15} can be calculated according to the voltage at both ends of the resistor R_{15}. As mentioned earlier, there is almost no current flowing into the operational amplifier U1A, so the current flowing through R_{15} is the current flowing through the memristor. By combining the equations (1.25)–(1.32), the expression of the current $i(t)$ can be obtained as follows:

$$i(t) = \frac{v_1 - v_9}{R_{15}} = \frac{v_1 - v_2}{R_{15}} + \frac{v_1 - v_2}{10R_5C_1R_{15}}\phi(t) + \frac{v_1 - v_2}{100R_5^2C_1^2R_{15}}\phi^2(t). \tag{1.33}$$

Comparing (1.25) with (1.33), the expressions for the important parameters α, β, and γ can be derived as:

$$\begin{cases} \alpha = \frac{1}{R_{15}} \\ 2\beta = \frac{1}{10R_5C_1R_{15}} \\ 3\gamma = \frac{1}{100R_5^2C_1^2R_{15}} \end{cases}. \tag{1.34}$$

1.3.3.3 Simulation Analysis of Third-Order Flux-Controlled Memristor Circuit

The volt-ampere characteristics of the above third-order flux-controlled memristor are simulated and analyzed using Cadence (PSPICE) software. Based on the relationship between α, β, and γ in (1.22) and with some tuning, the resistor-capacitor settings are as follows: $C_1 = 200\,\text{nF}$ and $R_5 = 10\,\text{k}\Omega$. It can be calculated as $\alpha = 0.1$ ms, $2\beta = 5$ ms/Wb, and $3\gamma = 0.25$ S/Wb^2.

Fig. 1.10 Test circuit of the equivalent circuit of the third-order flux-controlled memristor

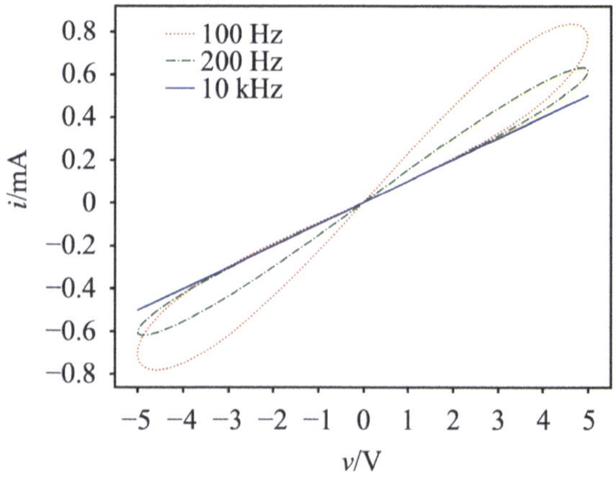

Fig. 1.11 Volt-ampere characteristic curves with different input frequencies

To test the volt-ampere of the equivalent circuit of the third-order flux-controlled memristor, the test circuit shown in Fig. 1.10 is used. Both ends of the sinusoidal power supply V_3 are connected to the MEM_PLUS end and the MEM_MINUS end of the memristor equivalent circuit. In order to reflect the floating characteristic of the equivalent circuit of the memristor, both ends of the voltage source are not grounded. At the same time, a current probe is marked at the sinusoidal voltage source to measure the current passing through the memristor, and the volt-ampere characteristic curve of the equivalent circuit of the memristor can be drawn by changing the Abscissa to the amplitude of the voltage source.

The amplitude of the sinusoidal voltage source is fixed at 5V, and the frequency is set to 100 Hz, 200 Hz, and 10 kHz, respectively. The volt-ampere characteristic curve is shown in Fig. 1.11. It can be seen that the memristor model conforms to the three characteristics of a memristor [41]: When periodic signals are applied at both ends of the memristor, the volt-ampere characteristic curve of the memristor is a hysteretic curve and passes through the origin. With the increase in frequency, the area of the hysteretic curve decreases. When the frequency approaches infinity, the hysteretic curve degenerates into a single-valued function and the memristor degenerates into a linear resistance.

The frequency of the voltage source is fixed at 100 Hz, and the amplitude is set to 3V, 5V, and 7V, respectively. The volt-ampere characteristic curve is shown in

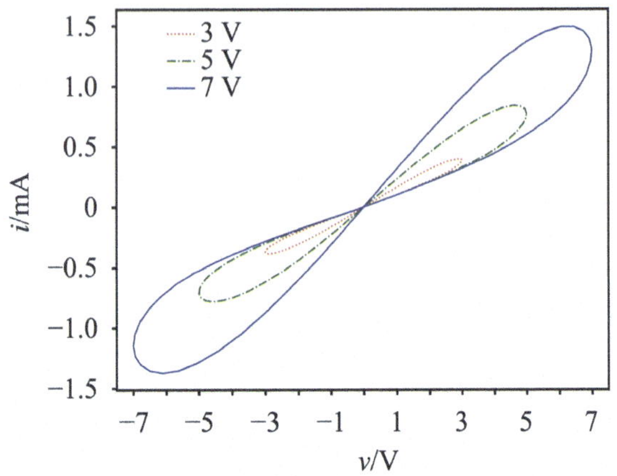

Fig. 1.12 Volt-ampere characteristic curves with different input amplitudes

Fig. 1.12. It can be seen that with the increase of the voltage at both ends of the memristor equivalent circuit, the area surrounded by the hysteretic curve will also increase.

1.3.4 Actual Circuit Analysis

The actual circuit of the equivalent circuit of the third-order flux-controlled memristor is shown in Fig. 1.13. The actual circuit adopts an AD712JN dual-channel operational amplifier and AD633JN multiplier. To stabilize the voltage of the voltage source of the operational amplifier, a capacitor is connected in parallel at the input of the positive and negative voltage source to eliminate interference.

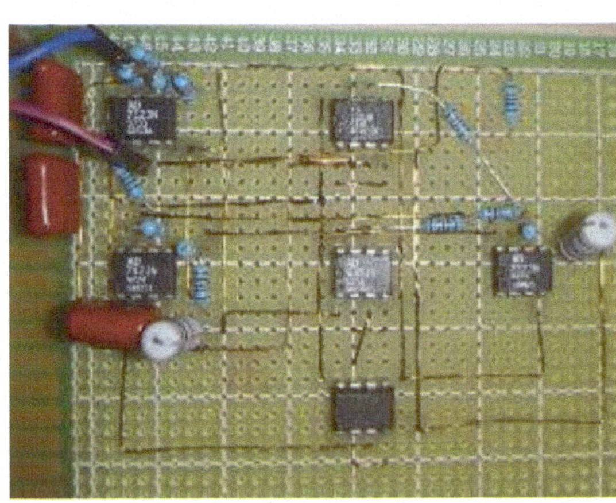

Fig. 1.13 The actual circuit of the equivalent circuit of the third-order flux-controlled memristor

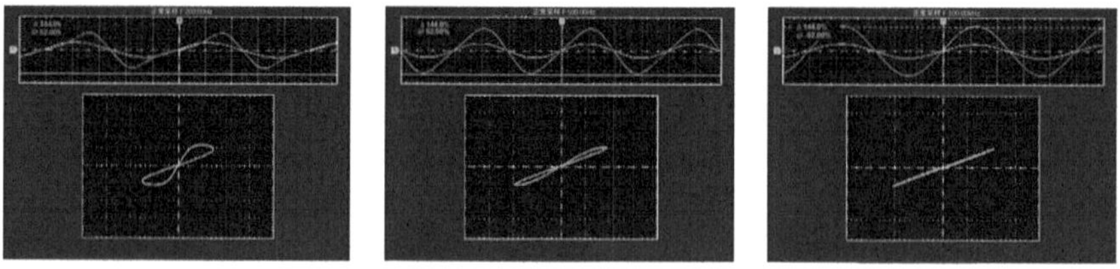

(a) (b) (c)

Fig. 1.14 The waveforms in time domain and the volt-ampere characteristic curves of the third-order flux-controlled memristor. (a) The waveform in time domain and the volt-ampere characteristic curve when frequency is 200 Hz. (b) The waveforms in time domain and the volt-ampere characteristic curves when frequency is 500 Hz. (c) The waveforms in time domain and the volt-ampere characteristic curves when frequency is 10 kHz

1.3.4.1 Volt-Ampere Characteristic Analysis of Equivalent Circuit of Third-Order Flux-Controlled Memristor

Since the current flowing through the memristor equivalent circuit is proportional to the voltage at both ends of the resistance R_{15}, the subtractor can be used to calculate the voltage at both ends of R_{15} so as to indirectly obtain the current waveform flowing through the memristor equivalent circuit.

The voltage amplitude of the sinusoidal input signal is fixed at 5V and the frequency is 200 Hz, 500 Hz, and 10 kHz, respectively. The voltage and current waveforms and the volt-ampere characteristic curves at both ends of the third-order flux-controlled memristor are shown in Fig. 1.14. Among them, yellow represents the voltage waveform and blue represents the current waveform.

It can be seen from the time domain waveform that there is a certain harmonic component in the current flowing through the memristor under the excitation of the low-frequency signal. With the increase of the excitation signal frequency, the current waveform gradually shows a standard sinusoidal signal, which is linear and consistent with the theoretical analysis and simulation results.

From the volt-ampere characteristic curve, it can be seen that the volt-ampere characteristic curve of the memristor is an oblique "8" hysteretic curve in the first and third quadrants and intersects itself at the origin; with the increase of frequency, the area of the hysteretic curve decreases gradually; when the frequency approaches infinity, the hysteretic curve degenerates into a straight line.

1.3.5 Conclusion

This section presents a new third-order flux-controlled memristor model characterized by a cubic nonlinear polynomial describing the charge-flux relationship, with its novelty lying in the inclusion of a quadratic term. An equivalent circuit for this memristor is developed using analog operation circuits, featuring a "floating" two-port design to overcome limitations of grounded models. The model's feasibility and the equivalent circuit's performance are verified through theoretical analysis, PSPICE-based simulations, and actual circuit implementations. Results confirm that the proposed model exhibits the three fundamental characteristics of memristors: nonlinear behavior under low-frequency excitation, pinched hysteresis loops passing through the origin, and convergence to linear behavior at high frequencies. This work contributes to the advancement of memristor modeling and circuit design, potentially enabling more accurate representations of memristive systems in various applications.

1.4 A Simple Memristor Emulator

1.4.1 Introduction

Memristor simulations, a crucial aspect of research in this field, typically rely on either memristor models or emulators. Memristor models, extensively discussed in [46, 47] and other literature, enhance our understanding of memristor characteristics and potential applications. However, these models cannot accurately represent memristor performance in actual circuits without physical fabrication.

Memristor emulators address this limitation by providing a more tangible representation of memristor behavior. They are prevalent due to their easier realization without requiring nanotechnology fabrication techniques. Emulators utilizing operational amplifiers (OPAMP) and microcontroller units (MCUs) have been thoroughly explored in [23]. While these emulators adequately mimic memristor characteristics to some degree, their complex implementations pose challenges in constructing circuits with multiple memristors.

Consequently, the development of simpler emulators becomes imperative. This section proposes a simple varactor-based memristor emulator composed of discrete components, which can be considered a generalized memristor under specific conditions. The simplicity of this emulator lies not only in its basic components with well-established models (such as SPICE models) but also in its straightforward physical realization and testing procedures. The emulator's generic properties as a memristor, including its current-voltage ($i - v$) relationship, are discussed later in this letter.

The proposed emulator facilitates the development of further applications in memory devices. Following the introduction of memcapacitors (capacitors with memory) and meminductors (inductors with memory) in [48], their emulators can be established using similar principles. This approach extends our understanding of memory device properties and simplifies research by enabling easier setup and testing of circuits with these emulators.

Furthermore, memristors are renowned for their applications in chaotic circuits, demonstrating their nonlinear properties. To validate the application of the proposed emulator, the simplest chaotic circuit is tested, consisting only of a linear passive inductor, a linear passive capacitor, and the emulator functioning as a nonlinear memristor. The resulting chaotic attractor is observed and thoroughly analyzed, further confirming the emulator's utility in practical applications.

1.4.2 Memristor Emulation

A varactor is known for its nonlinear property of variable capacitance under changing voltage, thus it can provide us the variable impedance when we apply a voltage bias. Using this character, we proposed a physical emulator of a two-terminal memristor. The equivalent circuit of the emulator is shown in Fig. 1.15a, which is simply composed of a common varactor in series with an inductor and a varactor. The varactor provides the varied impedance when the applied voltage changes, while the common diode serves as a bypass path when its applied voltage (same as the varactor) is positive.

To derive a generic form of memristance expression for the emulator, we proceed with the mathematical analysis starting from the memristor expression in [7, 46, 49]:

$$v(t) = \left(R_{ON} \frac{w(t)}{D} + R_{OFF} \left(1 - \frac{w(t)}{D} \right) \right) i(t), \tag{1.35}$$

where the lower resistance R_{ON} and the much higher resistance R_{OFF} are dominated by the maximum and the minimum value of the memristor, respectively. The ratio $x(t) = \frac{w(t)}{D}$ stands for the internal state of the memristor, which is restricted in between (0, 1).

When applying an external voltage, the total impendence of the emulator can be written as:

$$Z = Z_R + Z_L + Z_C = R + j\left(\omega L - \frac{1}{\omega C}\right), \tag{1.36}$$

where Z_R and Z_L are the resistance and the inductance, ω is the frequency of the applied voltage. When changing (1.36) into the form of (1.35), we can derive the equation for calculating the equivalent R_{ON}, R_{OFF} and the initial state $x(t)$, as shown below:

$$R_{ON} = Z_{\min} = Z|_{C_{\min}} = R + j\left(\omega L - \frac{1}{\omega C_{\min}}\right), \tag{1.37}$$

$$R_{OFF} = Z_{\max} = Z|_{C_{\max}} = R + j\left(\omega L - \frac{1}{\omega C_{\max}}\right), \tag{1.38}$$

$$x(t) = \frac{w(t)}{D} = \frac{C_{\min} \cdot (C_{\max} - C(t))}{(C_{\max} - C_{\min}) \cdot C(t)}, \tag{1.39}$$

where R represents impedance, $C_{\min}$ and $C_{\max}$ are determined by the absolute value of voltage on varactor $|v_C|_{\min}$ and $|v_C|_{\max}$. Here, R stands for impedance. Then we can finally get the generic memristance expression of this system:

$$M(q) = x(t) \cdot R_{ON} + (1 - x(t)) \cdot R_{OFF}. \tag{1.40}$$

This q-dependent expression is the crucial contribution to the memristance, and it changes with the absolute value of input voltage. R_{ON} and R_{OFF} are highly depended on the inherent characteristics of the varactor. SPICE simulation is used to verify the effectiveness of the proposed memristor emulator.

Applying a sinusoidal voltage with amplitude $v_0 = 1\,\text{V}$ upon the emulator, the $i - v$ curve is shown similarly as the HP Lab's experiment. Figure 1.15b shows the voltage and current as functions of time. It can be seen that unlike the regular resistor, the emulator's responds here to such a driving signal is a non-sinusoidal waveform.

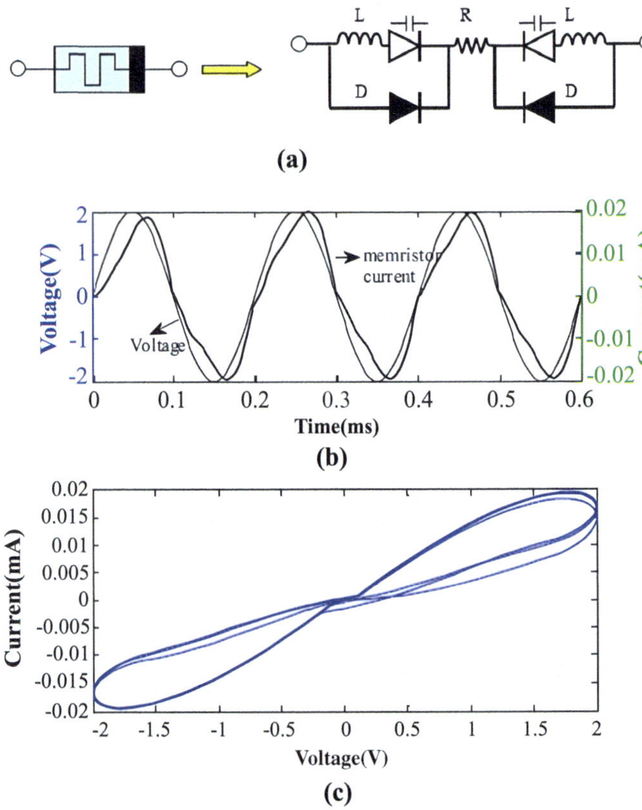

Fig. 1.15 (a) Memristor to equivalent circuit. (b) The applied sinusoidal voltage and the current flows through the emulator. (c) $i - v$ curve of the emulator when sinusoidal stimulus is applied

Figure 1.15c shows the $i - v$ curve produced by the emulator, which features the double-loop butterfly shape.

Here, the conclusion can be drawn that the emulator can only be seen as a generalized memristor. Firstly, the emulator behaves well as a memristive system: Fig. 1.15a shows us that it is passive, and the "pinched hysteretic loop" shown in Fig. 1.15c tells us about its history-related property [2], the emulator can be seen as a charge-controlled nonlinear resistor with memory, and the hysteretic loop does have the tendency to be a straight line when in the limit of infinite frequency. On the other hand, the memristor can only be called "generalized": as the existence of the inductor and varactor in the emulator, the calculated memristance is an impendence contains not only resistance but also inductance and capacitance, the physical unit cannot simply be ohm. During the existence of the varactor, the emulator can only act as a memristive system with the circuit while running, but it cannot store the information after the power is turned off, like an ideal memristor. However, the emulator can still help us develop and verify memristor applications. Our first application is focused on the simplest chaotic circuit.

1.4.3 Simplest Chaotic Circuit

It is widely reported that a memristor is extensively used in chaotic circuits [2, 23, 50, 51]. The simplest chaotic circuit has been discovered as the three-element circuit with the memristor to be the only one locally active element and the nonlinear device [23]. With the memristor emulator, the simplest three-element circuit as in Fig. 1.16 is established and tested.

In Fig. 1.16, L and C are the linear passive inductor and the linear passive capacitor, respectively, and MR stands for the memristor emulator. The circuit dynamic equations can be written as:

$$\begin{cases} V_M(t) = M(q) \cdot i(t) \\ i_C(t) = C\frac{dV_C(t)}{dt} \\ V_L(t) = L\frac{di_L(t)}{dt}. \end{cases} \quad (1.41)$$

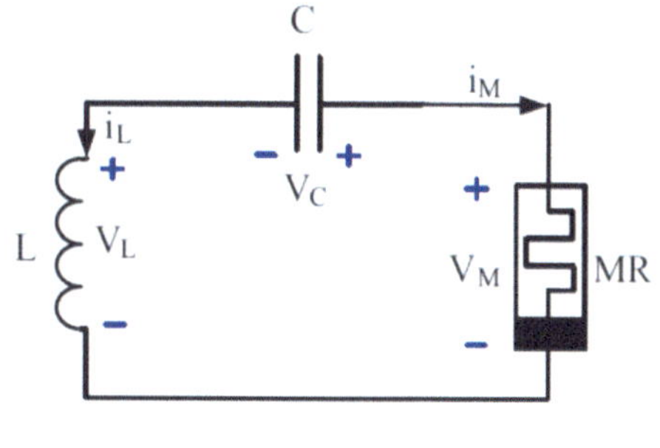

Fig. 1.16 Three-element circuit for chaos

The emulator here is the only nonlinear device, which provides the circuit with the necessary nonlinearity and the third state variable. It can be seen as a more generalized memristor, because its impendence contains not only resistance but also inductance and capacitance. The ideal *charge*-controlled memristor does not give rise to a third state variable according to [23]. With the emulator, the circuit exhibits a chaotic attractor as shown in Fig. 1.17, and current i flows through L as a function of v_C shows in Fig. 1.17a. As the numerical evidence of chaos, the Lyapunov exponent can also be found through calculation. This chaotic attractor in Fig. 1.17b has one positive Lyapunov exponent $\lambda = 0.0127$. The positive Lyapunov exponent implies that the circuit has an expanding direction in phase space. Also, for the three-dimensional

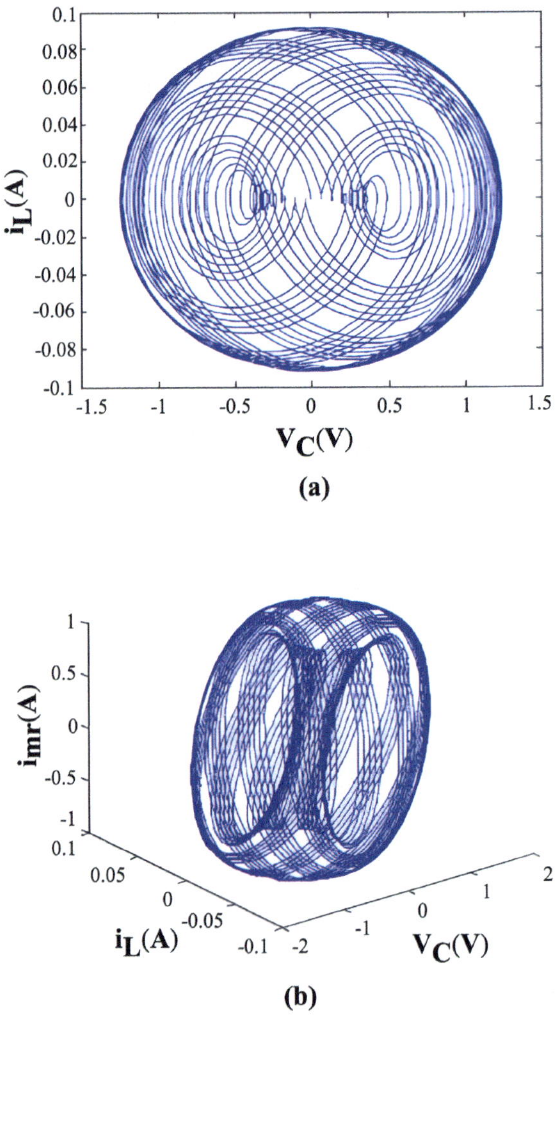

Fig. 1.17 (a) The $V_C(t)$ versus $i_L(t)$ chaotic attractor for circuit in Fig. 1.16. (b) The 3D chaotic attractor for circuit in Fig. 1.16. Here, the circuit is driven by a 5 KHz sin wave with the amplitude of $2V$

autonomous continuous time system in Fig. 1.17b, there should be three Lyapunov exponents. The *sum* of the exponents also indicates the chaotic behavior.

1.4.4 MATLAB Simulation

In this section, we use MATLAB to figure out the dynamic equations' numerical solutions with several different sets of parameters, and take the Jacobian method to calculate the biggest positive Lyapunov exponent to confirm that this circuit has chaotic properties. $x(t)$ represents the internal state of the memristor, which according to [23], is defined as follows:

$$R(x) = \beta \cdot \left(x^2 - 1\right). \tag{1.42}$$

So, the dynamic equations can be rewritten into this form:

$$\begin{cases} \frac{dV_C(t)}{dt} = i_L(t) \\ \frac{di_L(t)}{dt} = \frac{-V_C(t) - i_L(t)\cdot\beta\cdot\left(x(t)^2-1\right)}{L} \\ \frac{dx(t)}{dt} = -i_L(t) - \alpha \cdot x(t) - i_L x(t). \end{cases} \tag{1.43}$$

Therefore, the Jacobian Matrix of this system is:

$$\mathbf{J} = \begin{bmatrix} 0 & \frac{1}{C} & 0 \\ \frac{-1}{L} & -\frac{\beta\cdot\left(x^2-1\right)}{L} & -\frac{2\cdot\beta\cdot i_L\cdot x}{L} \\ 0 & -x-1 & -\alpha - i_L. \end{bmatrix} \tag{1.44}$$

The following figures show that a series of chaotic attractors and Lyapunov Exponents Spectrums, the positive Lyapunov Exponents also demonstrate that the emulator can generate chaos. The positive Lyapunov Exponents for Figs. 1.18, 1.19, 1.20, and 1.21 are with sequence as 0.032, 0.002, 0.0008, and 0.0032.

1.4.5 Conclusion

In this section, an emulator composed of conventional discrete components is proposed. The emulator shows most characters as a memristor, which plays two roles in the circuit: the nonlinearity and the essential state variable. With the help of this emulator, the simple chaotic circuit has been demonstrated, which can be taken as the simplest chaotic circuit, because there are only three existing components in the circuit, and only one nonlinear device serves as the localactive element as well. The chaotic attractors shown in the letter can help us learn more about the chaotic and the

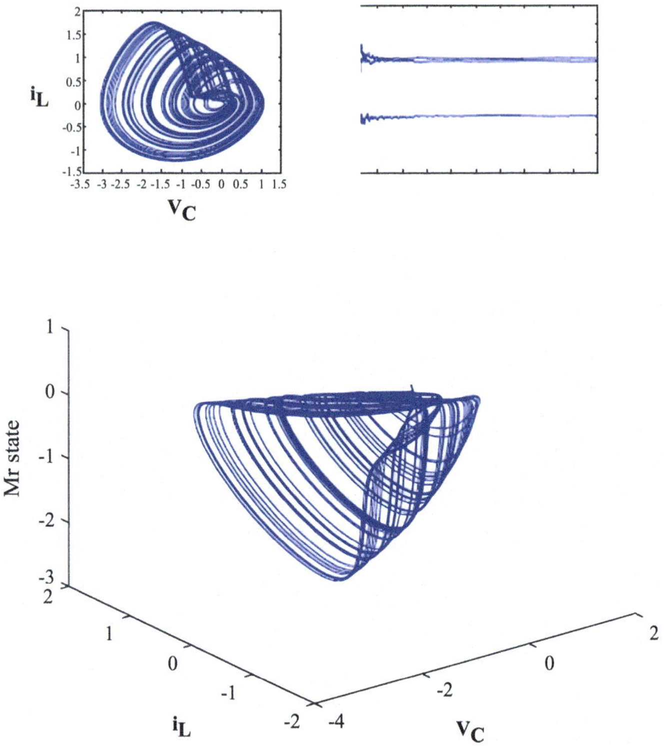

Fig. 1.18 With the parameters $C = 1$; $L = 3$; $\beta = 3/2$; $\alpha = 0.6$

memristor behavior characteristics. Our future work will be focused on pushing forward the research on the chaotic circuits and the further application of the memristor emulator.

1.5 Comments

This chapter delves into the modeling, simulation, and implementation techniques of memristors, proposing novel charge-controlled and flux-controlled memristor models, and designing corresponding equivalent circuits and emulators. These studies not only advance the theoretical foundations of memristors but also validate the feasibility of the models through practical circuit implementations and testing, laying the groundwork for memristor applications in nonvolatile memory, neuromorphic computing, and chaotic circuits.

Despite these advancements, memristor research still faces several challenges and opportunities. Future research can be expanded in the following areas:

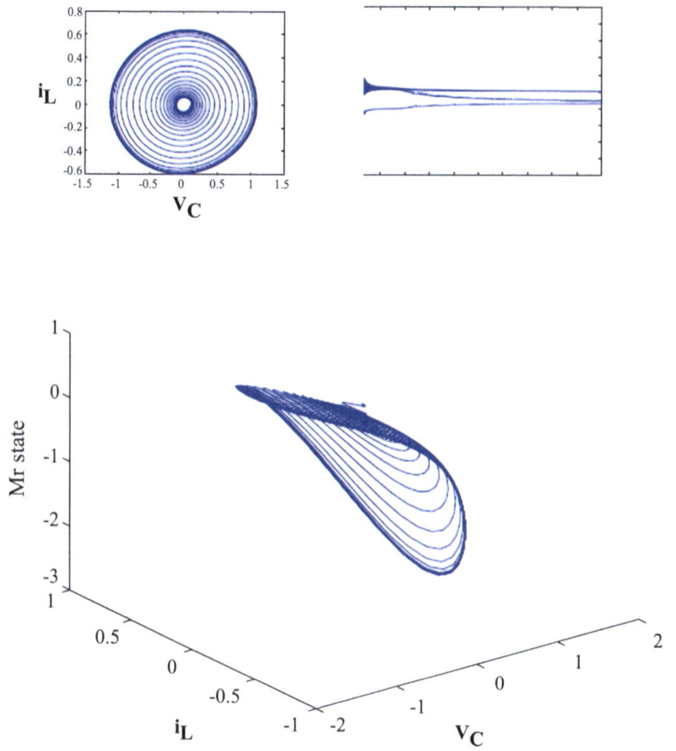

Fig. 1.19 With the parameters $C = 1$; $L = 3.3$; $\beta = 0.1$; $\alpha = 0.1$

(i) Developing high-precision, multi-physics coupled memristor models. Current models are primarily based on electrical characteristics. Future models need to comprehensively consider the influence of multiple physical fields such as thermal, mechanical, electrical, and magnetic, to establish more accurate multi-scale memristor models. This involves the integration of knowledge from nanomaterial science, solid-state physics, and computational electromagnetics. Specifically, research can focus on the effects of temperature, stress, and external magnetic fields on memristor switching characteristics, establishing dynamic models that consider these factors, and optimizing model parameters through molecular dynamics simulations and experimental validation.

(ii) Developing reliability and stability assessment methods for memristor devices. This includes establishing aging models and failure mechanism analysis methods for memristor devices and studying the impact of environmental factors (such as temperature, humidity, and radiation) on the long-term performance of memristors. This involves developing accelerated aging test methods, establishing mathematical models for memristor reliability, and designing self-healing memristor circuits to improve system reliability and lifespan.

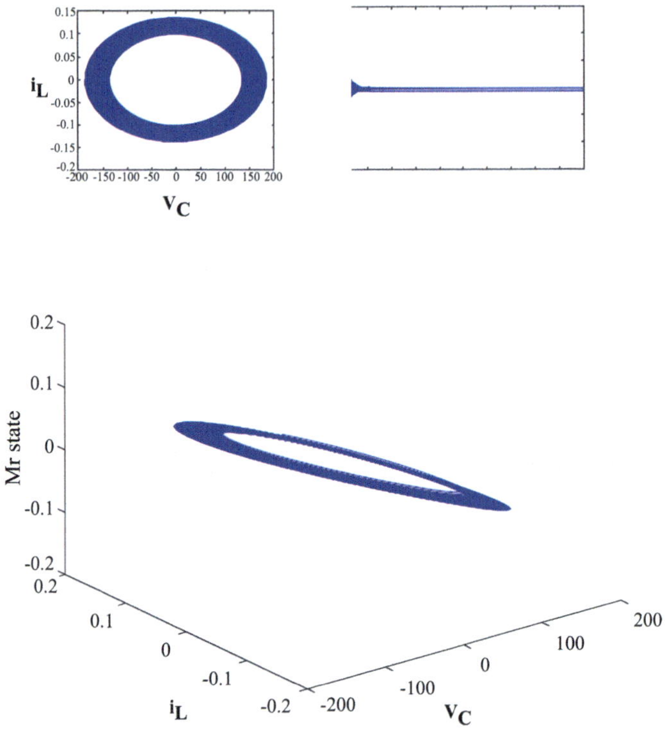

Fig. 1.20 With the parameters $C = 0.0006$; $L = 1100$; $\beta = 1.5$; $\alpha = 0.1$

(iii) Investigating memristor-based applications in complex analog and digital circuits. This involves exploring the potential of memristors in high-order filters, oscillators, analog-to-digital converters, and other complex circuits. The focus is on how to utilize the nonlinear characteristics and memory effects of memristors to simplify circuit structures, reduce power consumption, and improve performance. For example, designing memristor-based adaptive filters that use the dynamic characteristics of memristors to achieve online adjustment of filter parameters, or developing memristor-based chaotic communication systems to explore their applications in secure communications.

(iv) Analyzing the dynamics of memristor-based neural networks. This includes an in-depth study of the dynamic characteristics of memristor-based artificial neural networks, analyzing the evolution laws of memristor-based synaptic weights, studying the impact of network topology on learning performance, and exploring chaos and bifurcation phenomena in memristor-based neural networks. Specifically, this could involve establishing mathematical models of

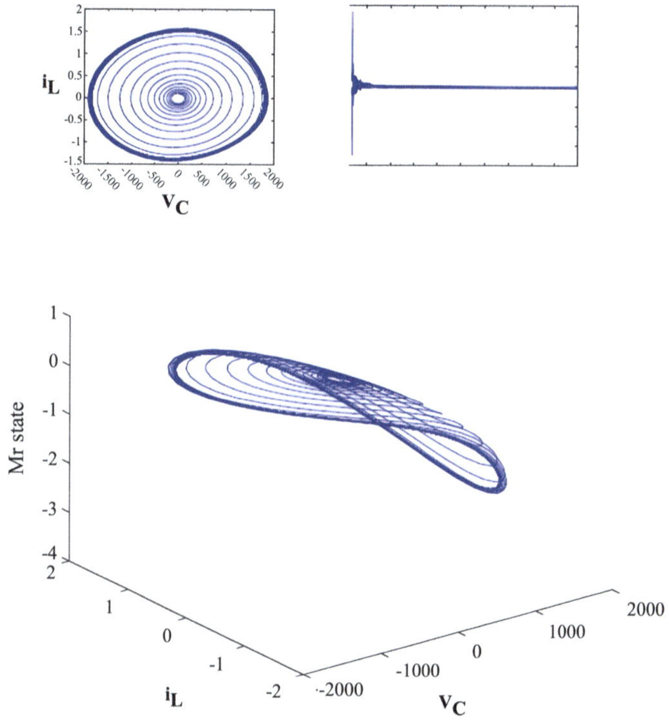

Fig. 1.21 With the parameters $C = 0.0006$; $L = 1000$; $\beta = 100$; $\alpha = 0.1$

memristor-based neurons to analyze their nonlinear dynamic behavior, studying the plasticity mechanisms of memristor-based synapses including short-term and long-term plasticity, and exploring synchronization phenomena in memristor-based neural networks and their role in information processing.

(v) Exploring memristor-based applications in novel computing architectures. This involves researching the role of memristors in emerging computing paradigms such as brain-like computing, edge computing, and quantum-classical hybrid computing. Specifically, this could include designing neural network hardware accelerators based on memristor arrays, studying how to use the analog characteristics of memristors to achieve efficient matrix-vector multiplication, or exploring the application of memristors as classical control units in quantum annealing algorithms, investigating how to leverage the nonlinear characteristics of memristors to enhance the computational capabilities of quantum-classical hybrid systems.

These research directions will help deepen the understanding of memristor systems, expand their applications in artificial intelligence, complex system analysis, and novel computing architectures, and promote the transition of memristor technology from theoretical research to practical applications, providing crucial support for future efficient, intelligent, and low-power electronic systems.

References

1. Li, Y., Wang, Z., Midya, R., Xia, Q., Yang, J.J.: Review of memristor devices in neuromorphic computing: materials sciences and device challenges. J. Phys. D Appl. Phys. **51**(50), 503002 (2018)
2. Chua, L.O.: Memristor-the missing circuit element. IEEE Trans. Circuit Theory **18**(5), 507–519 (1971)
3. Chua, L.O., Kang, S.M.: Memristive devices and systems. Proc. IEEE **64**(2), 209–223 (1976)
4. Chua, L.O.: The fourth element. Proc. IEEE **100**(6), 1920–1927 (2012)
5. Chua, L.O.: How we predicted the memristor. Nat. Electron. **1**(5), 322–322 (2018)
6. Chua, L.O.: Memristor: remembrance of things past. IEEE Micro **38**(5), 7–12 (2018)
7. Strukov, D.B., Snider, G.S., Stewart, D.R., Williams, R.S.: The missing memristor found. Nature **453**(7191), 80–83 (2008)
8. Williams, R.S.: How we found the missing memristor. IEEE Spectr. **45**(12), 28–35 (2008)
9. Chua, L.O.: 10 things you didn't know about memristors. In: Information Storage System and Technology, pp. IS1B–1. Optica Publishing Group (2017)
10. Chua, L.O.: If it's pinched it'sa memristor. Handbook of Memristor Networks, pp. 15–88 (2019)
11. Chua, L.O.: Everything you wish to know about memristors but are afraid to ask. Handbook of Memristor Networks, pp. 89–157 (2019)
12. Chua, L.O.: Resistance switching memories are memristors. Handbook of Memristor Networks, pp. 197–230 (2019)
13. Chua, L.O.: Hodgkin–huxley equations implies edge of chaos kernel. Japanese J. Appl. Phys. **61**(SM), SM0805 (2022)
14. Chua, L.O.: Local activity is the origin of complexity. Int. J. Bifurc. Chaos **15**(11), 3435–3456 (2005)
15. Mainzer, K.: Local activity principle: the cause of complexity and symmetry breaking. In: Chaos, CNN, Memristors and Beyond: A Festschrift for Leon Chua With DVD-ROM, composed by Eleonora Bilotta, pp. 146–159. World Scientific (2013)
16. Chua, L.O.: Five non-volatile memristor enigmas solved. Appl. Phys. A **124**, 1–43 (2018)
17. Ascoli, A., Demirkol, A.S., Tetzlaff, R., Chua, L.O.: Edge of chaos is sine qua non for turing instability. IEEE Trans. Circuits Syst. I Regul. Pap. **69**(11), 4596–4609 (2022)
18. Ascoli, A., Demirkol, A.S., Tetzlaff, R., Chua, L.O.: Edge of chaos theory resolves smale paradox. IEEE Trans. Circuits Syst. I Regul. Pap. **69**(3), 1252–1265 (2022)
19. Yi, W., Tsang, K.K., Lam, S.K., Bai, X., Crowell, J.A., Flores, E.A.: Biological plausibility and stochasticity in scalable vo2 active memristor neurons. Nat. Commun. **9**(1), 4661 (2018)
20. Kumar, S., Williams, R.S., Wang, Z.: Third-order nanocircuit elements for neuromorphic engineering. Nature **585**(7826), 518–523 (2020)
21. Itoh, M., Chua, L.O.: Memristor oscillators. Int. J. Bifurc. Chaos **18**(11), 3183–3206 (2008)
22. Muthuswamy, B.: Implementing memristor based chaotic circuits. Int. J. Bifurc. Chaos **20**(05), 1335–1350 (2010)
23. Muthuswamy, B., Chua, L.O.: Simplest chaotic circuit. Int. J. Bifurc. Chaos **20**(05), 1567–1580 (2010)
24. Xu, Q., Lin, Y., Bao, B., Chen, M.: Multiple attractors in a non-ideal active voltage-controlled memristor based chua's circuit. Chaos, Solitons & Fractals **83**, 186–200 (2016)
25. Bao, B., Wang, Q., Xu, J.: On memristor based five-order chaotic circuit. J. Circuits Syst. **16**(2), 66 (2011)

26. Bao, B., Ma, Z., Xu, J., Liu, Z., Xu, Q.: A simple memristor chaotic circuit with complex dynamics. Int. J. Bifurc. Chaos **21**(09), 2629–2645 (2011)
27. Song, Y., Shen, Y., Chang, Y.: Chaos control of a memristor-based chua's oscillator via backstepping method. In: International Conference on Information Science and Technology, pp. 1081–1084. IEEE (2011)
28. Iu, H.H.C., Yu, D., Fitch, A.L., Sreeram, V., Chen, H.: Controlling chaos in a memristor based circuit using a twin-t notch filter. IEEE Trans. Circuits Syst. I Regul. Pap. **58**(6), 1337–1344 (2011)
29. Ishaq Ahamed, A., Srinivasan, K., Murali, K., Lakshmanan, M.: Observation of chaotic beats in a driven memristive chua's circuit. Int. J. Bifurc. Chaos **21**(03), 737–757 (2011)
30. Ascoli, A., Tetzlaff, R., Corinto, F., Gilli, M.: Pspice switch-based versatile memristor model. In: 2013 IEEE International Symposium on Circuits and Systems (ISCAS), pp. 205–208. IEEE (2013)
31. Wong, H.S.P., Raoux, S., Kim, S., Liang, J., Reifenberg, J.P., Rajendran, B., Asheghi, M., Goodson, K.E.: Phase change memory. Proc. IEEE **98**(12), 2201–2227 (2010)
32. Raoux, S., Burr, G.W., Breitwisch, M.J., Rettner, C.T., Chen, Y.C., Shelby, R.M., Salinga, M., Krebs, D., Chen, S.H., Lung, H.L., et al.: Phase-change random access memory: a scalable technology. IBM J. Res. Dev. **52**(4.5), 465–479 (2008)
33. Sah, M.P., Yang, C., Kim, H., Chua, L.: A voltage mode memristor bridge synaptic circuit with memristor emulators. Sensors **12**(3), 3587–3604 (2012)
34. Ebong, I.E., Mazumder, P.: Cmos and memristor-based neural network design for position detection. Proc. IEEE **100**(6), 2050–2060 (2011)
35. Chua, L.: Resistance switching memories are memristors. Handbook of Memristor Networks, pp. 197–230 (2019)
36. Zhong, G.Q.: Implementation of chua's circuit with a cubic nonlinearity. IEEE Trans. Circuits Syst. I: Fundam. Theory Appl. **41**(12), 934–941 (1994)
37. Bao, B.C., Hu, W., Xu, J.P., Liu, Z., Zou, L.: Analysis and implementation of memristor chaotic circuit. Acta Physica Sinica **60**(12), 63–70 (2011)
38. Yang, L., Pu, Z., Huang, L., Deng, P., Su, J.: Active flux-controlled memristor emulator and the frequency response analysis of memristor based filter. Electronic Components Mater. **34**(5-10+14) (2015). 10.14106/j.cnki.1001-2028.2015.07.002
39. Ou, Q., Zhao, P., You, M.: Active low-pass filter characteristics based on the cubic nonlinear memristor. Appl. Sci. Technol. **41**(6–10) (2014)
40. Dang, B., Yu, Y., Ma, H., Yang, N., Nyima, T.: A novel charge-controlled memristor model and its emulator circuit. In: 2018 IEEE 18th International Conference on Communication Technology (ICCT), pp. 586–590. IEEE (2018)
41. Adhikari, S.P., Sah, M.P., Kim, H., Chua, L.O.: Three fingerprints of memristor. Handbook of Memristor Networks, pp. 165–196 (2019)
42. Liu, W., Wang, F.Q., Ma, X.K.: Exponential flux-controlled memristor model and its floating emulator. Chin. Phys. B **24**(11), 118401 (2015)
43. Sánchez-López, C., Mendoza-Lopez, J., Carrasco-Aguilar, M., Morales-Lopez, F.: A simple floating memristor emulator circuit based on current conveyors. In: 2013 10th International Conference on Electrical Engineering, Computing Science and Automatic Control (CCE), pp. 445–448. IEEE (2013)
44. Sánchez-López, C., Mendoza-Lopez, J., Carrasco-Aguilar, M., Muñiz-Montero, C.: A floating analog memristor emulator circuit. IEEE Trans. Circuits Syst. II Express Briefs **61**(5), 309–313 (2014)
45. Kim, H., Sah, M.P., Yang, C., Cho, S., Chua, L.O.: Memristor emulator for memristor circuit applications. IEEE Trans. Circuits Syst. I Regul. Pap. **59**(10), 2422–2431 (2012)
46. Zhang, X., Huang, Z., Yu, J.: Memristor model for spice. IEICE Trans. Electron. **93**(3), 355–360 (2010)
47. Zhang, Y., Zhang, X., Yu, J.: Approximated spice model for memristor. In: 2009 International Conference on Communications, Circuits and Systems, pp. 928–931. IEEE (2009)

48. Di Ventra, M., Pershin, Y.V., Chua, L.O.: Circuit elements with memory: memristors, memcapacitors, and meminductors. Proc. IEEE **97**(10), 1717–1724 (2009)
49. Tour, J.M., He, T.: The fourth element. Nature **453**(7191), 42–43 (2008)
50. Lindberg, E.: Sinusoidal excitation on the chua's circuit simulation of limit cycles and chaos. In: Proceedings of IEEE International Symposium on Circuits and Systems-ISCAS'94, vol. 6, pp. 5–8. IEEE (1994)
51. Muthuswamy, B., Kokate, P.P.: Memristor-based chaotic circuits. IETE Tech. Rev. **26**(6), 417–429 (2009)

Chapter 2
Chaotic Circuits Based on Memristor

Chaos theory has found significant applications across various fields, particularly in secure communications, encryption systems, and neuromorphic computing. The discovery of memristors—the fourth fundamental circuit element—has opened up new possibilities for developing chaotic circuits with enhanced performance and novel properties. This chapter explores several innovative approaches to designing and analyzing memristor-based chaotic circuits, advancing our understanding of these complex systems and their potential applications.

Section 2.1 investigates chaotic circuits based on two memristors. Through numerical simulations, this study demonstrates the system's chaotic attractors and provides an in-depth analysis of the circuit's dynamical behavior using the Lyapunov exponents and bifurcation diagrams. This work not only expands our understanding of multi-memristor dynamics but also lays a foundation for designing more sophisticated chaotic systems.

Section 2.2 examines memristor-based chaotic circuits with uncertain parameters. By introducing an uncertain model, this research addresses practical issues such as component variability and environmental influences. Through defining the bounds of uncertain parameters and studying their impact on chaotic behavior, it provides crucial insights for designing robust circuits that maintain their chaotic properties under varying conditions.

Section 2.3 introduces the concept of time-delay feedback to further enhance the complexity and applicability of memristive chaotic circuits. The study demonstrates how to incorporate time-delay feedback into conventional memristive circuits and proves through numerical simulations that chaos attractors can be generated in this type of circuit. This innovation not only expands the design space for chaotic circuits but also potentially improves their performance in encryption and secure communication applications.

Section 2.4 delves deeper into the dynamics of time-delay memristive systems. By introducing a time-delay feedback circuit, this study establishes a mathematical model of the time-delay memristive system and derives its chaotic model through

Y. Yu et al., *Memristive Computing*,
https://doi.org/10.1007/978-981-96-0332-9_2

dynamics analysis and computer simulation. This work provides a foundation for understanding and modeling phenomena ranging from weather patterns to stock market fluctuations.

Section 2.5 presents a novel chaotic circuit based on memristive elements, containing only three components: a memristor, a memcapacitor, and a meminductor. The study describes the state equations of the memristive circuit and reveals dynamical behaviors such as equilibrium set, the Lyapunov exponents, and bifurcation diagram through theoretical analyses and numerical simulations. Simulation results demonstrate that this system exhibits rich phenomena of periodic and chaotic behavior. The most striking feature of this work is that it builds a high-dimensional dynamic equation using only three circuit elements and exhibits rich periodic and chaotic behaviors.

Through these diverse studies, this chapter advances our understanding of memristor-based chaotic circuits and lays the groundwork for their application in various fields. From multi-memristor systems to time-delay feedback and uncertain parameters, each section contributes to the development of more sophisticated, robust, and versatile chaotic circuits with potential implications for secure communications, encryption, and neuromorphic computing.

2.1 Chaotic Circuit Based on Two Memristors

2.1.1 Introduction

The physical realization of the memristor, a previously missing fundamental circuit element, has unlocked new horizons across diverse domains including chaos theory, nanoelectronics, and neuromorphic computing. Groundbreaking research by Chua and colleagues has catapulted the study of memristors to the vanguard, conclusively demonstrating the prevalence of chaotic dynamics in nonlinear memristive circuits and underscoring the immense potential of memristor-based chaotic circuits, particularly in secure communications applications.

Contemporary literature in this field endeavors to harness the distinctive attributes of memristors to construct novel chaos generators. By leveraging the second-order characteristics of memristors, a memristor connected to a power source can inherently exhibit chaotic behavior [1]. One of the seminal memristor-based chaotic circuits was pioneered by Itoh and Chua, who elegantly derived memristor oscillators from Chua's oscillators by substituting Chua's diodes with memristors [2]. Similarly, memristive circuits can attain chaotic dynamics by adopting a modified Chua's circuit topology [3, 4]. Experimental observations of chaos in memristor-based chaotic circuits constructed using analog components have further validated these theoretical frameworks [5]. Notably, Muthuswamy and Chua presented the simplest chaotic circuit comprising a linear passive inductor, a linear passive capacitor, and a nonlinear active memristor [6].

However, the majority of these proposed chaotic circuits either overlook or insufficiently explore the rich dynamics arising from the interaction of two or more memristors in nonlinear electronic systems. Consequently, this section delves into the intricate chaotic behavior of circuits incorporating a pair of memristors, unveiling a tapestry of complex dynamical phenomena that emerge from their synergistic interplay.

2.1.2 Memristive Circuit

The memristor defined as the following is a passive two-terminal electronic element which imposes a relationship between the charge q and the flux ϕ. If the constitutive relation can be written as $\phi = \phi(q)$, it is called a charge-controlled memristor, that is memristance $M(q)$. On the other hand, it is called a flux-controlled memristor, that is memductance $W(\phi)$, if the constitutive relation can be written as $q = q(\phi)$. In this section, we design the memristive circuit using two flux-controlled memristors.

$$\begin{aligned} v &= M(q)i,\ M(q) \triangleq \frac{d\phi(q)}{dq} \\ i &= W(\phi)v,\ W(\phi) \triangleq \frac{dq(\phi)}{d\phi}. \end{aligned} \tag{2.1}$$

According to circuit laws, a general class of memristive systems can be designed by using the three traditional basic circuit elements and memristor. Without loss of generality, we can utilize two memristors, two capacitors, two resistors, and one inductor to design the memristive circuit. Figure 2.1 shows the model graphically, where M_1 and M_2 are two flux-controlled memristors, two capacitors C_1 and C_2, one inductor L, two resistors r and R. Applying circuit laws to memristor, we can obtain a relation between the two fundamental circuit variables of the charge and flux. As for capacitor, inductor, and resistor, we can use the voltage and the current to get the relation. When we apply Kirchhoff's circuit laws to the memristive circuit, we can obtain the following equations:

$$\begin{aligned} \frac{d\phi_1}{dt} &= v_3 \\ \frac{d\phi_2}{dt} &= \frac{v_4 - v_3}{RW_2 + 1} \\ \frac{dv_3}{dt} &= \frac{1}{C_1}\left[(-W_1)v_3 - \frac{W_2}{RW_2 + 1}(v_3 - v_4)\right] \\ \frac{dv_4}{dt} &= \frac{1}{C_2}\left[\frac{W_2}{RW_2 + 1}(v_3 - v_4) + i_5\right] \\ \frac{di_5}{dt} &= -\frac{1}{L}v_4 - \frac{r}{L}i_5. \end{aligned} \tag{2.2}$$

Fig. 2.1 Memristive circuit

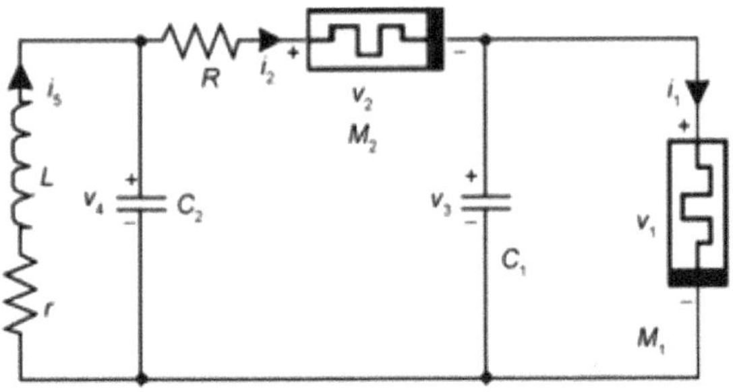

Equation (2.2) defines the relation among the four fundamental circuit variables of voltage, current, charge, and flux. Let $x = \phi_1$, $y = \phi_2$, $z = v_3$, $m = v_4$, $n = i_5$, $\alpha = \frac{1}{C_1}$, $\beta = \frac{1}{L}$, $\gamma = \frac{r}{L}$, $g = R$, and $C_2 = 1$. When $q = q(\phi) = \phi + \phi^3$ is designed, memductance $W(\phi)$ is defined as $W(\phi) \triangleq 1 + 3\phi^2$, and W_1 and W_2 in (2.2) are, respectively, defined as $W_1 = 1 + 3x^2$, $W_2 = 1 + 3y^2$. So, equation (2.2) can be transformed into the following form:

$$\begin{aligned}
\dot{x} &= z \\
\dot{y} &= \frac{m - z}{g(1 + 3y^2) + 1} \\
\dot{z} &= \alpha\left[(-1 - 3x^2)z - \frac{1 + 3y^2}{g(1 + 3y^2) + 1}(z - m)\right] \\
\dot{m} &= \frac{1 + 3y^2}{g(1 + 3y^2) + 1}(z - m) + n \\
\dot{n} &= -m\beta - n\gamma.
\end{aligned} \tag{2.3}$$

Equation (2.3) are a set of five first-order differential systems, which describe the dynamics of the memristive circuit shown in Fig. 2.1.

2.1.3 Numerical Evidence of Chaos

The Lyapunov exponents, which characterize the rate of separation of infinitesimally close trajectories in state space, provide empirical evidence of chaotic behavior. A positive Lyapunov exponent implies an expanding direction in phase space. Furthermore, the negative sum of the Lyapunov exponents indicates contracting volumes in phase space. These two seemingly contradictory properties are characteristic of chaos in a dynamical system. On the other hand, a bifurcation is a period doubling, quadrupling, and so on, that accompanies the onset of chaos in a dynamical system. So, the Lyapunov exponent and bifurcation are studied to indicate chaotic behavior of the memristive circuit in this section.

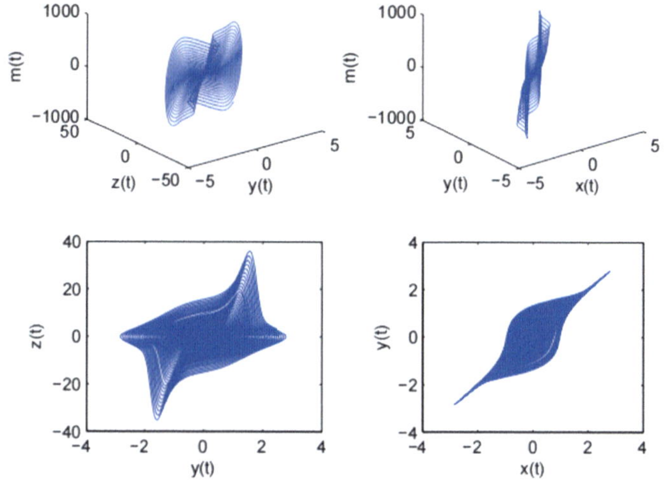

Fig. 2.2 Memristive circuit

2.1.3.1 Chaotic Attractor

Let $\alpha = 8$, $\beta_2 = 10$, $\gamma = 0$, $g = -12$, and initial conditions $[x, y, z, m, n] = [0, 0, 0, 0.08, 0]$, chaotic attractor is shown in Fig. 2.2. In the dynamical memristive circuit, the shape structure of strange attractor has three scrolls, which is different from the double-scroll Chua's attractors.

By computing the values of the Lyapunov exponents, the biggest Lyapunov exponents are positive, and the sum of the Lyapunov exponents is negative. These values of the Lyapunov exponents are indications of chaotic behavior in the memristive circuit.

2.1.3.2 Bifurcation

Bifurcation theory has intensively investigated the topic that bears on chaotic dynamics. In the memristive circuit, bifurcation is a qualitative change in its dynamics produced by varying parameters. The bifurcation parameter from (2.3) is g. When the circuit parameter makes change in $g > 0$, plots of bifurcation are shown in Fig. 2.3. On the other hand, plots of bifurcation corresponding to $g < 0$ are presented in Fig. 2.4 by numerical simulation. These figures illustrate the route to chaos and characterize the existence of chaotic attractors.

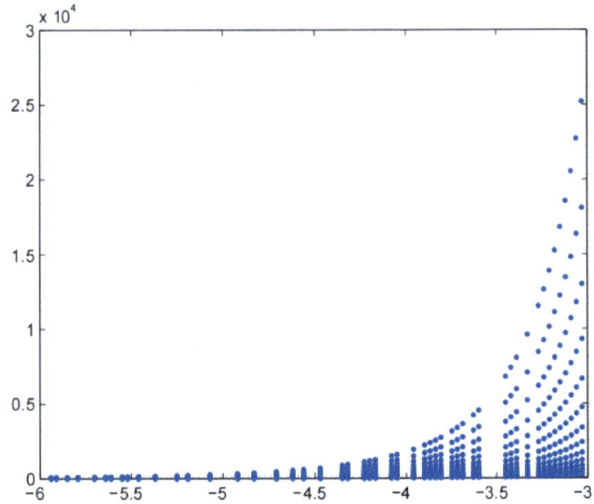

Fig. 2.3 Plots of bifurcation when $g < 0$

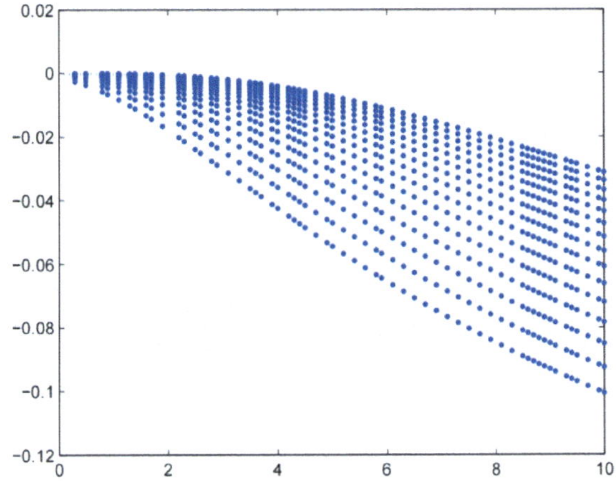

Fig. 2.4 Plots of bifurcation when $g > 0$

2.1.3.3 Lyapunov Exponent Spectrum

By computer simulation, we plot the Lyapunov exponents spectrum of the memristive circuit. Figure 2.5 shows the result from the memristive circuit. When the circuit parameter g is chosen, a plot of the Lyapunov exponent obtained by making simulation is shown in Fig. 2.5. The number of the Lyapunov exponents is same as the number of dimensions of the memristive circuit. So for a five-dimensional system described by the (2.3), we will have five Lyapunov exponents, namely L_1, L_2, L_3, L_4, and L. In Fig. 2.5, there exists a positive Lyapunov exponent, and the sum of the Lyapunov exponents is negative. So, these properties of the Lyapunov exponents are to verify chaotic behavior in the memristive circuit.

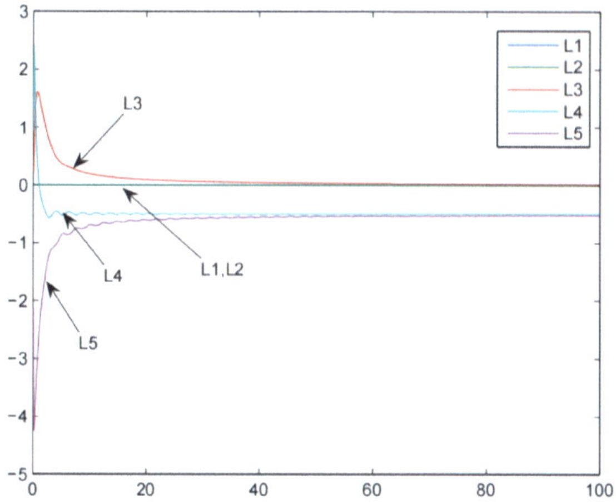

Fig. 2.5 Lyapunov exponent spectrum

2.1.4 *Conclusion*

In this section, chaotic circuit based on two memristors is presented. According to the numerical simulation of the Lyapunov exponent and bifurcation, chaotic behavior of the memristive circuit is studied. An interesting future direction would be to further investigate the dynamics and stability analysis of memristive circuit.

2.2 Memristive Chaotic Circuit with Uncertain Parameters

2.2.1 *Introduction*

Recent research on memristors has yielded significant advancements, positioning memristive devices as promising components for a wide range of applications, including digital memory, analog circuits, and neuromorphic systems. Borghetti et al. established objectives to identify novel physical state variables, devices, and architectures capable of providing both memory and logical functions [7]. Iu et al. designed and implemented a Twin-T notch filter feedback controller to regulate chaos in a memristor-based circuit, demonstrating the potential for controlled chaotic behavior in these systems [8]. Rose et al. explored the use of memristive circuits in realizing artificial synapses, paving the way for neuromorphic computing hardware implementation [9]. Furthermore, certain memristive systems, including memristor-based chaotic circuits (MCCs), show promise for integration into next-generation computing systems with brain-like "neural" capabilities. These diverse applications highlight the versatility and potential impact of memristive devices across multiple domains of electronic and computational systems. The unique properties of memristors, particularly their ability to retain state information and emulate synaptic behavior, position them as key components in the development of advanced computing architectures that more closely mimic biological neural networks.

However, there are some uncertain states and unknown traits in memristor and memristive circuit, which are defined as uncertain parameters of dynamical system. According to [10, 11], uncertain system is a dynamical system described by uncertain parameters. In the section, an uncertain system in MCC is designed, and the relationship between uncertain parameters and chaotic behavior of the memristive circuit is built. To the best of our knowledge, memristor-based chaotic circuit with uncertain parameters is challenging. This section introduces the uncertain model to probe into the nature of memristor and memristive circuit.

2.2.2 Uncertain System

In this section, we introduce uncertain parameters and design the uncertain model in MCC. Based on the model, the equilibrium point and its feature are discussed. The memristor used in this work is a flux-controlled memristor that is characterized by its incremental memductance function $W(\phi)$ describing the flux-dependent rate of change of charge [5]:

$$W(\phi(t)) = \frac{dq(\phi)}{d\phi}. \tag{2.4}$$

Building a circuit with a flux controlled memristor, as shown in Fig. 2.6, using x, y, z, and u instead of $\varphi(t)$, $v_1(t)$, $v_2(t)$, and $i_L(t)$, $\varphi(t)$ denotes the integral of v_1 with integral time constant $-\tau$.

Let a denote C_1, b denote C_2, c denote R and d denote L. The magnetic flux between two terminals of a memristor is $\varphi(t)$.

Then $\dot{x}$, $\dot{y}$, $\dot{z}$, and $\dot{u}$ can be denoted as follows:

$$\begin{cases} \dot{x} = \frac{dx}{d(t)} \\ \dot{y} = \frac{dy}{d(t)} \\ \dot{z} = \frac{dz}{d(t)} \\ \dot{u} = \frac{du}{d(t)}. \end{cases} \tag{2.5}$$

Next, the equations for this system are given below:

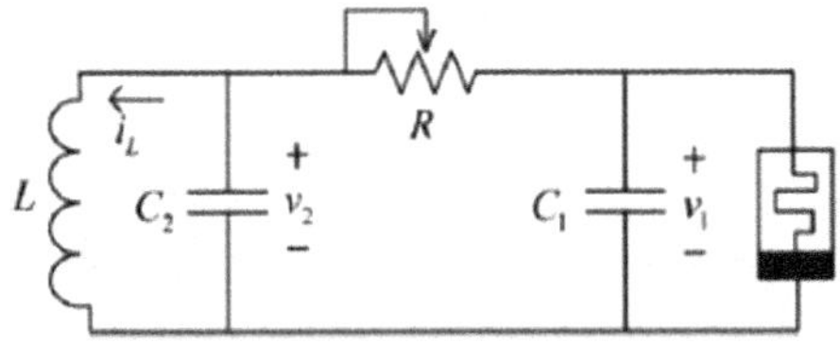

Fig. 2.6 Memristor-based chaotic circuit

$$\begin{cases} \dot{x} = -\tau y \\ \dot{y} = \frac{1}{a}(\frac{z-y}{c} - W(x)y) \\ \dot{z} = \frac{1}{b}(\frac{y-z}{c} - u) \\ \dot{u} = \frac{z}{d}. \end{cases} \tag{2.6}$$

where $\varphi(t)$ denotes the integral of v_1 with integral time constant $-\tau$.

$W(x)$ is the memductance function which given by [11]. Among the equation, m and n are the uncertain parameters.

$$W(\phi(t)) = m + n\varphi(t)^2. \tag{2.7}$$

When setting the $\dot{x}$, $\dot{y}$, $\dot{z}$, and $\dot{u}$ to zero, we obtain the equilibrium points of (2.6).

$$P_0 = (x^0, y^0, z^0, u^0) = (x_0, 0, 0, 0). \tag{2.8}$$

Equation (2.8) is one equilibrium of the system. We can see x_0 is uncertain but constant. The Jacobian matrix of the system at equilibrium point P_0 can be written as:

$$J(P_0) = \begin{bmatrix} 0 & -\tau & 0 & 0 \\ 0 & \frac{-1-c(m+nx_0^2)}{ca} & \frac{1}{ca} & 0 \\ 0 & \frac{1}{cb} & -\frac{1}{cb} & -\frac{1}{b} \\ 0 & 0 & \frac{1}{d} & 0 \end{bmatrix}. \tag{2.9}$$

The system characteristic equation can be written as:

$$\lambda(\lambda^3 + \alpha_2\lambda^2 + \alpha_1\lambda + \alpha_0) = 0, \tag{2.10}$$

where

$$\alpha_2 = \frac{1}{cb} + \frac{1 + c(m + nx_0^2)}{ac}, \tag{2.11}$$

$$\alpha_1 = \frac{1}{db} + \frac{(m + nx_0^2)}{cba}, \tag{2.12}$$

$$\alpha_0 = \frac{1 + c(m + nx_0^2)}{cdab}. \tag{2.13}$$

Figures 2.7 and 2.8 display the phase portraits and surface curved of α_0, α_1 and α_2. In these graphics, we set the bounds of m and n into $[m_{min}, m_{max}]$ and $[n_{min}, n_{max}]$, in Figs. 2.7 and 2.8, both the bounds of m and n chose [0, 0.002]. The bounds of α_0, α_1 and α_2 is $[0.0067, 1.5000e + 006]$, [3.333, 3.3333], [150000, 450000].

After observing the graphics of α_0, α_1, and α_2, the value of λ will be calculated. If no root of (2.10) has a real part greater than zero, P_0 is a stable equilibrium, else P_0 is an unstable equilibrium. So, the next section will find the bounds of some uncertain

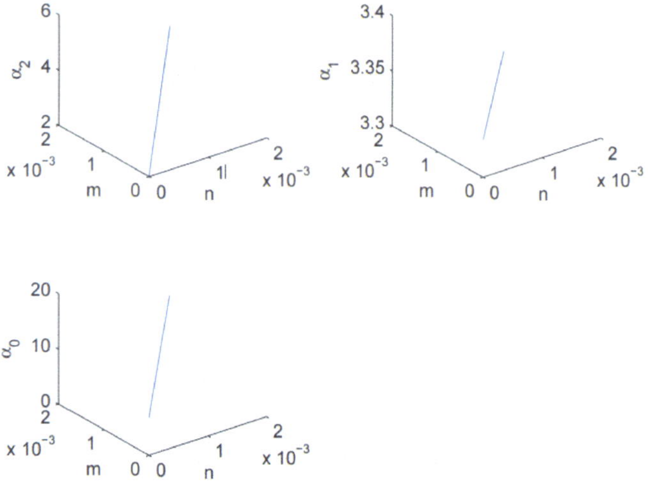

Fig. 2.7 Phase portraits of α_0, α_1, and α_2

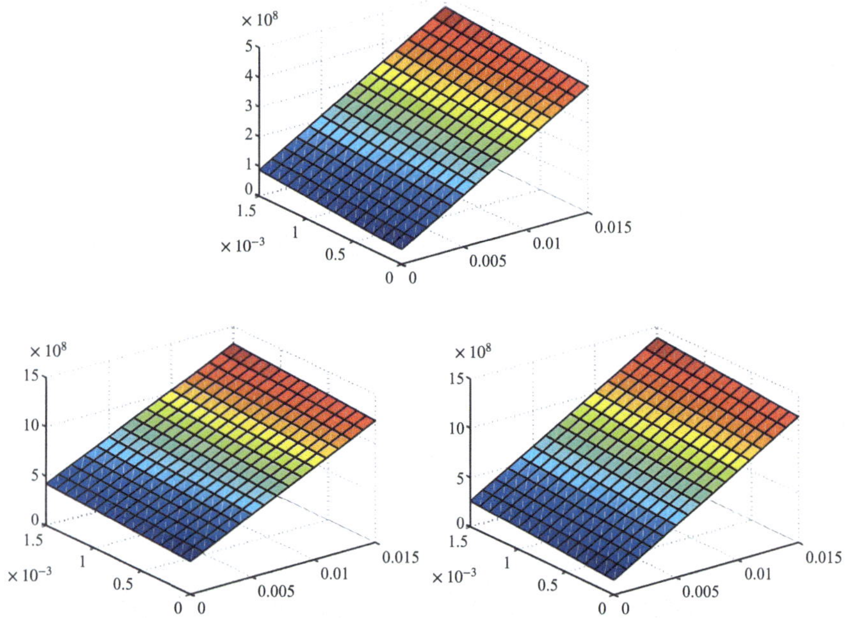

Fig. 2.8 Curve surface of α_0, α_1, and α_2

parameters after calculating the Lyapunov exponents of this uncertain system and depending on the several results.

2.2.3 *Numerical Evidence of Chaos*

According to the equations in section II, we calculate the uncertain parameters m and n in this section. By setting $d = 1, a = 0.001, b = 0.3, C = 500$, and $\tau = 1000$, we obtain an uncertain system for m and n. The initial values of state variables are $x = 0$, $y = 0$, $z = 1$, $u = -1$. By setting $m = 0.01$ and $n = 0.0003$, computer simulation shows that has a chaotic attractor as shown in Fig. 2.9. As known, if one Lyapunov exponent is greater than zero and sum of all Lyapunov exponents is less than zero, the system is known to exhibit chaos. In Fig. 2.10, we can see the $\lambda_1 = 0.58637$, $\lambda_2 = -0.05786$, $\lambda_3 = -2.3238$, and $\lambda_4 = -27.0186$. So, it can be clearly seen that the equilibrium point P_0 is not stable and Eq. (2.6) is in chaotic state. Next, we can change the uncertain parameters m and n into different values to test the stability of (2.6). If changing the value of n into 0.3 and 0.000000003. The phase portraits of chaotic attractor will become different and the stability of (2.6) changes too. Figure 2.11 shows the strange portraits of (2.6) when $n = 0.1$, while Fig. 2.12 shows the $n = 0.000000001$. By setting the value of n into 0.1 and 0.000000001, it can be seen that none of Lyapunov exponents is greater than zero but sum of all Lyapunov exponents is less than zero (Maybe the portraits of Fig. 2.12 looks chaotic). That

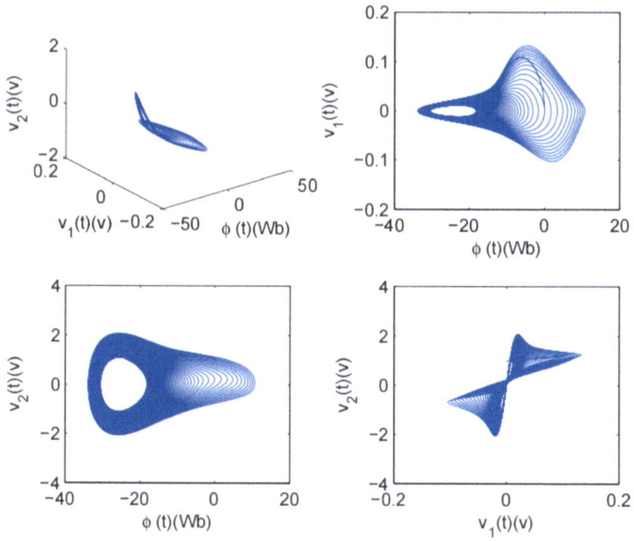

Fig. 2.9 Chaotic attractor of System

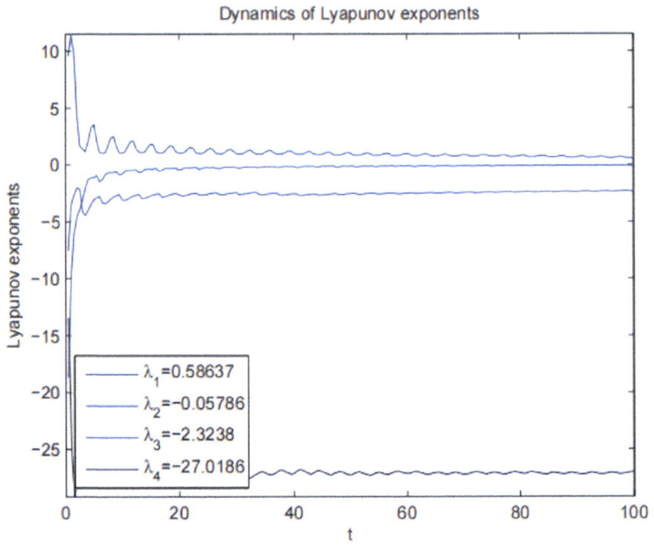

Fig. 2.10 Lyapunov exponent

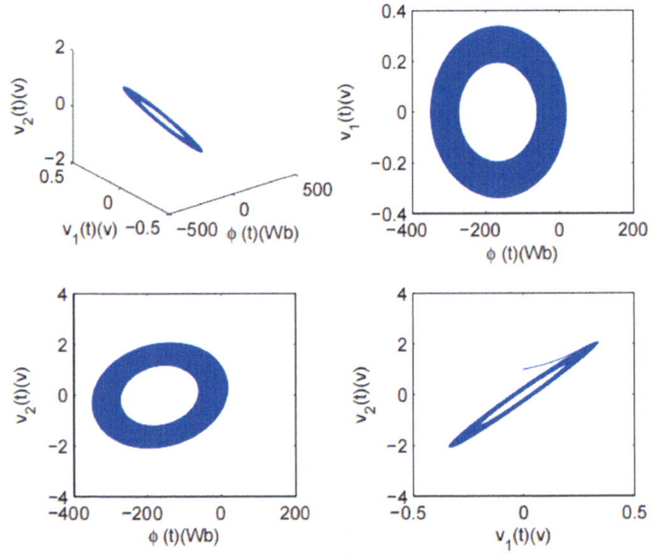

Fig. 2.11 Phase portraits of chaotic attractor, $n = 0.1$

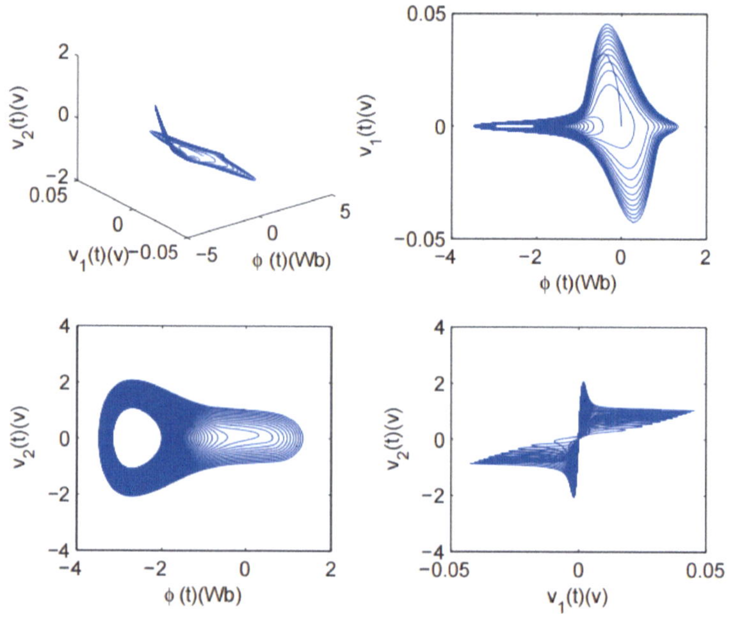

Fig. 2.12 Phase portraits of chaotic attractor, $n = 0.000000001$

means when $n = 0.1$ and $n = 0.000000001$, Eq. (2.6) is not chaotic, so $n = 0.1$ and $n = 0.000000001$ don't belong to $[n_{min}, n_{max}]$. As the same to the value of m, it can be known that not every value of uncertain parameters made the system become chaotic. By calculating the Lyapunov exponents in (2.6), we can get the chaos when $n \in (0.0000051, 0.0015)$ and $m \in (0, 0.0142)$ with the method in [8]. That's the range of the uncertain parameters.

2.2.4 Conclusion

This section introduces the uncertain model to study the nature of memristor and memristive circuit. We design an uncertain system in MCC to find the relationship between uncertain parameters and chaotic behavior. By the analysis of dynamical behavior, we make numerical simulation of the proposed memristive circuit with uncertain parameters, and present the bounds of the uncertain parameters.

2.3 Memristive Chaos Generator with Time-Delay Feedback

2.3.1 Introduction

The incorporation of time-delay feedback, a powerful nonlinear control technique, has emerged as a promising approach to enrich the dynamics and broaden the applicability of memristive chaotic circuits. Time-delay feedback, which involves introducing a time-delayed version of the system's output as an input, has been shown to induce or enhance chaotic behavior in various systems, including electronic circuits, lasers, and neural networks. The combination of memristive circuits and time-delay feedback offers an attractive framework for generating high-dimensional, tunable, and robust chaotic signals with potential applications in secure communication, chaos-based computing, and neuromorphic engineering.

In this context, this section focuses on the design and analysis of a novel chaos generator based on a memristive circuit with time-delay feedback.

2.3.2 About Memristor

The flux-controlled memristor in Fig. 2.13 is described by:

$$i = W(\varphi)v \tag{2.14}$$

and function $W(\varphi)$ called memductance, is defined by:

$$W(\varphi) \triangleq \frac{dq(\varphi)}{d\varphi}. \tag{2.15}$$

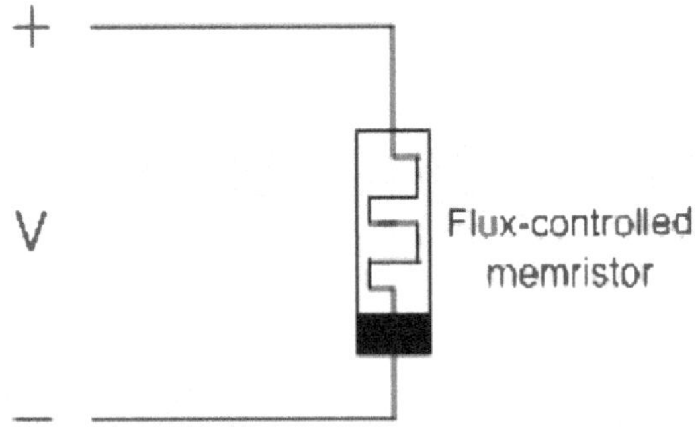

Fig. 2.13 Flux-controlled memristor

Fig. 2.14 Active memristor

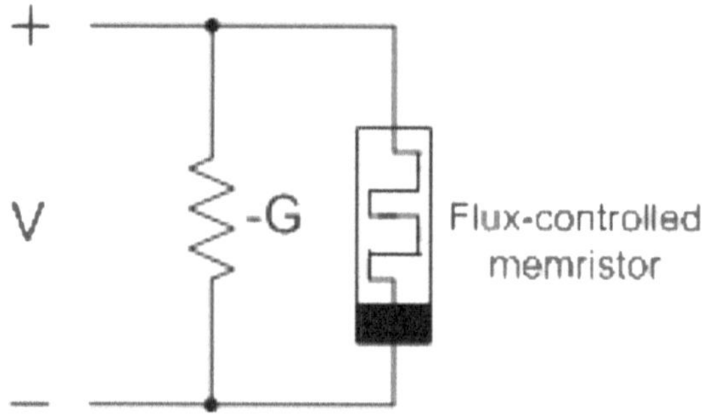

In this section, we assume that scalar function $q(\varphi)$ [2] is defined by:

$$q(\varphi) = b\varphi + 0.5(a - b)(|\varphi + 1| - |\varphi - 1|), \tag{2.16}$$

where $a, b > 0$. Thus, we obtain the memductance, as given below:

$$W(\varphi) = \frac{dq(\varphi)}{d\varphi} = \begin{cases} a, |w| < 1, \\ b, |w| > 1. \end{cases} \tag{2.17}$$

Since the memristor in Fig. 2.13 is passive, we can get an active memristor consisting of a negative conductor and a passive memristor, which is shown in Fig. 2.14.

2.3.3 *Time-Delay Memristive Circuit*

We use two resistors, one inductor, two capacitors, and a flux-controlled memristor to compose a conventional memristive circuit shown in Fig. 2.15. Now, we add the time-delay feedback into the memristive circuit, and obtain the circuit of Fig. 2.16. First, we analyze this circuit based on a passive memristor. The dynamical equation of this circuit can be described by:

$$\left.\begin{aligned} C_1\frac{dv_1}{dt} &= G(v_2 - v_1) - W(\varphi)v_1 \\ C_2\frac{dv_2}{dt} &= i_3 - G(v_2 - v_1) \\ L\frac{di_3}{dt} &= -(r_0 i_3 + v_2 + \varepsilon \sin(\sigma v_1(t - \tau))) \\ \frac{d\varphi}{dt} &= v_1. \end{aligned}\right\} \tag{2.18}$$

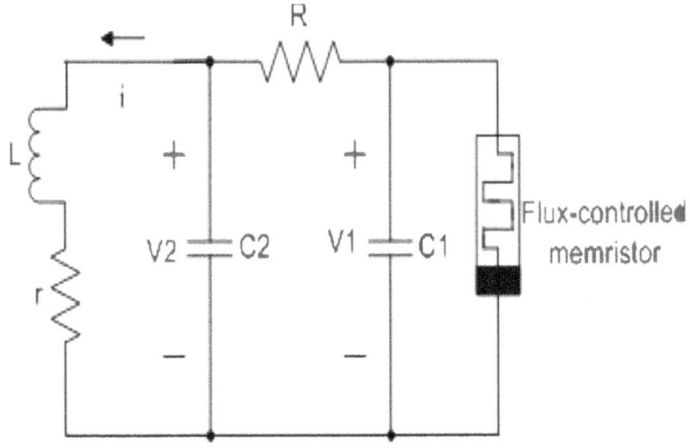

Fig. 2.15 Memristor circuit

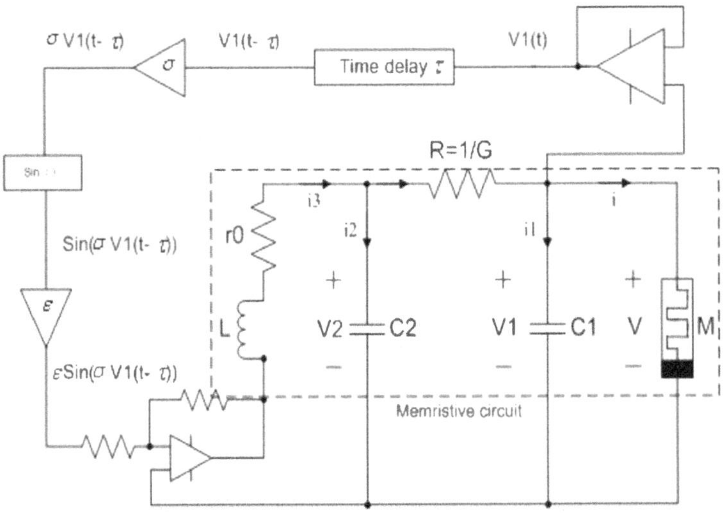

Fig. 2.16 Memristor circuit with time-delay feedback based on a passive memristor

The time-delay memristive circuit with an active memristor is shown in Fig. 2.17. Then, we give the dynamical equation of the circuit based on an active memristor:

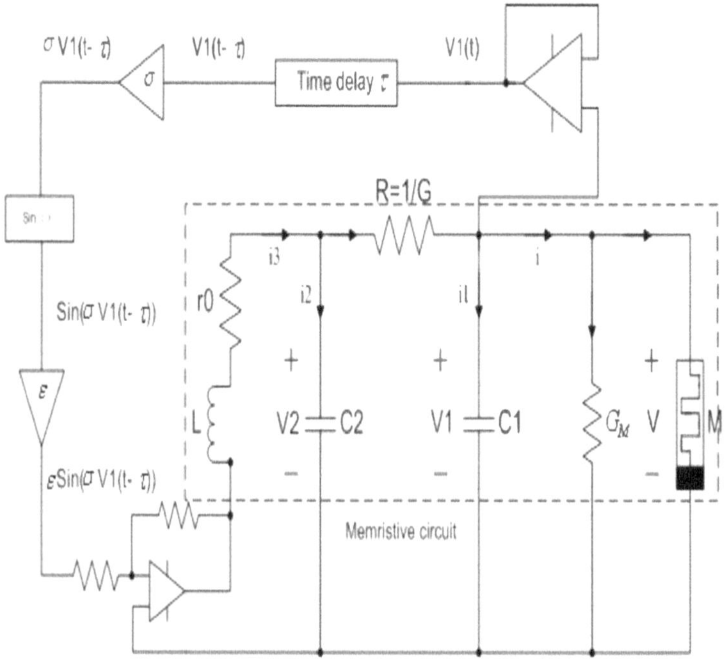

Fig. 2.17 Memristor circuit with time-delay feedback based on an active memristor

$$\left.\begin{aligned} C_1\frac{dv_1}{dt} &= G(v_2 - v_1) + G_M v_1 - W(\varphi)v_1 \\ C_2\frac{dv_2}{dt} &= i_3 - G(v_2 - v_1) \\ L\frac{di_3}{dt} &= -(r_0 i_3 + v_2 + \varepsilon \sin(\sigma v_1(t-\tau))) \\ \frac{d\varphi}{dt} &= v_1. \end{aligned}\right\} \tag{2.19}$$

2.3.4 Chaos Attractor

Computer simulation demonstrated that chaos attractor can be generated in the circuit of Fig. 2.16 when

$$\begin{aligned}
&C1 = 0.000001, C2 = 0.000001, L = 18, r0 = 14, \tau \\
&\quad = 0.001, G = 0.0000025, \varepsilon = 11, \sigma = 5, a = 0.0004, b \\
&\quad = 0.0005; \\
&C1 = 0.0001, C2 = 0.000001, L = 6, r0 = 1, \tau = 0.01, G = \\
&\quad 0.000003, \varepsilon = 1500, \sigma = 5, a = 0.0004, b = 0.0005; \\
&C1 = 0.0001, C2 = 0.000001, L = 1, r0 = 0.01, \tau \\
&\quad = 0.001, G = 0.0000025, \varepsilon = 3000, \sigma = 5, a = 0.0004, \\
&\quad b = 0.0005;
\end{aligned} \tag{2.20}$$

as shown in Fig. 2.18.

In the circuit of Fig. 2.17, chaos attractor can be generated when

$$\begin{aligned}
&C1 = 0.0000001, C2 = 0.0000001, L = 18, r0 = 14, \tau \\
&\quad = 0.001, G = 0.0000015, \varepsilon = 11, \sigma = 5, a = 0.0004, b \\
&\quad = 0.0005, GM = 0.000001; \\
&C1 = 0.00001, C2 = 0.0000001, L = 1, r0 = 1, \tau = 0.001, G = \\
&\quad 0.000001, \varepsilon = 30000, \sigma = 5, a = 0.0008, b = 0.0008, GM \\
&\quad = 0.000008; \\
&C1 = 0.001, C2 = 0.00001, L = 0.01, r0 = 0.01, \tau = 0.001, G \\
&\quad = 0.000001, \varepsilon = 35000, \sigma = 4, a = 0.0008, b = 0.0008, GM \\
&\quad = 0.00001;
\end{aligned} \tag{2.21}$$

as shown in Fig. 2.19.

2.3.5 Conclusion

Based on the time-delay feedback, conventional memristive circuit is used to design a new kind of circuit, simulation results demonstrated that this kind of memristive circuit can generate chaos attractor.

2.4 Chaotic Modeling of Time-Delay Memristive System

2.4.1 Introduction

Time-delay frequently appears in linear or nonlinear systems. Typical examples of time-delay systems are electronic networks, communication circuits, chemical processes, teleoperation systems, biosystems, underwater vehicles, and so on, where the

Fig. 2.18 Chaos attractor

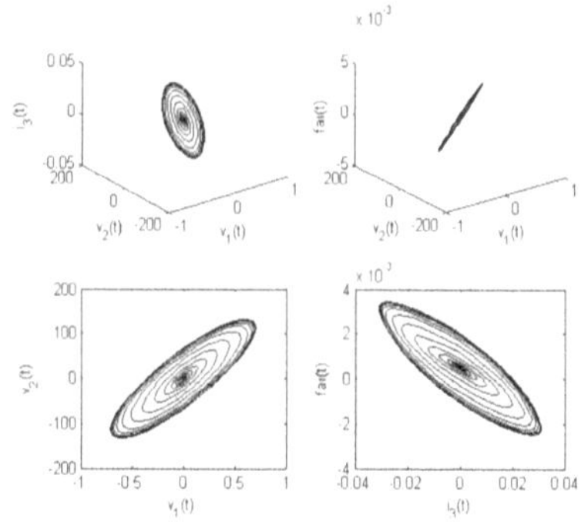

(a) chaos attractor of the first set of parameters

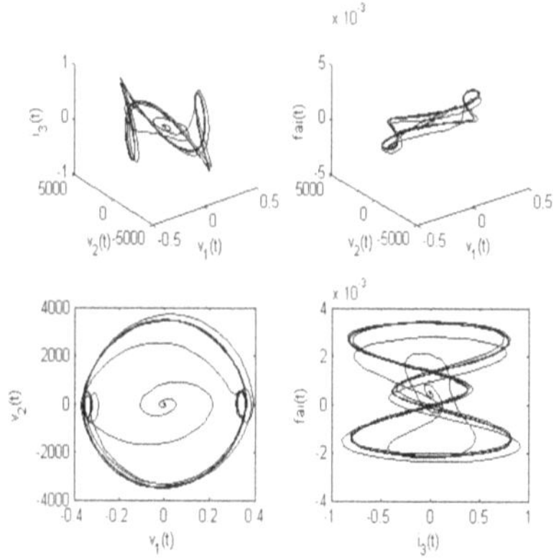

(b) chaos attractor of the second set of parameters

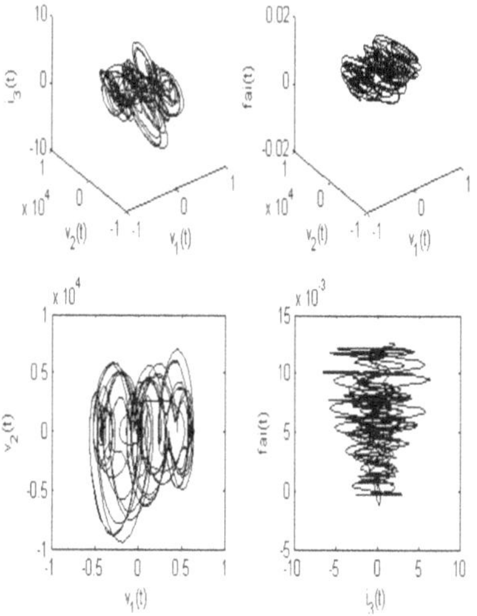

(c) chaos attractor of the third set of parameters

Fig. 2.19 Chaos attractor

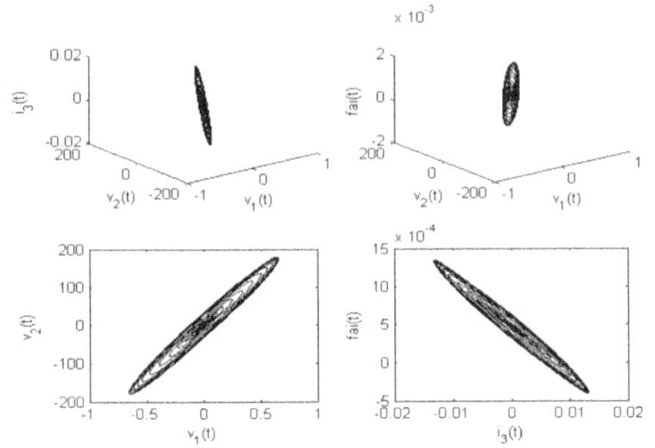

(a) chaos attractor of the first set of parameters

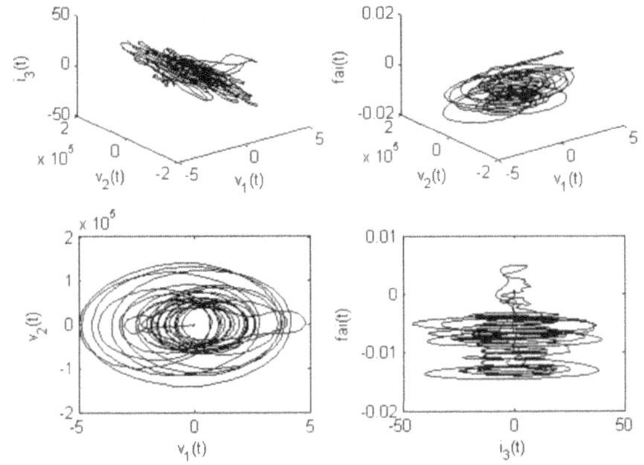

(b) chaos attractor of the second set of parameters

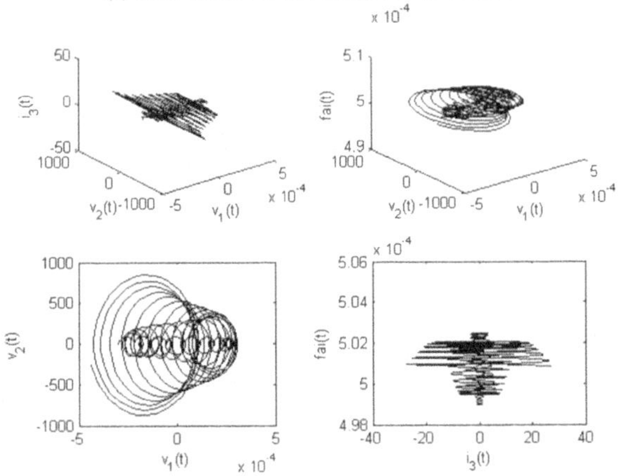

(c) chaos attractor of the third set of parameters

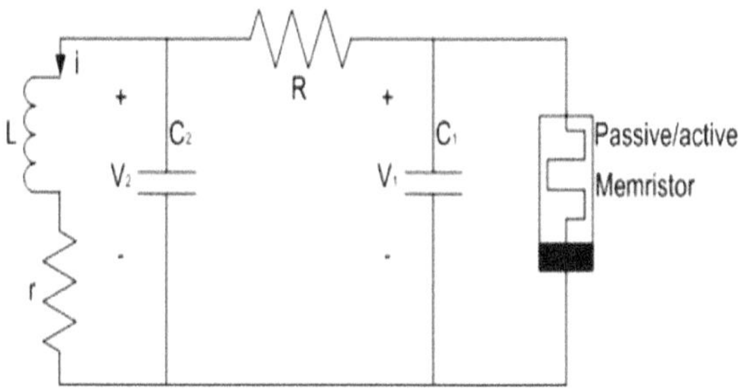

Fig. 2.20 Memristive system with *R*, *L*, *C*, and *M*

state change rate depends not only on the current state of the system but also its state at some time in the past. Delay effects may introduce instability, oscillators, chaos, and other complex dynamical behaviors. Moreover, the presence of time delay makes system design and analysis much more complicated. Thus, time-delay dynamical systems have been the subject of extensive research during the past decades [12–16].

This section investigates chaotic effects on memristive systems with time-delay. By introducing time-delay circuit, we build mathematical model of time-delay memristive system. Furthermore, we analyze the time-delay dynamical system and present some new chaotic behaviors of the time-delay memristive system with feasible and effective simulation results. To the best of our knowledge, the proposed work in this section on memristive systems with delayed and chaotic effects is new in the literature.

2.4.2 *Memristive System*

By using passive or active memristors and the three traditional basic circuit elements, we can design a general class of memristive systems. Without loss of generality, we study the flux-controlled memristive system in Fig. 2.20, which consists of two resistors, a inductor, two capacitors, and a passive memristor or an active memristor. Replaced passive or active memristor by Chua's diodes, the memristive system is transformed into a Chua's oscillator [2].

As for passive memristor, applying Kirchhoff's circuit laws to the nodes and the loop of the circuit in Fig. 2.20, we obtain:

$$\begin{cases} C_1\frac{dv_1}{dt} = \frac{v_2-v_1}{R} - W(\varphi)v_1 \\ C_2\frac{dv_2}{dt} = i - \frac{v_2-v_1}{R} \\ L\frac{di}{dt} = -ri - v_2 \\ \frac{d\varphi}{dt} = v_1. \end{cases} \tag{2.22}$$

When a flux-controlled memristor connects in parallel with a negative conductance ($-G_M$), an active memristor is obtained. At the same time, the dynamic equations of memristive system with an active memristor are presented below:

$$\begin{cases} C_1\frac{dv_1}{dt} = \frac{v_2-v_1}{R} + G_M v_1 - W(\varphi)v_1 \\ C_2\frac{dv_2}{dt} = i - \frac{v_2-v_1}{R} \\ L\frac{di}{dt} = -ri - v_2 \\ \frac{d\varphi}{dt} = v_1, \end{cases} \tag{2.23}$$

where the nonlinear function $W(\varphi)$ is given by:

$$W(\varphi) = \frac{dq(\varphi)}{d\varphi} = \frac{d[b\varphi + 0.5(a-b)(|\varphi+1| - |\varphi-1|)]}{d\varphi} = \begin{cases} a, |\varphi| < 1 \\ b, |\varphi| > 1 \\ . \end{cases}$$

According to the above-mentioned memristive systems, time delay happens frequently because of transmitting trait of electronic signal and processing of electronic devices. Hereby, we introduce time delay to memristive system and study dynamical behaviors of time-delay memristive systems.

2.4.3 *Main Results*

In this section, we introduce time delay in the memristive system and design a time-delay circuit. Then, we analyze the time-delay dynamical system, and study chaotic behaviors of the time-delay memristive system with numeral simulation.

2.4.3.1 Design of Time-Delay Memristive System

According to memristor characteristic and circuit theory, we can design time-delay memristive systems. By adding a small-amplitude voltage time-delay feedback in the memristive system, the modified memristive system is obtained. Figure 2.21 shows time-delay memristive system based on a passive memristor and an active memristor. When the memristor is passive, the dynamics of time-delay system is governed by the following state equations:

$$\begin{cases} C_1\frac{dv_1}{dt} = \frac{v_2-v_1}{R} - W(\varphi)v_1 \\ C_2\frac{dv_2}{dt} = i - \frac{v_2-v_1}{R} \\ L\frac{di}{dt} = -ri - v_2 - \varepsilon\sin(\sigma v_1(t-\tau)) \\ \frac{d\varphi}{dt} = v_1. \end{cases} \tag{2.24}$$

At the same time, the corresponding form of time-delay memristive system with an active memristor is presented:

$$\begin{cases} C_1\frac{dv_1}{dt} = \frac{v_2-v_1}{R} + G_M v_1 - W(\varphi)v_1 \\ C_2\frac{dv_2}{dt} = i - \frac{v_2-v_1}{R} \\ L\frac{di}{dt} = -ri - v_2 - \varepsilon\sin(\sigma v_1(t-\tau)) \\ \frac{d\varphi}{dt} = v_1. \end{cases} \tag{2.25}$$

We will show that the chaotic model of memristive system with time delay can be achieved for the given parameters of $C_1, C_2, L, r, R, a, b, \sigma, \varepsilon, \tau$.

Remark 2.1 Time-delay term is taken as $\varepsilon\sin(\sigma v_1(t-\tau))$. Obviously, the maximum amplitude of the time-delay term is ε, which is an important parameter in the dynamical model.

Remark 2.2 In the time-delay feedback circuit, time delay τ is small, which means the total delay of transmission line and circuit elements such as memristor, amplifier, and so on.

Remark 2.3 According to memristor theory and circuit law, we can implement circuit design of time delay.

2.4.3.2 Chaos Attractor

For memristors and standard circuit elements, their combination is a typical nonlinear system for nonlinear memristor. By the means of dynamics analysis and computer simulation, we study chaotic modeling of the nonlinear memristive system with time delay.

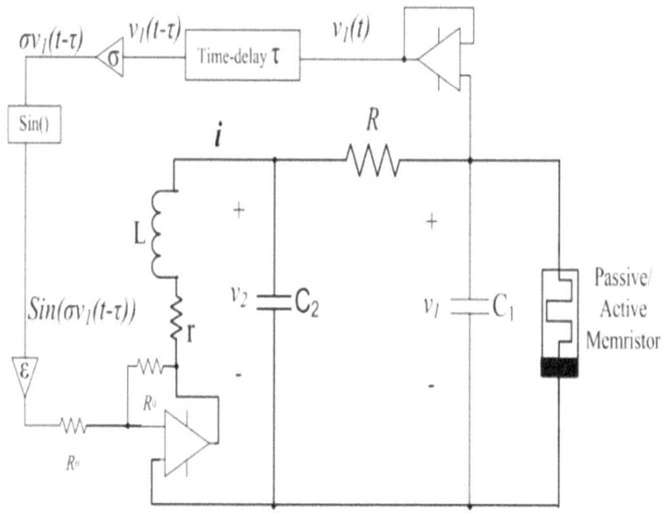

Fig. 2.21 Memristive system with time delay

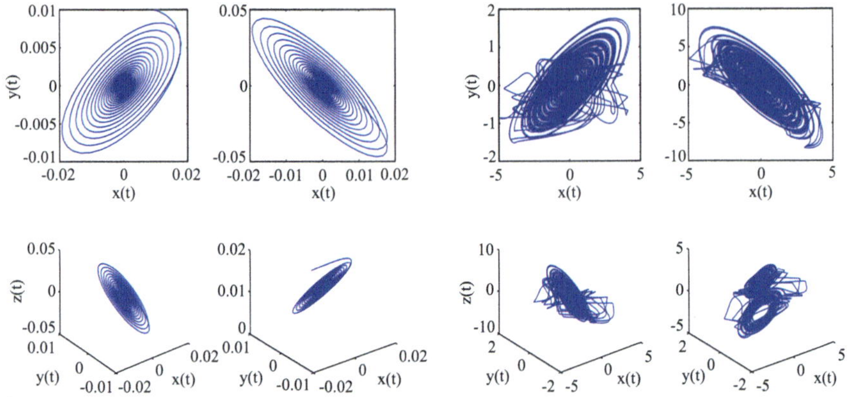

Fig. 2.22 The dynamical behavior of memristive system with a passive memristor (left) or an active memristor (right)

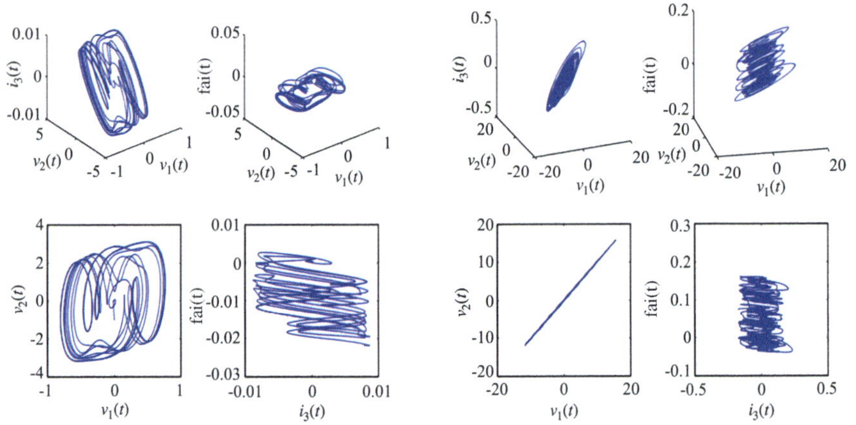

Fig. 2.23 The dynamical behavior of time-delay memristive system with a passive memristor (left) or an active memristor (right)

First, we investigate the dynamical behaviors of memristive system without time delay. When $C_1 = 0.1$, $C_2 = 1$, $L = 1/18$, $r = 1/45$, $R = 1$, $a = -0.7$, and $b = -10$, the dynamical behavior of memristive system with a passive memristor is shown in Fig. 2.22 (left). At the same time, while $C_1 = 0.1$, $C_2 = 1$, $L = 1/13$, $r = 7/260$, $G_M = 1.5$, $a = 0.3$, and $b = 0.8$, the memristive system with an active memristor has a double scroll attractor as shown in Fig. 2.22 (right). Figure 2.22 demonstrates that chaos happens in the memristive system. Obviously, the memristive system shown in Fig. 2.20 can be regarded as chaos generator.

Note that time delay may bring complex dynamical behaviors with memristive system. Consequently, we pay attention to the dynamics of a time-delay system based on a passive memristor or an active memristor. If we set $C_1 = 0.0001$, $C_2 =$

0.000001, $L = 1$, $r = 0.01$, $R = 1000/3$, $a = 0.0004$, $b = 0.0005$, $\varepsilon = 3$, $\sigma = 5$ and time delay $\tau = 0.01$, the computer simulation shows that time-delay dynamical system with a passive memristor has a chaotic attractor as shown in Fig. 2.23 (left). At the same time, the following parameter values of time-delay system with an active memristor are fixed for our computer simulations:

$$\begin{cases} C_1 = 0.00001, C_2 = 0.00001, L = 0.1, r = 1, R = 1, a = 0.01, b = 0.01 \\ G_M = 0.00001, \varepsilon = 10, \sigma = 1, \tau = 0.01 \end{cases}$$

and a strange chaotic attractor is obtained, as shown in Fig. 2.23 (right).

By adding the time-delay feedback with different values of σ, ε and τ, Fig. 2.23 shows distinct chaotic attractors. Firstly, Fig. 2.23 has multiple scroll attractor compared with a double scroll attractor shown in Fig. 2.22. The contour of simulation result shows that the introduced time delay may lead to dynamical complication of memristive system. Secondly, the parameters of σ, ε, and τ are sensitive to chaotic behaviors of the time-delay memristive system. Finally, the intent on study chaotic model of time-delay memristive system is to probe into the nonlinear characteristic of memristors.

2.4.4 *Conclusion*

This section summarizes comprehensively memristor traits and investigates memristive systems with chaotic and delayed effects. The contribution of this section lies in the following aspects. Firstly, we find an efficient way to introduce time delay and implement time-delay circuit in memristive system. Furthermore, the proposed time-delay design can be extended to other nonlinear systems. Secondly, we study chaotic behaviors of time-delay memristive system, and present a novel chaos generator based on a passive memristor and an active memristor. Finally, this work helps understand the nonlinear characteristic of memristor.

2.5 Chaotic Circuit with Memristive Elements

2.5.1 *Introduction*

Chaotic circuit is widely used to various fields encryption and decryption systems [17, 18] and neural networks [19–21]. The performance of the encryption and decryption system will be better if the dimension of the chaotic circuit is higher or the behavior is richer [22, 23]. Moreover, the memristive elements are dynamical elements with nonlinear characteristics. Therefore, there is no need designing extra nonlinear components when build to chaotic circuit. In this section, a high-dimensional chaotic circuit only using three circuit elements is presented, from which periodic and chaotic behaviors are exhibits.

In this section, by parallel connection circuit of the memristive elements, the novel chaotic circuit is designed. In this cases, complex nonlinear dynamic behaviors are investigated by using the Phase Diagrams, equilibrium set, Lyapunov exponent, bifurcation diagram. Furthermore, the electronic circuit is simulated to verify dynamical chaotic behaviors of the memristive elements based chaotic circuit.

2.5.2 *Preliminaries*

The models of the memristive elements, memristor [24], memcapacitor [25] and meminductor [26]. The mathematical model of memristor is proposed as follows:

$$\begin{cases} W(z) = a + bz^2 \\ \dot{z} = -v_M - \alpha z + v_M z^2, \end{cases} \tag{2.26}$$

where the $W(z)$ denotes memductance of memristor, a and b are the internal constant of $W(z)$, v_M denotes voltage of memristor, z is the internal state of memristor and α is the constant.

The memcapacitance which represents the relation of flux and charge is configured as charge controlled as follows:

$$\varphi = a\sigma + b\sigma^3. \tag{2.27}$$

The derivative equation form (2.27) with respect to time t can be obtained as follows:

$$u(t) = (a + b\sigma^2)q(t), \tag{2.28}$$

where the parameters a and b are constant, $\sigma = \int_{t_0}^{t} q(\tau)d\tau$.

A smooth continuous cubic monotone-increasing nonlinearity meminductor is described by:

$$q = c\rho + b\rho^3. \tag{2.29}$$

The derivative equation form (2.29) with respect to time t can be obtained as follows:

$$i(t) = (c + d\rho^2)\varphi(t), \tag{2.30}$$

where c and d are constant, $\varphi(t) = d\rho/dt$. The three models are applied into this section to build a new chaotic circuit.

Table 2.1 Comparisons of memristive element-based chaotic circuit

Item	Ref. [16]	Ref. [14]	Ref. [18]	Proposed circuit
Linear element	1 resistor 1 capacitor 1 Negative conductance	1 resistor 1 capacitor 1 Negative conductance	2 resistor 2 capacitor	No
Memristive elements	1 memristor 1 memcapacitor	1 memcapacitor 1 meminductor	1 meminductor	1 memristor 1 memcapacitor 1 meminductor
The number of components	Limit cycle Chaotic attractors	Five	Five	Three
Dimension	Five	Five	Four	Five
Dynamic behavior	Limit cycle Chaotic attractors Two kinds of chaotic attractors	Limit cycle Chaotic attractors	Limit cycle Chaotic attractors	Limit cycle Chaotic attractors

2.5.3 The Chaotic Circuit Memristive Element

The comparisons of some chaotic circuit of memristive elements are described in Table 2.1. It is obvious from the table that the chaotic circuit proposed in this section has a simple circuit structure only using three memristive elements to build a high-dimensional equation. This system has rich dynamic behavior. In the next work, the chaotic circuit based memristive elements is designed, the dynamical behaviors are revealed by theoretical analysis and numerical simulations.

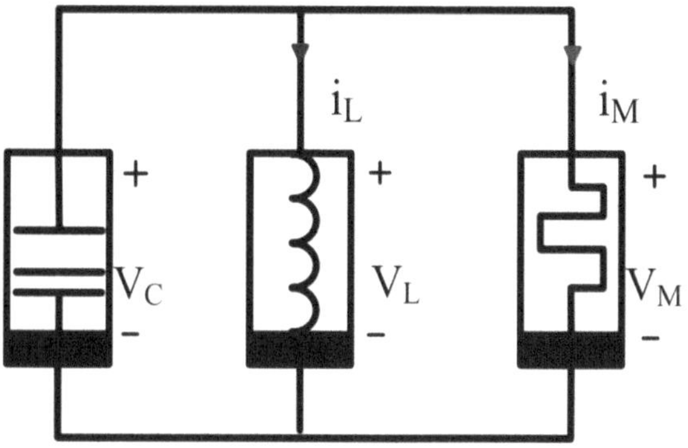

Fig. 2.24 Memristive elements chaotic circuit

2.5.3.1 Design of Memristive Elements Chaotic Circuit

The idea of designing the memristive elements chaotic circuit is inspired by the published discussion in [26]. By replacing the normal element with memory element, the memristive elements chaotic circuit can be obtained, as shown in Fig. 2.24. By applying Kirchhoff's laws to the circuit in Fig. 2.24, the five state equations of the memristive elements chaotic circuit are described by:

$$\begin{cases} \frac{dq_c}{dt} = (-c + d\rho_L^2)\varphi_L - (a + bz^2)(e + g\sigma_c^2)q_c \\ \frac{d\sigma_c}{dt} = q_c \\ \frac{d\varphi_L}{dt} = (e + g\sigma_c^2)q_c \\ \frac{d\rho_L}{dt} = \varphi_L \\ \frac{dz}{dt} = -(e + g\sigma_c^2)q_c - az + ((e + g\sigma_c^2)q_c)^2 z. \end{cases} \tag{2.31}$$

The fifth-order dimension-less mathematical model can be obtained by (2.31)

$$\begin{cases} \dot{x} = -(c + du^2)z - (a + bv^2)(e + gy^2)x \\ \dot{y} = x \\ \dot{z} = (e + gy^2)x \\ \dot{u} = z \\ \dot{v} = -(e + gy^2)x - \alpha v + ((e + gy^2)x)^2 v, \end{cases} \tag{2.32}$$

where $x = q_c, y = \sigma_c, z = \varphi_L, u = \rho_L, v = z, dw/dt = \dot{w}(w \triangleq x, y, z, u, v)$.

If we set $\alpha = 0.7, a = -1.4, b = 1.4, c = 1, d = 1, e = 1, g = 1$ and initial conditions $(0.06, 0, -0.1, 0, 0)$, the novel chaotic attractors can be exhibited as shown in Fig. 2.25a, b and c, if we set $\alpha = 2$ and other values are unchanged the dynamical behaviors of the system shows a periodic state as shown in Fig. 2.25d and e.

2.5.3.2 Equilibrium and Its Stability

Through setting $\dot{x} = \dot{y} = \dot{z} = \dot{u} = \dot{v} = 0$ of the (2.32), we can obtain the equilibrium set of the memristive elements chaotic circuit.

$$E = (x, y, z, u, v)|x = z = v = 0, y = k_1, u = k_2, \tag{2.33}$$

where k_1 and k_2 are real constant, it signifies that any points on the $y - u$ plane is the equilibrium points of the system. The Jacobian matrix at this equilibrium set of memristive elements circuit can be derived as:

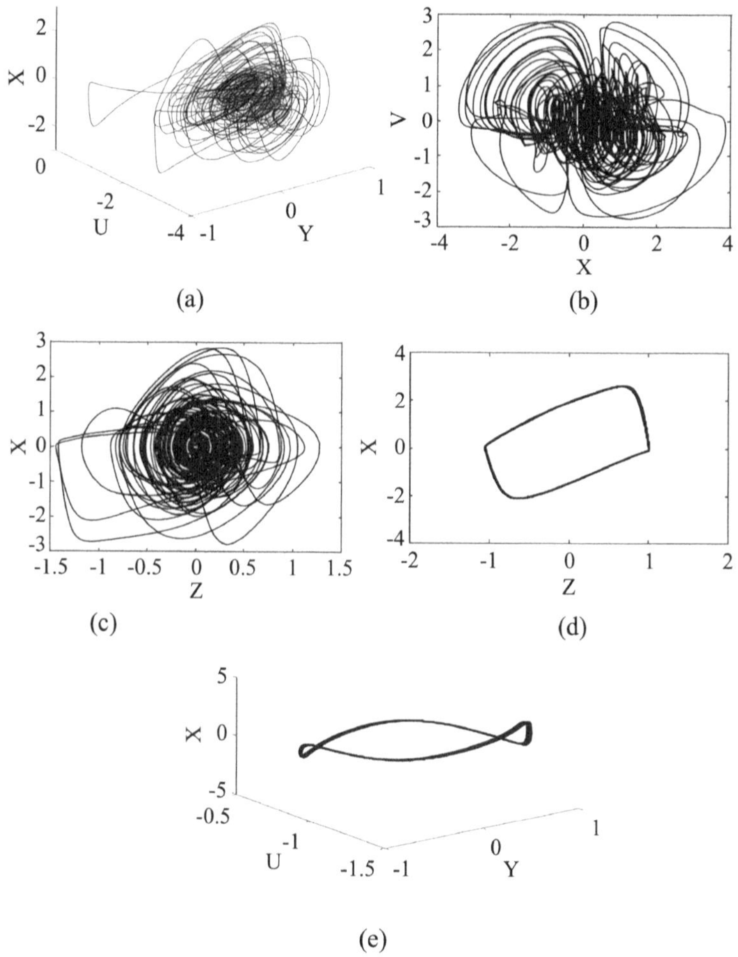

Fig. 2.25 Chaotic attractors of the memristive elements chaotic circuit: **a** The 3D chaotic attractor U-Y- X plane. **b** and **c** are the 2D chaotic attractor in Z-X and X-V plane respectively. **d** The 2D period limit cycle Z-X plane. **e** The 3D period limit cycle U-Y-X plane

$$\mathbf{J}_1 = \begin{bmatrix} -am & 0 & -n & 0 & 0 \\ 1 & 0 & 0 & 0 & 0 \\ m & 0 & 0 & 0 & 0 \\ 0 & 0 & 1 & 0 & 0 \\ -m & 0 & 0 & 0 & -\alpha. \end{bmatrix} \quad (2.34)$$

where $m = (e + 3gk_1^2)$, $n = (c + dk_2^2)$.

Then the characteristic equation can be described as follows:

$$\det(\lambda \mathbf{I} - \mathbf{J}_0) = \lambda^2(\lambda^3 + a_1\lambda^2 + a_2\lambda^1 + a_3), \tag{2.35}$$

$$\begin{aligned} h_1 &= 19998.99, h_2 = 20205.99 + 14k^2 \\ h_3 &= 119796.97 + 279992.93k^2 \\ h_4 &= 60600 - 141400k^2. \end{aligned} \tag{2.36}$$

If $k_1 = 1$ and $k_2 = 2$, the five eigenvalues $\lambda_i (i = 1, 2, 3, 4, 5)$ of the equilibrium set A are derived as follows:

$$\begin{aligned} &\lambda_1 = 0.0000, \lambda_2 = 0.0000, \lambda_3 = -0.7000 \\ &\lambda_4 = 0.7000 - 0.7141i, \lambda_5 = 0.7000 + 0.7141i. \end{aligned} \tag{2.37}$$

The equilibrium set A is an unstable saddle-focus because eigenvalues include a pair of conjugate complex root whose real part is positive.

2.5.3.3 Dynamical Behaviors

Calculations of Lyapunov exponents (LE) are widely used to indicate the existence of chaos or hyperchaos. The $LE_1 = 0.1773, LE_2 = 0.036, LE_3 = -0.0116, LE_4 = -0.1617, LE_5 = -0.8652$ and the Lyapunov dimension $dr = 4.0202$ can be gained by using the LET toolbox. The value of the time-varying Lyapunov exponents can be obtained by using wolf's method [27] in Fig. 2.26, which means that the sum of these five Lyapunov exponents is negative and the maximal Lyapunov exponent is

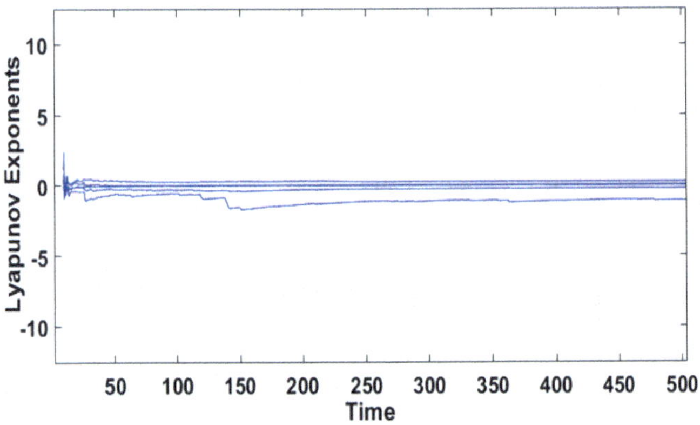

Fig. 2.26 Time-varying in the region [0, 500] of Lyapunov exponent

Fig. 2.27 **a** Bifurcation diagram and **b** Lyapunov exponent with respect to α

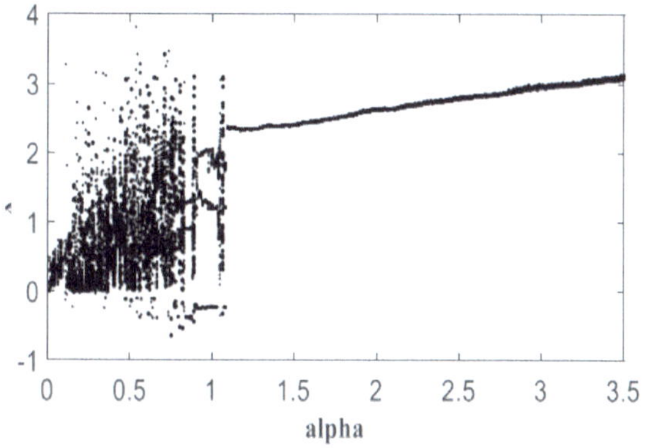

(a)

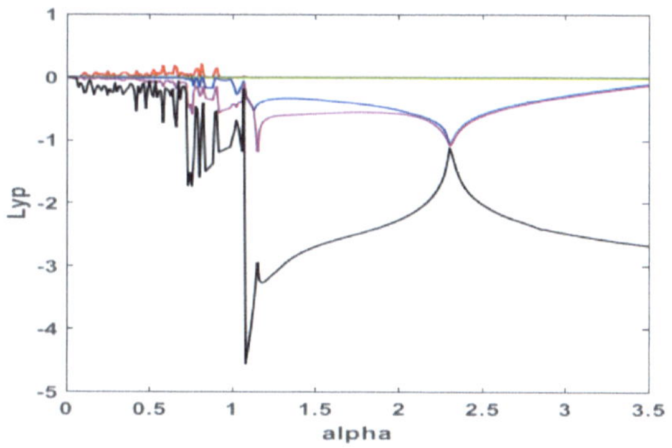

(b)

equal to 0.1773. We conclude that the proposed chaotic circuit satisfies the condition to produce chaotic oscillation.

The bifurcation diagram and Lyapunov exponent with respect to α is showed in Fig. 2.27a and b, it can be observed that the steady and unstable state of α goes along in the regions [0.01, 3.5], the system is consistent with each other (2.32). When the α goes along in the regions [0.05, 0.1], [0.9, 0.96], [1.01, 1.297], and [1.31, 3.5], the first Lyapunov exponent is mainly zero. The chaotic behavior mainly happens when α goes along in the regions [0.1, 0.9], [0.96, 1.01] and [1.29, 1.317]. These numerical simulations indicate the proposed circuit in Fig. 2.24 can generate two different topological structures of chaotic attractors.

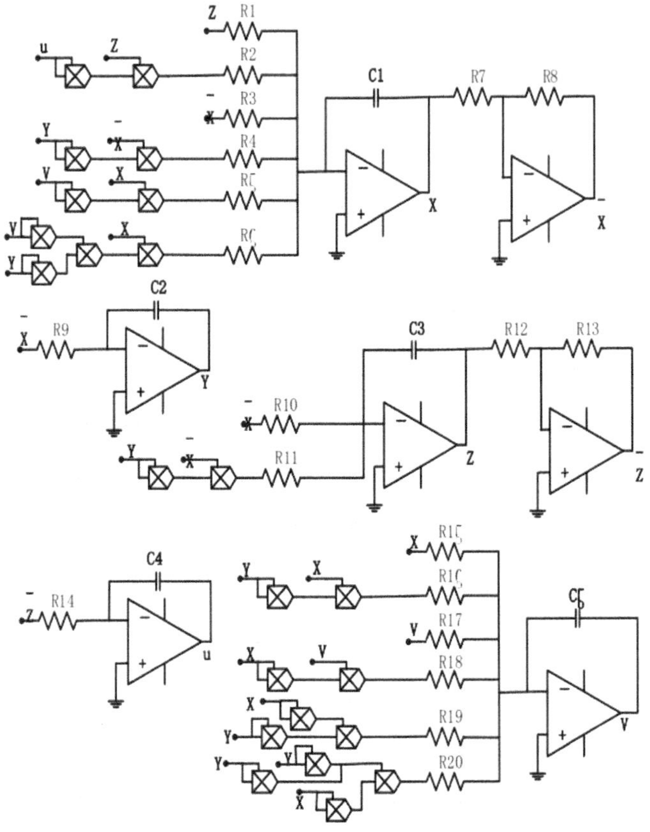

Fig. 2.28 Electronic circuit diagram

2.5.4 Circuit Simulation Result

In this section, the electronic circuit is simulated to verify the chaotic behavior of Eq. (2.32). The schematic of the electronic circuit as shown in Fig. 2.28. The main elements of the circuit have including op-amp OPO7CD, multiplier AD633IN, linear resistances and linear capacitors. The circuit (2.38) can be obtained according to the circuit schematic and Kirchhoff's circuit laws.

The main parameters are selected as $R_1 = R_2 = R_7 = R_8 = R_9 = R_{10} = R_{12} = R_{13} = R_{14} = R_{15} = R18 = 100\,\text{k}\Omega$, $R_3 = R_5 = 71.429\,\text{k}\Omega$, $R_4 = R_6 = 23.809\,\text{k}\Omega$, $R_{11} = R_{16} = 33.333\,\text{k}\Omega$, $R_{17} = 142.857\,\text{k}\Omega$, $R_{19} = 16.667\,\text{k}\Omega$, $R_{20} = 11.111\,\text{k}\Omega$, $C_1 = C_2 = C_3 = C_4 = C_5 = 100\,\text{nF}$. The circuit simulation results on the Multisim platform are shown in Fig. 2.29a and b.

$$\begin{cases} \dot{x} = \frac{1}{RC}[-\frac{R}{R_1}z - \frac{R}{R_2}zu^2 + \frac{R}{R_3}x + \frac{R}{R_4}xy^2 - \frac{R}{R_5}xv^2 - \frac{R}{R_6}xy^2v^2] \\ \dot{y} = \frac{1}{RC}[\frac{R}{R_9}x] \\ \dot{z} = \frac{1}{RC}[\frac{R}{R_{10}}x + \frac{R}{R_{11}}xy^2] \\ \dot{u} = \frac{1}{RC}[\frac{R}{R_{14}}z] \\ \dot{v} = \frac{1}{RC}[-\frac{R}{R_{15}}x - \frac{R}{R_{16}}xy^2 - \frac{R}{R_{17}}v + \frac{R}{R_{18}}vx^2 + \frac{R}{R_{19}}vx^2y^2 + \frac{R}{R_{20}}vx^2y^4]. \end{cases} \tag{2.38}$$

2.5.5 *Conclusion*

In this section, a novel chaotic circuit is presented using the memristor, memcapacitor, and meminductor. The nonlinear dynamical behaviors are investigated by using the Phase Diagrams, equilibrium set, Lyapunov exponent, and bifurcation diagram. The theoretical analyses and numerical simulation result shows that this circuit can produce two different dynamical behaviors including chaotic oscillation and period. This system has potential application values in the fields of encryption and decryption system and neural network because of complex dynamical behavior. This type of circuit will contribute to researching the physical characteristics of memristive elements and expanding nonlinear electronic circuits.

2.6 Comments

This chapter explores various memristor-based chaotic circuit configurations and their applications, demonstrating the versatility and potential of memristors in generating complex chaotic behavior, with significant implications for secure communication, encryption systems, and neuromorphic computing. Building upon single-memristor systems, the chapter focuses on investigating chaotic circuits based on multiple memristors, particularly the dual-memristor configuration. Through methods such as Lyapunov exponents and bifurcation diagrams, the rich dynamical behavior of these systems is thoroughly analyzed, not only expanding our understanding of multi-memristor dynamics but also laying the foundation for designing more sophisticated chaotic systems with enhanced security features. Furthermore, this

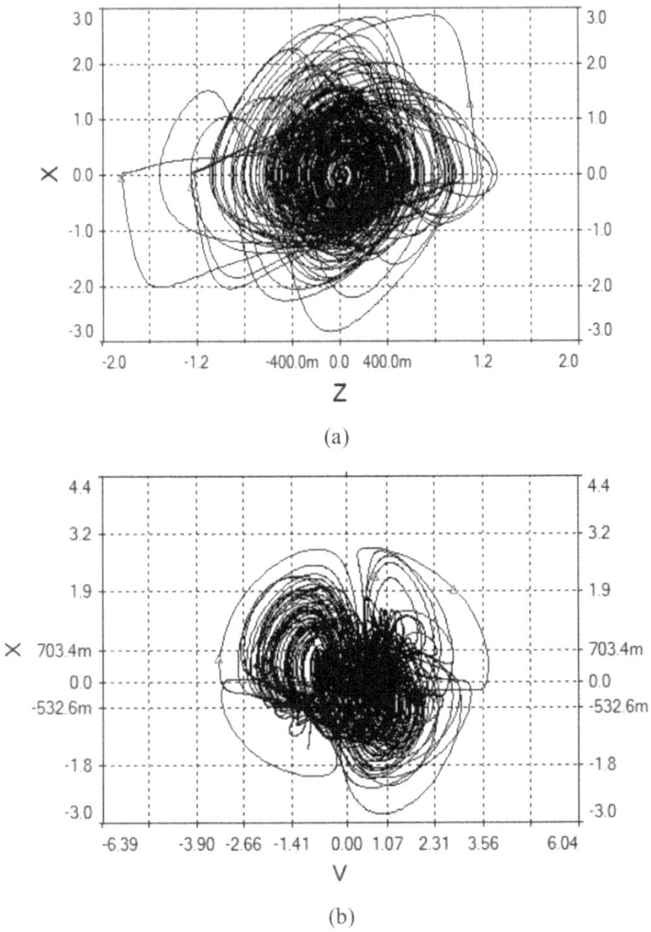

Fig. 2.29 The circuit simulation result **a** Phase portrait in the Z-X plane **b** Phase portrait in the V-X plane

chapter innovatively introduces time-delay feedback and uncertain parameters, further enhancing the complexity and applicability of memristor-based chaotic circuits. Overall, the novel chaotic circuits presented in this chapter open up new possibilities for applications requiring complex and unpredictable signals.

Despite the substantial contributions made by the research in this chapter, there are still limitations and room for improvement. Several key research directions could be focused on in the future:

Experimental validation using physical memristor devices. Although theoretical models and simulations provide valuable insights, it is crucial to validate these findings using physical memristor devices. Experimental studies can help uncover additional complexities and non-ideal behaviors that the models may not capture.

Future research could focus on fabricating and testing memristor-based chaotic circuits, comparing their performance with theoretical predictions, and refining the models based on experimental results.

Development of application-specific chaotic circuits. This chapter highlights the potential applications of memristor-based chaotic circuits in secure communication, encryption systems, and neuromorphic computing. Future research could focus on designing and optimizing chaotic circuits for specific applications, considering factors such as signal complexity, synchronization, and hardware constraints. This work can leverage the insights gained from the multi-memristor, time-delay feedback, and compact circuit designs presented in the chapter.

Development of automated design and optimization tools. As the complexity of memristor-based chaotic circuits increases, manual design and optimization become more challenging. Future research could focus on developing automated tools and algorithms for designing and optimizing these circuits, considering factors such as desired chaotic behavior, hardware constraints, and robustness to uncertainties. These tools can leverage machine learning and evolutionary algorithms to efficiently explore the vast design space and identify optimal circuit configurations.

Exploration of novel memristor devices and materials. The chapter focuses on the application of memristors, memcapacitors, and meminductors in chaotic circuits. However, the field of memristor devices is rapidly evolving, with new materials and device structures continuously emerging. Future research could explore the potential of these novel memristor devices in generating chaotic behavior, potentially leading to more compact and efficient chaotic circuits.

By pursuing these research directions, scientists and engineers can continue to advance the development of memristor-based chaotic circuits, unlocking new applications and deepening our understanding of these complex systems. The insights provided in this chapter lay a solid foundation for future work, highlighting the immense potential of memristor devices in the realm of chaotic circuits and beyond.

References

1. Driscoll, T., Pershin, Y., Basov, D., Di Ventra, M.: Chaotic memristor. Appl. Phys. A **102**, 885–889 (2011)
2. Itoh, M., Chua, L.O.: Memristor oscillators. Int. J. Bifurc. Chaos **18**(11), 3183–3206 (2008)
3. Muthuswamy, B., Kokate, P., et al.: Memristor-based chaotic circuits. IETE Tech. Rev. **26**(6), 417 (2009)
4. Bao, B., Liu, Z., Xu, J.: Steady periodic memristor oscillator with transient chaotic behaviours. Electron. Lett. **46**(3), 228 (2010)
5. Muthuswamy, B.: Implementing memristor based chaotic circuits. Int. J. Bifurc. Chaos **20**(05), 1335–1350 (2010)
6. Muthuswamy, B., Chua, L.O.: Simplest chaotic circuit. Int. J. Bifurc. Chaos **20**(05), 1567–1580 (2010)
7. Borghetti, J., Snider, G.S., Kuekes, P.J., Yang, J.J., Stewart, D.R., Williams, R.S.: Memristive switches enable stateful logic operations via material implication. Nature **464**(7290), 873–876 (2010)

8. Iu, H.H.C., Yu, D., Fitch, A.L., Sreeram, V., Chen, H.: Controlling chaos in a memristor based circuit using a twin-t notch filter. IEEE Trans. Circuits Syst. I Regul. Pap. **58**(6), 1337–1344 (2011)
9. Rose, G.S., Pino, R., Wu, Q.: Exploiting memristance for low-energy neuromorphic computing hardware. In: 2011 IEEE International Symposium of Circuits and Systems (ISCAS), pp. 2942–2945. IEEE (2011)
10. Hashimoto, T., Amemiya, T.: Controllability and observability of linear time-invariant uncertain systems irrespective of bounds of uncertain parameters. IEEE Trans. Autom. Control **56**(8), 1807–1817 (2010)
11. Lin, T.C., Lee, T.Y.: Chaos synchronization of uncertain fractional-order chaotic systems with time delay based on adaptive fuzzy sliding mode control. IEEE Trans. Fuzzy Syst. **19**(4), 623–635 (2011)
12. Wang, X.F., Zhong, G.Q., Tang, K.S., Man, K.F., Liu, Z.F.: Generating chaos in chua's circuit via time-delay feedback. IEEE Trans. Circuits Syst. I: Fundam. Theory Appl. **48**(9), 1151–1156 (2001)
13. Mahmoud, M.S., Ismail, A.: New results on delay-dependent control of time-delay systems. IEEE Trans. Autom. Control **50**(1), 95–100 (2005)
14. Niu, Y., Ho, D.W., Lam, J.: Robust integral sliding mode control for uncertain stochastic systems with time-varying delay. Automatica **41**(5), 873–880 (2005)
15. Yu, Y.B., Zhong, Q.S., Liao, X.F., Yu, J.B.: Exponential stability of takagi sugeno fuzzy systems with impulsive effects and small delays. Chin. Phys. B **17**(3), 842–846 (2008)
16. Yong-Bin, Y., Jing-Fu, B., Hong-Bin, Z., Qi-Shui, Z., Xiao-Feng, L., Jue-Bang, Y.: Impulsive control of nonlinear systems with time-varying delays. Chin. Phys. B **17**(7), 2377 (2008)
17. Yang, C., Hu, Q., Yu, Y., Zhang, R., Yao, Y., Cai, J.: Memristor-based chaotic circuit for text/image encryption and decryption. In: 2015 8th International Symposium on Computational Intelligence and Design (ISCID), vol. 1, pp. 447–450. IEEE (2015)
18. Ahmad, M., Alam, B., Farooq, O.: Chaos based mixed keystream generation for voice data encryption. Int. J. Cryptogr. Inf. Secur. **2**(1) (2014)
19. Liu, B., Wang, L., Jin, Y.H., Tang, F., Huang, D.X.: Improved particle swarm optimization combined with chaos. Chaos Solitons & Fractals **25**(5), 1261–1271 (2005)
20. Wu, A., Wen, S., Zeng, Z.: Synchronization control of a class of memristor-based recurrent neural networks. Inf. Sci. **183**(1), 106–116 (2012)
21. Yong, Y., Wanxing, S., Sunan, W.: Study of chaos genetic algorithms and its application in neural networks. In: 2002 IEEE Region 10 Conference on Computers, Communications, Control and Power Engineering. TENCOM'02. Proceedings., vol. 1, pp. 232–235. IEEE (2002)
22. Liu, H., Wang, X.: Color image encryption using spatial bit-level permutation and high-dimension chaotic system. Optics Commun. **284**(16–17), 3895–3903 (2011)
23. Min, L., Lan, X., Hao, L., Yang, X.: A 6 dimensional chaotic generalized synchronization system and design of pseudorandom number generator with application in image encryption. In: 2014 Tenth International Conference on Computational Intelligence and Security, pp. 356–362. IEEE (2014)
24. Tan, Z.: Parallel chaotic circuit based memristor. Electronic Components and Materials (2015)
25. Wang, G.Y., Cai, B.Z., Jin, P.P., Hu, T.L.: Memcapacitor model and its application in a chaotic oscillator. Chin. Phys. B **25**(1), 010503 (2015)
26. Zang, S., Wang, G., Shi, C.: Design of a chaotic oscillator circuit based on memcapacitor and meminductor. J. Hangzhou Dianzi Univ. (2017)
27. Fitch, A., Iu, H., Yu, D.: Chaos in a memcapacitor based circuit. In: 2014 IEEE International Symposium on Circuits and Systems (ISCAS), pp. 482–485 (2014)

Chapter 3
Memristor-Based En/Decryption System

This chapter focuses on the design, analysis, and application of memristor-based chaotic circuits for text and image encryption and decryption. By harnessing the nonlinear dynamics of memristors and exploiting the principles of chaotic cryptography, this chapter presents a comprehensive study of memristor-based encryption and decryption systems.

Section 3.1 delves into a novel application of memristors in the realm of cryptography. A new cryptosystem model based on a memristor-based chaotic circuit is proposed, taking into account the unique characteristics of memristors, nonlinear circuits, and chaotic cryptography. The feasibility and practicability of the proposed memristor-based chaotic circuit for text and image encryption and decryption are demonstrated through extensive simulations and analysis.

Section 3.2 presents two innovative image encryption algorithms based on a three-dimensional chaotic system. The section begins by introducing a novel three-dimensional chaotic circuit incorporating memristors and investigating its chaotic dynamics. Subsequently, two image encryption algorithms are developed utilizing the chaotic sequences generated by the memristor-based chaotic circuit. Rigorous theoretical and statistical analyses are conducted to assess the effectiveness and security of the proposed algorithms. The simulation results provided in this section further validate the robustness and efficiency of the memristor-based image encryption techniques.

The primary objective of this chapter is to showcase the immense potential of memristor-based chaotic systems in the field of cryptography and to highlight their applications in text and image encryption and decryption. By leveraging the unique properties of memristors and the complex dynamics of chaotic systems, the proposed encryption and decryption schemes offer enhanced security, improved performance, and increased resilience against various cryptographic attacks. The novel memristor-based chaotic circuits and encryption algorithms introduced in this

Y. Yu et al., *Memristive Computing*,
https://doi.org/10.1007/978-981-96-0332-9_3

chapter are expected to stimulate further research and development in the area of memristor-based cryptography and pave the way for practical implementations of secure communication systems.

3.1 Memristive Chaotic Text/Image En/Decryption Circuit

3.1.1 Introduction

Chaos theory has been extensively applied in cryptography since its introduction to cryptosystems in 1989 [1, 2]. The field of chaotic cryptography has subsequently experienced rapid development [3–9], with chaos-based cryptosystems becoming integral to encryption and decryption processes due to their inherent random properties and robust encryption performance [10].

A diverse array of chaotic encryption schemes has emerged, including wheel-switch chaotic schemes [11] and single chaotic map approaches [12]. These schemes have demonstrated remarkable versatility in encrypting various data types, encompassing plain text, color images, video files, and electronic mail [13–16]. In the realm of chaotic cryptography, the chaotic system serves as the fundamental component in the design of encryption and decryption algorithms, offering unique advantages such as high sensitivity to initial conditions and parameter variations.

Historically, cryptographic research has primarily focused on traditional chaotic systems such as the Logistic map, Lorenz system, and Chua circuit. These systems have been extensively studied and implemented in various encryption protocols. However, the advent of memristors has introduced new possibilities in the field of chaotic cryptography. Memristors, as the fourth fundamental circuit element, possess unique nonlinear characteristics that can potentially enhance the complexity and security of chaotic systems.

The integration of memristors into chaotic circuits presents an opportunity to develop novel encryption methodologies with potentially superior performance. Memristor-based chaotic circuits offer several advantages, including increased system dimensionality, enhanced sensitivity to initial conditions, and the potential for more complex dynamic behaviors. These properties could lead to more robust encryption schemes that are resistant to conventional cryptanalysis techniques.

Despite the potential benefits, the application of memristor-based chaotic circuits in data encryption remains largely unexplored in the literature. This gap in research presents a significant opportunity for innovation in the field of chaotic cryptography. By leveraging the unique properties of memristors, it may be possible to design encryption systems that offer improved security, efficiency, and adaptability compared to traditional chaotic cryptosystems.

This section aims to investigate the utilization of memristor-based chaotic circuits for text and image encryption and decryption. The research will explore the design and implementation of novel chaotic systems incorporating memristors, analyze their dynamic behaviors, and evaluate their performance in encryption tasks. This study represents an important step towards advancing the field of chaotic cryptography, potentially leading to the development of more secure and efficient encryption methods for various data types.

3.1.2 Memristor-Based Chaotic Circuit

A flux-controlled memristor circuit [17] is presented in Fig. 3.1, which assumes that the memristor is characterized by the "monotone increasing" and "piecewise-linear" nonlinearity described as follows:

$$q(\varphi) = d\varphi + 0.5(c - d)(|\varphi + 1| - |\varphi - 1|), \tag{3.1}$$

where $c > 0$, $d > 0$. Then,

$$W(\varphi) = \frac{dq(\varphi)}{d\varphi} = \begin{cases} c, & |\varphi| < 1 \\ d, & |\varphi| > 1. \end{cases} \tag{3.2}$$

According to Kirchhoff's circuit Laws, the memristor circuit's state equations are obtained as follows:

$$\begin{cases} C_1 \frac{dv_1}{dt} = i_3 - W(\varphi)v_1 \\ L\frac{di_3}{dt} = v_2 - v_1 \\ C_2 \frac{dv_2}{dt} = -i_3 + Gv_4 = -i_3 + Gv_2 \\ \frac{d\varphi}{dt} = v_1. \end{cases} \tag{3.3}$$

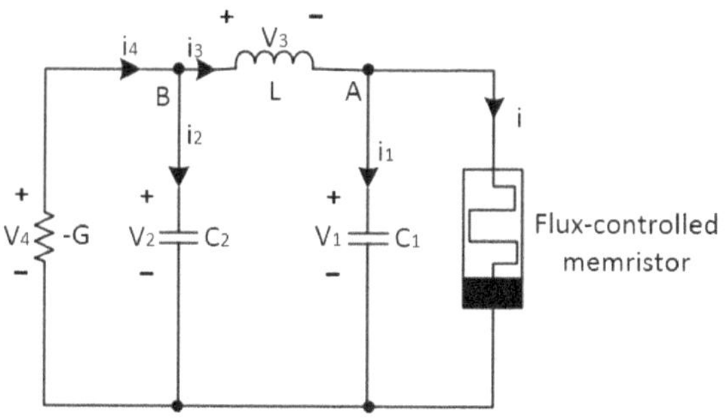

Fig. 3.1 Memristor-based chaotic circuit

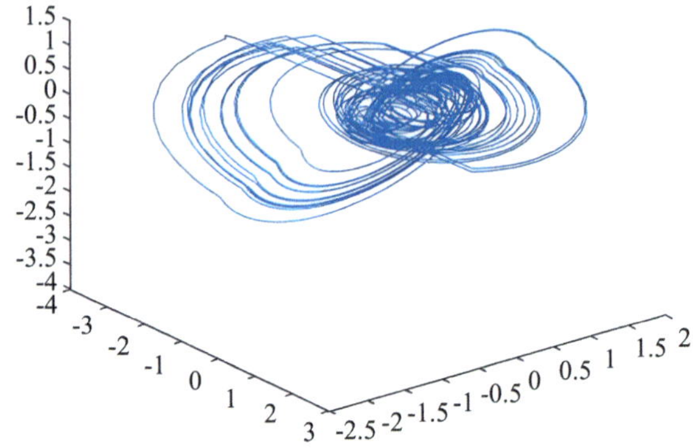

Fig. 3.2 Chaotic attractor

Given $x = v_1$, $y = i_3$, $z = v_2$, $\omega = \varphi$, $\alpha = \frac{1}{C_1}$, $\beta = \frac{1}{C_2}$, and $\gamma = \frac{G}{C_2}$, Eq. (3.3) is rewritten as follows:

$$\begin{cases} \frac{dx}{dt} = \alpha(y - W(\omega)x) \\ \frac{dy}{dt} = z - x \\ \frac{dz}{dt} = -\beta y + \gamma z \\ \frac{d\omega}{dt} = x. \end{cases} \tag{3.4}$$

When $t = 200$, $\alpha = 4$, $\beta = 1$, $\gamma = 0.65$, $a = 0.2$, and $b = 10$, we can obtain the chaotic attractor shown in Fig. 3.2.

The chaotic sequences $\{x_t, y_t, z_t, \omega_t | t = 1, 2, \ldots, N\}$ with aperiodic, non-convergent and the high sensitivity for initial values $\{x_0, y_0, z_0, \omega_0\}$ are generated, which are utilized in the following cryptosystem.

3.1.3 The Cryptosystem of Memristor-Based Chaotic Circuit

In this section, we design a novel model with the memristor-based chaotic circuit to encrypt and decrypt text and image information.

3.1.4 Cryptosystem Model

A memristor-based chaotic cryptosystem model consists of two parts shown in Fig. 3.3, which are chaotic encryption and chaotic decryption, respectively. In the cryptosystem model, the memristor-based chaotic circuit is critical to decide the chaotic encryption and decryption. When the User Key (UK), which is defined as the initial values resulted in chaos of memristor circuit, is given, the chaotic sequence (CS) is generated. Once the CS is obtained, the chaotic encryption and decryption

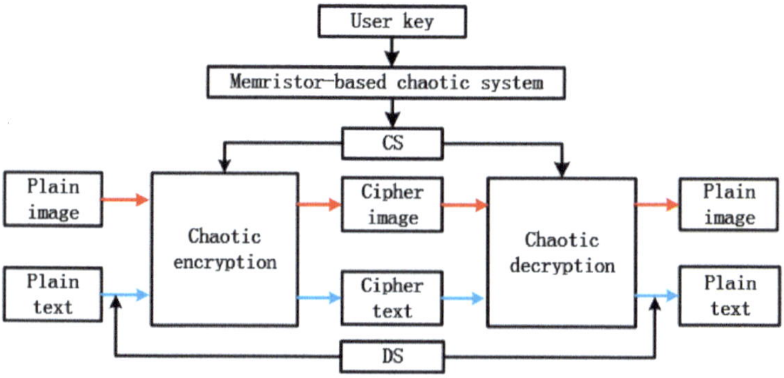

Fig. 3.3 Cryptosystem model

will be designed. According to the cryptosystem model, we can en/decrypt the text and image.

3.1.5 *Text En/Decryption*

According to the cryptosystem model shown in Fig. 3.3, the process of chaotic encryption of plain text is described as follows:

Step 1, the plain text is obtained as the input of the cryptosystem.
Step 2, with the input of UK, the memristor-based chaotic system generates the CS, so the UK maps CS $\{x_t, y_t, z_t, \omega_t | t = 1, 2, \ldots, N\}$.
Step 3, the CS is mapped to the value from zero to one.
Step 4, Fibonacci sequence (FS) is designed, as shown in Fig. 3.4a.
Step 5, by making the FS normalization with the value from zero to one, the disturbing sequence (DS) is generated, as shown in Fig. 3.4b.
Step 6, by mixing the CS and the DS, the key sequence (KS) is generated.
Step 7, the KS and plain text make operation, and obtain the cipher text sequence (CTS).

In light of the above-mentioned encryption process, we can design the corresponding decryption procedure.

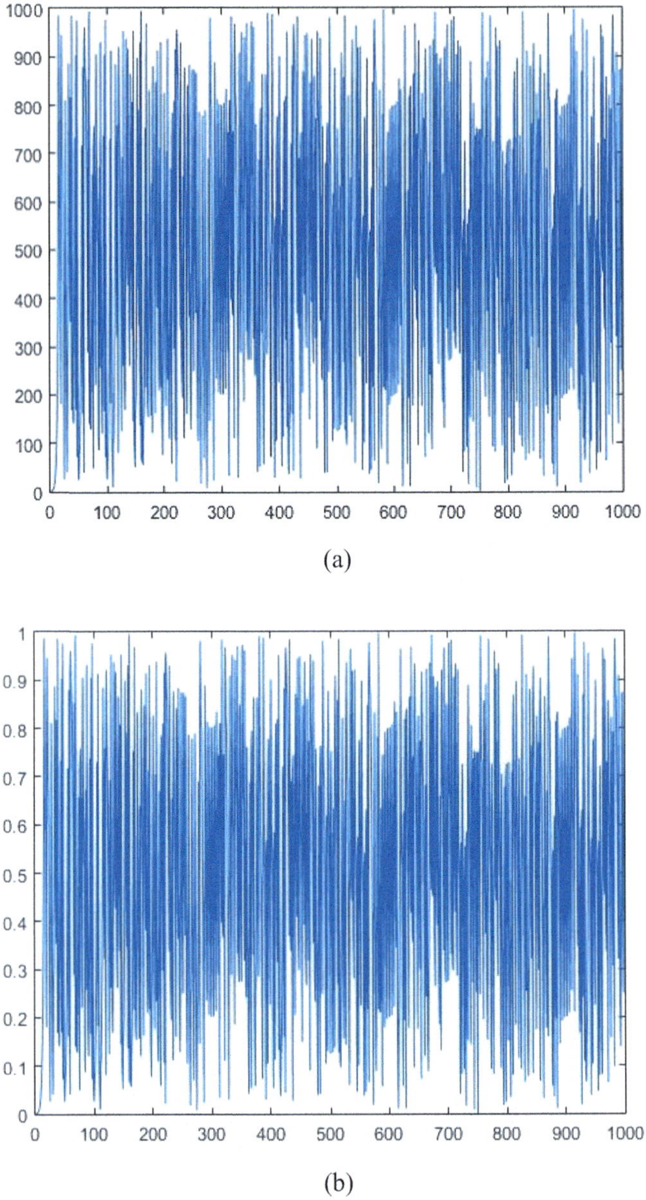

Fig. 3.4 The Fibonacci sequence and disturbing sequence: **a** the Fibonacci sequence. **b** the DS

3.1.6 Image En/Decryption

According to the cryptosystem model shown in Fig. 3.3, the process of chaotic encryption of plain image is presented as follows:

Step 1, the plain image is obtained as the input of the cryptosystem.
Step 2, with the input of UK, the memristor-based chaotic system generates the CS, so the UK maps CS $\{x_t, y_t, z_t, \omega_t | t = 1, 2, \ldots\}$.
Step 3, the CS is transformed as a new type of $\{x_t', y_t', z_t', \omega_t' | t = 1, 2, \ldots\}$.
Step 4, the pixel rows of image are changed by the CS.
Step 5, the pixel columns of image are changed by the CS.
Step 6, the values of image pixel are operated by the CS.
Step 7, the cipher image sequence (CIS) is obtained.

Similar to the encryption process, the decryption scheme is obtained.

3.1.7 Simulation Result

According to the cryptosystem model and the en/decryption procedure, we design the interface of text/image encryption and decryption, as shown in Fig. 3.5.

3.1.7.1 Simulation Results of Text En/Decryption

In this section, we choose the second dimension Y sequence $\{y_t | t = 1, 2, \ldots\}$ as shown in Fig. 3.6a. When the initial values ($x_0 = 0.01, y_0 = 0.0003, z_0 = 0.02, \omega_0 = 0$) resulted in chaos of the memristor circuit are chose, the UK is obtained, and the KS is shown in Fig. 3.6b is presented. Given the plain text, we make a program to implement the text en/decryption algorithms and gain the simulation results shown in Fig. 3.5.

3.1.7.2 Simulation Results of Image En/Decryption

According to the image en/decryption procedure, we make a program and simulation. The simulation results of the plain image encryption and corresponding decryption shown in Fig. 3.7 are presented.

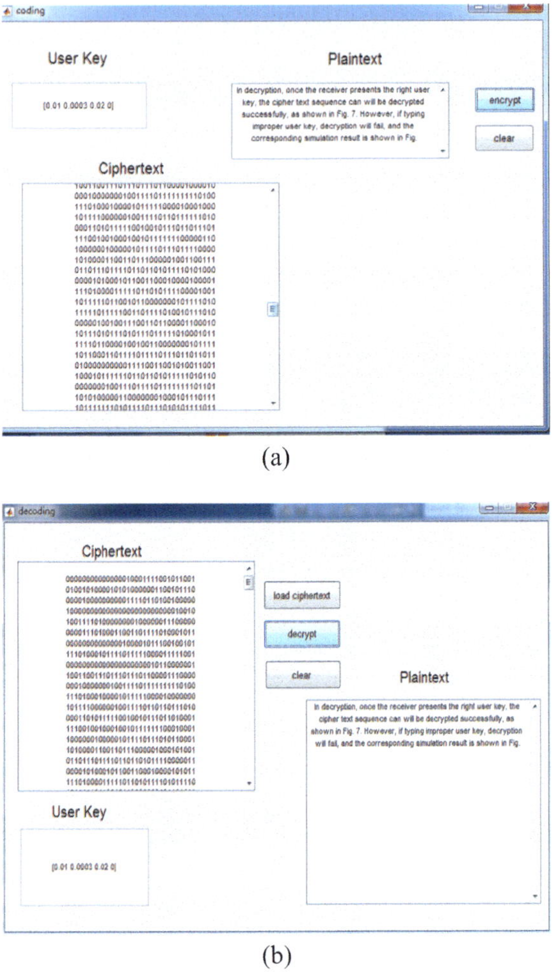

Fig. 3.5 The encryption and decryption interface

3.1.8 Conclusion

This section explores the application of memristor-based chaotic circuits for text and image encryption and decryption. By leveraging the principles of memristor circuits and chaos cryptography, a novel cryptosystem model based on memristor-based chaotic circuits is introduced. The proposed model is then utilized to investigate text and image encryption and decryption processes. Simulation results demonstrate the feasibility and practicability of employing memristor-based chaotic circuits for secure encryption and decryption tasks.

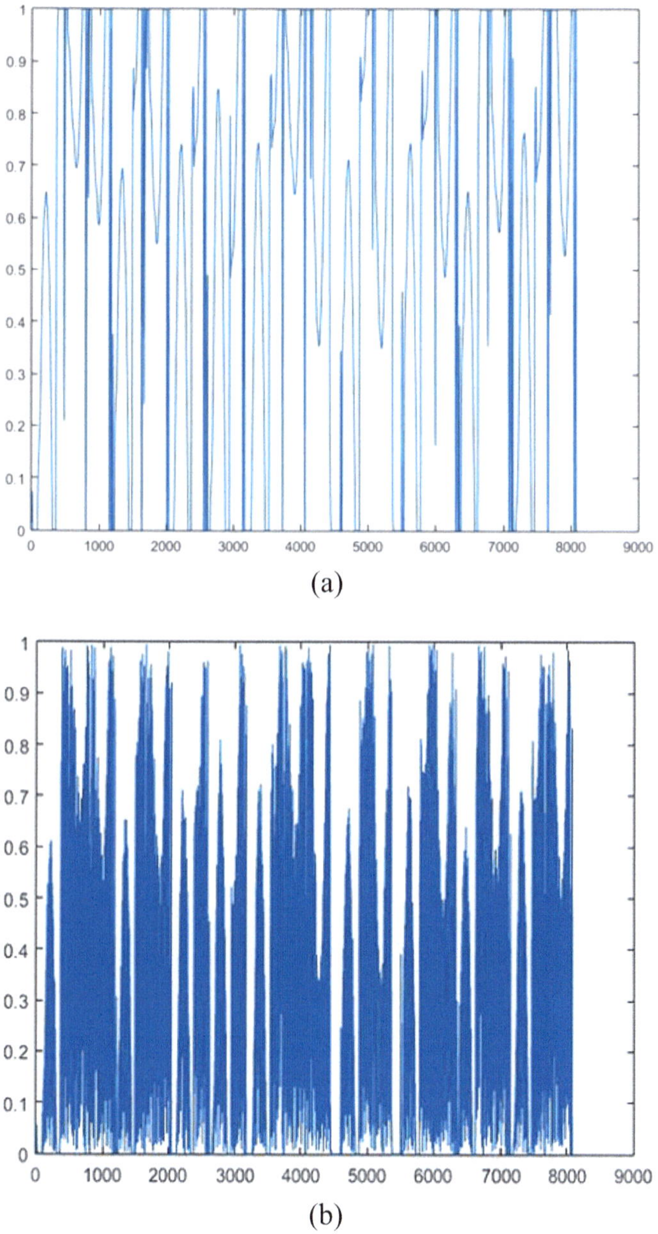

Fig. 3.6 Chaotic sequence and key sequence: **a** CS for text encryption. **b** KS for text encryption

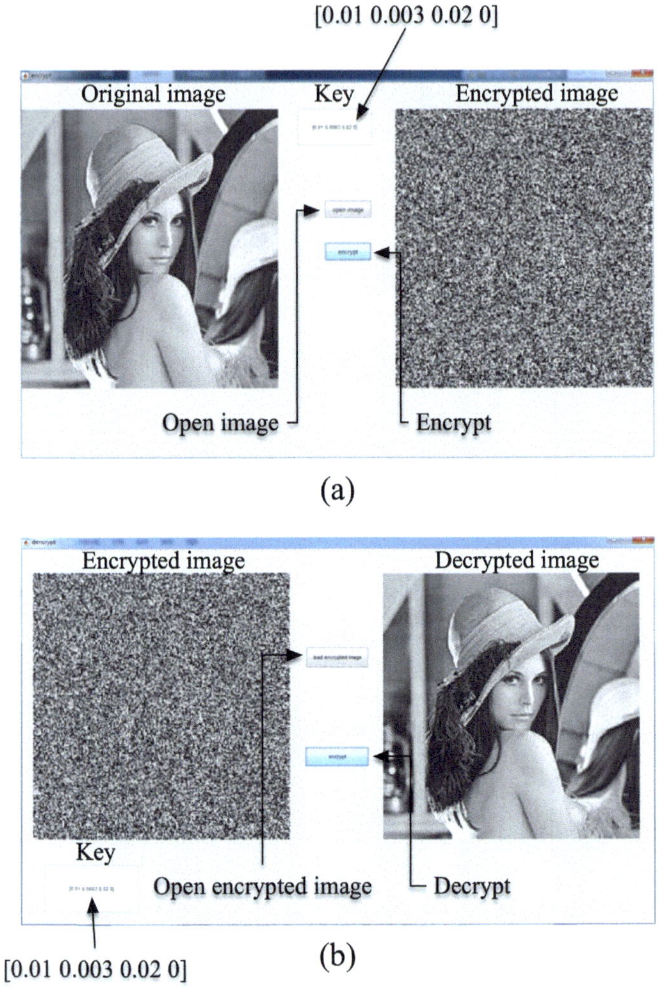

Fig. 3.7 Simulation result of image encryption and decryption

3.2 Memristive Chaotic Circuit and Its Application on Image En/Decryption

3.2.1 Introduction

In 1989, the British mathematician Matthews applied the chaos theory to the field of encryption [18]. Since chaos is sensitive to initial values, any slight changes to the parameters will result in a different output. Chaotic cryptography has developed rapidly [2, 5, 11, 14, 19]. Due to the fact that image data contains more redundant values than plain text, many researchers recently focused on image data encryption and are trying to implement different methods in encryption systems. The key space of the one-dimensional chaotic system is not large enough to resist the exhaustive attack, and the amount of chaotic system parameters and initial values is also an influence factor. In view of the questions above, many researchers have put forward to build a more complex and higher dimension chaos system, from one-dimensional logistic map encryption system developed to six-dimensional chaotic discrete system [5, 13, 20]. Besides, a lot of different algorithms, which are based on the pseudo-random chaos sequences, are presented, such as hybrid genetic algorithm [5], XOR operation [21], and linear congruence [22].

The dynamic properties of nonlinear chaos are more complicated than any other randomization technique, because the same parameters will present the same output in a particular chaotic system, even a small change in a parameter can cause a huge change. Cryptography researchers mostly focus on the traditional chaotic system, such as Logistic maps, Loren system Chua circuits, and so on. However, they rarely apply memristor chaos to encryption and decryption. Memristor is prone to produce chaos as a nonlinear element, with low power consumption, and high integration. It is very worth studying the use of memristor to design a high-dimensional chaotic circuit and applying the chaos to encryption and decryption.

3.2.2 Chaotic Circuit Based on Memristor

The chaotic circuit based on memristor is shown in Fig. 3.1. This circuit contains one memristor, one capacitance, one resistance, and one inductance. The circuit dynamics are described by:

$$\begin{aligned} C\frac{dV_C}{dt} &= i_L - \frac{V_C}{R}, \\ L\frac{di_L}{dt} &= V_C + \beta\left(z^2 - 1\right) i_M, \\ \frac{dz}{dt} &= i_L - \alpha z - i_L z, \end{aligned} \tag{3.5}$$

where the V_C denotes the voltage of capacitor and i_L denotes the inductor current, similarly, i_M is the memristor current. Z is the internal state of memtistor [11].

We know $i_M = i_L$, then we analyze the chaotic circuit characteristics. Equation (3.5) can be rewritten to this form:

$$\begin{aligned} C\frac{dx}{dt} &= y - \frac{x}{R}, \\ L\frac{dy}{dt} &= -x - \beta\left(z^2 - 1\right) y, \\ \frac{dz}{dt} &= -y - \alpha z + yz, \end{aligned} \tag{3.6}$$

where $x = V_C$, $y = i_L$. If we set $C = 0.5$, $R = 2$, $L = 4.45$, $\alpha = 1.9$, and $\beta = 0.5$. Through MATLAB simulation, we found the circuit has a chaotic attractor as shown in Fig. 3.8. Calculating the Lyapunov exponents, we got one positive Lyapunov exponent $\gamma_1 \approx 0.3397$, $\gamma_2 \approx 0$, and $\gamma_3 \approx -6.4157$. The bifurcation diagram with

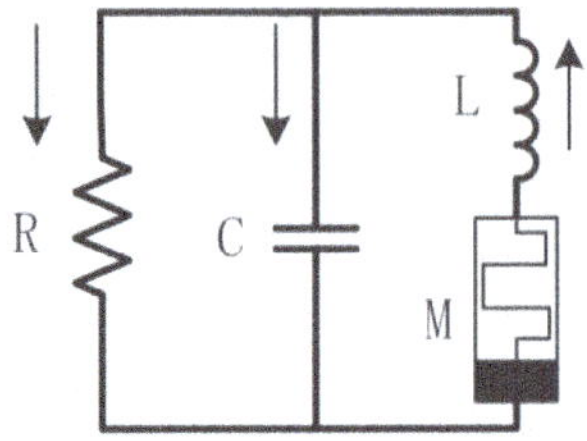

Fig. 3.8 Memristor-based chaotic circuit

(a) plain image

(b) cipher image

(c) decrypted image

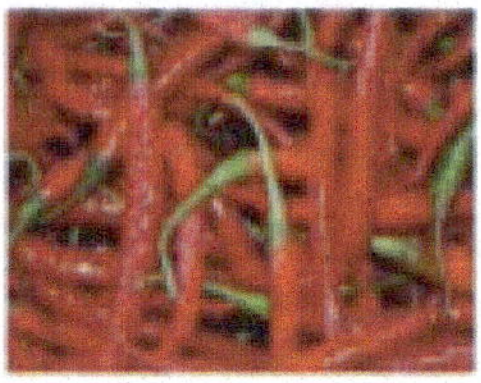

(d) RGB plain image

(e) RGB cipher image

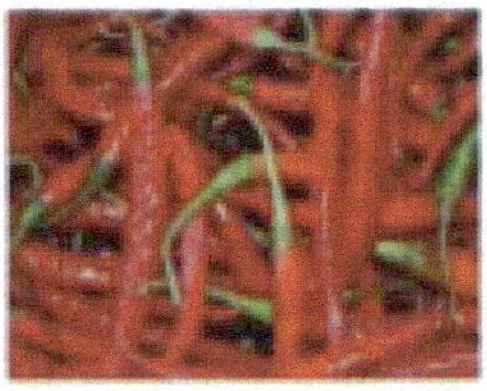

(f) RGB decrypted image

Fig. 3.9 IED2-MBCC

change in parameter β, Poincare cross diagram on the $z = 0$ section are shown in Fig. 3.11 (Figs. 3.9 and 3.10).

3.2.3 *Image En/Decryption Algorithms*

In the following works, we are working with Gray and RGB color images. Since three channels of gray image's pixel values are same, we just need to take either one of the channels pixel values, and store them in a matrix. But RGB color image is more complex, we separate the red, green, and blue channels from the image and store the pixel values for each channel in its corresponding matrixes. In the following texts, we take RGB color image as an example to show en/decryption algorithms,

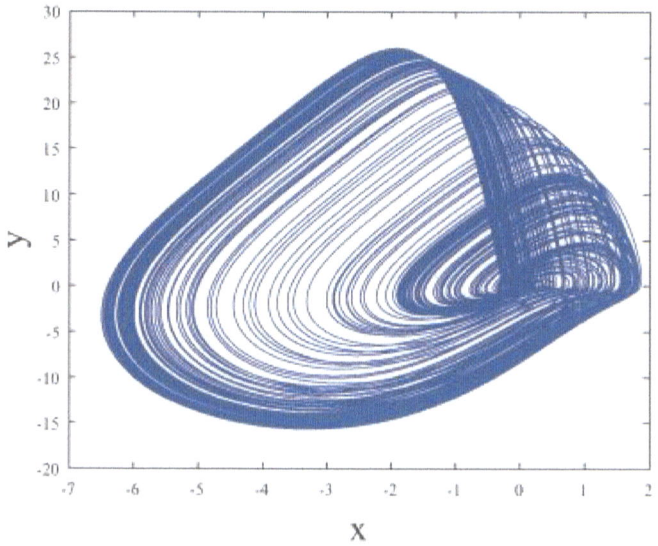

Fig. 3.10 Chaotic attractor

which is based on memristor chaos. Here IED1-MBCC is the first algorithm image en/decryption by using memristor-based chaotic circuit, similarly, IED2-MBCC is the second algorithm image en/decryption by using memristor-based chaotic circuit.

3.2.4 *IED1-MBCC*

The encryption process is divided into two parts: shuffling and substituting. We use discrete logarithm operation to shuffle three matrices and mod operation to substitute image pixels. The chaotic encryption process of images is described below [23].

Step 1, the plain image is the input matrices.
Step 2, separating three channels pixel values from images and storing them in RBG matrices, respectively.
Step 3, segmenting those three matrices into sub-blocks of size 8×8, and those sub-matrices are marked R_{sub_i}, G_{sub_i}, B_{sub_i}, $i = 1, 2, \ldots, m$, m is total numbers of sub-blocks.
Step 4, transforming those above sub-matrices $R^{'}_{sub_i}$, $G^{'}_{sub_i}$, $B^{'}_{sub_i}$ through discrete logarithm operation.
Step 5, chaotic sequences (CS) discussed in part two $\{x_t, y_t, z_t | t = 1, 2, \ldots, N\}$ are obtained.
Step 6, encrypting pixels in sub-matrices (Step 4), the cipher algorithm is formulated below, using the red channel as an example:

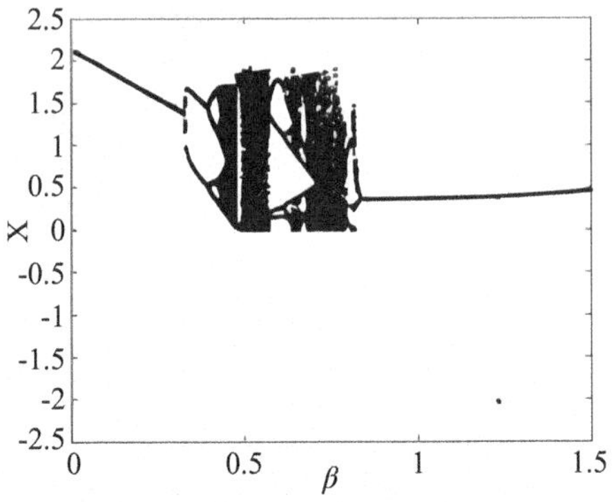

(a) bifurcation diagram

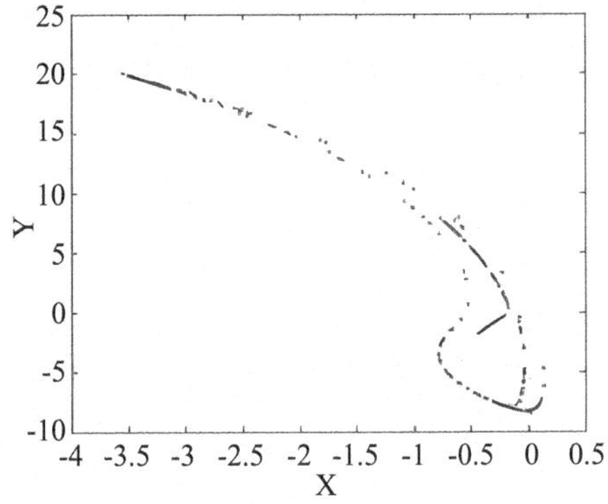

(b) poincare cross diagram

Fig. 3.11 Bifurcation and Poincare cross diagram

$$R_{cipher_{old_n}} = (R'_{sub_i} + R_{cipher_{old_{n-1}}} + x'_t) \bmod \lfloor \max(y'_t, z'_t) \rfloor \quad i = 1, 2, \ldots, m.$$

Step 7, combining $R_{cipher_{old}}$, $G_{cipher_{old}}$, $B_{cipher_{old}}$ into a matrix as cipher image sequence.

There are some illustrations about the above steps. The numbers of matrices have changed twice during the process of encryption, the first change takes place in Step 3, we need to increase matrices to finish the segmentation task. The second change is due to the discrete logarithm in Step 4. Both times, adding matrix's value is zero.

The decryption is the inverse process of encryption, but there is a slight change for the pixels in the decryption method, which is given below:

$$R_{sub_i} = (R_{cipher_{old_n}} - R_{cipher_{old_{n-1}}} - x_t^{'})$$
$$\bmod \lfloor \max(y_t^{'}, z_t^{'}) \rfloor \quad i = 1, 2, \ldots, m.$$

3.2.5 *IED2-MBCC*

The second encryption algorithms is based on the Hill algorithm. Plain image matrix is marked P, and the cipher image matrix is labeled C. K is encryption key and decryption requires matrix K, which is invertible. Next, we construct a reversible matrix K. In mod operation, inverse matrix is different from general inverse matrix, the inverse matrix is calculated as follows: $[K^{-1}]_{ij} = (det\ K)^{-1}(-1)^{i+j}D_{ij}$, D_{ij} is complement. For example, if determinant of A is equal to 9, where mod number is equal to 26, $9 \times 3 = 27$, $27 \bmod 26 = 1$, $(det\ A)^{-1} = 3$.

$$C = PK \bmod 256. \tag{3.7}$$

We set matrix A_1 as stated below, where $det(A) = 1$. By the same method, we construct $A_2, A_3, \ldots, A_{2n}$, then do tensor product to build encrypted matrixes as follows $A_1 \otimes A_2 = K_1, A_3 \otimes A_4 = K_2, \ldots, A_{2n-1} \otimes A_{2n} = K_n$, where r is chaotic sequences.

$$A_1 = \begin{bmatrix} r & r+1 \\ r-1 & r \end{bmatrix}.$$

The encryption process is described below:

Step 1, the plain image is the input matrices.
Step 2, separating three channels pixel values from images and storing them in RBG matrices, respectively.
Step 3, segmenting those three matrices into sub-blocks of size 4×4, and those sub-matrices are marked R_{sub_i}, G_{sub_i}, B_{sub_i}, $i = 1, 2, \ldots, m$, m is total numbers of sub-blocks.
Step 4, transforming those above sub-matrices $R_{sub_i}^{'}$, $G_{sub_i}^{'}$, $B_{sub_i}^{'}$ through discrete logarithm operation.
Step 5, chaotic sequences (CS) we discuss in part two $\{x_t, y_t, z_t | t = 1, 2, \ldots, N\}$ are obtained.
Step 6, encryption key matrixes $K_1, K_2, \ldots, K_m$ are gotten by constructing $A_1, A_2, \ldots, A_{2m}$ and doing tensor product operation.
Step 7, encrypting pixels in sub-matrices (Step 4), the cipher algorithm as formula (3.7).

$$R_{cipher_{oldn}} = (R'_{sub_i} + R_{cipher_{old_{n-1}}} + x'_t) \bmod \lfloor \max(y'_t, z'_t) \rfloor \quad i = 1, 2, \ldots, m.$$

Step 8, combining $R_{cipher_{old}}$, $G_{cipher_{old}}$, $B_{cipher_{old}}$ into a matrix as cipher image sequence.

3.2.6 Simulation Results

The simulation results of two kinds of images: gray images and RGB color images, are shown in Fig. 3.12. As a basic property of a good encryption algorithm, it needs to provide sufficient superiority to against a variety of known attacks like secret-key attacks, plain text attacks, brute-force attacks, and differential attacks.

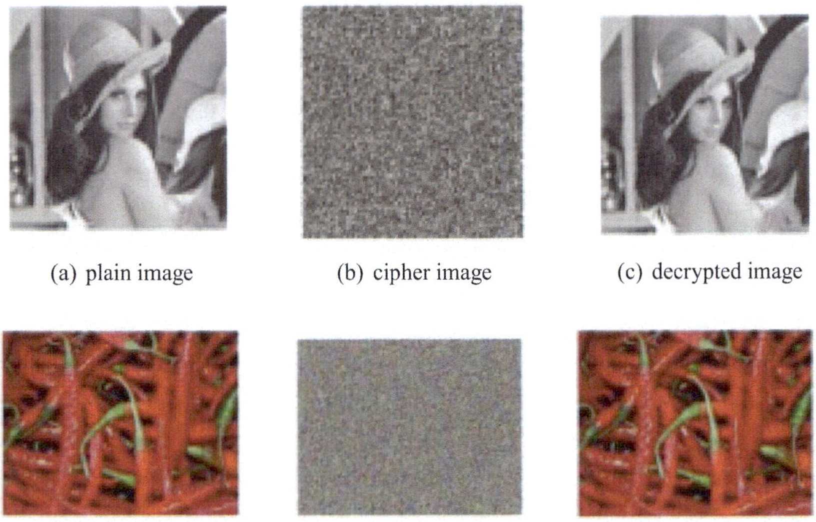

(a) plain image (b) cipher image (c) decrypted image

(d) RGB plain image (e) RGB cipher image (f) RGB decrypted image

Fig. 3.12 IED1-MBCC

3.2.6.1 Secret-Key Analysis

In our proposed method, we use three-dimensional chaos as encryption and decryption keys, and total key space is $2^{192} \approx 6.28 \times 10^{57}$. For a size $M \times N$ encrypted image with G gradation, if using brute-force method of attack, calculation time is $(M \times N)!G^{M \times N}$. With the increase in image size and gradation, there is an icrease in the difficulty of the brute-force attack.

3.2.6.2 Statistical Analysis

Correlation values of comparative analysis are given in Tables 3.1 and 3.2. Figures 3.13 and 3.14 show plain and cipher image histograms, which are encrypted by two methods, the cipher image with a very desirable uniform distribution.

Table 3.1 The correlation and entropy values of IED1-MBCC

Method	Image	Correlation					
		Plain			Cipher		
		horizontal	vertical	diagonal	horizontal	vertical	diagonal
IED1-MBCC	Gray	0.9342	0.9016	0.8820	−0.0037	0.0053	−0.0233
	RGB	0.9394	0.9500	0.9279	−0.00075	0.0696	0.0059
[22]	Gray	0.92756	0.9613	–	0.0237	−0.0161	–
	RGB	0.8836	0.8813	–	0.01985	−0.0198	–

Table 3.2 The correlation and entropy values of IED2-MBCC

Method	Image	Correlation					
		Plain			Cipher		
		Horizontal	Vertical	Diagonal	Horizontal	Vertical	Diagonal
IED1-MBCC	Gray	0.9391	0.9161	0.8838	−0.0720	0.0552	−0.0011
	RGB	0.9484	0.9496	0.9251	−0.0019	0.0472	−0.0032
[18]	Gray	0.9271	0.9230	0.9847	−0.0574	−0.0035	0.0578
	RGB	0.9640	0.9501	0.9566	−0.0351	−0.0028	0.0475

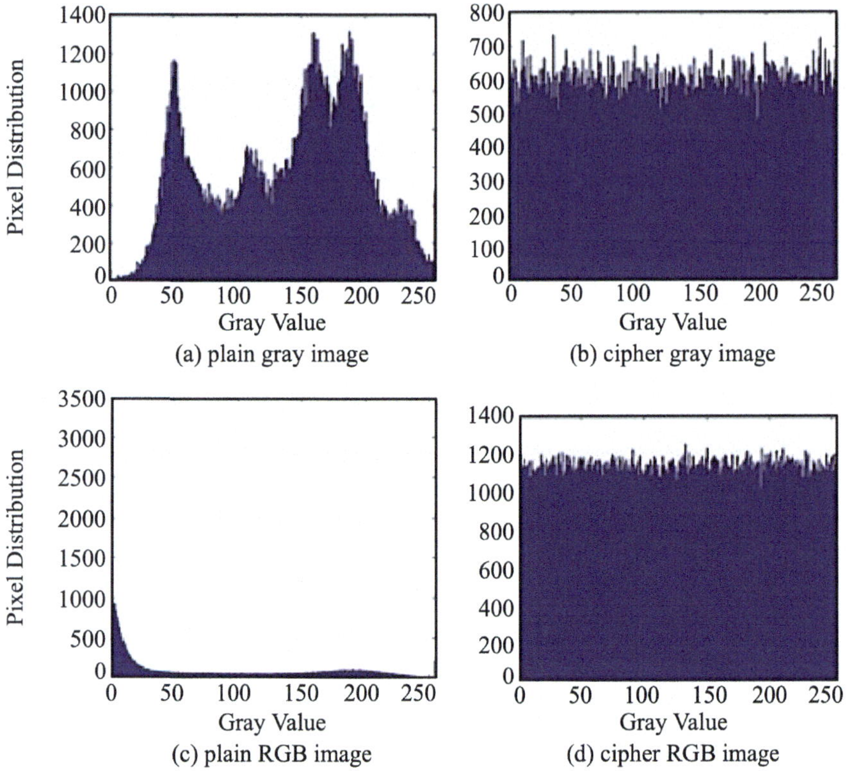

Fig. 3.13 Histogram by IED1-MBCC

3.3 Comments

Chaotic cryptography has emerged as a promising field in information security, leveraging the inherent properties of chaotic systems such as sensitivity to initial conditions, ergodicity, and pseudo-randomness. While traditional chaotic systems like logistic maps and Chua circuits have been extensively studied, the advent of memristors has opened new avenues in this field. This chapter explores the novel application of memristor-based chaotic circuits in text and image encryption and decryption, representing a significant advancement. The study proposes a novel encryption system model based on memristor chaotic circuits and develops two innovative image encryption algorithms: IED1-MBCC, employing discrete logarithm operations and modular arithmetic, and IED2-MBCC, based on the Hill cipher with tensor products. Through extensive simulation and rigorous analysis, including statistical tests and security evaluations, the feasibility and practicality of these methods are verified, demonstrating the potential of memristor-based circuits in enhancing the security and efficiency of cryptographic applications.

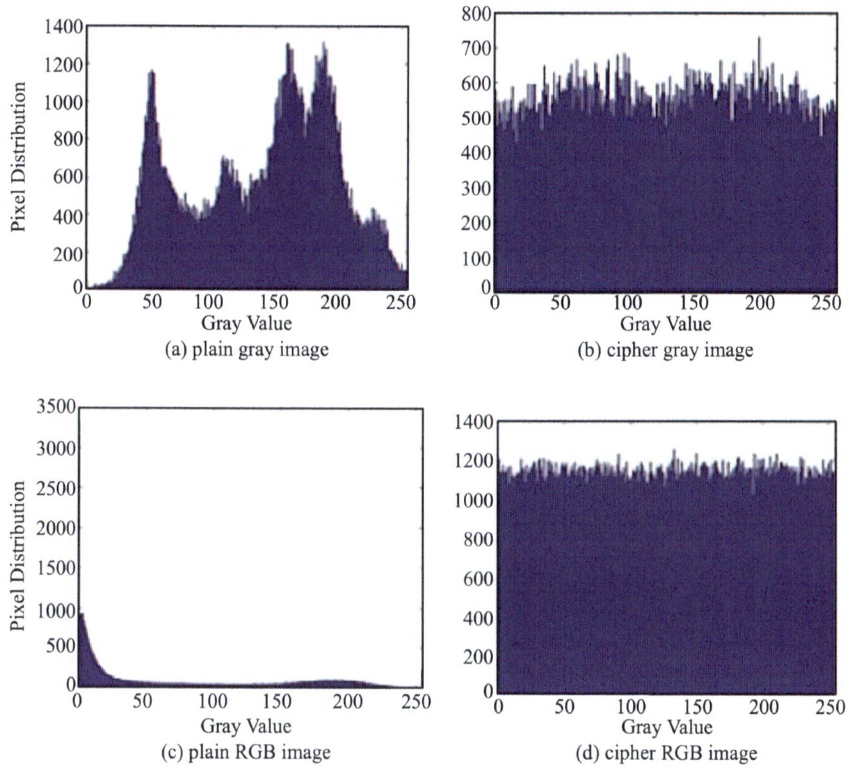

Fig. 3.14 Histogram by IED2-MBCC

Building on these initial findings, future research can further explore and expand the potential of memristor-based chaotic encryption systems. Specifically, the following directions may become research trends in this field:

Design of high-dimensional chaotic systems with multiple coupled memristors. Explore circuit topologies with three or more coupled memristors, such as star, ring, or fully connected structures. By adjusting the nonlinear characteristics of memristors and coupling strengths, more complex chaotic dynamics can be generated. For example, a six-dimensional memristor chaotic system could be designed, with a Lyapunov exponent spectrum featuring multiple positive exponents, significantly increasing the key space and enhancing encryption strength.

Investigation of memristor chaotic systems in post-processing of quantum key distribution. Combine memristor chaotic systems with quantum key distribution technology to design efficient key distillation and expansion algorithms. For instance, use memristor chaotic sequences to shuffle and expand raw quantum keys, improving the randomness and security of the final key. Study the impact of different memristor characteristics (such as nonlinearity degree, and on-off ratio) on key distillation efficiency to optimize algorithm performance.

Development of new image encryption methods based on memristor chaotic networks. Design complex networks composed of multiple memristor chaotic nodes, with each node responsible for encrypting a local area of the image. Achieve multi-scale image encryption by adjusting network topology and inter-node coupling strengths. Study the relationship between network synchronization characteristics and encryption strength, and propose an adaptive key update mechanism based on network synchronization degree to enhance the dynamic security of the encryption system.

References

1. Paul, A., Das, N., Prusty, A.K., Das, C.: Rgb image encryption by using discrete log with and lorenz's chaotic function. In: 2013 4th International Conference on Computer and Communication Technology (ICCCT), pp. 199–204 (2013)
2. Liangrui, T., Jing, L., Yi, S.: Image encryption based on a four-dimensional chaotic system. In: 2009 Fifth International Conference on Natural Computation, vol. 5, pp. 487–491. IEEE (2009)
3. Gao, T., Chen, Z.: A new image encryption algorithm based on hyper-chaos. Phys. Lett. A **372**(4), 394–400 (2008)
4. Liu, H., Wang, X.: Color image encryption based on one-time keys and robust chaotic maps. Comput. & Math. Appl. **59**(10), 3320–3327 (2010)
5. Seyedzadeh, S.M., Mirzakuchaki, S.: A fast color image encryption algorithm based on coupled two-dimensional piecewise chaotic map. Signal Proc. **92**(5), 1202–1215 (2012)
6. Wang, J., Jiang, G.p.: Cryptanalysis of a hyper-chaotic image encryption algorithm and its improved version. Acta Phys. Sin **60**, 060503(1)–060503(11) (2011)
7. Li, T.: Study on digital image encryption algorithm based on 3d chaotic sequences. J. Comput. Appl. **26**(7), 1652 (2006)
8. Wei, P.C.: Studies on the design and implementation of chaotic stream ciphers. Master Dissertation of Chongqing University, pp. 30–48 (2004)
9. Liu, C.b.: Study and implementation of the chaotic encryption algorithm. Master Dissertation of Beijing University of Posts and Telecommunications, pp. 28–37 (2008)
10. Qin, Z.G.: Cryptography algorithm-survey and trends. Comput. Appl. **24**, 1–4 (2004)
11. Wu, Y., Noonan, J.P., Agaian, S.: A wheel-switch chaotic system for image encryption. In: Proceedings 2011 International Conference on System Science and Engineering, pp. 23–27. IEEE (2011)
12. Bao, L., Zhou, Y., Chen, C.P., Liu, H.: A new chaotic system for image encryption. In: 2012 International Conference on System Science and Engineering (ICSSE), pp. 69–73. IEEE (2012)
13. Hao, R., Min, L., Han, D., Wang, X.: A 6 dimensional chaotic generalized synchronization system and design of pseudorandom number generator with application in text encryption. In: 2014 12th International Conference on Signal Processing (ICSP), pp. 1826–1831. IEEE (2014)
14. Corron, N.J., Reed, B.R., Blakely, J.N., Myneni, K., Pethel, S.D.: Chaotic scrambling for wireless analog video. Commun. Nonlinear Sci. Numer. Simul. **15**(9), 2504–2513 (2010)
15. Lin, Z., Yu, S., Lü, J., Cai, S., Chen, G.: Design and arm-embedded implementation of a chaotic map-based real-time secure video communication system. IEEE Trans. Circuits Syst. Video Technol. **25**(7), 1203–1216 (2015)
16. Wu, Q., Wang, G., Yuan, L.: E-mail encryption based on dual chaotic map. In: 2010 International Workshop on Chaos-Fractal Theories and Applications, pp. 159–163. IEEE (2010)
17. Itoh, M., Chua, L.O.: Memristor oscillators. Int. J. Bifurc. Chaos **18**(11), 3183–3206 (2008)
18. Matthews, R.: On the derivation of a "chaotic" encryption algorithm. Cryptologia **13**(1), 29–42 (1989)

19. Chen, G., Mao, Y., Chui, C.K.: A symmetric image encryption scheme based on 3d chaotic cat maps. Chaos, Solitons & Fractals **21**(3), 749–761 (2004)
20. Liu, H., Wang, X.: Color image encryption using spatial bit-level permutation and high-dimension chaotic system. Opt. Commun. **284**(16–17), 3895–3903 (2011)
21. Li, C., Li, S., Alvarez, G., Chen, G., Lo, K.T.: Cryptanalysis of two chaotic encryption schemes based on circular bit shift and xor operations. Phys. Lett. A **369**(1–2), 23–30 (2007)
22. François, M., Grosges, T., Barchiesi, D., Erra, R.: A new image encryption scheme based on a chaotic function. Signal Proc.: Image Commun. **27**(3), 249–259 (2012)
23. Paul, A., Das, N., Prusty, A.K.: An advanced gray image encryption scheme by using discrete logarithm with logistic and heh64 chaotic functions. In: 2013 3rd IEEE International Advance Computing Conference (IACC), pp. 1114–1120 (2013)

Chapter 4
Filter Design Based on Memristive Family

Memristive components, including memristors, memcapacitors, and meminductors, represent a paradigm shift in electronic circuit design. These novel elements, characterized by their ability to "remember" past states, offer unique nonlinear dynamics and history-dependent behavior. Memristors, first theorized by Leon Chua in 1971 and physically realized by HP Labs in 2008, exhibit a relationship between charge and magnetic flux. Memcapacitors and meminductors, extend this concept to capacitive and inductive elements, respectively. Their nonlinear characteristics and inherent memory effects provide new degrees of freedom in circuit design, potentially leading to more efficient and compact electronic systems. One promising application area for these memristive components is in the field of signal processing, particularly in filter design. Filters, essential elements in numerous electronic systems, are traditionally constructed using conventional passive and active components. However, the integration of memristive elements into filter circuits opens up exciting possibilities for enhancing performance and functionality. By leveraging the unique properties of memristive devices, such as their nonlinear response and state-dependent behavior, engineers can potentially create adaptive filters with improved frequency selectivity, reduced power consumption, and enhanced noise rejection capabilities.

Section 4.1 provides an overview of the fundamental characteristics of three memory circuit elements: memristors, memcapacitors, and meminductors, and presents their generalized mathematical models. Research has shown that these elements exhibit unique hysteresis characteristics under sinusoidal signal inputs. Notably, their equivalent resistance, capacitance, and inductance are not fixed, but closely related to internal state variables, thus demonstrating a dependence on historical states. Furthermore, these elements possess a significant zero-crossing property: when the input quantity is zero, the corresponding output quantity is zero. This property results in a characteristic curve that forms a hysteresis loop passing through the origin of the coordinate system when subjected to periodic zero-crossing signal inputs. These findings not only deepen our understanding of memory circuit elements but also lay a theoretical foundation for their future applications in electronic circuit design.

Y. Yu et al., *Memristive Computing*,
https://doi.org/10.1007/978-981-96-0332-9_4

Section 4.2 introduces a novel approach to incorporating memristors and memcapacitors into low-pass filter (LPF) designs. Inspired by both traditional RC-LPFs and Memristor-C-LPFs, a new Memristor-Memcapacitor-LPF is proposed. The transfer function of this LPF is derived, and its amplitude-frequency and phase-frequency responses are analyzed. Simulation results validate the theoretical findings, highlighting the potential extension of memristive elements to other filter circuits.

Building upon the concepts introduced in the previous section, Sect. 4.3 presents a pioneering model for integrating memristive components into high-pass filter (HPF) designs. By combining the principles of Memristor-Capacitor-HPFs and traditional RC-HPFs, a novel Memristor-Memcapacitor-HPF is developed. The study examines the HPF's transfer function, amplitude-frequency response, and phase-frequency response, with simulation results confirming the theoretical analysis. This section emphasizes the versatility of memristive elements across different filter circuits.

Section 4.4 delves into the realm of active filters based on memristors and memcapacitors (AFMMs), encompassing designs for low-pass, high-pass, band pass, and band stop filters. By replacing resistors and capacitors with memristors and memcapacitors, novel AFMM configurations are developed. The study demonstrates the transfer functions, cut-off frequencies, and Bode plots of these filters, with experimental results illustrating the time-varying characteristics of amplitude-frequency and phase-frequency responses. This section highlights the potential of memristive components to enhance the performance and flexibility of active filters.

Finally, Sect. 4.5 introduces a groundbreaking electrical circuit that replaces traditional resistors, capacitors, and inductors with their memristive counterparts: memristors, memcapacitors, and meminductors. This innovative approach gives rise to a new type of RLC circuit with unique characteristics. The section explores the properties and applications of band pass and band stop filters using these mem-devices, emphasizing their time-varying nature, which enables enhanced control and usability in signal processing applications.

Throughout this chapter, the integration of memristive components into filter circuits is thoroughly investigated, showcasing their potential to improve filter performance, flexibility, and adaptability. The theoretical analyses, simulation results, and experimental findings presented in each section contribute to a comprehensive understanding of memristive filters and their applications in signal processing. By harnessing the unique properties of memristors, memcapacitors, and meminductors, this chapter paves the way for the development of advanced filter designs that can revolutionize various fields, from telecommunications to biomedical signal processing.

4.1 Memristive Family

4.1.1 The Proposal of Memristors, Memcapacitors, and Meminductors

In 1971, Chua [1], grounded in the completeness of circuit theory, introduced the memristor concept, highlighting its role as the fourth fundamental circuit component. He proposed that the memristor could directly describe a specific relationship between charge q and magnetic flux ϕ. This concept is mathematically defined as:

$$G(\phi, q) = 0. \tag{4.1}$$

When Eq. (4.1) is expressed solely as a single-valued function of charge, it is termed a charge-governed memristor, with the following mathematical formulation:

$$v_M(t) = M_R(q_M(t))i_M(t), \tag{4.2}$$

where $M_R(q_M) = \frac{d\phi_M(q_M)}{dq_M}$.

Conversely, when Eq. (4.1) is represented only by a single-valued function of magnetic flux, it is designated as a flux-governed memristor, expressed mathematically as:

$$\begin{aligned} i_M(t) &= G_M(\phi_M(t))v_M(t), \\ G_M(\phi_M) &= \frac{dq_M(\phi_M)}{d\phi_M}, \end{aligned} \tag{4.3}$$

where M_R denotes memristance; G_M represents memductance; $i_M(t)$ and $q_M(t)$ are the current and charge traversing the device; $v_M(t)$ and $\phi_M(t)$ are the voltage and magnetic flux across the device; the subscript M indicates variables associated with the memristor.

In 2009, Chua et al. [2] broadened the scope of memory-bearing elements from memristors to memristive systems by introducing two new variables, thus establishing a comprehensive system of memory-bearing elements analogous to traditional circuit components [3], as shown in Fig. 4.1.

$$\begin{aligned} \xi &= \int_{t_0}^{t} q(\tau)d\tau, \\ \psi &= \int_{t_0}^{t} \phi(\tau)d\tau. \end{aligned} \tag{4.4}$$

Akin to the categorization of memristors, memelements can be classified into flux-governed memcapacitors and charge-governed memcapacitors based on distinct control quantities [4], represented by the following equations:

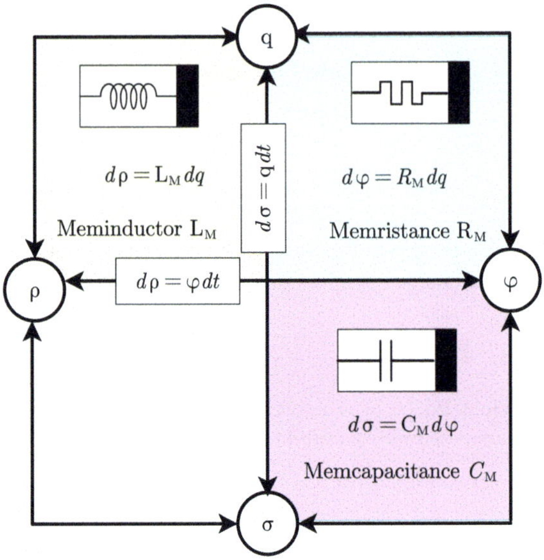

Fig. 4.1 Relationship among Memristor, Memcapacitor, and Meminductor

$$
\begin{aligned}
q_C(t) &= M_C(\phi_C(t)) \cdot v_C(t) = M_C \int_{t_0}^{t} v_C(\tau)d\tau \cdot v_C(t), \\
M_C(\phi_C) &= \frac{d\xi_C(\phi_C)}{d\phi_C},
\end{aligned}
\tag{4.5}
$$

$$
\begin{aligned}
v_C(t) &= M_C^{-1}(\xi_C(t)) \cdot q_C(t) = M_C^{-1} \int_{t_0}^{t} q_C(\tau)d\tau \cdot q_C(t), \\
M_C^{-1}(\xi_C) &= \frac{d\phi_C(\xi_C)}{d\xi_C}.
\end{aligned}
\tag{4.6}
$$

Meanwhile, meminductors can similarly be characterized as current-governed and flux-governed types, as illustrated in the following equations:

$$
\begin{aligned}
\phi_L(t) &= M_L(q_L(t)) \cdot i_L(t) = M_L \int_{t_0}^{t} i_L(\tau)d\tau \cdot i_L(t), \\
M_L(q_L) &= \frac{d\psi_L(q_L)}{dq_L},
\end{aligned}
\tag{4.7}
$$

$$
\begin{aligned}
i_L(t) &= M_L^{-1}(\psi_L(t)) \cdot \phi_L(t) = M_L^{-1} \int_{t_0}^{t} \phi_L(\tau)d\tau \cdot \phi_L(t), \\
M_L^{-1}(\psi_L) &= \frac{dq_L(\psi_L)}{d\psi_L},
\end{aligned}
\tag{4.8}
$$

where M_C represents memcapacitance; M_C^{-1} denotes inverse memcapacitance; $q_C(t)$ is the charge traversing the memcapacitor; $v_C(t)$ and $\phi_C(t)$ are the voltage and magnetic flux across the memcapacitor; M_L signifies meminductance; M_L^{-1} indicates inverse meminductance; $q_L(t)$ and $i_L(t)$ are the charge and current flowing through the meminductor; $\phi_L(t)$ is the magnetic flux across the meminductor; subscripts C and L denote variables corresponding to memcapacitors and meminductors, respectively.

4.1.2 General Mathematical Models of Memristors, Memcapacitors, and Meminductors

In 2009, Di Ventra et al. [5] proposed that elements with memory characteristics are not limited to memristors, but can actually be extended to memcapacitors and meminductors. They pointed out that if x represents a set of n-dimensional state variables of a memory element, $u(t)$ and $y(t)$ represent any two complementary constitutive variables (i.e., current, charge, voltage, or magnetic flux) as the input and output of the system, and g represents a generalized response function, then a general nth-order memory element can be defined by:

$$y(t) = g(x, u, t)u(t), \tag{4.9}$$

$$\dot{x}(t) = f(x, u, t), \tag{4.10}$$

where f is a continuous n-dimensional vector function. Physically speaking, if an initial state $u(t = t_0)$ is given at time t_0, then Eq. (4.10) has a unique solution.

Based on the above theory, when $u(t)$ and $y(t)$ are taken as different complementary variables, the general mathematical expressions for memristors, memcapacitors, and meminductors can be obtained as follows.

4.1.2.1 General Mathematical Definition of Memristors

Current-controlled memristor:

$$\begin{aligned} v_M(t) &= R_M(x_M, i_M, t)i_M(t), \\ \dot{x}_M &= f(x_M, i_M, t). \end{aligned} \tag{4.11}$$

Voltage-controlled memristor:

$$\begin{aligned} i_M(t) &= W_M(x_M, v_M, t)v_M(t), \\ \dot{x}_M &= f(x_M, v_M, t). \end{aligned} \tag{4.12}$$

4.1.2.2 General Mathematical Definition of Memcapacitors

Voltage-controlled memcapacitor:

$$\begin{aligned} q_C(t) &= C_M(x_C, v_C, t)v_C(t), \\ \dot{x}_C &= f(x_C, v_C, t). \end{aligned} \tag{4.13}$$

Charge-controlled memcapacitor:

$$\begin{aligned} v_C(t) &= C_M^{-1}(x_C, q_C, t)q_C(t), \\ \dot{x}_C &= f(x_C, q_C, t). \end{aligned} \tag{4.14}$$

4.1.2.3 General Mathematical Definition of Meminductors

Current-controlled meminductor:

$$\begin{aligned} \phi_L(t) &= L_M(x_L, i_L, t)i_L(t), \\ \dot{x}_L &= f(x_L, i_L, t). \end{aligned} \tag{4.15}$$

Flux-controlled meminductor:

$$\begin{aligned} i_L(t) &= L_M^{-1}(x_L, \phi_L, t)\phi_L(t), \\ \dot{x}_L &= f(x_L, \phi_L, t). \end{aligned} \tag{4.16}$$

Here, x_M, x_C, and x_L represent the dimensionless state variables of the memristor, memcapacitor, and meminductor systems, respectively.

Memristors, memcapacitors, and meminductors share the typical circuit characteristics: under sinusoidal signal input, there exists a hysteresis loop relationship between their respective inherent internal variables. This unique relationship is specifically manifested as the v_M-i_M relationship for memristors, the q_C-v_C relationship for memcapacitors, and the i_L-ϕ_L relationship for meminductors.

Moreover, from Eqs. (4.1)–(4.8) above, it can be seen that the equivalent resistance, capacitance, and inductance of memristors, memcapacitors, and meminductors are related to their respective internal variables of magnetic flux, charge, and their integrals over time. In other words, they are related to the historical states of voltage, magnetic flux applied across them, and current and charge flowing through them. From these equations, it can be deduced that any memristive element should have a zero-crossing property. For example, for a memristor, when $i_M(t) = 0$, regardless of the value of $R_M(q_M(t))$, $v_M(t) = 0$ must hold; for a memcapacitor, when $v_C(t) = 0$, $q_C(t) = 0$ must hold; similarly, a meminductor also has the property that $i_L(t) = 0$ and $\phi_L(t) = 0$. Therefore, when the input signal to a memory element is a zero-crossing periodic signal of a certain frequency, its corresponding characteristic curve must be a hysteresis loop passing through the origin of the coordinate system.

4.2 A Low-Pass Filter Based on Memristor and Memcapacitor

4.2.1 Introduction

Many researchers have investigated filter circuits based on memristors and analyzed their time-varying characteristics. By replacing conventional resistors with memristors, as illustrated in Fig. 4.2b, these studies have established time-varying systems where the low-pass filter (LPF) characteristics become functions of time. Consequently, the circuit's performance varies at different time instants and states, as evidenced by the analysis of transfer function and cut-off frequency simulation results.

Inspired by the Memristor-C-LPF, it is hypothesized that memcapacitors, which possess memory properties and belong to the memristive system family like memristors, can also be incorporated into such circuits. This integration aims to further enhance the time-varying nature of the filter and potentially introduce novel functionalities. Figure 4.2c presents an innovative RC circuit where all basic components are replaced by memristive elements. The Memristor-Memcapacitor-LPF is a serial circuit comprising two memristive elements, necessitating independent consideration and analysis of both the memristor and memcapacitor.

In contrast to traditional LPF circuits with constant, time-invariant properties, the incorporation of memristive elements yields dynamic, time-dependent characteristics. To realize a time-varying LPF circuit, the relationship between the memristor

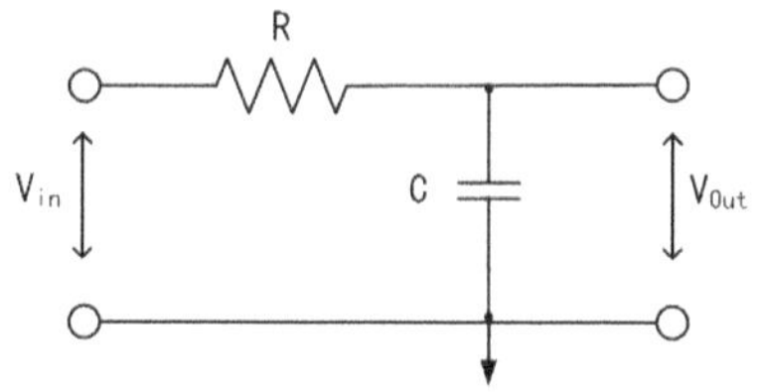

(a) The conventional RC-LPF circuit.

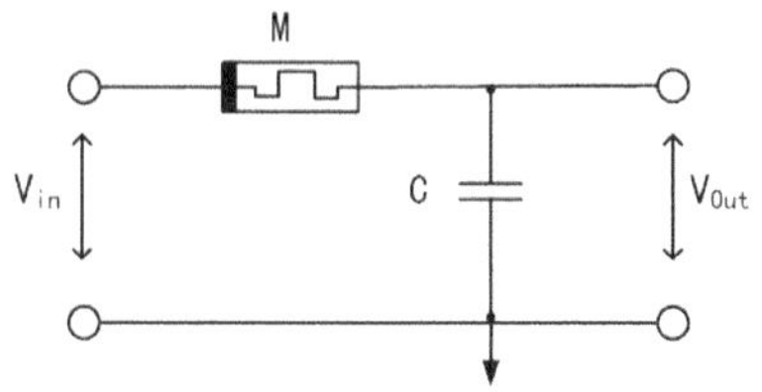

(b) The Memristor-C-LPF circuit.

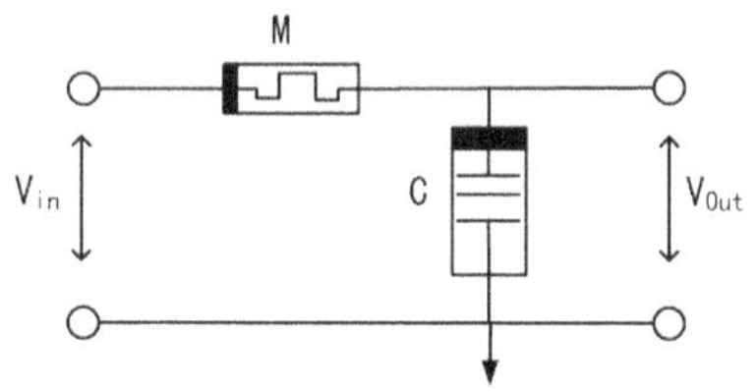

(c) The M-Memcapacitor-LPF circuit.

Fig. 4.2 **a** First-order RC, **b** Memristor-C filter and **c** The Memristor-Memcapacitor-based low-pass filter and memcapacitor should be independent and related to an offset signal, denoted as V_t. It is crucial to note that V_t determines the state of the memristive system.

For practical implementation, the circuit requires V_t to be applied separately to both the memristor and memcapacitor. This configuration allows the system's corresponding features to be modulated by altering the offset signal's amplitude and frequency. Consequently, at different times t, the system exhibits distinct states. The system state can also be manipulated by adjusting the duration of the applied signal. Importantly, the current applied signal is independent of the signal to be filtered.

The nonlinear nature of memristive elements and their inherent threshold voltage characteristics necessitate careful signal processing. When the amplitude of V_t is below the threshold voltage, the memristive elements behave similarly to linear elements. Therefore, the filtering signal cannot be directly applied to the circuit; instead, amplitude processing must be performed to ensure proper operation.

This novel approach to filter design, leveraging the unique properties of memristors and memcapacitors, opens up new possibilities for creating adaptive and reconfigurable filter circuits. The time-varying characteristics of these memristive element-based filters could potentially lead to enhanced performance in various signal processing applications, particularly in scenarios where dynamic filtering requirements are essential.

4.2.2 Analysis of Memristor-Memcapacitor LPF

The first-order passive low-pass filter is a impedance series circuit. From the vector form of Kirchhoff's voltage law (KVL) [6, 7]:

$$\sum \dot{U}_k = 0 \tag{4.17}$$

and the vector form of Ohm's law:

$$\dot{U} = Z\dot{I} \tag{4.18}$$

we obtain the following relation:

$$\dot{U} = \dot{U}_M + \dot{U}_C = Z_M\dot{I} + Z_C\dot{I} = Z\dot{I}. \tag{4.19}$$

RC series connection circuit constitutes a two-port network. Considering a dynamic circuit under the ω frequency excitation, the ratio of sinusoidal steady-state response vector to activation vector is called Sine steady-state network function, mathematically represented as:

$$H(j\omega) = \frac{V_{\text{out}}(j\omega)}{V_{\text{in}}(j\omega)}, \tag{4.20}$$

When the load end is open, capacitor voltage to input voltage transfer ratio is:

$$H(j\omega) = \frac{V_{\text{out}}(j\omega)}{V_{\text{in}}(j\omega)} = \frac{\frac{1}{j\omega C}}{R + \frac{1}{j\omega C}} = \frac{1}{1 + j\omega RC}. \tag{4.21}$$

RC is the circuits time constant. Let $\tau = RC$, we obtain:

$$\omega_H = \frac{1}{\tau}, \tag{4.22}$$

$$f = \frac{\omega_H}{2\pi} = \frac{1}{2\pi\tau} = \frac{1}{2\pi RC}. \tag{4.23}$$

Then, Eq. (4.23) can be written as:

$$f = \frac{1}{1 + j\frac{\omega}{\omega_H}} = \frac{1}{1 + j\frac{f}{f_H}}. \tag{4.24}$$

Where R_m denotes the resistance of the memristor and the R_c denotes the capacitive reactance of the memcapacitor. Given a sinusoidal waveform as input, we conclude from simulation results that the memristor resistance can be described as follows:

$$R_m = R_{\text{avg}} \pm \Delta R\sin(\omega t + \varphi). \tag{4.25}$$

And the memcapacitor capacitive reactance for sinusoidal input signal is represented as:

$$R_c = R_c\sin(\omega t + \varphi). \tag{4.26}$$

If we replace the corresponding elements with the memristive system elements, the transfer function and the cut-off frequency can be given by the following equations:

$$\frac{V_{\text{out}}}{V_{\text{in}}} = \frac{1}{1 + j\omega(R_{\text{avg}} \pm \Delta R\sin(\omega_1 t + \varphi_1))R_c\sin(\omega_2 t + \varphi_2)}, \tag{4.27}$$

$$f = \frac{1}{2\pi(R_{\text{avg}} \pm \Delta R\sin(\omega_1 t + \varphi_1))R_c\sin(\omega_2 t + \varphi_2)}. \tag{4.28}$$

where ω denotes the frequency of V_t and different values of ω corresponds to different results.

4.2.3 Simulation Result

4.2.3.1 Memristor Simulation

The memristor model parameters given in the same study are used: $R_{init} = 80\,\mathrm{K}\Omega$, $R_{on} = 1\,\mathrm{K}\Omega$, $R_{off} = 160\,\mathrm{K}\Omega$, $D = 10\,\mathrm{nm}$, $\mu_v = 10\,\mathrm{Fm}^2/\mathrm{s}\,Vp = 10$, where R_{init} and p parameters denote the initial resistance and exponent of window function, respectively. Plotting the Lissajous figure of $(i(t), v(t))$ in the $V - I$ plane in Fig. 4.3 yields the pinched hysteresis loop, which is one of the features of a memristor. Figure 4.4 shows the simulation result when a sinusoidal signal $V(t) = V_0 \sin(2\pi \Omega t)$ is added, with $V_0 = 1\,\mathrm{V}$ and $\Omega = 1\,\mathrm{Hz}$.

4.2.3.2 Memcapacitor Simulation

To get the memcapacitor simulation result, we use the piecewise-linear function model of memcapacitor [8]. The memcapacitor model parameters are: $\alpha = 50\,\mu\mathrm{F}(\mathrm{V}\cdot\mathrm{s})$, $\beta = 2\alpha$, $V_t = 0.5\,\mathrm{V}$, and $C_{init} = 1\,\mu\mathrm{F}$. Figures 4.5 and 4.6 show the simulated results when a sinusoidal signal $V(t) = V_0 \sin(2\pi \omega t)$ is added, where $V_0 = 1\,\mathrm{V}$, $\omega = 1\,\mathrm{Hz}$.

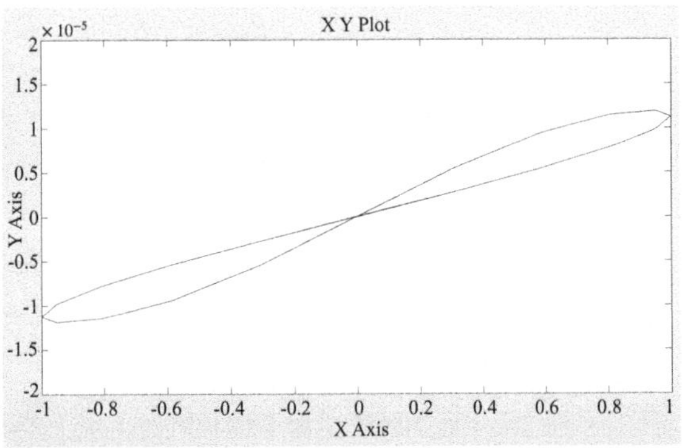

Fig. 4.3 The pinched hysteresis loop of memristor

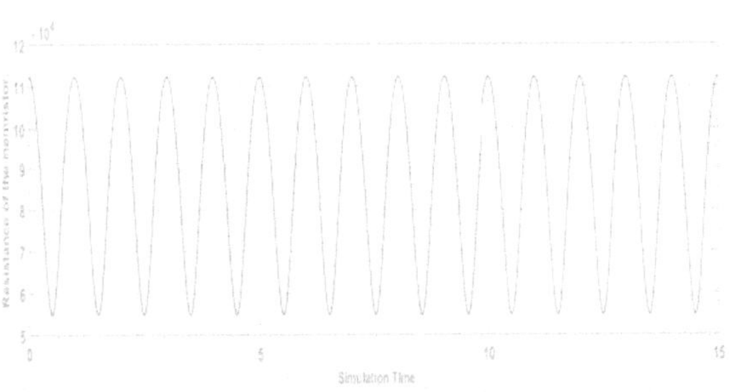

Fig. 4.4 Resistance of the memristor under different time

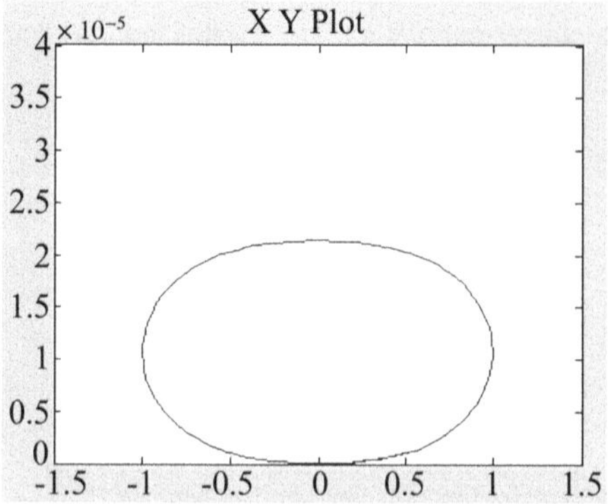

Fig. 4.5 The voltage-capacitance curve

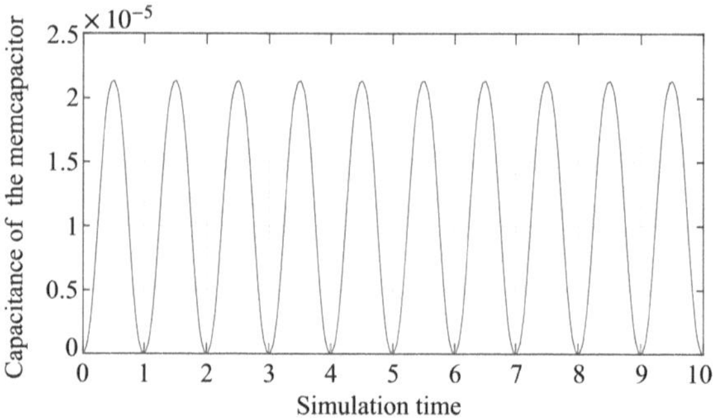

Fig. 4.6 The capacitance of memcapacitor change over time

4.2.4 *LPF Based on Memristive System*

Different signals arrive at different transfer functions and cut-off frequencies. Figure 4.7 shows the curve of the transfer function changes over time and Fig. 4.8 shows the cut-off frequency. Here, the offset signal is $V(t) = V_0 \sin(2t)$, where $V_0 = 1$ V and $\omega = 1$ Hz. In the research on frequency response of amplifying circuit, the frequency of input signal ranges from a few hertz to hundreds of megahertz, meanwhile the magnification can be from several times to millions of times. Thus, logarithmic coordinate is widely used in frequency characteristic curve referred to as Bode plot. Figures 4.9 and 4.10 show the simulation result ploted in Bode plot. Select $t_1 = 1.5$ s and $t_2 = 0.25$ s, at t_1 and t_2 the system has different resistance and capacitance. From Figs. 4.9 and 4.10, we can see that at different times t_1 and t_2, the LPF circuit has different amplitude-frequency response and phase-frequency response. This verifies the time-varying system which is in our expectation.

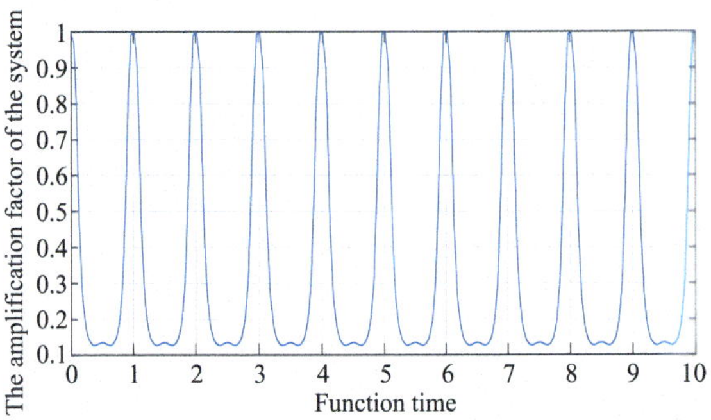

Fig. 4.7 The curve of the transfer function changes over time

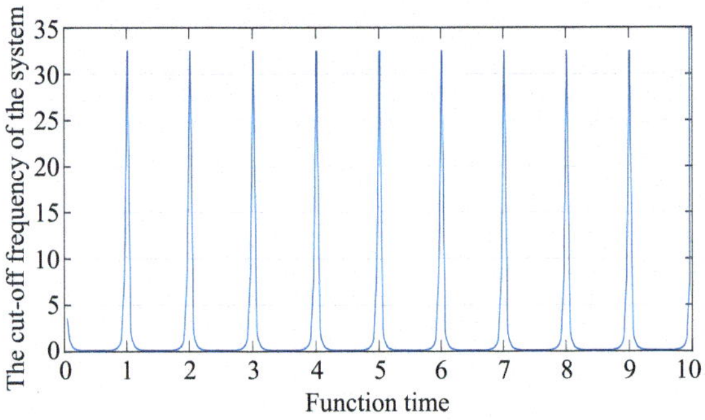

Fig. 4.8 The curve of the cut-off frequency changes over time

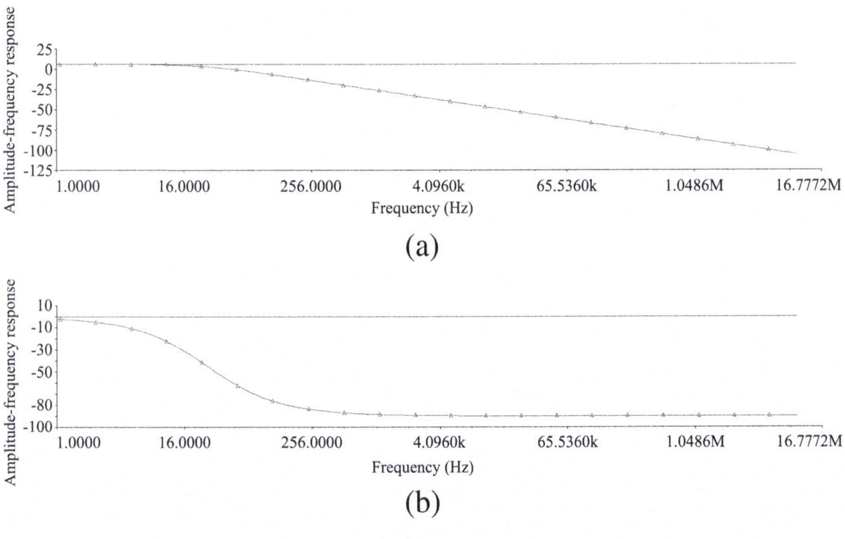

Fig. 4.9 Amplitude-frequency response and phase-frequency response at t_1

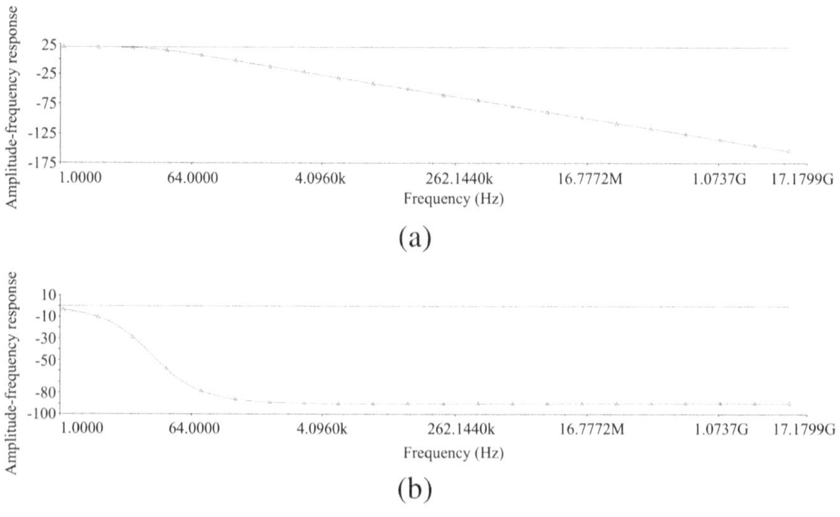

Fig. 4.10 Amplitude-frequency response and phase-frequency response at t_2

4.2.5 Conclusion

From the upper simulation result we can find the transfer function and cut-off frequency are time-variant which is in our expectation. At any given time, the circuit has a different amplitude-frequency response and phase-frequency response. Thus, it is assumed that this circuit is a time-variant and memristive system. The range of the change is limited and periodic. This is associated with our offset signal. Different signal corresponds to the different transfer function and cut-off frequency. Finding the most suitable offset signal is the next phase of task. The filter circuit can be classified into low-pass, band pass and high-pass filter. The memristive elements can also be applied into those type circuit to change circuits performance.

4.3 A High-Pass Filter Based on Memristor and Memcapacitor

4.3.1 Introduction

Filter invented by Campbell and Wagner [9], is of importance in the field of signal processing such as communication engineering, digital image processing, and so on. In the point of view of pass band and stop band, filter can be classified as low-pass filter (LPF), high-pass filter, band pass filter and band stop filter [9]. In the point of view of presence or absence of power supply, filter can be classified as active filter and passive filter. In this section, the research on low-pass filtering, high-pass filter, active filter, band pass, and band stop filter circuit based on memristor and memcapacitor will be provided.

For all kinds of filters, many scholars have focused on filter circuit design and its application [10–18]. On one hand, a compact second-order band pass filter is presented with its adjustable center frequency [17], and a three-step method for the synthesis and design of such a filter is proposed to solve the problem of mounting high-temperature superconductor, which inserts into the body of the metal waveguide in order to ensure good electrical contact [18]. On the other hand, the deep neural network filter can be used to improve the reliability of the speaker verification system [12], and the adaptive filter can be designed for image sample classification [13]. In light of these research results, filter circuit design and its application are very important and challenging problems.

For the first-order passive RC HPF circuit, Memristor-C filter circuit is presented by replacing the resistor with memristor [10, 11]. Based on RC LPF, we have studied a new Memristor-Memcapacitor LPF [19]. In this study, a novel HPF circuit model named Memristor-Memcapacitor-HPF is presented by replacing both resistor and capacitor with memristive elements in first-order RC HPF circuit and which is rarely researched before. With the analysis of the amplitude-frequency response and phase-frequency response by the Simulink model, a time-varying system such that its performance and characters can be changed with time, is obtained.

4.3.2 Design of Memristor-Memcapacitor-HPF

The time-variation characteristic of the memristor-based filter circuit has been researched by replacing the ordinary resistance with memristor as shown in Fig. 4.11a and b. With the analysis of the transfer function and cut-off frequency simulation results, a time-varying system where all the characters of the HPF circuit are the functions of time parameter is presented. Based on this, the performance of the HPF can be diverse when the instants of time and states are different. Inspired by the memristor-based HPF, the capacitor had been replaced with the memcapacitor and a novel HPF circuit named memristor-memcapacitor-HPF is obtained as shown in Fig. 4.11c.

Remark 1: with the consideration of the relationship between memristive elements and RC, new HPF circuit can be designed.

In this Memristor-Memcapacitor-HPF circuit, the nonlinear dopant drift model of memristor and the piecewise-linear model of memcapacitor are introduced [20, 21]. Based on these models, the definition of the memristance and the memcapacitance is presented in (4.29), where RM denotes the resistance of the memristor, and RC denotes the capacitive reactance of the memcapacitor.

$$\begin{cases} R_M = M_{init} + \Delta M \sin(\omega_1 t + \varphi_1) \\ R_C = C_{init} + \Delta C \sin(\omega_2 t + \varphi_2). \end{cases} \tag{4.29}$$

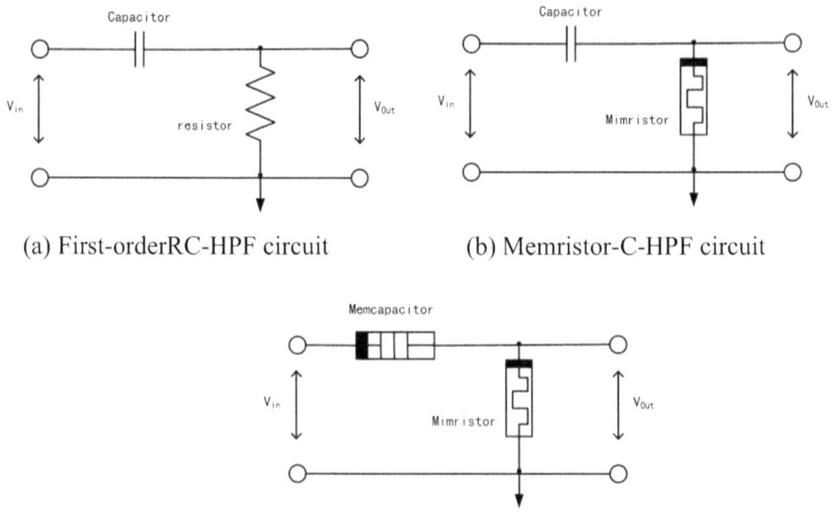

(a) First-orderRC-HPF circuit

(b) Memristor-C-HPF circuit

(c) Memristor-Memcapacitor-HPF circuit

Fig. 4.11 Three types of HPF circuit

4.3.3 Analysis of Memristor-Memcapacitor-HPF

RC series connection circuit constitutes a two-port network. Considering a dynamic circuit under the frequency excitation, the ratio of sinusoidal steady-state response vector to activation vector is called Sine steady-state network function as follows:

$$H(j\omega) = \frac{V_{\text{out}}(j\omega)}{V_{\text{in}}(j\omega)}. \tag{4.30}$$

The transfer ratio of resistor voltage and input voltage is

$$H(j\omega) = \frac{V_{\text{out}}(j\omega)}{V_{\text{in}}(j\omega)} = \frac{R}{R + \frac{1}{j\omega C}} = \frac{1}{1 + \frac{1}{j\omega RC}} \tag{4.31}$$

by setting $\tau = RC$, the following formula is obtained:

$$\omega_L = \frac{1}{\tau}. \tag{4.32}$$

The cut-off frequency is determined by

$$f = \frac{\omega_L}{2\pi} = \frac{1}{2\pi RC}. \tag{4.33}$$

Furthermore, the models of the Memristor and the Memcapacitor shown in (4.29) are adopted. The transfer function and the cut-off frequency of the Memristor-Memcapacitor-HPF circuit can be obtained as follows:

$$H(j\omega) = \frac{1}{1 + \frac{1}{j\omega R_M R_C}}, \tag{4.34}$$

$$f = \frac{1}{2\pi R_M R_C}, \tag{4.35}$$

where ω denotes the frequency of $V_O S$. It's important to state that the sinusoidal input signal added to the Memristor is different from the one added to the Memcapacitor.

4.3.4 Simulation Result

For this study, simulation was done using the dynamic and analytic models. First, the memristor model and the memcapacitor model are presented, respectively, with their simulation results. Then, the transfer function and cut-off frequency curve of the Memristor-Memcapacitor-HPF circuit are provided. Finally, different amplitude-frequency response and phase-frequency response are provided at different time, which means that the new HPF circuit is time-varying. All simulation results are solved by Matlab and Multisim.

In this case, the Memristor parameters are taken as follows:

$M_{init} = 80\,\text{K}\Omega$, $R_{on} = 1\,\text{K}\Omega$, $R_{off} = 160\,\text{K}\Omega$, $D = 10\,\text{nm}$, $\mu_v = 10\,\text{Fm}^2/\text{sV}$, $p = 10$, $\omega_1 = \pi$, $\varphi_1 = \frac{\pi}{2}$, in which the p denotes the exponent of window function. The Lissajous figure is shown in Fig. 4.12a, which is one of the features of the Memristor. Based on the (4.29), the time-varying resistance of the Memristor is presented in Fig. 4.12b with the added sinusoidal signal: $V(t) = V_1 \sin(2\pi\omega t)$, $V_1 = 1v$, $\omega = 1\,\text{Hz}$.

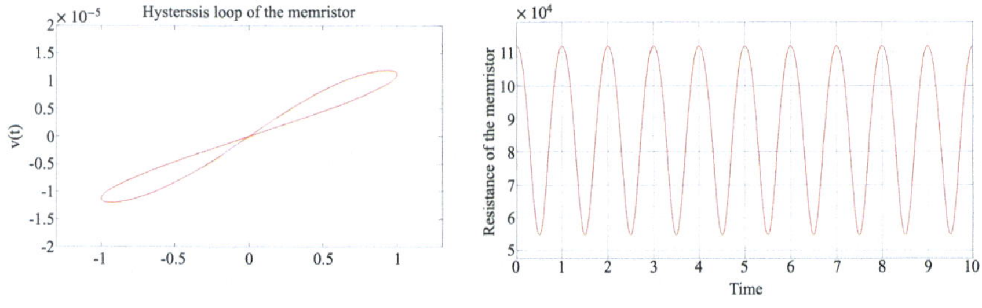

(a) (b)

Fig. 4.12 The simulation of the Memristor. **a**: the hysteresis loop of Memristor. **b**: the resistance of Memristor

On the other hand, the parameters for the Memcapacitor are chosen as $\alpha = 50\,\mu\text{F}(\text{V}\cdot\text{s})$, $\beta = 2\alpha$, $V_O S = 0.5v$, $\omega_2 = \pi$, $\varphi_2 = -\pi/2$, $C_i nit = 11\,\mu\text{F}$. The results are provided in Fig. 4.13a and b, with the added sinusoidal signal: $V(t) = V_2 \sin(2\pi\omega t)$, $V_2 = 1v$, $\omega = 1$ Hz.

Based on memristive models, Fig. 4.14 shows the curve of transfer function changed over time, and the time-varying cut-off frequency is presented in Fig. 4.15. Simultaneously, Figs. 4.16 and 4.17 demonstrate the Bode plot, which shows different amplitude-frequency response and phase-frequency response to three different times. Given a period of time, the TIME-BODE-DIAGRAM (i.e., a traditional bode-diagram with a time shaft) shown in Figs. 4.18 and 4.19, is presented, which intuitively present the variation of the new Memristor-Memcapacitor-HPF's parameters including cut-off frequency, Amplitude-Frequency response, Phase-Frequency response etc. Figures 4.18 and 4.19 verify that the new circuit is HPF circuit.

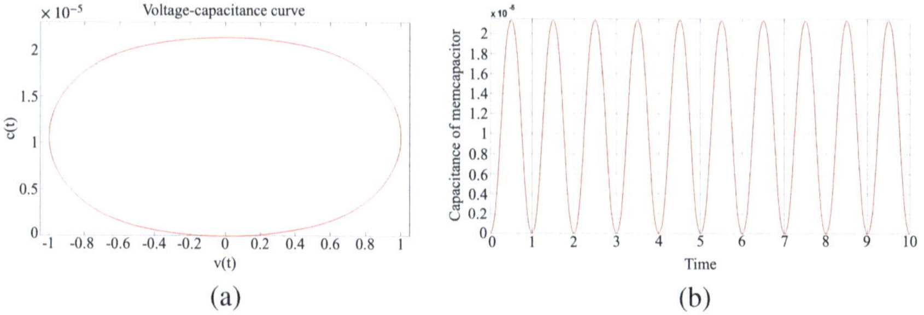

Fig. 4.13 The simulation of the Memcapacitor. **a**: the V-C curve of Memcapacitor. **b**: the capacitance of Memcapacitor

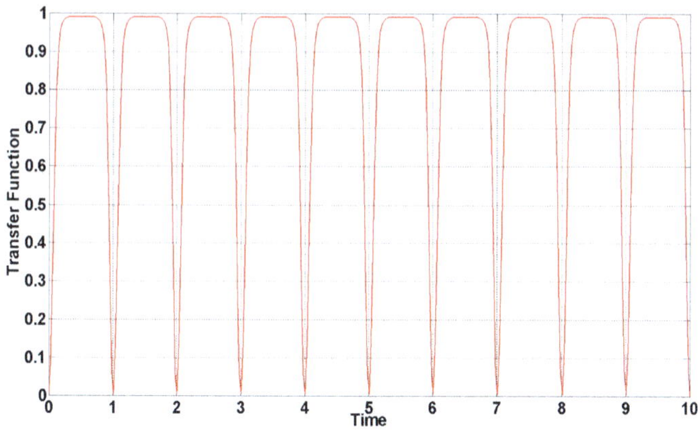

Fig. 4.14 Time-varying transfer function

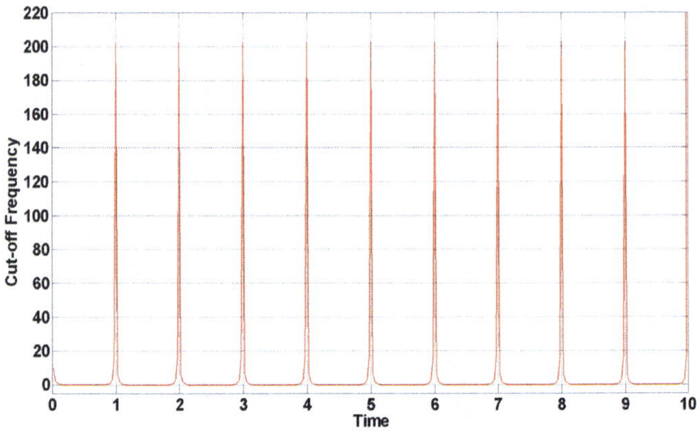

Fig. 4.15 Time-varying cut-off frequency

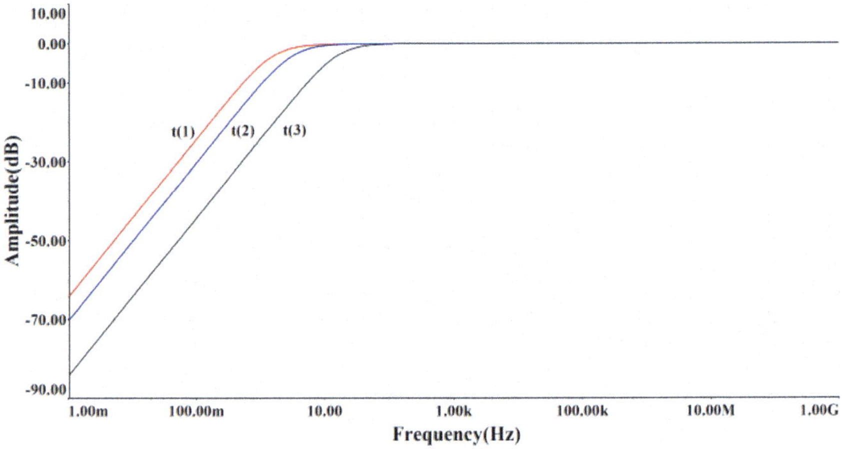

Fig. 4.16 Different Amplitude-Frequency respond at three different moments

4.3.5 Conclusion

In this section, a Memristor-Memcapacitor-HPF circuit rarely researched is designed and analyzed. Simulation results verify both analytical and dynamical models of the circuit. It is shown that the presented system is time-variant as expected and the circuit has different amplitude-frequency response and phase-frequency response to different times. Compared with traditional HPF, there are more possibilities and opportunities for practical application. Obviously, memristive elements can be utilized in other filter circuits.

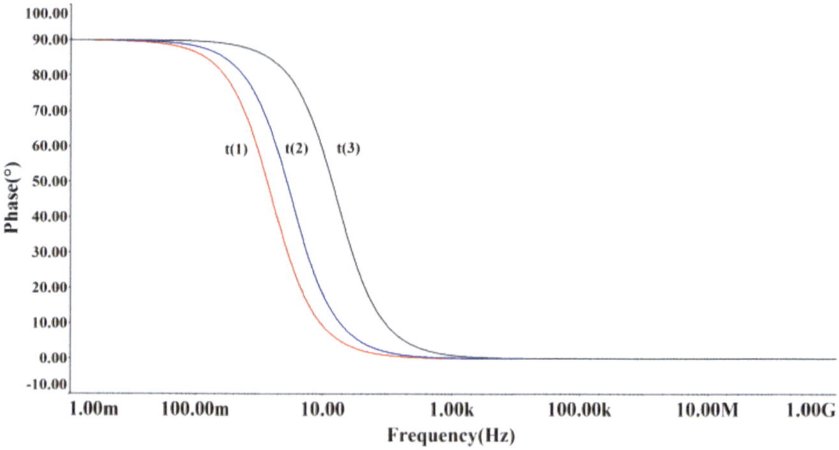

Fig. 4.17 Different Phase-Frequency responds at three different moments

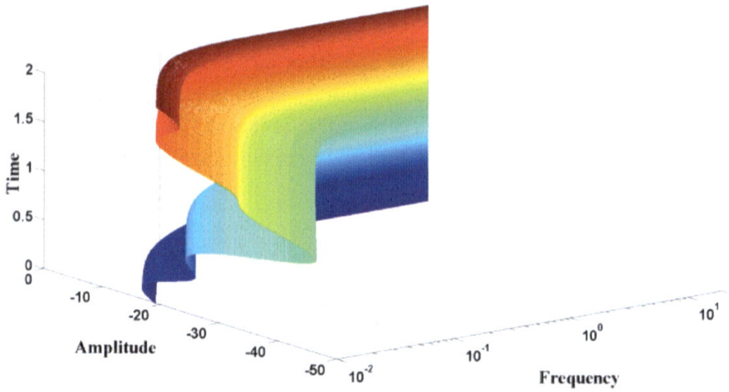

Fig. 4.18 Time-Amplitude-Frequency response

4.4 An Active Filter Based on Memristor and Memcapacitor

4.4.1 Introduction

Active filters, comprising resistors, capacitors, and amplifiers, are fundamental components in signal processing and can be categorized into low-pass, high-pass, band pass, and band stop filters [22]. Conventional filters, once designed with specific resistances and capacitances, have fixed frequency bands and cut-off frequencies. This static nature limits their adaptability in dynamic signal processing environments.

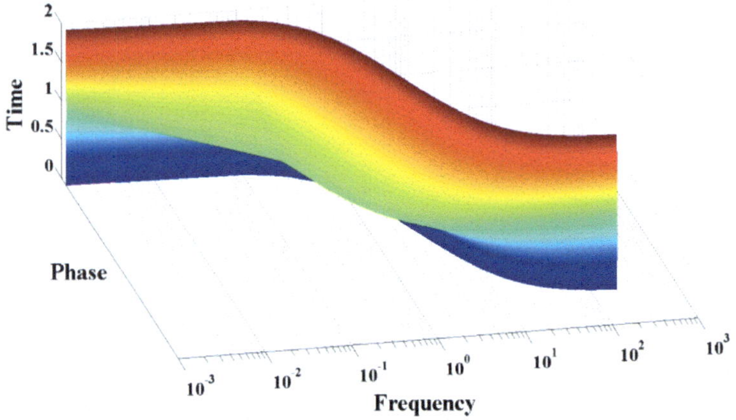

Fig. 4.19 Time-Phase-Frequency response

The advent of memristors and memcapacitors has introduced new possibilities in filter design. These devices exhibit nonlinear, continuous changing characteristics and can be conveniently controlled by pulse voltages. This property allows for dynamic adjustment of memristance and memcapacitance, potentially leading to filters with time-varying parameters.

This section introduces a novel class of active filters based on memristors and memcapacitors. By replacing conventional resistors and capacitors with their memristive counterparts, the proposed filter designs allow for cut-off frequencies that can be dynamically altered over time. This approach represents a significant advancement in active filter technology, offering enhanced flexibility and adaptability.

The integration of memristors and memcapacitors into active filter circuits presents several advantages. The ability to adjust memristance and memcapacitance in real-time allows for on-the-fly modification of the filter's frequency response. These filters can potentially adapt to changing signal environments, optimizing their performance based on input characteristics. Memristive devices can lead to more compact filter designs, as they can replace multiple conventional components. The state-dependent nature of memristive elements enables the creation of programmable filters, where specific frequency responses can be stored and recalled. Additionally, memristive devices often operate with lower power requirements compared to traditional circuit elements.

The proposed active filters leverage these unique properties of memristors and memcapacitors to create versatile, time-varying filter structures. Various configurations, including low-pass, high-pass, and band pass filters, will be explored, demonstrating how the incorporation of memristive elements can enhance their functionality and performance.

This research not only advances the field of active filter design but also paves the way for new applications in adaptive signal processing, cognitive radio systems, and reconfigurable electronics. By harnessing the nonlinear dynamics of memristive devices, this section aims to develop a new generation of filters capable of meeting the evolving demands of modern communication and signal processing systems.

4.4.2 Design of AFMM

In [23], a memristor-based second-order active low-pass filter is researched. Traditional resistances are replaced by memristors, as shown in Fig. 4.20a. The cut-off frequency can be continuously changed by controlling the duration of the programming voltage applied to the memristor. Based on the memristor-based circuit, a novel active low-pass filter circuit, which is called active low-pass filter based on memristor and memcapacitor (ALPFMM), is presented as shown in Fig. 4.20b. In this circuit, the memristor and the memcapacitor are utilized to substitute the resistance and capacitor. Meanwhile, active high-pass filter based on memristor and memcapacitor (AHPFMM), active band pass filter based on memristor and memcapacitor (ABPFMM) and active band stop filter based on memristor and memcapacitor (ABSFMM) are shown in Fig. 4.20c, d, and e, respectively.

In these circuits, which are based on memristor and memcapacitor, the memristor use the nonlinear dopant drift model [1, 24] and the memcapacitor use the piecewise-linear model [25]. The memristor and memcapacitor can be defined as (4.36), where R_M denotes the memristance and R_C refers to the memcapacitance.

$$\begin{aligned} R_M &= R_{init} + \triangle R sin(\omega t + \omega) \\ R_C &= C_{init} + \triangle C sim(\omega t + \omega). \end{aligned} \tag{4.36}$$

4.4.3 Analysis of AFMM

In active filter circuit, the voltage and current are always transformed into $U(s)$ and $I(s)$ by Laplace transform. Therefore, the $R(s)$ of resistance is equal to R and $Z_c(s)$ of capacitor is equal to $\frac{1}{sC}$. Meanwhile, the ratio of output to input is called the transfer function as follows:

$$A_u(s) = \frac{U_o(s)}{U_i(s)}. \tag{4.37}$$

By transforming s into $j\phi$, the amplification factor can be obtained.

In this section, by using the model of memristor and memcapacitor shown in (4.36), the transfer function and the cut-off frequency of the active filter circuit can be obtained as follows.

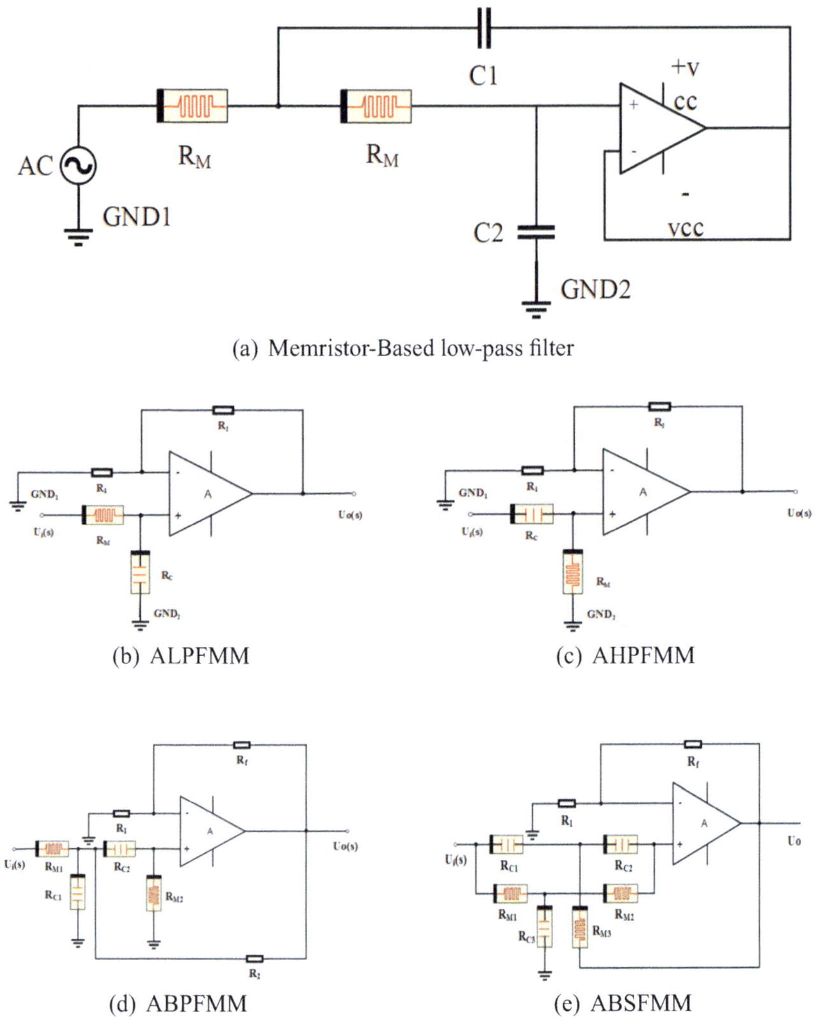

(a) Memristor-Based low-pass filter

(b) ALPFMM

(c) AHPFMM

(d) ABPFMM

(e) ABSFMM

Fig. 4.20 Active filter circuit

For the ALPFMM as Fig. 4.20b, the transfer function is shown below:

$$A_u(s) = \frac{U_o(s)}{U_i(s)} = \left(1 + \frac{R_2}{R_1}\right)\frac{1}{1 + sR_M R_C}. \tag{4.38}$$

In addition, the cut-off frequency is shown below:

$$f = \frac{1}{2\pi R_M R_C}. \tag{4.39}$$

For the AHPFMM as Fig. 4.20c, the transfer function can be obtained as follows:

$$A_u(s) = \frac{U_o(s)}{U_i(s)} = \left(1 + \frac{R_2}{R_1}\right)\frac{sR_M R_C}{1 + sR_M R_C}. \tag{4.40}$$

For the ABPFMM as Fig. 4.20d, when $R_{C1} = R_{C2} = R_C$, $R_{M1} = R_M$, $R_{M2} = 2R_M$, the transfer function can be expressed as:

$$A_u(s) = A_{uf}(s) \cdot \frac{sR_M R_C}{1 + [3 - A_{uf}(s)]sR_M R_C + (sR_M R_C)^2}, \tag{4.41}$$

where $A_{uf}(s) = 1 + \frac{R_f}{R_1}$. The band pass is represented as:

$$f_{bw} = |3 - A_{uf}| f_0, \tag{4.42}$$

where $f_0 = f$.

For ABSFMM, when $R_{C1} = R_{C2} = 2R_{C3}$, $R_{M1} = R_{M2} = \frac{1}{2}R_{M3}$, the transfer function is given by:

$$A_u(s) = A_{up}(s) \cdot \frac{1 + (sR_M R_C)^2}{1 + 2[2 - A_{up}(s)]sR_M R_C + (sR_M R_C)^2}, \tag{4.43}$$

where $A_{up} = 1 + \frac{R_f}{R_1}$. The stop width is given by:

$$BW = 2|2 - A_{up}| f_0. \tag{4.44}$$

4.4.4 Simulation Results

In this section, the simulations have been implemented. The memristor model and the memcapacitor model are simulated by Simulink as Fig. 4.21. The parameters of memristor and memcapacitor are taken as provided in Table 4.1.

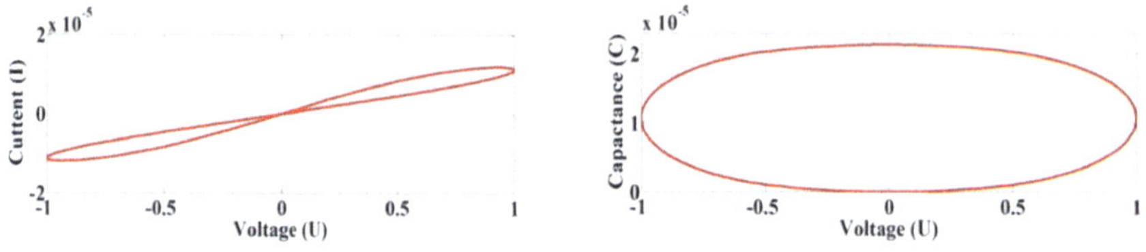

(a) The hysteresis loop of Memristor

(b) The V-C curve of Memcapacitor

Fig. 4.21 The simulation of Memristor and Memcapacitor

Table 4.1 The parameters used in the simulation of Memristor and Memcapacitor

Memristor		Memcapacitor	
Symbol	*Value*	*Symbol*	*Value*
R_{init}	80 KΩ	C_{init}	11 μF
R_{on}	1 KΩ	α	50 μF(V · s)
R_{off}	160 KΩ	β	2α
D	10 nm	V_{os}	0.5 V
μ_v	$10\,\mathrm{Fm^2s^{-1}V^{-1}}$	ϕ_2	π
p	10	φ_2	$-\pi/2$
ϕ_1	π		
φ_1	$\pi/2$		

Notations: R_{init} denotes the initial value of memristance. R_{on} is the resistance value of ON state. R_{off} is the resistance value of OFF state. D is the thickness of the film. μ_v denotes the average ion mobility in memristor. p denotes the exponent of window function. C_{init} is the initial value of memcapacitance. α, β, V_{os} denotes constancies.

Based on the memristive models, the curves of transfer functions and cut-off frequencies will vary over time. Figure 4.22a–d, demonstrates the time-varying transfer functions of ALPFMM, AHPFMM, ABPFMM, and ABSFMM, respectively. The time-varying cut-off frequency of ALPHMM and AHPFMM, band pass of ABPFMM, and stop width of ABSFMM are presented in Fig. 4.23.

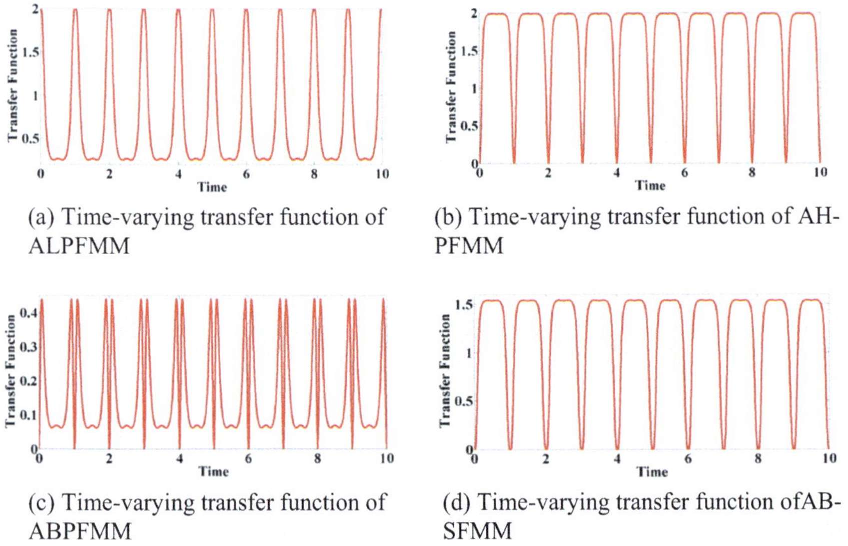

(a) Time-varying transfer function of ALPFMM

(b) Time-varying transfer function of AH-PFMM

(c) Time-varying transfer function of ABPFMM

(d) Time-varying transfer function ofAB-SFMM

Fig. 4.22 Time-varying transfer functions

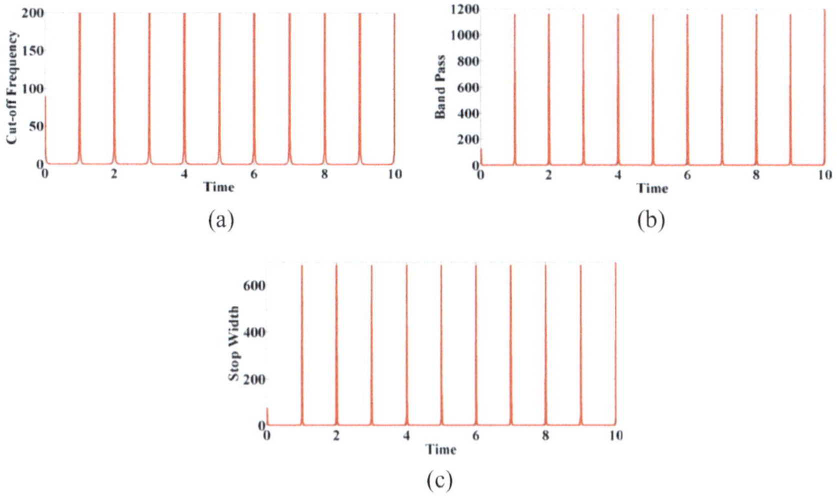

Fig. 4.23 Time-varying curves

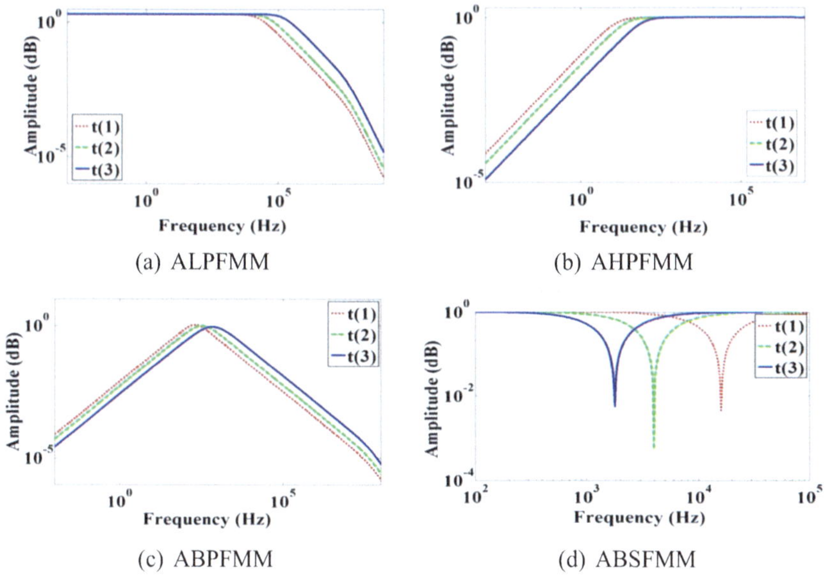

Fig. 4.24 Different AFRs of active filters at three different units of time

Since the cut-off frequencies of the low-pass and high-pass filter circuits are the same, we provide one curve, as seen in Fig. 4.23a. Figure 4.23b is the time-varying band pass of ABPFMM, while Fig. 4.23c is the time-varying stop width of ABSFMM.

Figures 4.24 and 4.25 show the Bode plots of active filters. Because memristance and memcapacitance change over time, there are different AFR

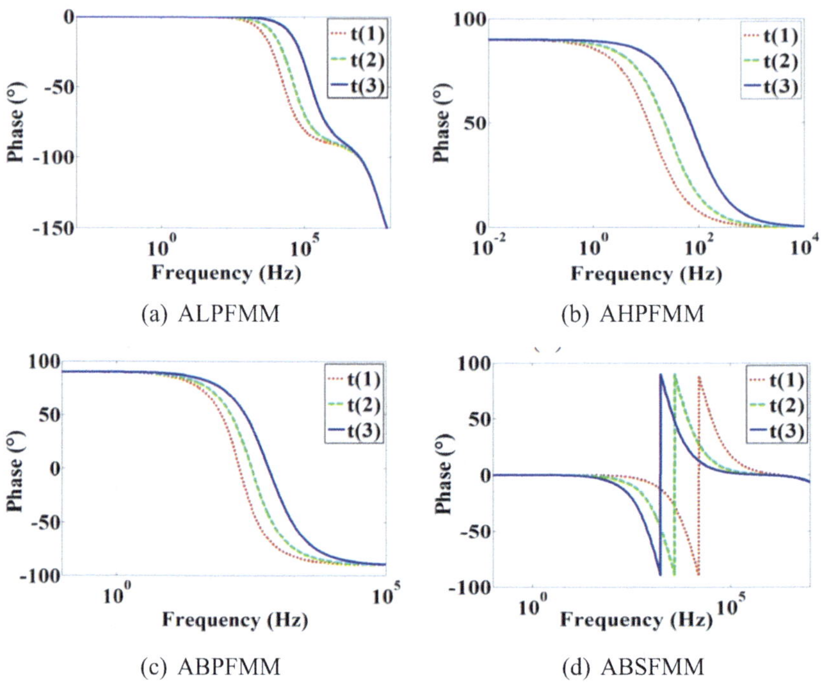

Fig. 4.25 Different PFRs of active filters at three different units of time

(Amplitude-Frequency Response) and different PFR (Phase-Frequency Response) in different units of time. Without losing the generality, this section chooses three different moments. The AFRs of ALPFMM, AHPFMM, ABPFMM, and ABSFMM are presented in (a), (b), (c), (d) of Fig. 4.24. Furthermore, the PFRs of ALPFMM, AHPFMM, ABPFMM, and ABSFMM are presented in (a), (b), (c), (d) of Fig. 4.25, respectively. For all the figures, the dotted line is generated at $t(1)$. The imaginary line is generated at $t(2)$, and the solid line is generated at $t(3)$.

4.4.5 Conclusion

The main contributions are summarized as follows. First, memristor and memcapacitor are applied into four kinds of representative active filters, which are low-pass, high-pass, band pass, and band stop filter. Second, simulation experiments show that amplitude-frequency and phase-frequency responses vary over time.

4.5 Band Pass and Band Stop Filters Based on Three Memory Devices

4.5.1 Introduction

This section investigates the application of three memory devices—memristors, memcapacitors, and meminductors—in the design of band pass and band stop filters. These mem-devices, characterized by their nonlinear, time-varying properties and memory effects, present new possibilities for filter design.

The study is divided into two main parts. The first part analyzes the characteristics of these three mem-devices, providing insight into their behavior in circuit configurations and how they differ from conventional components. The second part explores the practical application of these mem-devices in constructing band pass and band stop filters. While previous research has examined memristors in low-pass and high-pass filters [26, 27], this study replaces all three traditional electrical components (resistor, capacitor, and inductor) with their memristive counterparts.

The research examines whether circuits composed entirely of mem-devices can replicate or alter the functionality of traditional band pass and band stop filters. This investigation extends the application of mem-devices in filter design and provides insights into the behavior of circuits composed of memory elements. Mathematical models, circuit designs, and simulation results are presented to demonstrate the characteristics of these mem-device-based filters. The performance of these novel filters is compared with traditional counterparts, highlighting the differences and potential advantages of incorporating mem-devices in filter design.

4.5.2 Memory Devices

The circuit in this section will use three mem-devices, memristor, memcapacitor and meminductor. They have memory and they are time-varying. The introduction has mentioned that these three mem-devices can be analyzed as traditional component. To realize how to analyze the mem-devices in traditional way, the internal variable x is used [25]. The basic formula of memristor is shown below [25]:

$$\begin{aligned} V_M(t) &= R(x, I, t)I(t) \\ \dot{x} &= f(x, I, t)^{\cdot} \end{aligned} \tag{4.45}$$

For the formula above, x is internal state variable, V and I are voltage and current across memristor, respectively. Considering physical model of memristor implemented by HP [26, 28], the model shows that two layers of TiO_2 thin film are located between Platinum [26, 29]. One of the layers is oxygen vacancies, acting as a semiconductor, the other layer is vacancies, acting as an insulator [26].

According to HP model, W is thickness of oxygen vacancies, D is total thickness of memristor, x can be defined as W/D [26, 29]. HP physical model is helpful for MATLAB SIMULINK model of memristor and it will be explained in the "simulation" section. Because the value of memristor is time-varying, the $V - I$ relationship is not linear, and the $V - I$ relationship is shown as pinched hysteresis loop [30].

Like memristors, memcapacitors and meminductors can also be expressed in traditional way. The formula of the memcapacitor is shown below [25]:

$$q(t) = C(x, V_c, t)V_c(t)\dot{x} = f(x, V_c, t), \tag{4.46}$$

$$\phi(t) = L(x, I, t)I(t)\dot{x} = f(x, I, t). \tag{4.47}$$

Like the formula of memristor, x is internal variable, and all of variables are time-varying, the relationship is shown as pinched hysteresis loop. The figures of relationship for each mem-devices will be shown in the "simulation" section.

4.5.3 *Memristive Circuits*

The memristive circuits in this section are similar to the traditional circuits of band pass and band stop filter, however, the resistor, capacitor and inductor are replaced with memristor, memcapacitor and meminductor. Figures 4.26 and 4.27 show the circuit of band pass filter and band stop filter in the memristive system.

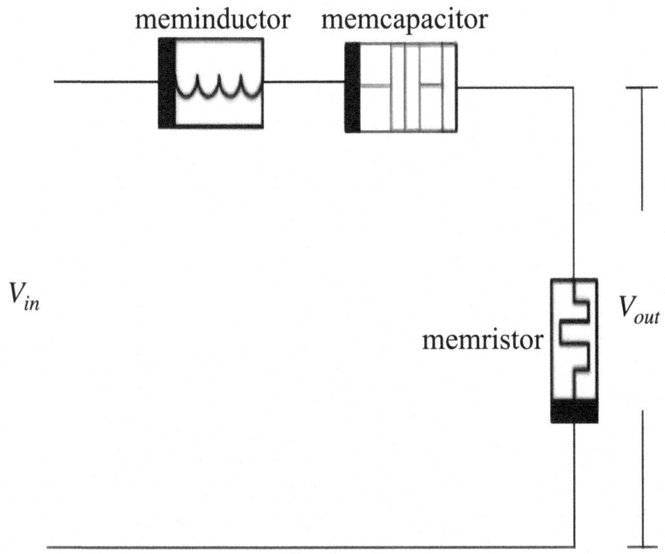

Fig. 4.26 Band pass filter circuit

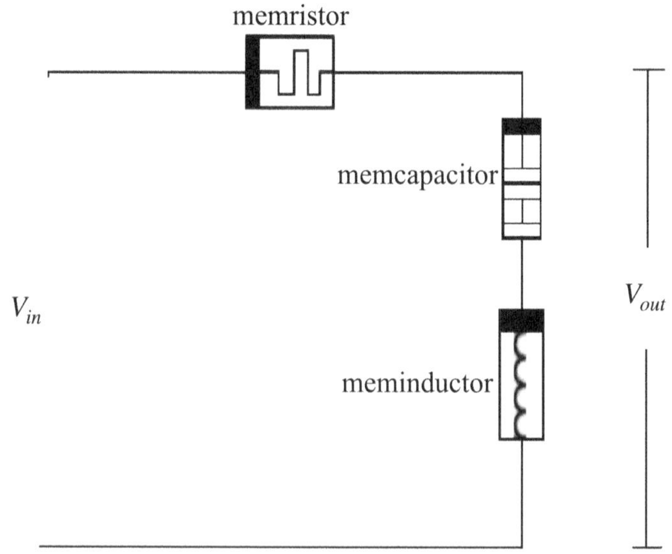

Fig. 4.27 Band stop filter circuit

Like the traditional band pass and band filter circuit, the output voltage is the voltage across memristor for band pass filter, on the other hand, the output voltage is the total voltage across meminductor and memcapacitor for band stop filter. Because the values of mem-devices are time-varying, the sinusoidal wave is applied. The value of parameters will be introduced in the "simulation" section. Because the output voltage of band pass filter and band stop filter is mentioned above, the transfer function of two filters can be expressed as follows:

$$H(jw)_{BPF} = \frac{V_{out}}{V_{in}} = \left(\frac{R_m}{R_m + \frac{1}{jwC} + jwL} \right), \tag{4.48}$$

$$H(jw)_{BSF} = \frac{V_{out}}{V_{in}} = \left(\frac{\frac{1}{jwC} + jwL}{R_m + \frac{1}{jwC} + jwL} \right). \tag{4.49}$$

Equations (4.48) and (4.49) are transfer functions of band pass and band stop filter, respectively. What's more, to analyze the characteristic of the filter, it is necessary to find the phase of filters.

$$\theta_{BPF} = -arctan \left(\frac{wL - \frac{1}{wC}}{R_m} \right), \tag{4.50}$$

$$\theta_{BSF} = \frac{\pi}{2} - arctan\left(\frac{wL - \frac{1}{wC}}{R_m}\right). \tag{4.51}$$

Equations (4.50) and (4.51) represent phases of transfer functions of band pass filter and band stop filter, respectively. The other important parameter to analyze filter is cut-off frequency. Based on the concept of cut-off frequency, there are two possible solutions for frequencies, and the same mem-devices are used for the cut-off frequencies of band pass filter and band stop filter.

4.5.4 *Simulation*

Comparing with traditional RLC circuits, all electric components are replaced with mem-devices. Before analyzing the characteristics of filter, we need to simulate memristor, memcapacitor and meminductor first. The value of mem-devices are time-varying, the simulation result of memristor, memcapacitor, and meminductor will be introduced below, respectively.

4.5.4.1 Simulation of Memristor

Based on the HP physic model in the Introduction section, the simulation of the memristor is based on several formulas below [26, 31]:

$$R_m = R_{on}x + R_{off}(1 - x), \tag{4.52}$$

$$\frac{dx}{dt} = kf(x)i(t),\ k = \frac{u_v + R_on}{D^2}, \tag{4.53}$$

$$f(x) = 1 - (2x - 1)^{2p}. \tag{4.54}$$

R_{on} and R_{off} are bond value of the memristor, and $f(x)$ is window function to find internal value x. Setting parameter following: $R_{on} = 100$ ohm, $R_{off} = 16000$ ohm, $R_{initial} = 11000$ ohm, $p = 10$, $D = 10$ nm, $u_v = 10^{-14}\,m^2s^{-1}V^{-1}$. As a result, two figures are obtained for memristor, Fig. 4.28 is Lissajous figure of relationship between $v(t)$ and $i(t)$, the result shows a pinched hysteresis loop relationship. The value of memristor is time-varying, so the sinusoidal signal is added in the simulation of the memristor, and the signal function is represented as:

$$V(t) = V_0 sin(2\pi wt). \tag{4.55}$$

Where $V_0 = 1$V, $w = 1$ Hz. The values of memristor in the first 10 s are shown in Fig. 4.29.

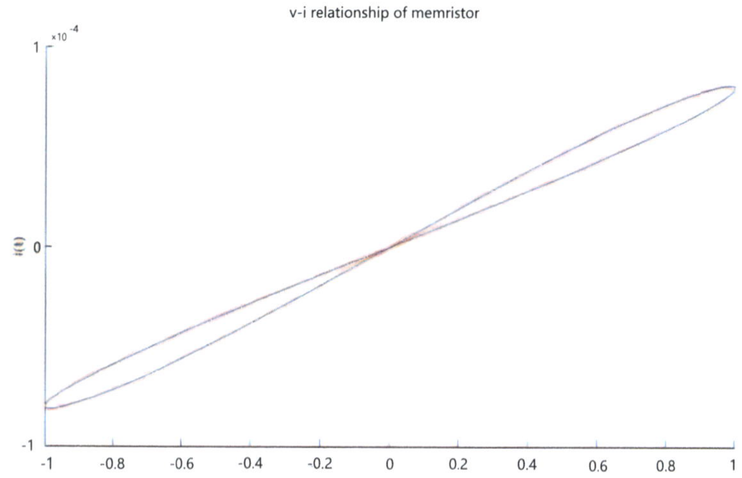

Fig. 4.28 Lissajous figure of relationship between $v(t)$ and $i(t)$

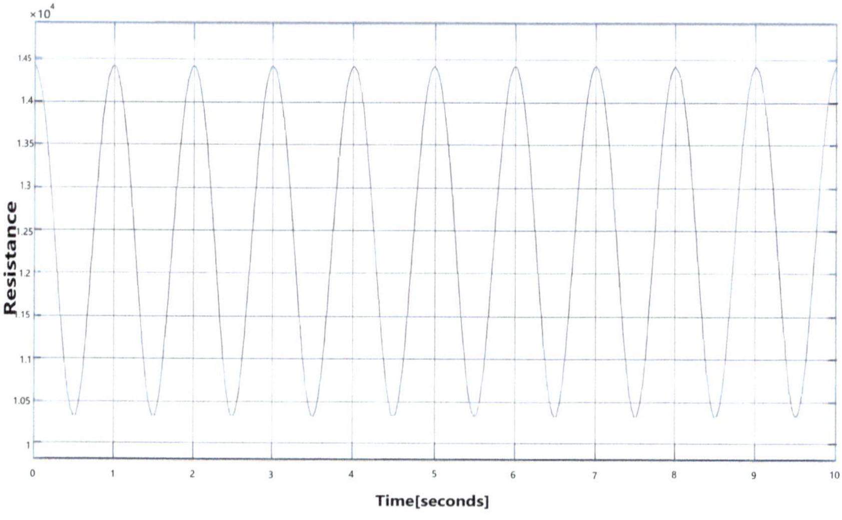

Fig. 4.29 The value of memristor in first 10 seconds

4.5.4.2 Simulation of Memcapacitor

In the Introduction section, it is mentioned that the value of memcapacitor can be expressed in a traditional way, and the value is time-varying and depends on internal value x, and $C = x$. For the memcapacitor simulation, the mathematical model in Simulink model is based on following (4.50)

$$\frac{dx}{dt} = \beta V_c + 0.5(\alpha - \beta)[|V_C + V_T| - |V_C - V_T|]. \tag{4.56}$$

For the formula above, α and β are constant. In memcapacitor simulation, $\alpha = 50\,\text{uF}/(\text{V} * \text{s})$, $\beta = 2\alpha$, $V_T = 0.5\,\text{V}$. Like memristor, the sinusoidal signal is added in simulation of memcapacitor, the value of VC is same is (4.55) with $w = 1\,\text{Hz}$ and $V_0 = 1\,\text{V}$.

Figure 4.30 shows the relationship between voltage and value of memcapacitor, and Fig. 4.31 shows the value of memcapacitor in first 10 s.

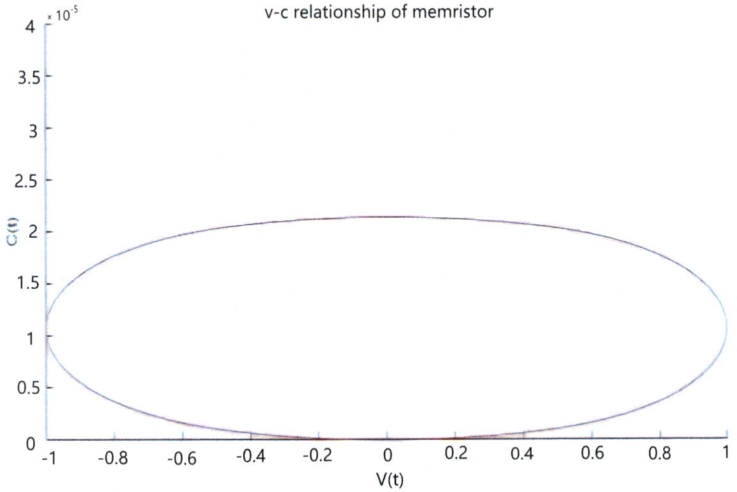

Fig. 4.30 V-C relationship for memcapacitor

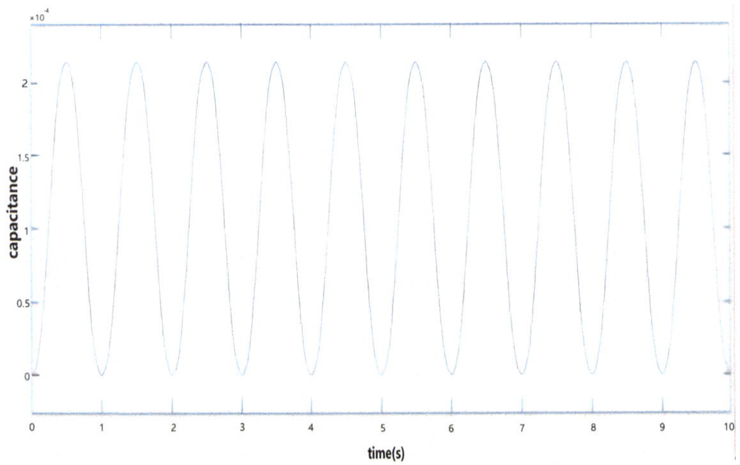

Fig. 4.31 The value of memcapacitor for first 10 s

4.5.4.3 Simulation of Meminductor

Like memristors and memcapacitors, the meminductor can also be expressed in traditional way, and whose value depends on internal value x. For the mathematics model of meminductor, the internal value x can be expressed by maximum and minimum value of meminductor l_{max} and l_{min} based on electromechanical model of meminductor [28]:

$$x = \frac{l - l_{min}}{l_{max} - l_{min}}, \tag{4.57}$$

$$L_m(t) = [\sqrt{L_{min}} + x(t)(\sqrt{L_{max}} - \sqrt{L_{min}})]^2. \tag{4.58}$$

The internal value in meminductor simulation is similar to memristor simulation, the only difference is that the value k is directly set to 10. Other parameters are set as following: $L_{min} = 1\,\text{mH}$, $L_{max} = 20\,\text{mH}$, $p = 10$. Unlike the simulation of memristor and memcapacitor, the sinusoidal signal added is not voltage, the signal is the current output from subsystem of memristor. As a result, Fig. 4.32 is the relationship between current and flux, and the result shows the pinched hysteresis in Fig. 4.29. Figure 4.33 shows the relationship between current and value of meminductor.

4.5.5 *Filter Analysis*

The characteristic of three mem-devices are analyzed in the last section. To create and analyze the band pass and band stop filter, it is necessary to put memristor,

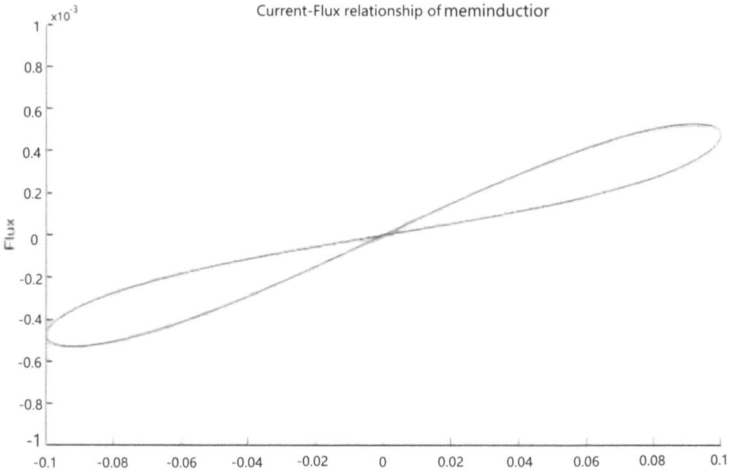

Fig. 4.32 Relationship between i and flux for memristor

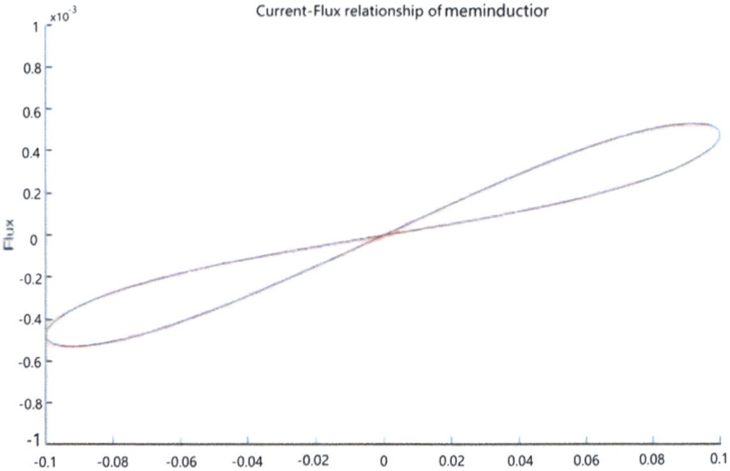

Fig. 4.33 $I - L$ relationship for meminductor

memcapacitor and meminductor together and analyze cut-off frequency and transfer function. Because the values of three mem-devices are time-varying, the cut-off frequency will also be time-varying. Figures 4.34 and 4.35 show two possible cut-off frequencies.

The other factor to analyze the characteristic of filter is transfer function. The purpose of this section is to find characteristic of band pass and band stop filters with mem-devices, and compare with traditional band pass and band stop filters. Because

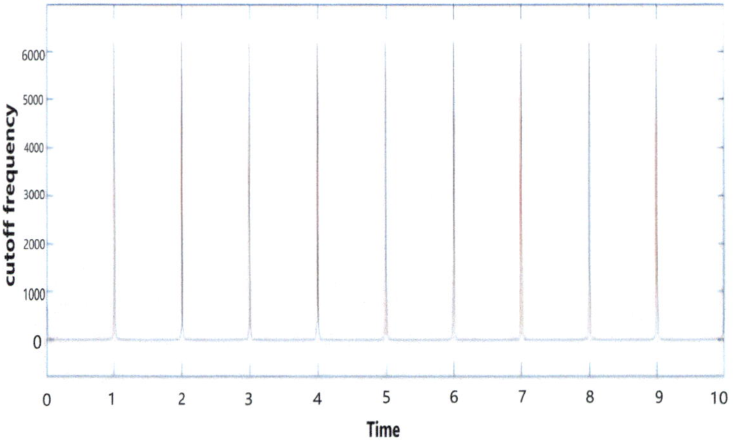

Fig. 4.34 Cut-off frequency of $f1$

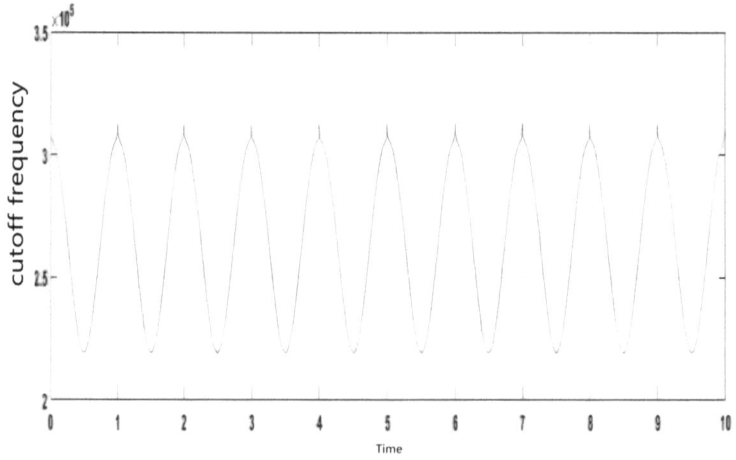

Fig. 4.35 Cut-off frequency of $f2$

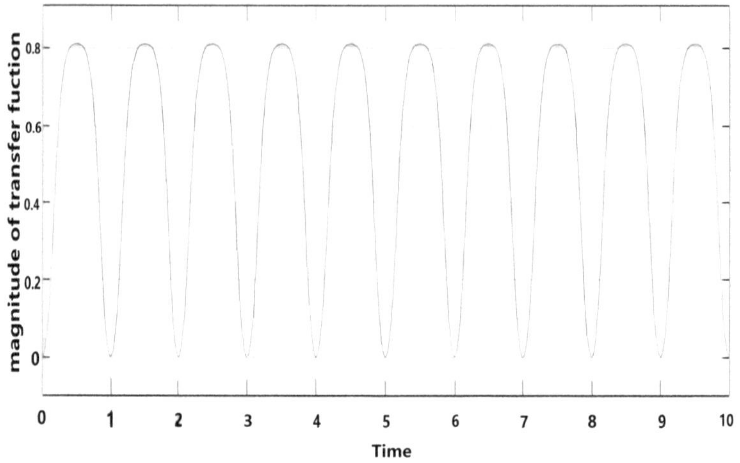

Fig. 4.36 Magnitude of transfer function of band pass filter

the value of three mem-devices are time-varying, the value of mem-devices in different time are adopted to find if both moment have same characteristic. For analysis below, $t_1 = 0.25$ s, $t_2 = 0.5$ s. Like cut-off frequency, the magnitude of transfer function is time-varying. Figures 4.36 and 4.37 are the transfer functions of band pass filter and band stop filter of band pass and band stop filter, respectively. Figures 4.34 and 4.35 show two possible cut-off frequencies.

According to Figs. 4.29 and 4.31, the values of mem-devices are available at time $t = 0.25$ and 0.5. For band pass circuit, the relationship between the magnitudes of

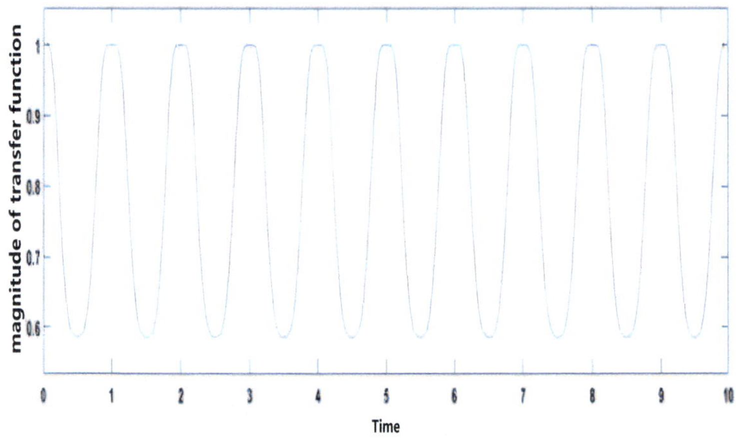

Fig. 4.37 Magnitude of transfer function of band stop filter

transfer function (dB) and frequencies (Hz) is shown in Fig. 4.38, the magnitude-frequency relationship of band stop circuit is shown in Fig. 4.39. For the phase-frequency relationship of band pass filter, the result is shown in Fig. 4.40. According to the figures above, the trend of magnitude and phase are similar to the traditional circuit, what's more, at different times, the value of magnitude and phase are different but the trend remains the same, which means after replacing the traditional electric

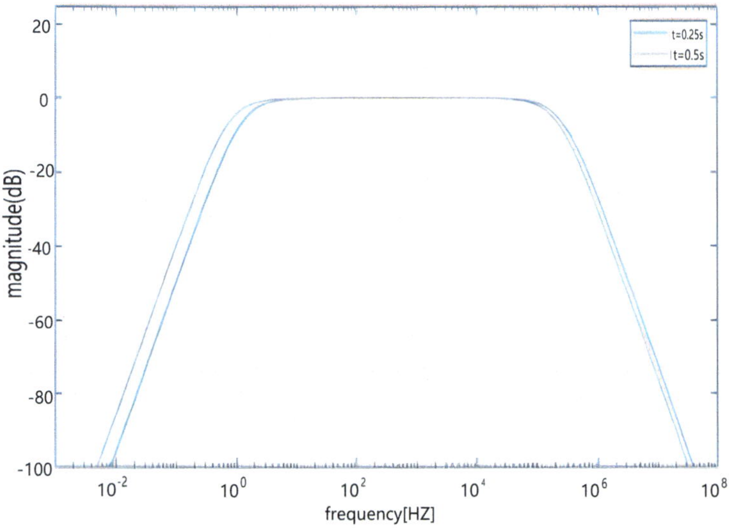

Fig. 4.38 Frequency-magnitude relationship for band pass filter

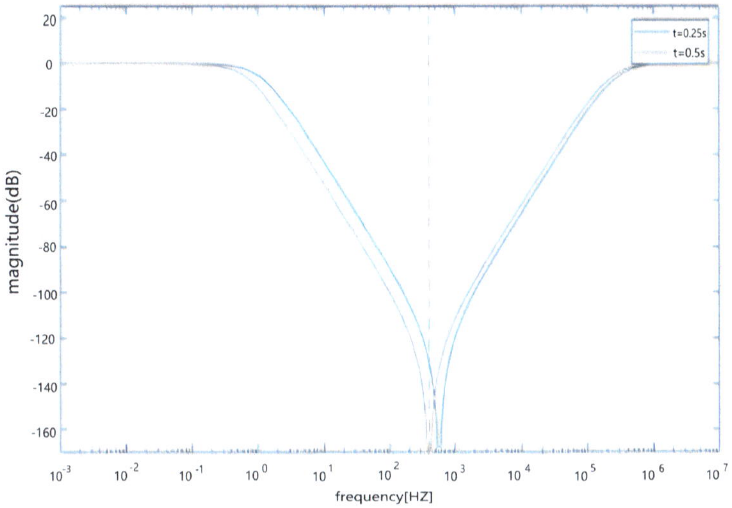

Fig. 4.39 Frequency-magnitude relationship for band pass filter

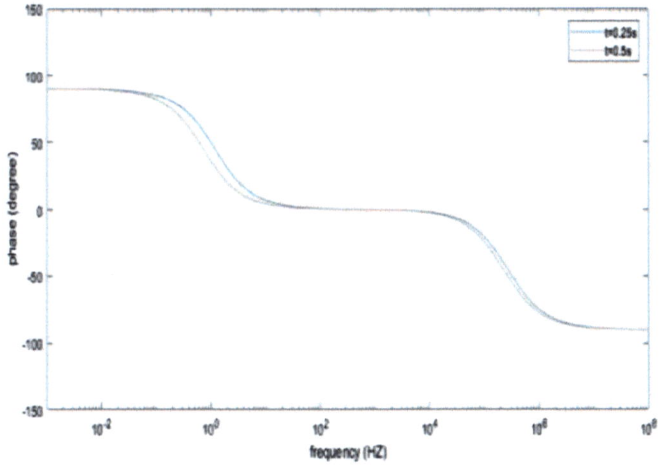

Fig. 4.40 Frequency-phase relationship of band pass filter

component with mem-devices, the circuit still shows the characteristic of band pass and band stop filter.

4.5.6 *Conclusion*

The result of filter analysis shows that the circuit replacing with mem-devices has similar characteristic as the traditional band pass and band stop circuits. What's more, the parameter value at different times is different, which shows the mem-devices are time-varying, the trend at different time has a similar trend, which shows the mem-devices can be applied in band pass and band stop filter.

4.6 Comments

This chapter explores the design of filters based on memristors and memcapacitors, demonstrating the immense potential of these memory devices in the field of signal processing. The research proposes several novel filter structures, including low-pass, high-pass, band pass, and band stop filters, and validates their feasibility through theoretical analysis and simulation. These works pave the way for developing adaptive, compact, and efficient filtering systems. Future research should focus on expanding and optimizing filter technologies based on three memory elements: memristors, meminductors, and memcapacitors.

A key direction is the development of more complex and flexible filter structures, fully utilizing the unique characteristics of each memory element. The exploration will involve designing and implementing advanced structures such as Extrapolated Impulse Response (EIR) filters and filter banks, leveraging memristor crossbar arrays, meminductor-based resonant circuits, and memcapacitor charge storage capabilities. These complex filters will employ innovative combinations of parallel, cascade, and feedback topologies, significantly enhancing filtering performance and flexibility.

The integration of all three memory elements in filter design will be a crucial area of study. The research will investigate how to synergistically combine the complementary characteristics of memristors, meminductors, and memcapacitors to create hybrid filters with enhanced functionality. This may include exploring memristor-meminductor oscillators for frequency-selective filtering, memcapacitor-memristor voltage-controlled filters, or triple memory element resonant circuits for sharp cut-off characteristics.

To support these advanced filter structures, research into synchronous sampling and delay circuits based on memory elements will be critical. Future works will focus on developing circuits capable of precise signal delay control, utilizing the unique properties of each memory element. For instance, memcapacitor-based sample-and-hold circuits might be combined with memristor-based programmable delay lines to achieve FIR filters with longer orders and improved accuracy.

Research on reconfigurable filter systems will exploit the programmable nature of all three memory elements. This will involve developing efficient weight update algorithms and flexible control circuits that can dynamically adjust the characteristics of memristors, meminductors, and memcapacitors. The focus will be on achieving smooth transitions and real-time reconstruction of filters without interrupting signal processing, optimizing memory element array structures for rapid and accurate parameter updates.

Additionally, the research will explore the nonlinear characteristics of memory elements for specialized filtering applications. This may include using the hysteresis properties of memristors for adaptive noise cancellation, the frequency-dependent behavior of meminductors for wideband matching networks, or the voltage-dependent capacitance of memcapacitors for voltage-controlled filtering.

Through these interconnected research directions, the potential of memory elements in signal processing will be further unlocked. These advancements will not only enhance filter performance and adaptability but also revolutionize analog signal processing. As research progresses, memory element-based filtering technology is poised to become the cornerstone of next-generation signal processing solutions, offering more efficient, flexible, and powerful signal processing capabilities for future electronic systems.

References

1. Chua, L.O.: Memristor-the missing circuit element. IEEE Trans. Circuit Theory **18**(5), 507–519 (1971). https://doi.org/10.1109/TCT.1971.1083337
2. Di Ventra, M., Pershin, Y.V., Chua, L.O.: Putting memory into circuit elements: memristors, memcapacitors, and meminductors [point of view]. Proc. IEEE **97**(8), 1371–1372 (2009)
3. Xiao-Yuan, W., Jun, Y., Guang-Yi, W.: Simulink modeling of memristor, memcapacitor, meminductor and their characteristics analysis. Acta Physica Sinica **67**(9), 098501 (2018)
4. Xiao-Yuan, W., Jun, Y., Guang-Yi, W.: Simulink modeling of memristor, memcapacitor, meminductor and their characteristics analysis. Acta Physica Sinica **67**(9), 098501 (2018). https://doi.org/10.7498/aps.67.20172674
5. Di Ventra, M., Pershin, Y.V., Chua, L.O.: Circuit elements with memory: memristors, memcapacitors, and meminductors. Proc. IEEE **97**(10), 1717–1724 (2009)
6. Guillemin, E.A.: Introductory Circuit Theory. Wiley, New York (1953)
7. Valkenburg, M.E.V.: Newwork Analysis. Prentice-Hall, Englewood Cliffs, NJ (1974)
8. Di Ventra, M., Pershin, Y.V., Chua, L.O.: Circuit elements with memory: Memristors memristors memcapactiors and meminductors. In: Proc. IEEE, vol. 97, pp. 0018–9219
9. Campbell, G.A.: Physical theory of the electric wave-filter. Bell Syst. Techn. J. **1**(2), 1–32 (1922). https://doi.org/10.1002/j.1538-7305.1922.tb00386.x
10. Sözen, H., Çam, U.: On the realization of memristor based rc high pass filter. In: 2013 8th International Conference on Electrical and Electronics Engineering (ELECO), pp. 45–48 (2013). https://doi.org/10.1109/ELECO.2013.6713933
11. Sozen, H., Cam, U.: First-order memristor-capacitor filter circuits employing hp memristor. J. Circuits Syst. Comput. **23**(08), 1450116 (2014)
12. Yu, H., Tan, Z.H., Zhang, Y., Ma, Z., Guo, J.: Dnn filter bank cepstral coefficients for spoofing detection. IEEE Access **5**, 4779–4787 (2017). https://doi.org/10.1109/ACCESS.2017.2687041
13. Karczewicz, M., Zhang, L., Chien, W.J., Li, X.: Geometry transformation-based adaptive in-loop filter. In: 2016 Picture Coding Symposium (PCS), pp. 1–5 (2016). https://doi.org/10.1109/PCS.2016.7906346
14. Dannehl, J., Fuchs, F.W., Thøgersen, P.B.: Pi state space current control of grid-connected pwm converters with lcl filters. IEEE Trans. Power Electron. **25**(9), 2320–2330 (2010). https://doi.org/10.1109/TPEL.2010.2047408

15. Parker, S.G., McGrath, B.P., Holmes, D.G.: Regions of active damping control for lcl filters. IEEE Trans. Ind. Appl. **50**(1), 424–432 (2014). https://doi.org/10.1109/TIA.2013.2266892
16. Viani, F., Robol, F., Salucci, M., Azaro, R.: Automatic emi filter design through particle swarm optimization. IEEE Trans. Electromagn. Compat. **59**(4), 1079–1094 (2017). https://doi.org/10.1109/TEMC.2017.2652416
17. Thanapitak, S., Pawarangkoon, P., Sawigun, C.: A flipped voltage follower second-order bandpass filter. J. Circuits Syst. Comput. **26**(07), 1750112 (2017)
18. He, Y.S., Barannik, A., Cherpak, N., Sun, L., Skresanov, V., Bian, Y., Wang, J., Natarov, M., Zolotaryov, V.: Novel design of band-pass waveguide filter with hts e-plane insert. IEEE Trans. Appl. Supercond. **27**(4), 1–4 (2017). https://doi.org/10.1109/TASC.2017.2654350
19. Li, Y., Yang, C., Yu, Y., Fernández Díez, F.: Research on low pass filter based on memristor and memcapacitor. In: 2017 36th Chinese Control Conference (CCC), pp. 5110–5113 (2017). https://doi.org/10.23919/ChiCC.2017.8028162
20. Strukov, D.B., Snider, G.S., Stewart, D.R., Williams, R.S.: The missing memristor found. Nature **453**(7191), 80–83 (2008). https://doi.org/10.1038/nature06932
21. Mazumder, P., Li, S.R., Ebong, I.E.: Tunneling-based cellular nonlinear network architectures for image processing. IEEE Trans. Very Large Scale Integr. (VLSI) Syst. **17**(4), 487–495 (2009). https://doi.org/10.1109/TVLSI.2009.2014771
22. Linvill, J.G.: Rc active filters. Proc. IRE **42**(3), 555–564 (1954)
23. Shen, Y., Liu, Y., Wang, X.: Realization of a memristor-based second-order active low-pass filter. In: 2015 5th International Conference on Information Science and Technology (ICIST), pp. 351–355 (2015). https://doi.org/10.1109/ICIST.2015.7288995
24. Chua, L.O.: Resistance switching memories are memristors. Appl. Phys. A (2011)
25. Chua, L.O., Kang, S.M.: Memristive devices and systems. Proc. IEEE **64**(2), 209–223 (1976). https://doi.org/10.1109/PROC.1976.10092
26. Mahvash, M., Parker, A.C.: A memristor spice model for designing memristor circuits. In: 2010 53rd IEEE International Midwest Symposium on Circuits and Systems, pp. 989–992 (2010). https://doi.org/10.1109/MWSCAS.2010.5548803
27. Driscoll, T., Quinn, J., Klein, S., Kim, H.T., Kim, B., Pershin, Y.V., Di Ventra, M., Basov, D.: Memristive adaptive filters. Appl. Phys. Lett. **97**(9) (2010)
28. Xiao-Yuan, W., Jun, Y., Guang-Yi, W.: Simulink modeling of memristor, memcapacitor, meminductor and their characteristics analysis. Acta Physica Sinica **67**(9) (2018)
29. Kim, H., Sah, M.P., Yang, C., Cho, S., Chua, L.O.: Memristor emulator for memristor circuit applications. IEEE Trans. Circuits Syst. I Regul. Pap. **59**(10), 2422–2431 (2012)
30. Wang, W., Yu, Q., Xu, C., Cui, Y.: Study of filter characteristics based on pwl memristor. In: 2009 International Conference on Communications, Circuits and Systems, pp. 969–973. IEEE (2009)
31. Abunahla, H., Mohammad, B., Abunahla, H., Mohammad, B.: Memristor device modeling. Memristor Technology: Synthesis and Modeling for Sensing and Security Applications pp. 93–104 (2018)

Chapter 5
Memristive Filter for Signal Processing

Signal processing plays a crucial role across various disciplines, from telecommunications to biomedical engineering and beyond. It involves manipulating and analyzing signals to extract useful information, enhance quality, or achieve specific goals. The importance of signal processing in extracting meaningful information from signals makes it foundational in modern technology and scientific research.

Building on the foundations established in previous chapters, which explored memristor-based chaotic circuits and filter designs, this chapter delves into the application of memristive devices in more advanced signal processing tasks. The focus shifts from basic filter designs to specialized applications in signal segmentation and image processing, demonstrating the versatility and potential of memristor-based circuits in real-world scenarios.

Section 5.1 introduces a novel approach to filter design by integrating memristors into band pass filter (BPF) and band stop filter (BSF) circuits. These filters are characterized by their time-varying properties, a unique feature enabled by the use of memristive components. The development of these circuits, alongside a signal division model, has the potential to enhance digital signal processing, particularly in the underserved area of filter design. This section demonstrates the application of these circuits through signal division and explores their extension to various signal processing tasks, potentially enhancing traditional processing methods.

Further advancing the application of memristive devices in signal processing, Sect. 5.2 presents a key contribution in the form of a memristor bridge-based low-pass filter (LPF). This innovative design exhibits cut-off frequency, transfer function, amplitude-frequency response (AFR), and phase-frequency response (PFR) that vary with time and memristance. Building on this foundation, an adaptive Gaussian filter is developed by incorporating the memristor bridge's changeable parameters into the energy function. This approach leads to a significant increase in peak signal-to-noise ratio (PSNR) for denoising compared to the traditional Gaussian filters. Additionally, the variable parameter of the memristor bridge is integrated into the Gaussian pyramid

Y. Yu et al., *Memristive Computing*,
https://doi.org/10.1007/978-981-96-0332-9_5

to enhance an existing Scale-Invariant Feature Transform (SIFT) algorithm, resulting in increased feature point matches and improved accuracy compared to the traditional SIFT method.

Through these advancements, this chapter demonstrates the practical applications of memristor-based circuits in signal processing, moving beyond theoretical filter designs to tackle real-world challenges in signal segmentation and image processing. The research presented here not only validates the feasibility and effectiveness of the proposed algorithms but also opens new avenues for applying memristive devices in complex signal processing tasks.

5.1 Memristive Band Pass and Band Stop Filters for Signal Segmentation

5.1.1 Introduction

Filter is presented by Campbell and Wagner [1], which attracts much interest of researchers among all the analog circuits [1–6]. With the increasing attention on memristive elements, the filter circuit usher in a new development [7–9]. A piecewise-linear memristor-based proportional-integral controller is proposed for the design of the loop filter of the charge-pump phase-locked loop [8], and a memristor-based HPF is investigated in [9]. In addition, in the field of signal processing, much attention has been paid to filter for its unique feature [10–12]. The possible application of the complex BPF in Coriolis mass flow meter signal processing was explored in [11]. Hardware implementation of edge detection in real-time video signals is explained by using Laplacian filters [12]. In light of the filtering specialty of filter circuits, the memristive BPF and BSF are studied for signal division in this section.

Based on the classical RLC BPF and BSF circuits, the new memristive BPF and BSF circuits are obtained by replaying the resistor and capacitor with memristor and memcapacitor. Due to the filtering effects of BPF and BSF, the cut-off frequencies can divide the input signal into some parts. With the adjustable quality of memristors and memcapacitors, this signal division processing is also adjustable, which is presented in this section as an application of the new memristive BPF and BSF circuits. The Simulink simulation has demonstrated the validity of new BPF and BSF, and the software experiment has realized the aforesaid signal division model.

5.1.2 Memristive Filter Circuits

In this section, the new memristive BPF circuit and BSF circuit are provided by replacing resistor and capacitor with the memristor and memcapacitor, as shown in Figs. 5.1 and 5.2. The nonlinear dopant drift model of memristors [13, 14] and the

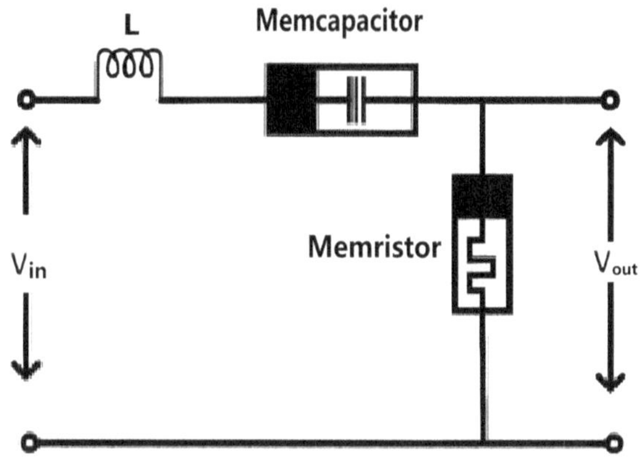

Fig. 5.1 Memristive BPF circuit

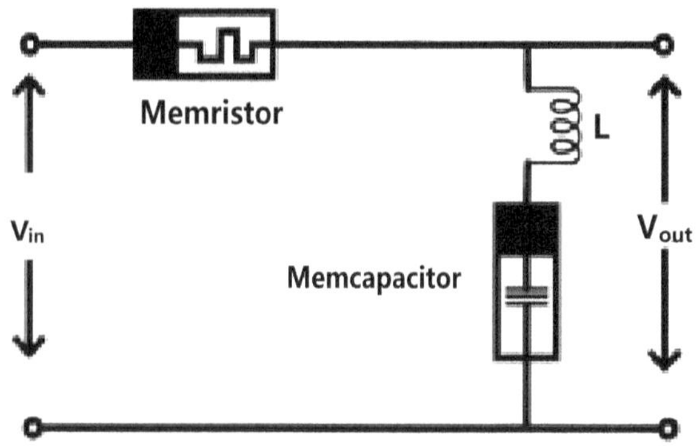

Fig. 5.2 Memristive BSF circuit

piecewise-linear model of memcapacitors [15] are introduced to build memristive BPF and BSF circuits. When a sinusoidal signal is added to the memristor and memcapacitor, the R_M denoting the resistance of the memristor and the R_c denoting the capacitive reactance of the memcapacitor can be presented as follows:

$$\begin{cases} R_M = M_{\text{init}} + \Delta M \sin(\omega_1 t + \varphi_1) \\ R_C = C_{\text{init}} + \Delta C \sin(\omega_2 t + \varphi_2) \end{cases}, \tag{5.1}$$

where $\sin(\omega_1 t + \varphi_1)$ is the signal added to the memristor and $sin(\omega_2 t + \varphi_2)$ is the signal added to the memcapacitor.

In the new memristive BPF, the series group of memcapacitor and inductor is treated as output shown in Fig. 5.1, and its transfer function is provided as shown in (5.2). If the memristor is treated as output shown in Fig. 5.2, the transfer function of the BSF circuit can be described in (5.3).

$$H(j\omega)_{BPF} = \frac{V_{out}}{V_{\text{in}}} = \frac{R_M}{R_M + \frac{1}{j\omega R_C} + j\omega L}, \tag{5.2}$$

$$H(j\omega)_{BSF} = \frac{V_{\text{out}}}{V_{\text{in}}} = \frac{\frac{1}{j\omega R_C} + j\omega L}{R_M + \frac{1}{j\omega R_C} + j\omega L}. \tag{5.3}$$

The two phases are presented as follows:

$$\theta_{BPF} = -\arctan\left(\frac{\omega L - 1/\omega R_C}{R_M}\right), \tag{5.4}$$

$$\theta_{BSF} = \mathbf{90} - \arctan\left(\frac{\omega L - 1/\omega R_C}{R_M}\right). \tag{5.5}$$

Due to the same memristor model, memcapacitor model, and inductor which are introduced for the BPF and BSF, the same cut-off frequencies which are applied to signal division can be obtained as follows:

$$\begin{cases} f_1 = \left[-\frac{R_M}{2L} + \sqrt{\left(\frac{R_M}{2L}\right)^2 + \frac{1}{LR_C}}\right]/2\pi \\ f_2 = \left[\frac{R_M}{2L} + \sqrt{\left(\frac{R_M}{2L}\right)^2 + \frac{1}{LR_C}}\right]/2\pi. \end{cases} \tag{5.6}$$

5.1.3 Signal Division Model

In this section, a signal division model based on memristive filter circuits is presented in Fig. 5.3. The model uses cut-off frequencies to divide the input signal into some parts by setting up suitable parameters. The details can be described as follows:

- Step 1: The original signal with its maximum frequency f_{max} and minimum frequency f_{min} are obtained in input terminal.
- Step 2: By employing suitable parameters of memristive filter circuits, the correlations, $f_1 > f_{min}$ and $f_2 < f_{max}$, are installed.
- Step 3: Based on the filtering effects, the original signal is divided into two parts which are the P-signal and the S-signal.
- Step 4: By adjusting the parameters of memristive filter circuits, the P-signal and S-signal are, respectively, processed like step I-step I.
- Step 5: The P-P-signal and P-S-signal are obtained from P-signal, and the S-P-signal and S-S-signal are gained from S-signal.
- Step 6: In the output terminal, the original signal can be restored by allying P-P-signal, P-S-signal, S-P-signal, and S-S-signal.

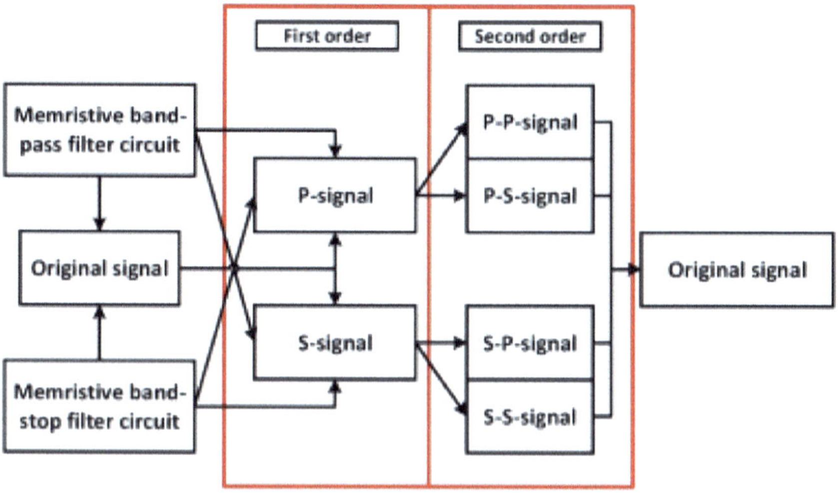

Fig. 5.3 Signal division model

During transmitting, the original signal is divided into four parts, as shown in Fig. 5.3. In first-order division, the frequency band $[f_1, f_2]$ named P-signal is passed by BPF and the frequency band $[f_{\min}, f_1] \cup [f_2, f_{\max}]$ named S-signal is passed by BSF. Then, in second-order division, the P-signal is treated as the "new original signal" which has the new maximum frequency f_1 and the new minimum frequency f_2. By adjusting the parameters for filter circuits, the new cut-off frequencies $f_1 > f_1'$ and $f_2 < f_2'$ are obtained; and these are used to divide the P-signal into P-P-signal and P-S-signal. The division of S-signal is similar to P-signal. In this model, due to the time-varying memresistance of memristor and capacitive reactance of memcapacitor, the filtering effects of memristive BPF and BSF circuits are time-varying too, which means the cut-off frequencies are time-varying. Based on this performance, the disparate result of signal division can be obtained by setting up different parameters for filter circuits, and the original signal cannot be speculated by incomplete signal parts.

5.1.4 *Simulation*

For circuit simulation, the parameters for the memristor are adopted as follows: initial resistance $R_{\text{init}} = 80\,\text{k}\Omega$, on-state resistance $R_{\text{on}} = 1\,\text{k}\Omega$, off-state resistance $R_{\text{off}} = 160\,\text{k}\Omega$, doping layer thickness $D = 10\,\text{nm}$, mobility $\mu = 10\,\text{Fm}^2/\text{sV}$, and exponent of the window function $p = 10$, with $\omega = \pi$ and $\phi = -\pi/2$. The parameters for the memcapacitor are employed as initial capacitance $\alpha = 50\,\mu\text{F(V.s)}$, $\beta = 2\alpha$, offset voltage $V_{\text{os}} = 0.5\,\text{V}$, angular frequency $\omega_2 = \pi$, phase $\phi_2 = -\pi/2$,

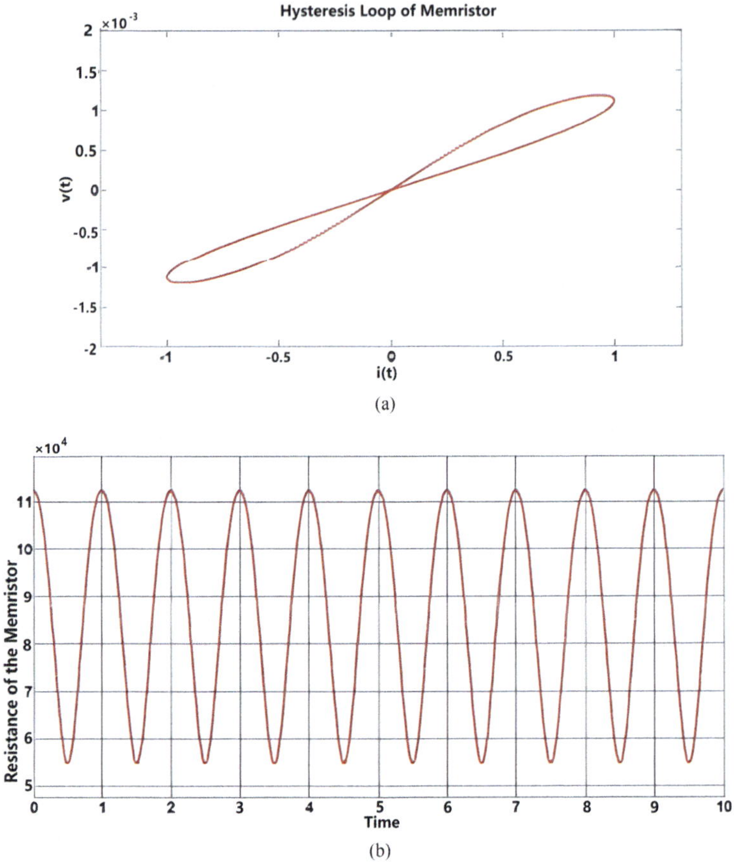

Fig. 5.4 The simulation of the memristor. **a** The hysteresis loop of memristor. **b** The resistance of memristor

and $C_{\text{init}} = 1.1\,\mu\text{F}$. The Lissajous figure is shown in Fig. 5.4a, which is one feature of the memristor. Based on (5.1), the time-varying resistance of the memristor is presented in Fig. 5.4b. Meanwhile, the memcapacitor simulation is presented in Fig. 5.5. The same added signal for the memristor and memcapacitor is given by:

$$V(t) = V\sin(2\pi f t), \quad V = 1\,\text{V}, \quad f = 1\,\text{Hz}.$$

On this basis, the cut-off frequencies f_1 and f_2 are obtained in Fig. 5.6. Meanwhile, with those simulation results of the memristor and memcapacitor, the Bode plots of the BPF and BSF in Figs. 5.1 and 5.2 are provided in Figs. 5.7 and 5.8, respectively. This verifies that the presented memristive filter circuits are indeed BPF and BSF circuits and show different amplitude-frequency responses and phase-frequency responses at three different times. Furthermore, the filter characteristics

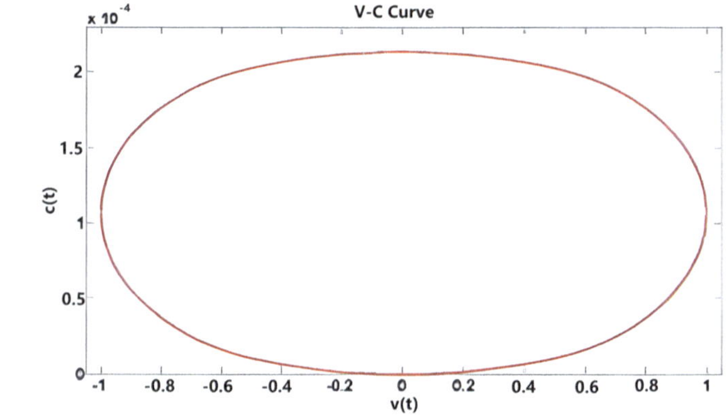

(a)

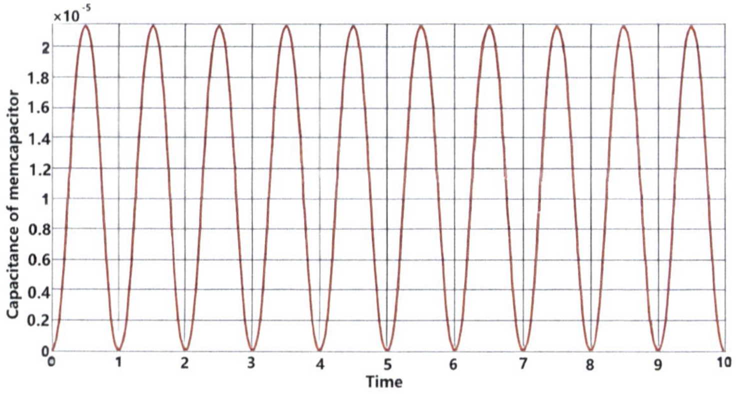

(b)

Fig. 5.5 The simulation of the memcapacitor. **a** The V-C curve of memcapacitor. **b** The capacitance of memcapacitor

are time-varying, which means the Bode characteristics can be easily controlled by changing the memristance of the memristor and the capacitance of the memcapacitor. The memristive filter circuits are more flexible than traditional filters, which implies that more applications can be designed for these new filters.

As shown in Fig. 5.6, the cut-off frequencies f_1 and f_2 are time-varying, which means the frequency band passed by the BPF is also time-varying. For simulation, the parameters for the analog signal are set as: $f_{\min} = 1$ Hz and $f_{\max} = 10^4$ Hz, and the division result is shown in Fig. 5.9. This software interface is divided into two parts: the figure part and the control part. The figure part shows the division results (from top to bottom: original signal, restoring signal, P-signal, P_P-signal, P_S-signal, S-signal, S_P-signal, S_S-signal) and the control part presents the interaction to adjust the

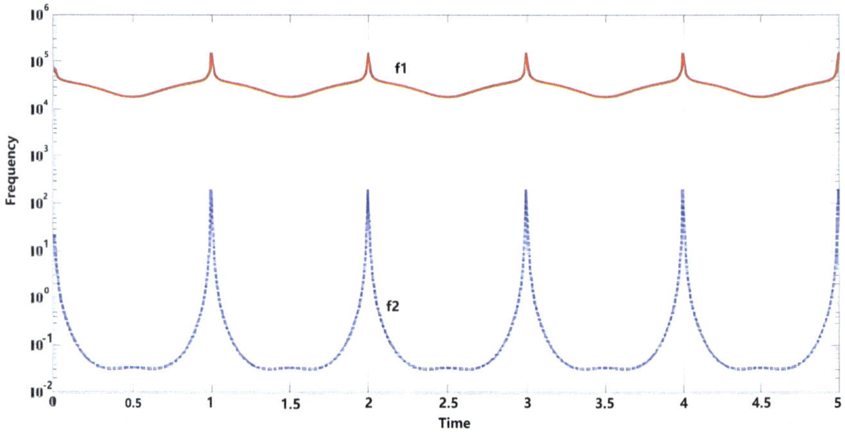

Fig. 5.6 The cut-off frequencies f_1 and f_2 of filter circuits

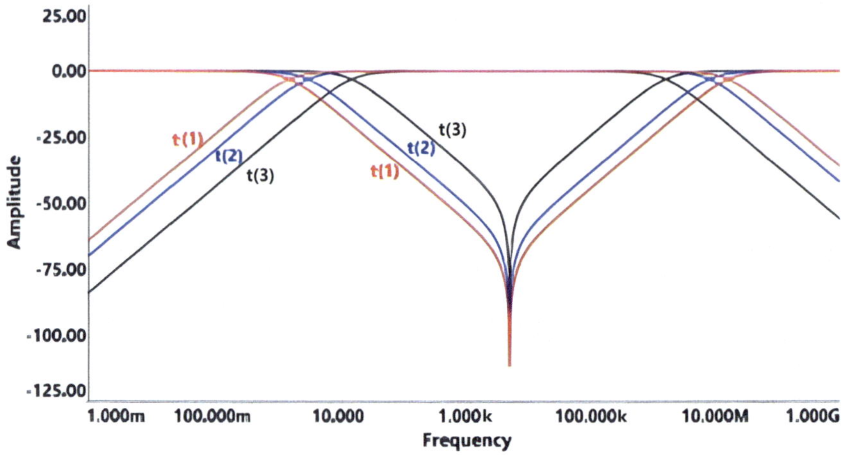

Fig. 5.7 The amplitude-frequency responses of BPF circuit and BSF circuit at three different moments

parameters for the original signal. Comparing with some traditional signal division models, some advantages like more flexibility can be found in the new one. With the controllable cut-off frequencies, signal division can become some new processing like signal selection, which can be used in intelligent algorithms such as genetic algorithm, the periphery filtering of neural networks, and so on.

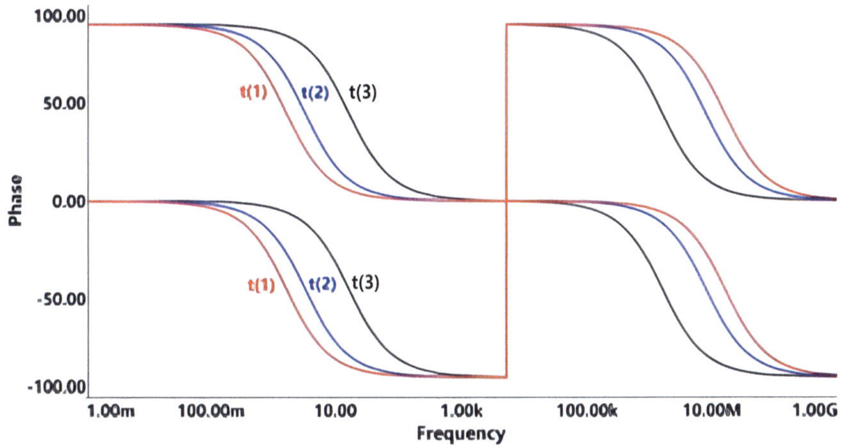

Fig. 5.8 The phase-frequency responses of BPF circuit and BSF circuit at three different moments

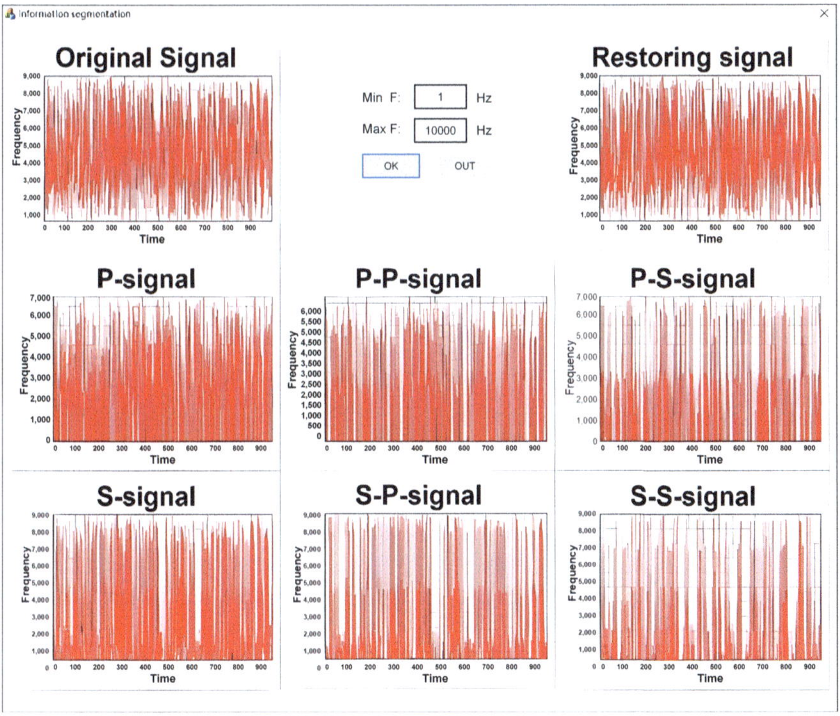

Fig. 5.9 Signal division result

5.1.5 Conclusion

This section presents the new memristive BPF and BSF circuits. Circuit analysis and numerical simulation are performed, which shows that the novel filter circuits are BPF and BSF with the time-varying feature. In light of this feature, a signal division model is provided with software experiments. With the realization of new filter circuits, the field of filter is growing gradually, which is important and scarce in the digital signal processing. This section has researched the application of signal division as an example. By employing the new filter circuits, many signal processing applications can be studied, which may improve the performance of traditional processing.

5.2 Memristor Bridge-Based Low-Pass Filter for Image Processing

5.2.1 Introduction

In this section, the memristor bridge is introduced to design the LPF for image denoising and scale-invariant feature transform (SIFT) algorithm design. On the one hand, design and analysis of memristor bridge-based LPF are presented. On the other hand, an adaptive Gaussian filter for the denoising image and a new Gaussian pyramid for SIFT are studied.

5.2.2 Design and Analysis of LPF Based on Memristor Bridge

5.2.2.1 Design of LPF Based on Memristor Bridge

In 2008, HP lab proposed a nanoscale double-ended resistor with memory [16], which is defined as:

$$R_W = R_{\text{on}}(1 - \alpha), \tag{5.7}$$

where R_{m} denotes the memristance.

Another expression of memristance for sinusoidal input voltage $V(t) = V_{\text{m}} \sin(2\pi\omega t)$ is provided below [7]:

$$R_{\text{M}} = R_{\text{avg}} \pm \Delta R_{\text{M}} \sin(2\omega t + \phi). \tag{5.8}$$

In [17], the memristor bridge was proposed as shown in Fig. 5.10. The memristor bridge is a Wheatstone bridge-like circuit composed of four identical memristors with the polarities. The memristor model is the same as in (5.7).

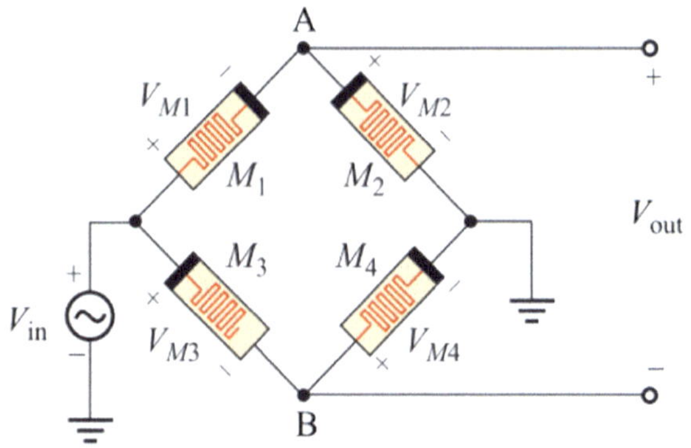

Fig. 5.10 Memristor bridge

The relationship of V_{out} and V_{in} is given by

$$V_{\text{out}} = \frac{V_{\text{A}} - V_{\text{B}}}{R_{\text{M1}} + R_{\text{M2}}} \cdot \left(\frac{1}{R_{\text{M3}} + R_{\text{M4}}} \right) \quad (3). \tag{5.9}$$

The cut-off frequency of the LPF varies over time because of the memristance varying over time [18]. The memristance of the memristor bridge varies not only over time but also over the memristance of each memristor in the memristor bridge, which exhibits more dynamic behaviors. In this section, the memristor bridge is applied into a resistor-capacitance (RC) filter to replace the resistor. The circuit can be designed as shown in Fig. 5.11.

5.2.2.2 Analysis of LPF Based on Memristor Bridge

In the filter circuit of Fig. 5.11, the single port capacitor C can be disconnected as shown in Fig. 5.12a. According to the direction of V_{ab}, the voltage of V_{ab} is obtained as follows:

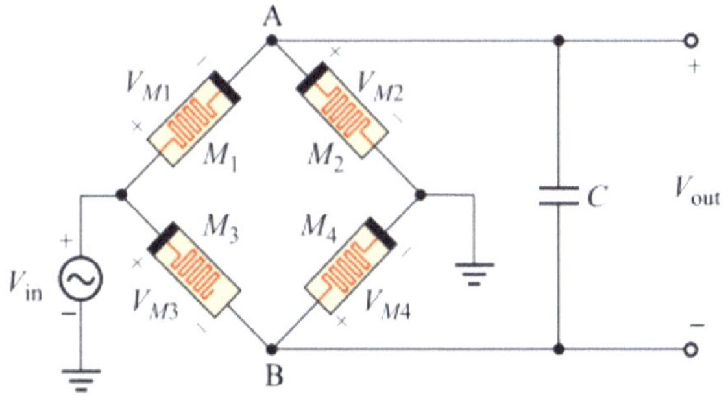

Fig. 5.11 New LPF based on memristor bridge

$$V_{ab} = \frac{V_{in}}{R_{M1} + R_{M2}} \left(\frac{1}{R_{M3} + R_{M4}} \right). \tag{5.10}$$

The independent voltage source is replaced by a short circuit as shown in Fig. 5.12b. R_{MB} can be calculated by (5.11). We use a single-port network of the Davidin equivalent circuit to replace the single-port network in Fig. 5.12c. Thus, the transfer function of the filter circuit (5.12) can be obtained as follows:

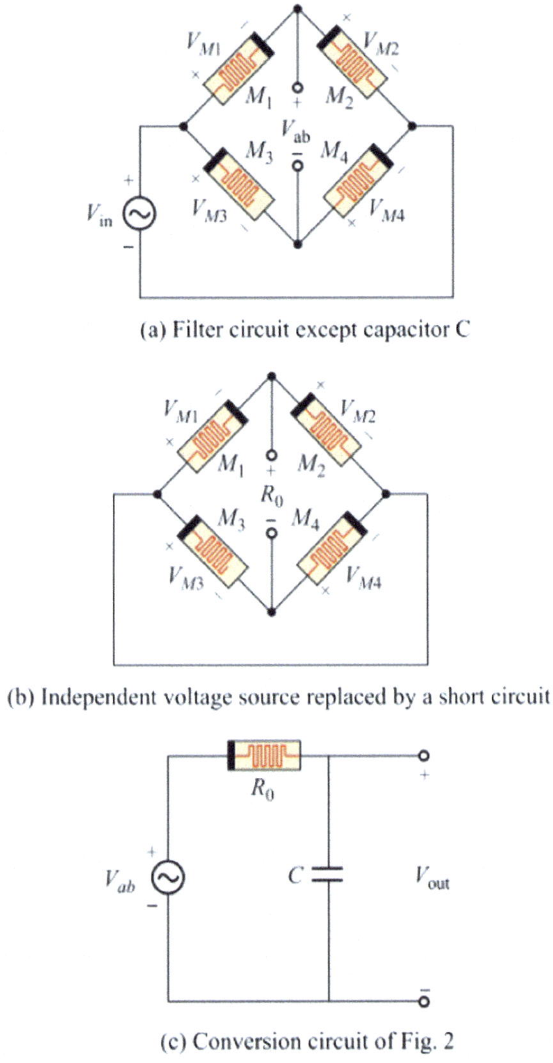

Fig. 5.12 Circuit conversion process

$$R_{\mathrm{MB}} = \frac{R_{\mathrm{M}} + R_{\mathrm{M2}}}{R_{\mathrm{M1}} + R_{\mathrm{M3}}} \cdot \frac{R_{\mathrm{M5}} + R_{\mathrm{M4}}}{R_{\mathrm{M4}}}, \tag{5.11}$$

$$1 + j\omega R_{\mathrm{MB}}C, \tag{5.12}$$

$$\frac{1}{2\pi R_{\mathrm{MB}}C}. \tag{5.13}$$

For this study, the memristance of the memristor bridge not only varies over time but also changes with each memristor's memristance of the memristor bridge. Compared to a single memristor, the advantage of the memristor bridge is that even if one of the memristors fails, the memristor bridge can still work well.

In Fig. 5.13a, the memristances of each memristor in the memristor bridge are the same. The memristor parameters are taken as follows:

$$\mathrm{Minit} = 110\,\mathrm{k}\Omega, \quad R_{\mathrm{on}} = 60\,\mathrm{k}\Omega, \quad R_{\mathrm{off}} = 160\,\mathrm{k}\Omega, \quad W = 91. \tag{5.14}$$

In Fig. 5.13b, the memristance of the memristor is not the same with each other because of the inequality of Minit. There is one Minit of 80 kΩ with $R_{\mathrm{on}} = 1\,\mathrm{k}\Omega$ and three Minit of 110 kΩ. The other variables are the same as the memristors in Fig. 5.13a.

Based on the memristor bridge models in Fig. 5.13, the curves of time-varying cut-off frequency are shown in 5.14a and b and the transfer functions which change over time are presented in Fig. 5.14c and d.

Figure 5.15 demonstrates the bode plot. Figure 5.15a and b show the amplitude-frequency responses which vary over time. Figure 5.15c and d provide the time-

Fig. 5.13 Simulation of the two memristor bridges with different memristance

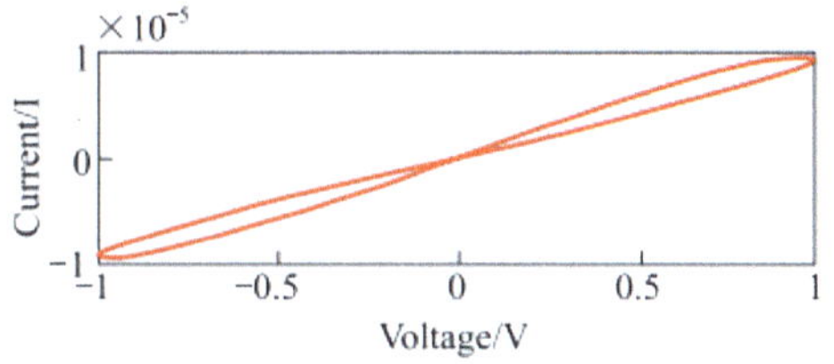

(a) Hysteresis curve of memristor bridge with same memristances

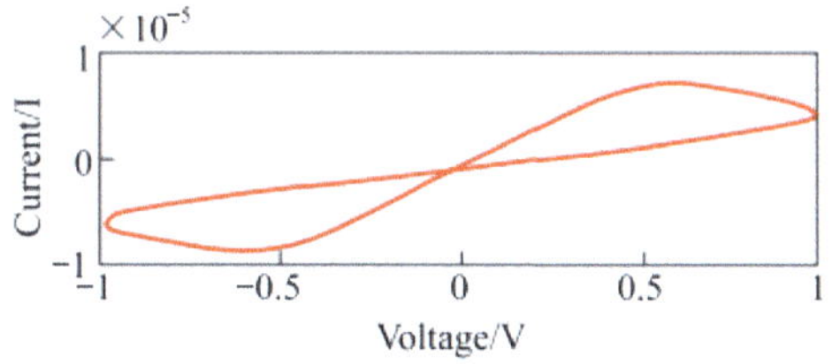

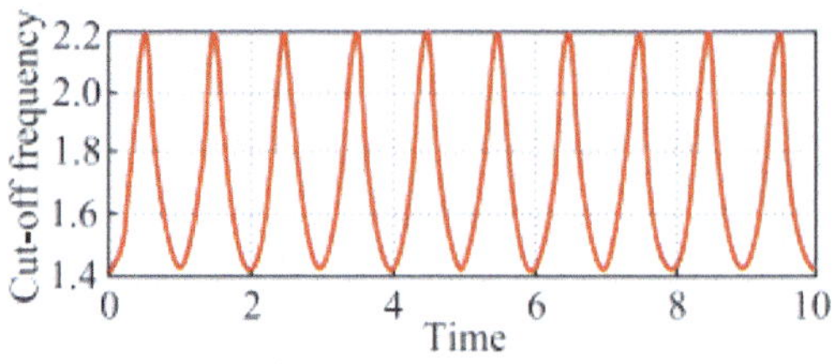

(a) Time-varying cut-off frequency based on memristor bridge model in Fig. 4(a)

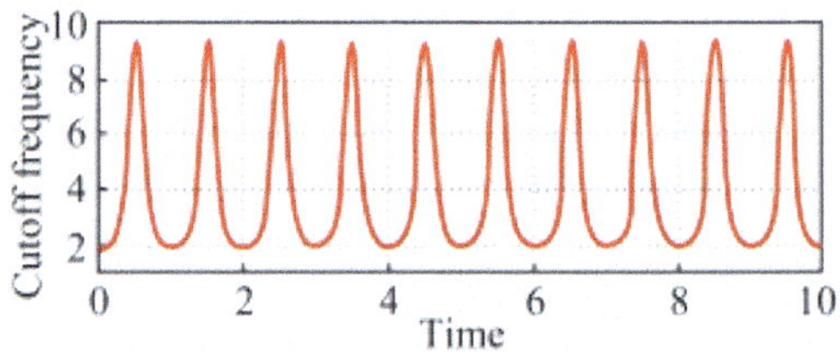

(b) Time-varying cut-off frequency based on memristor bridge model in Fig. 4(b)

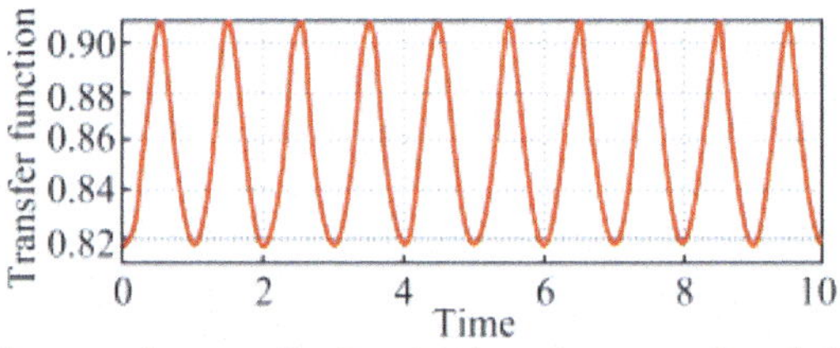

(c) Time-varying transfer function based on memristor bridge model in Fig. 4(a)

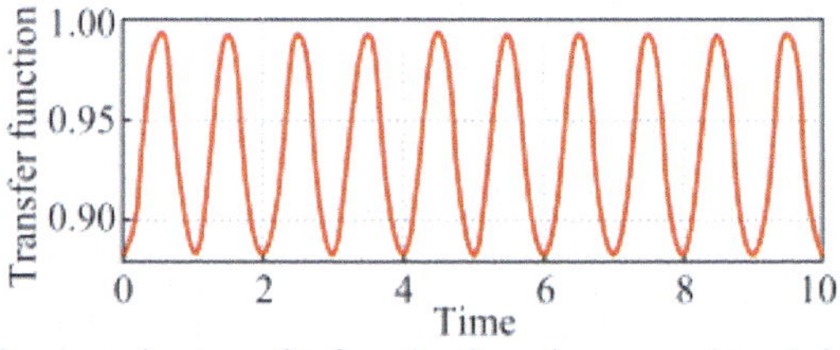

(d) Time-varying transfer function based on memristor bridge model in Fig. 4(b)

Fig. 5.14 Time-varying cut-off frequency and transfer function based on different memristor bridge models

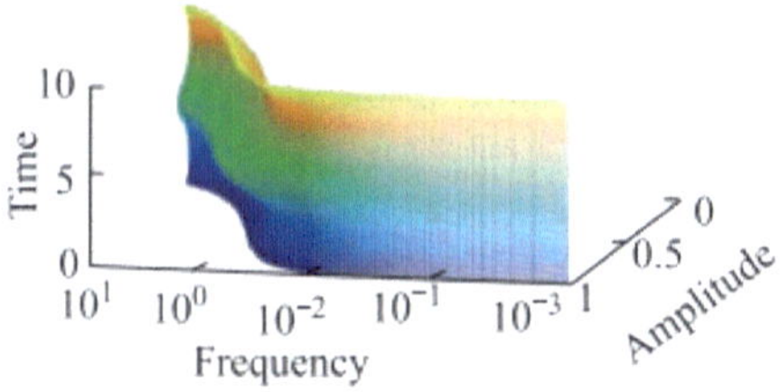

(a) Time-varying amplitude-frequency response based on memristor bridge model in Fig. 4(a)

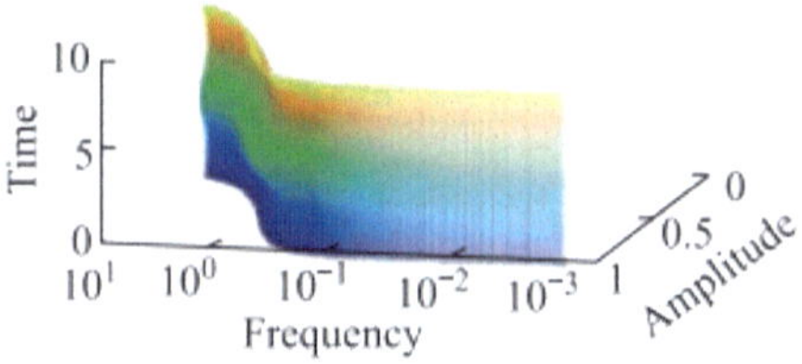

(b) Time-varying amplitude-frequency response based on memristor bridge model in Fig. 4(b)

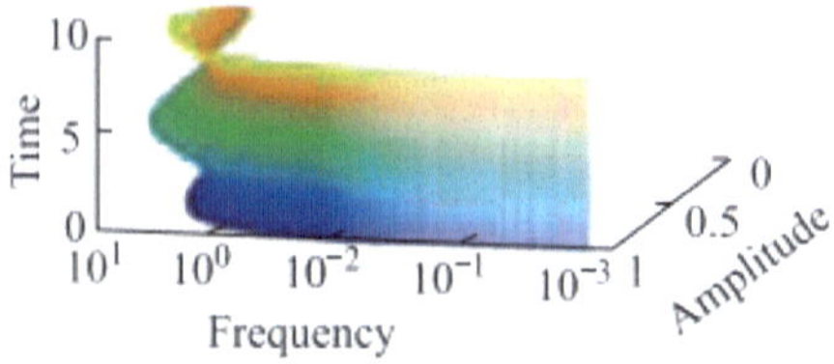

(c) Time-varying phase-frequency response based on memristor bridge model in Fig. 4(a)

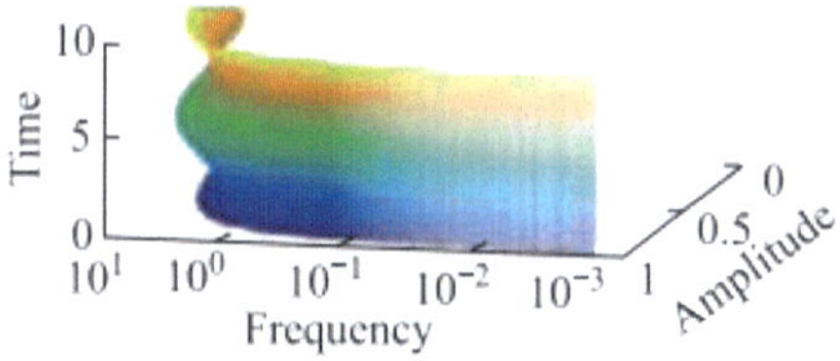

(d) Time-varying phase-frequency response based on memristor bridge model in Fig. 4(b)

Fig. 5.15 Time-varying amplitude-frequency response and phase-frequency response based on different memristor bridge models varying phase-frequency responses. Figure 5.15a is different from Fig. 5.15b, and c is not same as Fig. 5.15d because of the difference of memristor bridge models.

5.2.3 Image Processing

The memristor bridge-based LPF with its cut-off frequency varying over time is discussed. Inspired by the memristive filter and its typical characteristic, we design the memristive Gaussian filter and apply it to image processing. The adaptive Gaussian kernel will change in different situations, so the character of our LPF, which is variable cut-off frequency, can be combined with the Gaussian filter to denoise the image. Meanwhile, because of the scale of the Gaussian pyramid, each octave is gradually increased in SIFT, and the cut-off frequency of our LPF is increased during a specific time period, the new rules of the Gaussian pyramid scale change can be established by mapping the changeable cut-off frequency driven by the memristor bridge to SIFT.

The Gaussian blurring is a very typical example of image convolution. In essence, the Gaussian blurring is a convolution operation of a grayscale image I and a Gaussian kernel:

$$I(r, y) = I_0(t, y) \cdot G(t, y). \tag{5.15}$$

The convolution operation is denoted by the asterisk. The two-dimensional Gaussian kernel G_σ with standard deviation σ is defined as follows:

$$G_\sigma(r, y) = \frac{1}{2\pi\sigma^2}. \tag{5.16}$$

5.2.3.1 Adaptive Gaussian Filter for Denoising Image

We propose an adaptive Gaussian filter for denoising an image, which uses an improved energy function based on the function in [19] as follows:

$$\sigma = \arg\min\left(\sum_h + \frac{E_\sigma(r, y)}{h}\right). \tag{5.17}$$

Here, σ_M is defined by the cut-off frequency of the new LPF such that

$$\sigma_M = \frac{1}{2\pi R_{\mathrm{MB}} C}. \tag{5.18}$$

h is a normalizing parameter, and $E_\sigma(r, y)$ is the residual difference between the original pixel and the filtered pixel [20], which is

$$E_{\sigma}(r, y) = I_{o}(r, y) - I_{\sigma M}(r, y). \tag{5.19}$$

σ_M affects the smoothness. The image will be smoother with σ_M increasing, while the residual will increase and potentially lose image details. Therefore, the adaptive filter can select a proper σ by changing R_m for a compromise between smoothening results and holding details.

Then, the adaptive Gaussian filter and the Gaussian filter proposed in [21] are compared by denoising the images. The test images are standard clean images with equal size, which is 512 × 512 pixels for each image in Fig. 5.16a. All images are

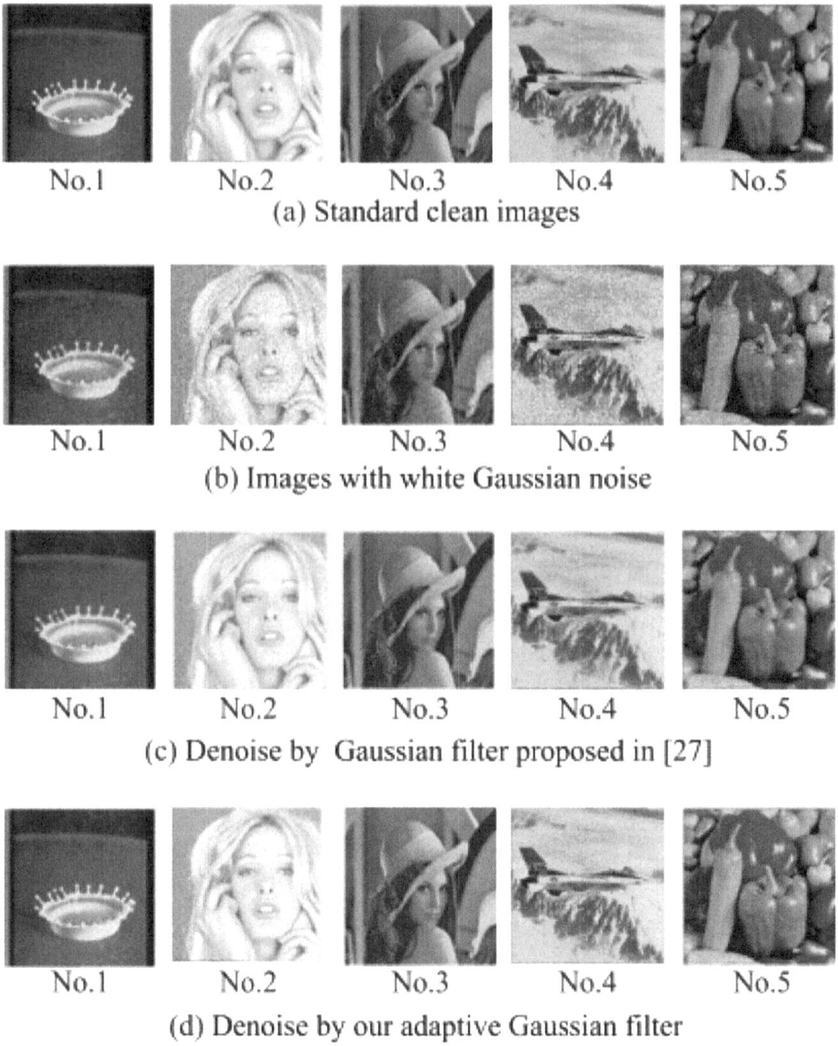

Fig. 5.16 Testing images and filter result

from the USC-SIPI image database. We generate noisy observations by adding the white Gaussian noise of standard deviation $\alpha = \frac{10}{255}$ to each clean image as shown in Fig. 5.16b. These noisy images are then denoised by the Gaussian filter proposed in [21] and our adaptive Gaussian filter, respectively. The size of the Gaussian template is set as 11×11. The cut-off frequency of the second low-pass filter model in Section 2 is used as σ_M for the adaptive Gaussian filter. Since the memristance changes with time, σ_M of the adaptive Gaussian filter also changes with time. We can select a proper σ_M for our filter, with $h = 1$. The average of all σ_M is set as σ for the Gaussian filter proposed in [21]. The filter results are as shown in Fig. 5.16c and d, respectively.

The filter results of our adaptive filter and the Gaussian filter [21] are compared by grayscale histogram, mean squared error (MSE), and peak signal-to-noise ratio (PSNR), which are shown in Fig. 5.17 and Tables 5.1 and 5.2, respectively. The histogram shows the number of pixels in an image at each different intensity value found in that image. Figure 5.17 shows the histogram of clean, noisy, Gaussian filtering [21] image, and our adaptive filtering Lena image. The variance of each histogram is calculated in Fig. 5.17. The different variance demonstrates that the variance of the original filtering image is the lowest because of the smooth effect of the Gaussian filter. However, the variance of our filtering image is higher for holding details. The MSE is a measure of the quality of an estimator. It is always nonnegative, and if the values are closer to zero, the results after processing perform better. From Table 5.1, the MSE of all five images, which are processed by our improved filter, is lower than the MSE that are processed by the Gaussian filter in [21]. PSNR is an engineering term for the ratio between the maximum possible power of a signal and the power of corrupting noise that affects the fidelity of its representation. From Table 5.2, we can note that the adaptive filter improves the PSNR by more than 2.77 dB.

From these three indicators, it is obvious that the Gaussian filter based on the memristor bridge-based low-pass filter is better than the Gaussian filter in [21], because the higher PSNR indicates lower noise and the greater variance represents more complete details.

For the algorithm of image denoising, the cost of time complexity is $O(n^2)$ because of the convolution operation, which is the core of the algorithm. Meanwhile, since σ_M is defined by the cut-off frequency of the new LPF, σ is time-varying and the Gaussian kernels are changed with the property of the cut-off frequency. Therefore, when more extendibility and reusability are needed such as different characters of the Gaussian kernels, the presented model can easily satisfy those needs with the introduction and improvement of memristor, which means the design of the Gaussian kernels does not need extra time which is more efficient than the design of the Gaussian kernel in [21].

5.2.3.2 New Gaussian Pyramid for SIFT

SIFT was proposed by Lowe [22]. The scale-space of the SIFT algorithm is represented by a Gaussian pyramid. The scale of each layer of the image [23] can be

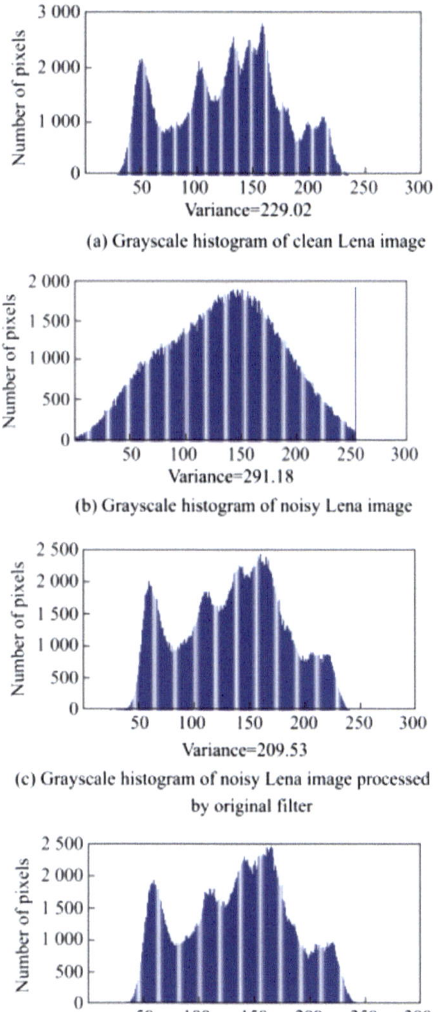

Fig. 5.17 Grayscale histogram of Lena image

(a) Grayscale histogram of clean Lena image

(b) Grayscale histogram of noisy Lena image

(c) Grayscale histogram of noisy Lena image processed by original filter

(d) Grayscale histogram of noisy Lena image processed by adaptive filter

uniformly expressed as follows:

$$\theta^{(\ell)}(c, r) = \theta^{(0)}(c + \ell, r), \tag{5.20}$$

where $\theta^{(\ell)}$ is the octave coordinate, $\ell \in \{0, \ldots, L-1\}$: $r \in \{0, \ldots, s+2\}$, with L being the number of octaves, c is the layer coordinate, $\theta^{(0)}$ is the base scale, and s is an integer number divided in each octave of the Gaussian pyramid.

Table 5.1 MSE of all five images in Fig. 5.16a

MSE	Images				
	No.1	No.2	No.3	No.4	No.5
Noisy	53.3923	53.8595	53.8858	54.2862	52.9164
Original filter	6.4781	10.2410	10.0236	13.0692	8.3866
Adaptive filter	2.4909	5.4001	3.0768	3.8197	3.0555
MSE reduced	3.9872	4.8409	6.9468	9.2450	5.3311

Table 5.2 PSNR of all five images in Fig. 5.17a

PSNR	Images				
	No.1	No.2	No.3	No.4	No.5
Noisy	30.8560	30.8182	30.8161	30.7839	30.8949
Original filter	40.0163	38.0274	38.1206	36.9683	38.8950
Adaptive filter	44.1672	40.8068	43.2498	42.3105	43.2799
PSNR improved	4.1509	2.7794	5.1292	5.3422	4.3849

In this section, the scale algorithm of the Gaussian pyramid is changed by using the normalized cut-off frequency. Because the scale of each octave is gradually increased in SIFT, the rise section of the cut-off frequency of the memristor bridge-based LPF is applied as the scale algorithm of the Gaussian pyramid, which is shown as follows:

$$\sigma = \frac{1}{2\sqrt{\pi} R_{\text{MB}} C}, \tag{5.21}$$

The test images are from the image database of the University of Oxford. Four groups of images are tested in this section as shown in Fig. 5.18.

In Lowe's SIFT algorithm, let $\sigma_0 = 1.6$, $s = 3$. In our algorithm, six consecutive values of the same time interval are set to the scale of each octave. The summary of the six scales is approximately the sum of $\sigma(\sigma_{\text{base}}, r)$ in SIFT. Other parameters are the same as SIFT. Figure 5.18c and d are used as representatives to show our results. Figure 5.19 is the result of our Gaussian pyramid algorithm for Fig. 5.18d. The two images in Fig. 5.20 are the feature extractions of Fig. 5.18c and d, respectively. Figure 5.21 is the result of the match. In this section, the value of the ratio is 0.6.

Figures 5.22 and 5.23 are the experimental results of SIFT and our algorithm for the four group images in Fig. 5.18 respectively. The number of feature points, matched number, and percent of correct matches of our algorithm are all more than SIFT. Therefore, the normalized cut-off frequency of the memristor bridge-based LPF can be used well in the Gaussian pyramid.

For the improved SIFT algorithm in this section, because most of the parameters are the same as SIFT, the original time complexity and spatial complexity will not

(a) One picture of the first group

(b) Another picture of the first group

(c) One picture of the second group

(d) Another picture of the second group

(e) One picture of the third group

(f) Another picture of the third group

(g) One picture of the fourth group

(h) Another picture of the fourth group

Fig. 5.18 Test images

Fig. 5.19 Our Gaussian pyramid algorithm for Fig. 5.18d

Fig. 5.20 Feature extraction

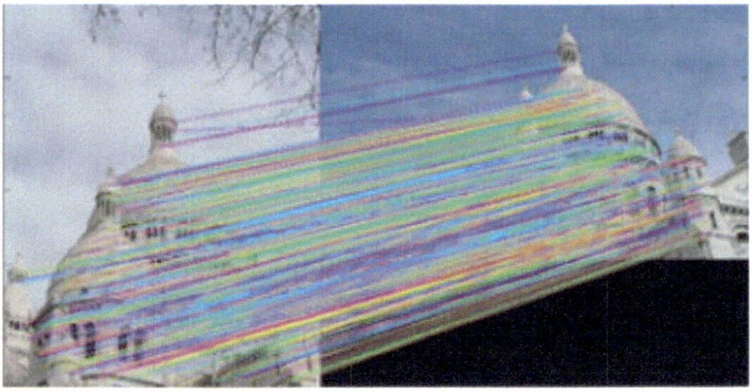

Fig. 5.21 Results of the match

Image	Number of feature points	Matched	Correct matches	Percent of correct matches/%
(a)	5 524	1 566	1 411	90
(b)	4 327			
(c)	4 111	1 239	828	67
(d)	2 845			
(e)	4 584	1 041	935	90
(f)	4 186			
(g)	9 763	772	754	98
(h)	6 854			

Fig. 5.22 Experimental results of SIFT

Image	Number of feature points	Matched	Correct matches	Percent of correct matches/%
(a)	12 787	4 164	4 085	97
(b)	10 640			
(c)	9 781	1 472	1 062	72
(d)	6 161			
(e)	6 305	1 480	1 410	95
(f)	6 176			
(g)	29 938	975	966	99
(h)	14 093			

Fig. 5.23 Experimental results of our algorithm

be increased. However, the rise section of the cut-off frequency of the memristor bridge-based LPF, which is applied as the scale algorithm of the Gaussian pyramid, is changed with time. So, we do not need to calculate the scale of the Gaussian pyramid. In other words, the efficiency of our algorithm improves to some extent.

5.2.4 *Conclusions*

This section introduces the memristor filter design and its application to image processing. The main contributions in this section are summarized as follows. First, a novel kind of LPF based on the memristor bridge is designed, whose characteristics are that the cut-off frequency, transfer function, amplitude-frequency response (AFR), and phase-frequency response (PFR) are demonstrated to change with different time and memristance. Second, the changeable parameter (for example, cut-off frequency) of the memristor bridge-based LPF is introduced to the energy function, and an improved adaptive Gaussian filter along with some experiments shows that the PSNR of our denoising increases by more than 2.77 dB compared to the corresponding Gaussian filter [21]. At the same time, the variable parameter of the memristor bridge-based LPF is introduced to the Gaussian pyramid, and an improved SIFT algorithm is presented. Experimental results show that the number of matched feature points and the percent of correct matches are increased compared to the traditional SIFT [14], which verifies the feasibility and effectiveness of our algorithm.

5.3 Comments

This chapter explores innovative applications of memristor-based filters in signal processing and image analysis, demonstrating the potential of these novel devices to enhance traditional filtering techniques. The research introduces memristor-based band pass and band stop filters with time-varying properties, as well as a memristor bridge-based low-pass filter, and applies them to tasks such as signal segmentation, image denoising, and feature extraction. This work not only advances the theoretical study of memristor circuits but also achieves significant results in practical applications, such as improving the peak signal-to-noise ratio in image denoising and increasing the accuracy of feature matching.

Despite these advancements, several challenges remain in fully realizing the potential of memristor-based filters. Current designs are limited in their ability to simultaneously adapt to multiple signal characteristics, potentially limiting their effectiveness in complex, rapidly changing environments. Additionally, the robustness and stability of these filters under varying operational conditions and device non-idealities have not been fully addressed, posing potential reliability issues in critical applications. To address these limitations, several key areas emerge as promising avenues for future research:

Developing multi-dimensional adaptive filters. Utilizing the nonlinear characteristics and time-varying nature of memristors to design filters capable of adaptive adjustment in multiple dimensions simultaneously, including frequency, amplitude, and phase. This research will involve constructing complex memristor network topologies, developing new control algorithms for dynamic parameter adjustment, and studying how to balance filtering performance under multi-objective optimization. These filters will adapt to complex, non-stationary signals, providing real-time superior performance in applications such as cognitive radio systems or advanced noise cancellation in dynamic environments.

Investigating the robustness and stability of memristor-based filters. Conducting an in-depth analysis of the impact of non-ideal characteristics of memristor devices (such as variability, noise, and temperature sensitivity) on filtering performance, and developing compensation techniques and fault-tolerance mechanisms. The research will include designing self-calibrating circuits, studying redundancy strategies, and developing control algorithms that maintain stable performance under device parameter drift. The goal of this work is to create robust memristor-based filters that maintain consistent performance over time and under varying environmental conditions, crucial for their application in critical systems.

Exploring the application of memristor-based filters in real-time video processing. Extending the image processing techniques presented in this chapter to the video domain, studying how to leverage the time-varying characteristics of memristor filters to process dynamic scenes. This involves developing new spatio-temporal filtering algorithms, designing adaptive mechanisms that can quickly respond to scene changes, and researching how to achieve efficient real-time processing with limited computational resources. These techniques will be applied to advanced surveillance systems, autonomous vehicle vision systems, or next-generation video compression algorithms.

Optimizing memristor-based filters by integrating machine learning techniques. Investigating how to combine deep learning or reinforcement learning technologies with memristor-based filters to achieve intelligent optimization and automatic adjustment of filter parameters. This includes designing new neural network structures to simulate and control memristor filters, developing online learning algorithms to adapt to different signal environments, and exploring how to utilize the physical properties of memristors to achieve more efficient learning processes. This approach will lead to highly adaptive filtering systems that continuously improve their performance based on input data, potentially revolutionizing areas such as adaptive signal processing and intelligent sensor networks.

These research directions will further advance the application of memristors in the field of signal processing and pave the way for the development of next-generation intelligent, efficient, and adaptive signal processing systems.

References

1. Campbell, G.A.: Physical theory of the electric wave-filter. Bell Syst. Tech. J. **1**, 1–32 (1922)
2. Dannehl, J., Fuchs, F.W., Thogersen, P.B.: Pi state space current control of grid-connected pwm converters with lcl filters. In: IEEE Energy Conversion Congress and Exposition (2009)
3. Parker, S.G., McGrath, B.P., Holmes, D.G.: Regions of active damping control for lcl filters. IEEE Trans. Ind. Appl. **50**(1), 424–432 (2014)
4. Viani, F., Robol, F., Salucci, M., Azaro, R.: Automatic emi filter design through particle swarm optimization. IEEE Trans. Electromagn. Compat. **59**, 1079–1094 (2017)
5. Thanapitak, S., Pawarangkoon, P., Sawigun, C.: A flipped voltage follower second-order band-pass filter. J. Circuits Syst. Comput. **26**(7) (2017)
6. He, Y., Barannik, A., Cherpak, N.: Novel design of band-pass waveguide filter with hts e-plane insert. IEEE Trans. Appl. Superconduct. **27**(4) (2017)
7. Cam, U.: First-order memristor-capacitor filter circuits employing hp memristor. J. Circuits Syst. Comput. **23**(8) (2014)
8. Zhao, Y.B., Tse, C.K., Feng, J.C.: Application of memristor-based controller for loop filter design in charge-pump phase-locked loops. Circuits Syst. Signal Process **32**, 1013–1023 (2013)
9. Sozen, H., Cam, U.: On the realization of memristor based rc high pass filter. In: 8th IEEE International Conference on Electrical and Electronics Engineering, pp. 45–48 (2013)
10. Ascoli, A., Tetzlaff, R., Corinto, F., Mirchev, M., Gilli, M.: Memristor-based filtering applications. In: 14th Latin American Test Workshop, pp. 1–6 (2013)
11. Li, M., Henry, M.: Complex bandpass filtering for coriolis mass flow meter signal processing. In: IECON 2016 - 42nd Annual Conference of the IEEE Industrial Electronics Society, pp. 4952–4957 (2016)
12. Onat, E.: Fpga implementation of real time video signal processing using sobel, robert, prewitt and laplacian filters. In: 25th Signal Processing and Communications Applications Conference (SIU), pp. 1–4 (2017)
13. Chua, L.O.: Memristor - the missing circuit element. IEEE Trans. Circuit Theory **CT-18**(5), 507–519 (1971)
14. Chua, L.: Resistance switching memories are memristors. Handbook of Memristor Networks, pp. 197–230 (2019)
15. Di Ventra, M., Pershin, Y.V., Chua, L.O.: Circuit elements with memory: memristors, memcapacitors, and meminductors. Proc. IEEE **97**(10), 1717–1724 (2009)
16. Strukov, D.B., Snider, G.S., Stewart, D.R., Williams, R.S.: The missing memristor found. Nature **453**(7191), 80–83 (2008)
17. Kim, H., Sah, M.P., Yang, C., Roska, T., Chua, L.O.: Memristor-bridge synapses. Proc. IEEE **100**(6), 2061–2070 (2012)

18. Li, Y., Yang, C., Yu, Y., Díez, F.F.: Research on low pass filter based on memristor and mem-capacitor. In: 2017 36th Chinese Control Conference (CCC), pp. 5110–5113. IEEE (2017)
19. Belyaev Y, S.H.: Mesh smoothing by adaptive and anisotropic gaussian filter applied to mesh normal. In: Proceedings of the Vision, Modeling, and Visualization, pp. 203–210 (2002)
20. G, G.: Local smoothness in terms of variance: the adaptive gaussian filter. In: Proceedings of the British Machine Vision Conference, pp. 1–10 (2000)
21. Wang, M., Zheng, S., Li, X., Qin, X.: A new image denoising method based on gaussian filter. In: 2014 International Conference on information science, electronics and electrical engineering, vol. 1, pp. 163–167. IEEE (2014)
22. Lowe, D.G.: Object recognition from local scale-invariant features. In: Proceedings of the 7th IEEE International Conference on Computer Vision, pp. 1150–1157 (1999)
23. Lowe, D.G.: Distinctive image features from scale-invariant keypoints. Int. J. Comput. Vis. **60**(2), 91–110 (2004)

Chapter 6
Memristor Network-Based Swarm Intelligence

Swarm intelligence, inspired by the collective behavior of natural systems such as ant colonies and fish schools, has emerged as a powerful paradigm for solving complex optimization problems. In recent years, the advent of memristor technology has opened up new avenues for the hardware implementation of swarm intelligence algorithms. Memristors, with their unique properties of non-volatility, nanoscale dimensions, and the ability to perform both storage and computation, provide an ideal platform for realizing swarm intelligence in a highly parallel and efficient manner.

The chapter begins by introducing a memristor crossbar array-based implementation of ant colony optimization (ACO) for image edge detection in Sect. 6.1. A nonlinear voltage-controlled memristor model with a relaxation term is proposed, along with an improved ACO algorithm incorporating a padding strategy. The memristor crossbar array, with its high device density and parallel computing capabilities, enables efficient deployment of ACO for edge detection tasks. Experimental results demonstrate the superiority of the proposed method compared to existing edge detection techniques.

Section 6.2 presents a memristive network-based genetic algorithm (MNGA) for image edge detection. By establishing a mapping between the memristive network (MN) and the genetic algorithm (GA), a mem-computing model is developed to facilitate the hardware implementation of the GA. The MNGA-based edge detection algorithm, integrating image filtering and GA operators deployed by MN, exhibits an improved figure of merit (FoM) compared to state-of-the-art memristor-based swarm intelligence methods.

In Sect. 6.3, the relationship between artificial fish swarm algorithm (AFSA) and memristive networks is investigated. Inspired by the correlation between ACO and the memristive networks, mapping rules between AFSA's four main behaviors and memristive networks are highlighted. Simulations of single fish motion and swarm behavior are conducted, and comparisons with conceptual models are made.

Y. Yu et al., *Memristive Computing*,
https://doi.org/10.1007/978-981-96-0332-9_6

Section 6.4 introduces a matrix-friendly genetic algorithm (MGA) that leverages the parallel matrix computing capabilities of memristor networks. By representing the population as a matrix and performing evolution through matrix operations, MGA achieves better convergence and faster execution compared to baseline GA. The deployment of MGA on memristor circuits enables parallelization and in-memory computing, leading to significant efficiency gains. The effectiveness of MGA is demonstrated through a feature selection experiment on logistic regression using Sonar datasets.

Finally, Sect. 6.5 explores the deployment of GA using memristors for image restoration. The similarity between memristor-based arrays and chromosomes in genetics, as well as the resemblance to image gray value matrices, motivates the development of a memristor-based GA for image restoration. Experiments with different parameter settings showcase the feasibility and effectiveness of implementing GA with memristors.

The integration of memristor networks and swarm intelligence algorithms presented in this chapter offers a promising approach to tackle complex optimization problems in various domains. By exploiting the unique properties of memristors, such as their ability to perform in-memory computing and parallel processing, the proposed methods demonstrate enhanced performance, scalability, and efficiency compared to traditional swarm intelligence implementations. The insights and techniques presented in this chapter pave the way for the development of highly efficient and intelligent memristor-based swarm intelligence systems, with potential applications in image processing, feature selection, optimization, and beyond.

6.1 Memristor Crossbar Array-Based Ant Colony Optimization

6.1.1 Introduction

The bio-inspired algorithm is stemmed from the principles that exist in natural systems. This algorithm provides problem-solving techniques with enhanced robustness and flexibility [1, 2]. Ant Colony Optimization (ACO), a widely used and highly effective bio-inspired algorithm, is originally designed for solving a well-known traveling salesman problem (TSP) [3]. In the field of image processing, ACO also shows high performance on image edge detection with regarding pixel intensity variation as the heuristic factor [4–6]. For the sake of using ACO more efficiently, researchers are exploring a novel electronic device to implement ACO and the memristor is one of the most potential devices.

For high-density and systematic deployment of memristor, crossbar architecture is introduced by researchers [7, 8]. Furthermore, memristor crossbar array is widely used in Spiking Neural Networks [9], pattern recognition [10] and so on [11–13]. Especially, similarities between ACO and memristors have been demonstrated by researchers [14]. By using a piecewise-linear memristor model, a memristor network is designed by researchers to implement ACO [15].

As one of the most important and useful research directions for Artificial Intelligence, computer vision includes many tasks, such as image recognition [16–18], object detection [19, 20], pose estimation [21, 22] and other tasks [23–25]. Image edge detection is a fundamental task in the computer vision field [26]. A common practice of ACO for image edge detection is taking the contrast of pixels as a heuristic factor and ants on random initial position will find local shortest edges [4–6].

In this section, an efficient memristor crossbar array deployment of ACO is proposed to realize image edge detection.

6.1.2 Memristor Model

Different from the piecewise-linear memristor model used by [15], a nonlinear model is proposed, and the similarities between this model and the ACO is explained. In order to understand the similarity between ACO and memristors intuitively, a simple shortest path problem with two paths was illustrated [14]. Ants go from point A to point B and there are two paths L2 = 2L1 as depicted in Fig. 6.1a. During the initialization phase of ACO, pheromones in L1 are equal to L2, and the heuristic factors of each path are the reciprocal of path lengths. When the first iteration starts, L1 will be selected by more ants because its heuristic factor is bigger. Therefore, more pheromones are added to L1 than L2. With the increment of iterations, the gaps of pheromones between L1 and L2 increase gradually. Finally, the probability of ant selecting L2 is close to zero, and it means that L1 is the shortest path obtained by ACO.

The parallel circuit of memristors in Fig. 6.1b is designed to solve the two-path problem. There are two parallel branches, and M_1 is a current-controlled memristor while M_2 consists of two same memristors. More current will go through M_1 than M_2 because of Kirchhoff's law when the switch is closed. The resistance of M_1 goes down faster as the result of more passing current. Thus, with the increment of time,

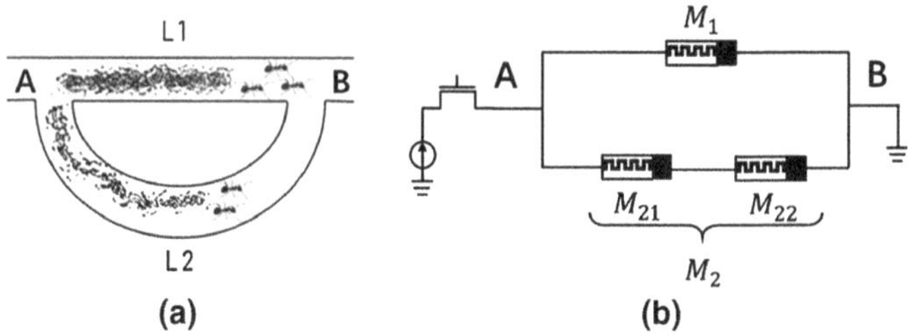

Fig. 6.1 **a** From A to B, there are two paths L1 and L2, such that L2 = 2L1. **b** A simple parallel circuit of memristor to implement ACO in (**a**)

the memristance of M_1 will decrease continuously. Finally, the same shortest path as ACO is obtained.

Zoha et al. [15] used a linear memristor model to detect image edges. However, the intrinsic mapping between this model and ACO is not obvious. In this chapter, we proposed a nonlinear memristor model with a one-to-one relationship to ACO defined by (6.1)–(6.3).

$$M = R_{off}(1 - x) + R_{on}x, \quad 0 \leq x \leq 1 \tag{6.1}$$

In (6.1), M is the memristance (resistance of memristor), R_{off} is the maximum memristance and R_{on} is the minimum memristance. The internal variable x determines the value of M and the derivative of x to t is described by (6.2) and (6.3).

$$\frac{dx}{dt} = -f_m(1 - x) \tag{6.2}$$

$$f_m = KV_m - \xi \tag{6.3}$$

This model is a nonlinear voltage-controlled memristor with a relaxation term. f_m is the window function where V_m is the voltage of memristor, ξ is the relaxation term and K is the drift constant of memristor.

$$\frac{dM}{dt} = -\frac{dx}{dt}(R_{off} - R_{on}) \tag{6.4}$$

$$dG = d\frac{1}{M} = -\frac{1}{M^2}dM \tag{6.5}$$

By plugging (6.2) in the derivative of (6.1) to time t, the derivative of M to t is obtained as shown in (6.4). The relationship between differentials of conductance (G) and memristance (M) is described by (6.5). When $R_{off} \gg R_{on}$, (6.2) can be replaced by (6.6) because $R_{off}/(R_{off} - R_{on}) \approx 1$. Finally, the derivative of conductance to t is obtained by (6.1) and (6.4)–(6.6), which is shown in (6.7). In (6.7), $I_m = V_m/M$ is the current goes through the memristor.

$$\frac{dx}{dt} \approx -f_m\left(\frac{R_{off}}{R_{off} - R_{on}} - x\right) \tag{6.6}$$

$$\frac{dG}{dt} \approx KI_m - \xi G \tag{6.7}$$

The memristor model is evaluated by MATLAB while using the necessary parameters listed in Table 6.1. The input voltage is a sinusoidal signal whose amplitude is equal to 1mV and frequency is equal to 1 Hz. As the $I - V$ curve shown in Fig. 6.2, a hysteretic curve is obtained, which is often used to distinguish whether a device is a memristor [27–29].

Table 6.1 Parameters of memristor model

Parameter	Value
R_{off}	1 MΩ
R_{on}	1 Ω
ξ	15.68
K	1.96×10^4

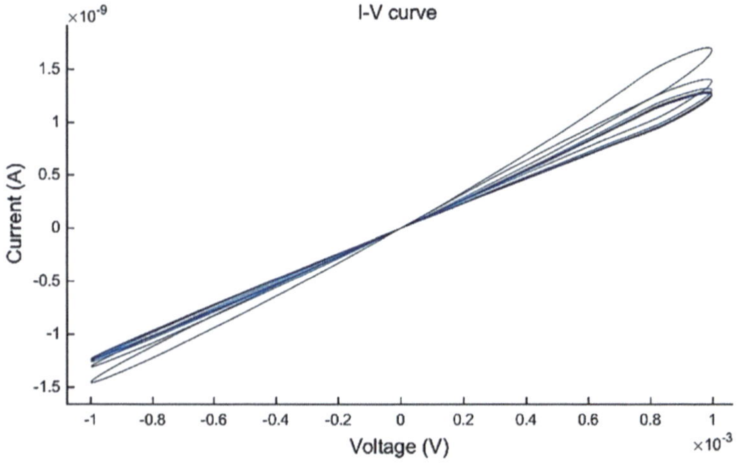

Fig. 6.2 $I-V$ curve of proposed memristor model

The ACO is based on the stochastic propagation of multiple moving agents (ants) through a graph [3]. Especially, based on [14], the probability p_i of ants selecting path i (for two-path problem,) shown in Fig. 6.1a is given by

$$p_i = \frac{\tau_i^{\alpha}\left(\frac{1}{L_i}\right)^{\beta}}{\sum_j \tau_j^{\alpha}\left(\frac{1}{L_j}\right)^{\beta}} \tag{6.8}$$

where τ_i is the pheromone of path i, and L_i is the length of path i, α and β are the importance of the pheromone and the heuristic factor, respectively.

Moreover, for each step, the amount of pheromone is updated according to:

$$\tau_i(k+1) = (1-\rho)\tau_i(k) + \frac{vQ}{L_i} \tag{6.9}$$

where ρ describes the pheromone evaporation, Q is a constant, $v = 1$ depicts one ant visits the path and otherwise $v = 0$.

Then, if the ant is added to the point A at a constant rate γ, the derivatives of τ_1 and τ_2 to t can be defined by

$$\frac{d\tau_i}{dt} = -\gamma\rho\tau_i + \frac{\gamma Q}{L_i} \frac{\tau_i^\alpha \frac{1}{L_i^\beta}}{\sum_j \tau_j^\alpha \left(\frac{1}{L_j}\right)^\beta} \tag{6.10}$$

When current of source in Fig. 6.1b is equal to I_0, we can rewrite Eq. 6.7 based on Kirchhoff's Current Law as:

$$\frac{d\hat{G}_i}{dt} = -\xi\hat{G}_i + \frac{K I_0}{G_{off,i}} \frac{\hat{G}_i G_{off,i}}{\sum_j \hat{G}_j G_{off,j}} \tag{6.11}$$

where $\hat{G} = G_i/G_{off}$, and $G_{off,i} = 1/M_{on,i}$ which represents the maximum of conductance. Comparing (6.10) with (6.11), one can notice that the memristor model can implement ACO with appropriate parameters when $\alpha = \beta = 1$. Then, we can get following relationships: (1) the change of pheromone $d\tau_i$ is similar to $d\hat{G}_i$; (2) the pheromone evaporation ρ corresponds to the relaxation term ξ; (3) the inspired factor $1/L_i$ can be realized by $G_{off,i}$.

In order to build and verify similarities between ACO and the proposed memristor model, we make a two-path problem experiment which is shown in Fig. 6.1. The result of the ACO and memristor circuit is shown in Fig. 6.3a, b respectively. Whether it's ACO or memristor circuit, the final shortest path is the same and the convergence process is similar. However, the initial values of those two methods are different. The initial values of τ_1 and τ_2 are equal, while the initial value of G_1 is twice that of G_2.

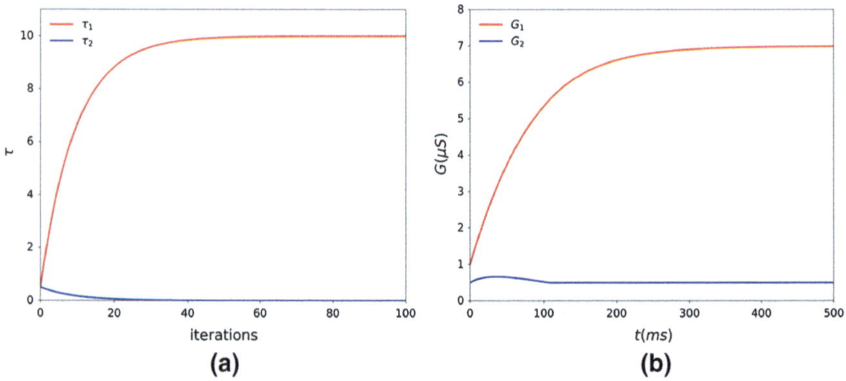

Fig. 6.3 Experiments on solving two-path problem by ACO. **a** The proposed memristor model. **b** These plots were obtained by using the following parameter values: $\tau_i(0) = 0.5$, $\alpha = \beta = 1$, $L_1 = L_2/2 = 1$, $\rho = 0.1$, $\gamma = 1$, $Q = 1$, $dt = 1\,\text{ms}$, $I_0 = 6 \times 10^{-3}\,\mu\text{A}$ and parameters of memristor model are shown in Table 6.1.

6.1.3 Simple Ant Colony Optimization

In this section, we introduce the hardware-friendly optimization of ACO and the padding strategy that is used for masking image margin. Before we calculate the heuristic factor of the image, a 1×1 reflection padding process should be done. Reflection padding can ensure that the heuristic factor in the circumference of the image can be calculated correctly. A reflection padding example is shown in Fig. 6.4 for an image with the size of 3×3. This kind of padding regards the head and tail pixel of each row and column as the symmetry center. For example, the pixel $I_{0,-1}$ is the symmetry pixel of $I_{0,1}$ when $I_{0,0}$ is symmetry center and the value of $I_{0,-1}$ is equal to $I_{0,1}$. But $I_{-1,-1}$, $I_{-1,3}$, $I_{3,-1}$ and $I_{3,3}$ are different, because its symmetry center is not the pixel of the image but the padding pixel. Specifically, the symmetry center of the $I_{-1,-1}$ is the pixel $I_{-1,0}$ and $I_{-1,1}$.

Then, we need to calculate the heuristic factor of image edge detection and it is defined by (6.12). I_{max} is a normalizing factor set to the maximum intensity of the whole image to limit $0 \leq \eta_{ij} \leq 1$. η_{ij} is determined by the contrast of vertical adjacent pixels and horizontal adjacent pixels. In the ACO algorithm, η_{ij} is the path length from other adjacent pixels to pixel I_{ij}. The next step is reflection padding, and the padding size is equal to $L - 1$ where L is the count of pixels of each ant traveled path. The reason why we use reflection padding is to set the boundary of ant travel. It should be noted that we don't calculate the heuristic factor of the first padding pixel in Fig. 6.4, and the reflection factor is a new padding after computing all the heuristic factors (η). The complete process is shown in Fig. 6.5.

$$\eta_{ij} = \frac{1}{I_{max}}(|I_{i,j-1} - I_{i,j+1}| + |I_{i-1,j} - I_{i+1,j}|) \tag{6.12}$$

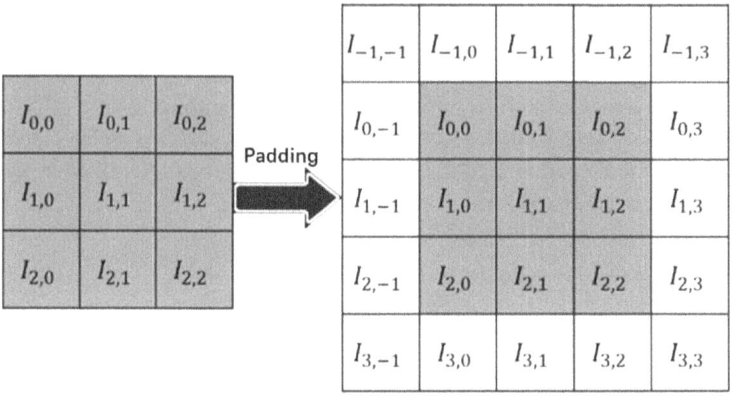

Fig. 6.4 1×1 reflect padding of a 3×3 image

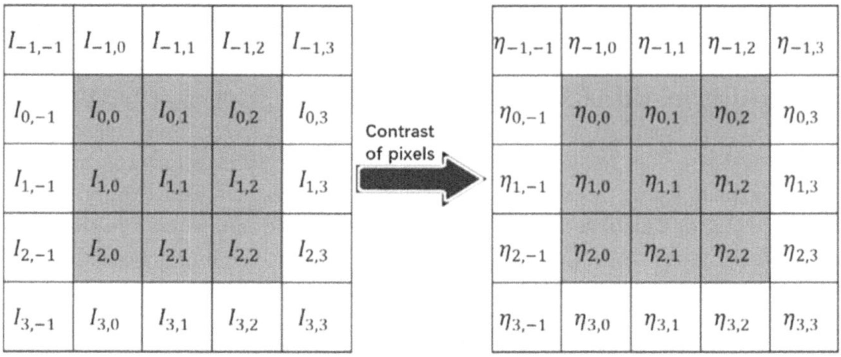

Fig. 6.5 Calculate of heuristic factor and reflection padding for the path length of ant travel is equal to 2

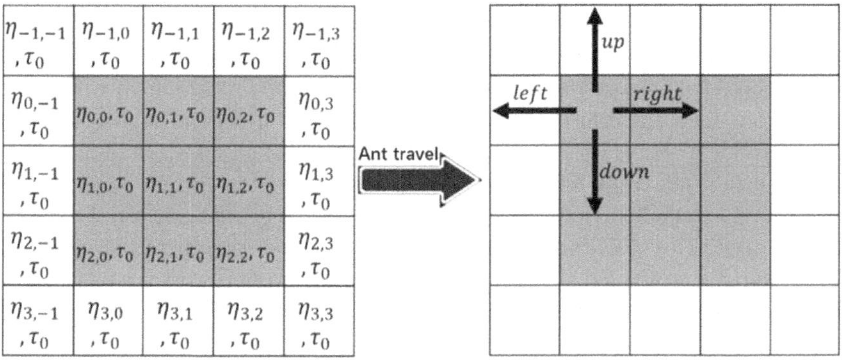

Fig. 6.6 Ant travel example for the path length of ant travel is equal to 2

Finally, we will start ant travel and pheromone update. Before that, the pheromone will all be initialized to τ_0. In particular, we just consider ants can only move vertically or horizontally and ants start from each pixel as the example shown in Fig. 6.6. In other words, every ant can only select four directions in a step and each path contains two pixels, given the path length $L = 2$. The selection probability of each path can be calculated by (6.13) and (6.14).

$$p_{path_m} = \frac{m_n}{\sum_{n \in \{up, down, left, right\}} m_n} \tag{6.13}$$

$$m_n = \prod_{(i,j) \in path_n} \tau_{ij}^{\alpha} \left(\frac{1}{\eta_{ij}} \right)^{\beta} \tag{6.14}$$

6.1.4 Memristor Crossbar Array

In this section, a memristor crossbar array is utilized to implement the ACO algorithm. Memristor crossbar array architecture offers high device density and fully Operationality [7, 8]. The memristor crossbar array is shown in Fig. 6.7 and it is the hardware deployment of ACO example shown in Fig. 6.6.

The size of crossbar array is equal to $(h + 2(L - 1)) \times (w + 2(L - 1))$, where h and w are the height and width of the image, respectively, and $L - 1$ is the padding size. Hence, the size of the crossbar array is equal to 5×5 while the path length is 2. There are two controlled circuits for selecting a memristor. The top one controls the selection of columns and the left one controls the selection of rows.

The pheromone updating implemented by a crossbar array of four direction paths starts from pixel $L_{0,0}$ is shown in Fig. 6.8. The switches on the column-select circuit and row-select circuit control the connection of memristors and the current path. For example, there are three switches in Fig. 6.8a: (1) S_I controls the inlet of current; (2) S_C controls the connect of adjacent columns; (3) S_G controls the exit of current. The difference between row updating and column updating is the location of S_C. For Fig. 6.8c, S_C is located on the row-select circuit and used to control the connect of adjacent rows.

The pulse of the current source for updating is shown in Fig. 6.9. It shows that there is only one pulse for each update and the value of current is 1 μA under 1 μs duration. In this case, the ideal time is 1 μs and the power is 1 μW for each update of a memristor.

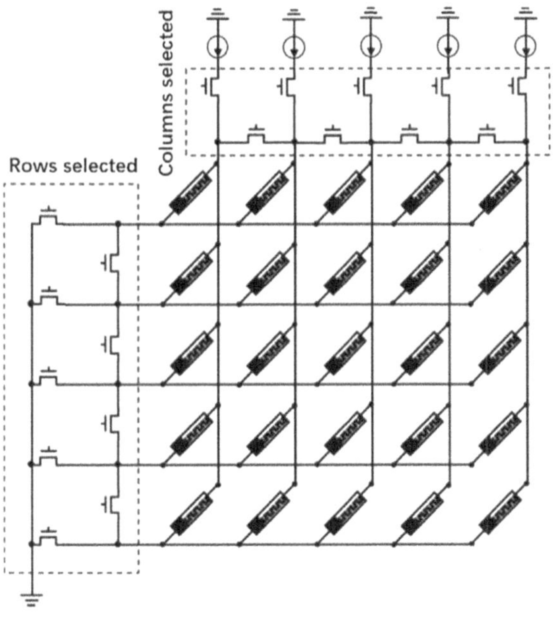

Fig. 6.7 Memristor crossbar array for implementing ACO

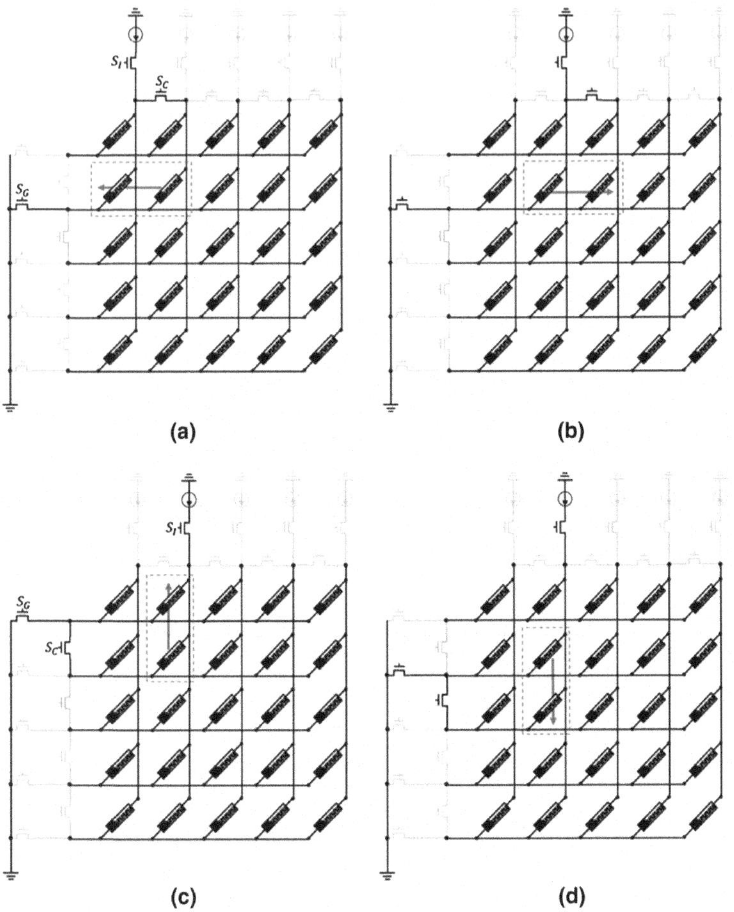

Fig. 6.8 Updating on crossbar array for left, right, up, and down paths

As shown in Fig. 6.10, the travel paths of four directions can be updated parallelly. If we don't use the memristor crossbar to parallelize ACO, there will be $4 \times w \times h$ paths in each iteration and the time complexity will become $O(wh)$. Using the parallelization strategy illustrated in Fig. 6.10, a row can be updated in the left or right direction simultaneously, and all columns can be updated in the up and down directions simultaneously. Consequently, the time complexity decreases to $O(h + w)$. Furthermore, the parallelism of up and down directions is higher since the current source is set on a column-select circuit.

As shown in Fig. 6.11, the flow of the method is summarized as follows: 1. analyzing the similarities between the memristor model and ACO; 2. designing the crossbar array circuit; 3. making simulation of image edge detection including image graying, computing contrast of pixel intensity, initializing memristance, updating, reading the

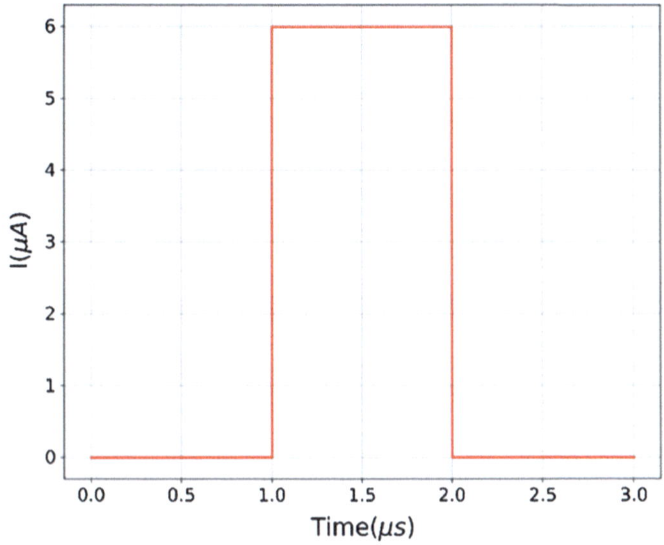

Fig. 6.9 The pulse of current source for memristor update

conductance matrix, and setting threshold. The "contrast" and "conductance matrix" have blurry edges, and the edges of the "conductance matrix" are more obvious.

6.1.5 Simulation and Results

The simulation process of implementing ACO by memristor crossbar array consists of three steps: initialization, multiple update iterations, and generation of edge detection results.

Firstly, the internal variable x_{ij} of memristor in row i and column j is initialized to the heuristic factors η_{ij}. As x_{ij} gets bigger, the memristance M_{ij} gets smaller and the conductance of memristor G_{ij} gets bigger. This will result in pixel I_{ij} with larger heuristic factors η being more likely to be judged as the edge pixel.

In each update iteration, paths with length $L(L \geq 2)$ in four directions is updated separately as shown in Fig. 6.8. The pulse of the current source for each update is shown in Fig. 6.9. Memristors in the same path get different voltages if their conductance G is unequal because of Kirchhoff's voltage law. This is the reason why they will get different increments of conductance.

After an appropriate iterative process, a matrix of memristive conductance is obtained and we use the symbol $\mathcal{G}$ to represent it. In order to output a binary matrix $\mathcal{R}(I_{ij}$ is an edge pixel given $\mathcal{R})$, a threshold should be set. It is interesting to note that appropriate edge detection results are obtained when the threshold is the mean value of the matrix $\mathcal{G}$. It means that $\mathcal{R}$ can be calculated by Eq. 6.15.

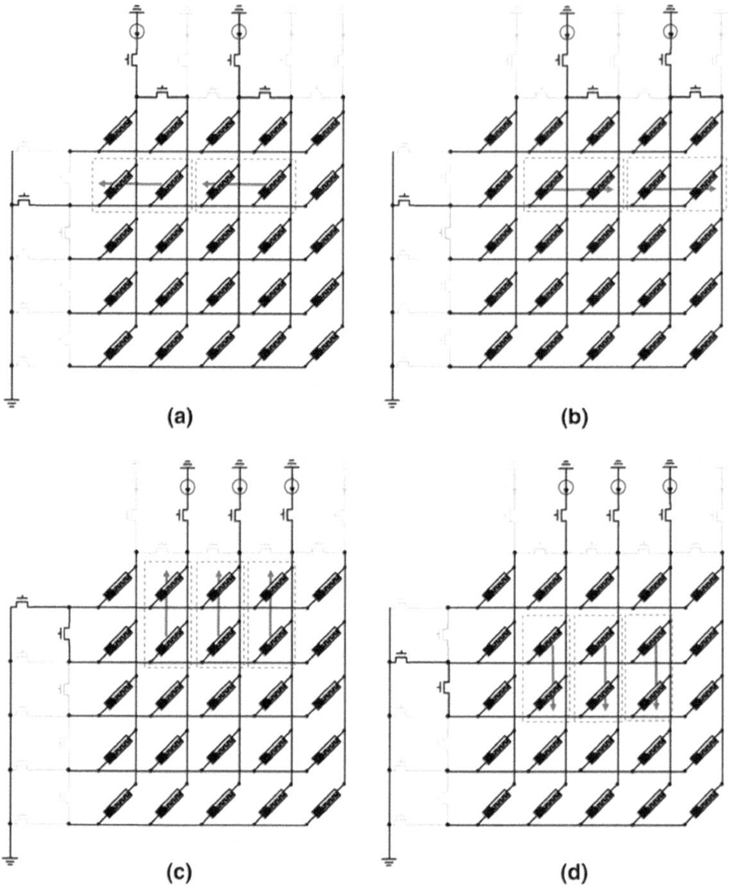

Fig. 6.10 Updating in parallel on crossbar array for four directions

$$\mathcal{R}_{ij} = \begin{cases} 1, & \mathcal{G}_{ij} > mean(\mathcal{G}) \\ 0, & else \end{cases} \tag{6.15}$$

Canny [26] and Sobel [30] are traditional operators for image edge detection. Holistically-nested Edge Detection (HED) is a neural network with multi-scale and multi-level feature learning [31]. The visual results of those three methods and the proposed method are shown in Figs. 6.12 and 6.13. Three original images, a large building, many small-looking elephants, and "Lena" image, are used for the experiments of image edge detection. The results of the proposed model are similar to those of traditional operators because they are all unsupervised methods.

For large-scale objects like the building in Fig. 6.12, HED can effectively obtain the outline of the object while ignoring the details in the object. As ACO tends to be trapped by local optimizations, the proposed model cannot ignore the details in

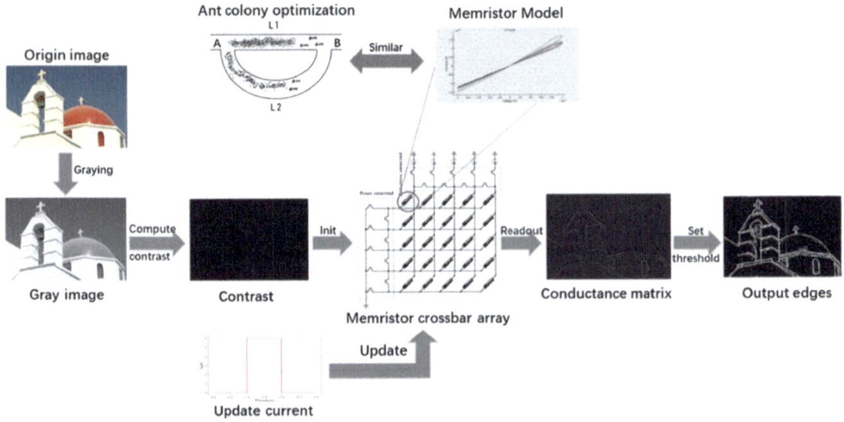

Fig. 6.11 Pipeline of proposed method

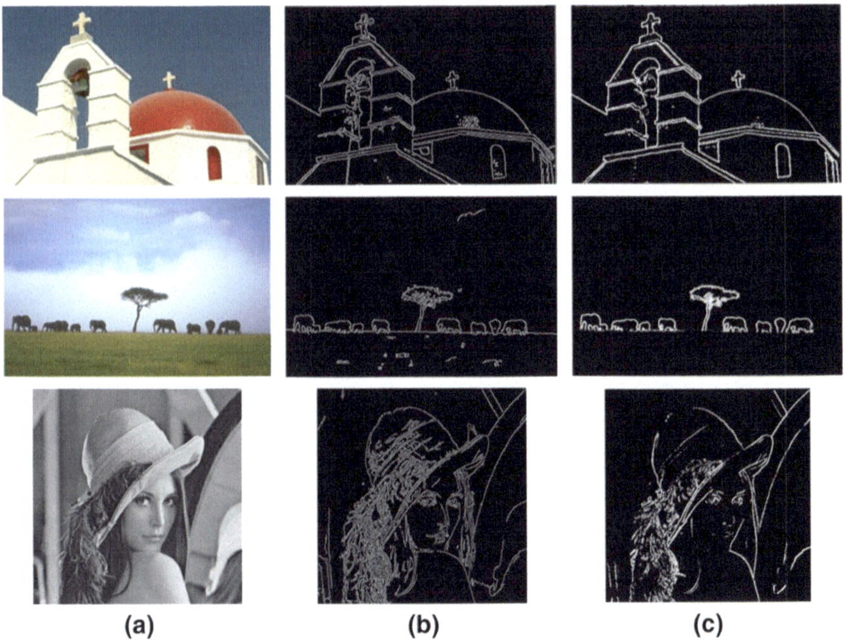

Fig. 6.12 Detected edges for source image (**a**) by Canny algorithm (**b**) and Sobel algorithm (**c**)

the object. However, for small-scale objects just like the elephants in Fig. 6.13, the edges generated by HED are rough and large while those generated by the proposed model are more delicate.

Furthermore, we analyze the influence of different path lengths and iterations on the edge detection results. As shown in Fig. 6.14, the detected edges are coarse, and

Fig. 6.13 Detected edges by HED (a) and the proposed method (b)

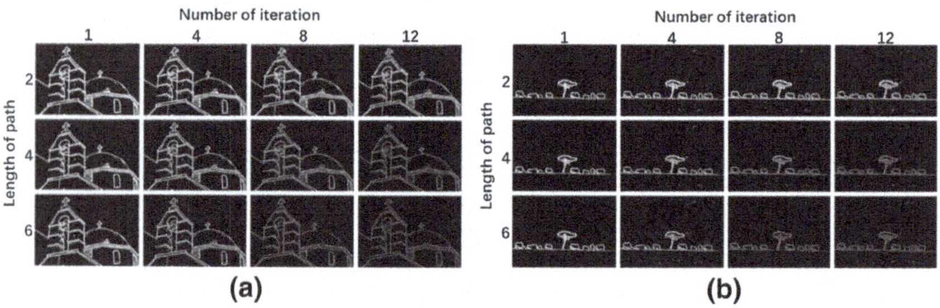

Fig. 6.14 Various results obtained by different iterations and lengths of path

even the close edges are merged together when the number of iterations and the lengths are small. With the increment of the iterations or the length, the detected edges become more delicate. However, due to the ceiling of the conductance, the memristor will be saturated when the number of iterations is too large. This will inadvertently make the edges discontinuous and blurry.

In Fig. 6.15, FoM [32], regarding the results of the Canny algorithm as actual edge images, is utilized to compare the results of the Sobel operator and the proposed method. When the length of the path and the number of iterations is appropriate, the FoM of the proposed model will be higher than the corresponding result of the Sobel operator which is the best result with different thresholds (2, 4, 8, 16, 32, 64, 128, 192).

We also analyze the change of memristive conductance of some memristors, which is shown in Fig. 6.16. For clarity, we use the value of memristive conductance minus the mean of all conductance of the memristor crossbar array as the Y-axis. The result shows that the gap between the conductance of the edge pixel and non-edge pixel will widen with the increasing of iterations. However, the edge pixel may be regarded as the non-edge pixel in some cases.

As shown in Fig. 6.17, x will eventually converge to one for both edge pixel and non-edge pixel, which indicates that the memristor is saturated. int the drop of FoM values in Fig. 6.15 and the blurry results in Fig. 6.14.

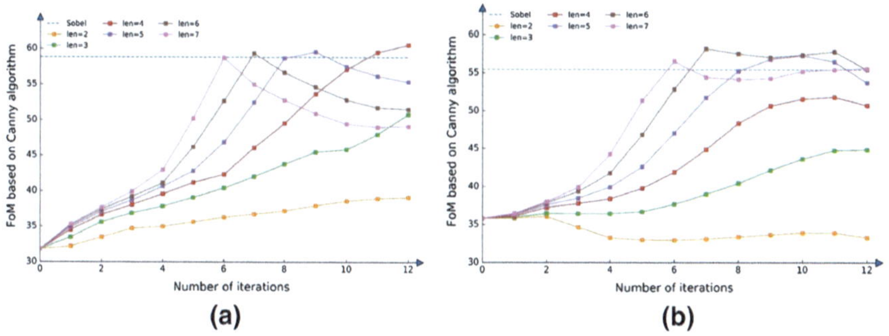

Fig. 6.15 Comparison of FoM between the proposed model with different path lengths and the Sobel algorithm

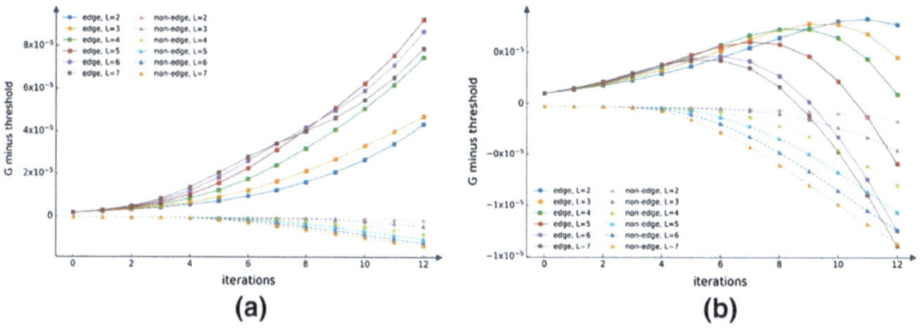

Fig. 6.16 An edge pixel (solid lines) and a non-edge pixel (dotted lines) are randomly selected. The value of the memristive conductance minus the mean of all conductance of the memristor cross bar array is analyzed. (a) Result for two pixels with a larger initial value difference (b) Result for two pixels with a smaller initial value difference

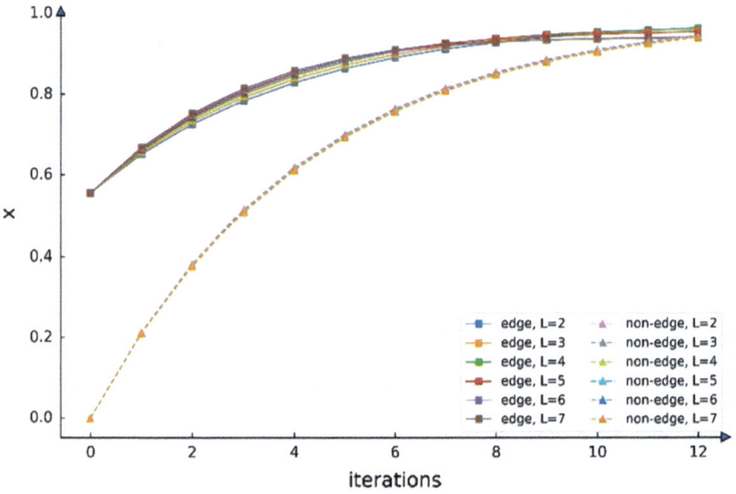

Fig. 6.17 The change of the internal variable x for an edge pixel (solid lines) and a non-edge pixel (dotted lines) under different lengths and iterations

6.1.6 *Conclusion*

In conclusion, the usage of a memristor crossbar array for image edge detection based on ACO is proposed. By using a nonlinear memristor model with a relaxation term, both analytical and numerical similarity between ACO and memristor circuit has been shown. The ACO is simplified for implementing it to the memristor network. Then, a memristor crossbar array is used for high device density. The results of the simulation show that delicate edges of the image are obtained by this memristor crossbar array. Furthermore, both the storing and computing abilities of memristor endow it with the potential for many applications. Furthermore, the use of memristors for other swarm intelligence is worth exploring, such as genetic algorithms, fish swarm algorithms, and novel swarm intelligence algorithms.

6.2 Memristive Crossbar Array-Based Genetic Algorithm

6.2.1 *Introduction*

This section explores the intersection of memristive networks (MNs) and genetic algorithms (GAs), two seemingly distinct yet potentially complementary approaches in computational intelligence. As units capable of both storing and computing information, memristors have paved the way for mem-computing, a novel paradigm inspired by the brain's architecture of interacting memory elements [14]. This approach has shown promise in various applications, including reservoir computing [33], sparse coding [12], and image processing [15].

The primary motivation of this research is to conduct a precise comparison between mem-computing and genetic algorithms, elucidating their analogies and differences. The aim is to bridge the gap between these two computational paradigms and explore potential synergies.

The focus lies in designing memristive networks that implement mem-computing principles within the framework of biologically inspired, nature-based meta-heuristic algorithms, specifically genetic algorithms. The result of this integration is the Memristive Network-based Genetic Algorithm (MNGA), which represents a significant innovation in the field.

A key advantage of the MNGA is its ability to efficiently compute and store fitness values, a crucial aspect of genetic algorithms. This capability marks a meaningful advancement over traditional software-based implementations, potentially offering improvements in speed and efficiency. To demonstrate and validate the effectiveness of the proposed MNGA, an application in image edge detection is presented. This application treats edge detection as an optimization problem, showcasing the versatility and practical utility of this approach.

By combining the adaptive, parallel processing capabilities of memristive networks with the robust optimization framework of genetic algorithms, this research aims to open new avenues in computational intelligence. The MNGA has the potential to address complex optimization problems more efficiently, particularly in scenarios where hardware implementation and real-time processing are crucial.

This section not only contributes to the theoretical understanding of mem-computing and genetic algorithms but also provides a practical framework for their integration, potentially impacting fields ranging from image processing to optimization in various domains.

6.2.2 Computing Model

6.2.3 GA

Nature-inspired meta-heuristic algorithms [14, 15, 34–48] focus on optimization problems by mimicking biological or physical phenomena, which can be classified into human-based algorithms, physics-based algorithms, swarm-based algorithms [14, 15, 34, 35], and evolutionary algorithms. Evolutionary algorithms inspired by natural evolution, especially the genetic algorithm (GA), are of importance to solve optimization problems. GA inspired by the genetic process of biological organisms, was developed by Holland [37]. Recently, increasing attention to GA has been paid by researchers and efforts have been made to improve GA for wide applications [38–48]. Especially, an effective genetic first-order statistical image segmentation algorithm was proposed in [47], which is the research field that we focus on. The GA premises

Table 6.2 Comparison of the proposed method with [14, 15]

Symbol	Meaning
G	Generations
P	Population
B	Individual
C	Chromosome
gl	Gene list
$S_r(b)$	Rule of selection
$C_r(b_1, b_2)$	Rule of crossover
$B_r(b)$	Rule of biomutation
$FF(b)$	Fitness function

that there are a certain number of individuals named "population" which perform biological behaviors in ways of natural selection, biomutation, and gene crossover. Every individual has a gene list and processes fitness based on its gene list. Fitness is the key to selection which eliminates low-level fitness individuals with high probabilities and high-level individuals with low probabilities. With the evolution of performance, more and more high-level individuals remain and the global fitness of the population can be promoted gradually. Finally, the optimal solution is obtained.

To simulate the GA, parameters need to be initialized as shown in Table 6.2. Figure 6.18 shows a common GA process which can be implemented with selected $S_r(b)$, $C_r(b_1, b_2)$, and $B_r(b)$.

Step 1 Initialize the MN and parameters of the GA and the fitness of the population can be calculated.

Step 2 If the algorithm is not terminated, $S_r(b)$ disposes individuals b one by one. If not, go to Step 6.

Step 3 After choosing two individuals b_1 and b_2 under the rule of S_r, $C_r(b_1, b_2)$ is completed to obtain a new individual.

Step 4 The process $B_r(b)$ is implemented to decide whether individuals b are variable or not.

Step 5 After processes $S_r(b)$, $C_r(b_1, b_2)$, and $B_r(b)$, the new population is obtained. Then, the algorithm goes back to Step 2.

Step 6 The final population is the result of the GA.

6.2.3.1 MN

In this section, MN is discussed. An MN includes basic elements (the memristor, the memcapacitor, the meminductor, and their emulators). It is a processor used to solve the shortest path problems and builds neural networks. Recent research [49] summarized that the type and functionality of memristive emulators are the key to network dynamics.

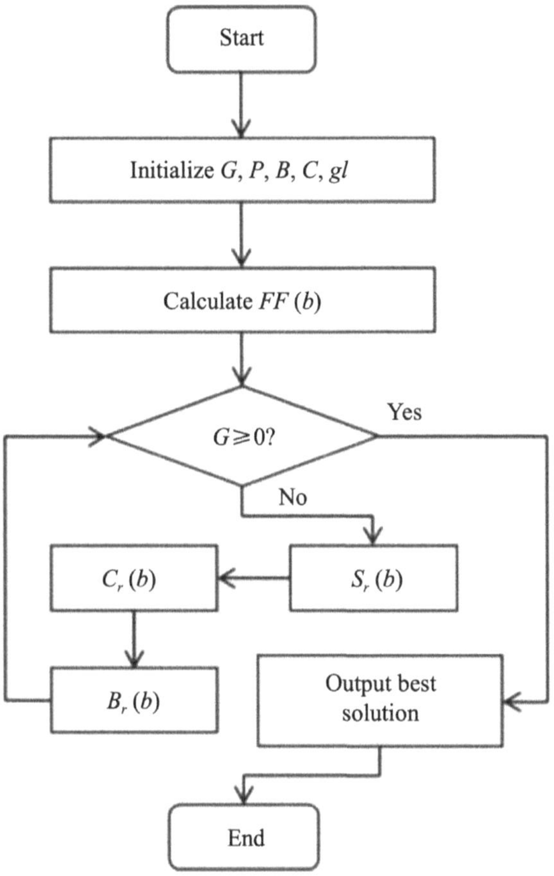

Fig. 6.18 Initialization and genetic operation of the GA process

Inspired by [14, 15], we find that there are still other traditional computing models which can be transformed with MN. Based on the structural, computing, and memory features of the MN, the MNGA is implemented to detect the edge information of images. If the memristive units like memristors are used to map the individuals of population, the MN is very suitable for GA which has a settled population. We also need a way to store the information of individuals for genetic processing. Finally, the MN can change the states of units expediently, which can map the changes of individuals naturally.

6.2.3.2 MNGA

Image segmentation can be classified into threshold-based methods, region-growing methods, hybrid methods, and edge detection methods. It aims to separate the desired foreground object from the background [47, 50]. By algorithm implementation and analysis [51, 52], we find that the process of detecting edges is very similar to the evolution in GA, where edges in images are seen as high-level fitness individuals in the population. In light of this design, we have implemented a new image edge detection method based on MN-based GA (MNGA) to obtain the edge information from gray images. The analogy and relation between the GA and the MN to solve optimization problems are then complete. To illustrate the mapping in Table 6.3, we assume an optimization problem $OP(x, y, z)$ where we find its maximum value with x, y and z changing from 0 to 10. Firstly, every value group $[x, y, z]$ for $OP(x, y, z)$ is a feasible solution while that for the MN is a memristive subnet (which has three memristors in this case). A single memristance for the memristor maps a single value coding of x, y or z for a feasible solution. Incidentally, in different GA models, the codings of the feasible solution are various such as binary coding, and floating coding, which require different memristor models. Finally, when $S_r(b)$, $C_r(b_1, b_2)$, and $B_r(b)$ are initialized, those approaches of solution in the GA maps the mem-computing approaches in Table 6.3. After this general discussion, we provide explicit applications based on MNGA.

The representation of chromosomes by memristors is shown in Fig. 6.19. As we can see, there are six genes and each of them is coded by floating-point code. The memristors (denoted by M_i, $i = 1, 2, ..., 6$) representation contains six genes and the value of each memristance is proportional to the corresponding gene. Furthermore, each memristor is connected to a current source which allows us to change the memristance independently and parallelly.

The mapping operators of the chromosome and the memristive network are shown in Table 6.3. In the proposed method, the memristor is mainly used as a memory to read and write gene values. The selection operator is implemented by reading the memristance of memristors and the selection rule is run on memristances for selecting memristors. The biomutation operator can randomly select some memristors and change memristances randomly. Finally, the crossover operation is performed by reading and exchanging the memristance of the corresponding memristors. Compared with traditional methods, the parallel gene update of each operator can be realized by memristor deployment.

Table 6.3 Mapping rules of GA and mem-computing

Nature	GA	Mem-computing
Population	Some feasible solutions	Memristive network
Individual	Feasible solution	Memristive subnet
Chromosome	Part of feasible solution	Memristor
Gene	Feasible solution coding	Memristance
Natural selection	Solution selection	Memristor selection
Biomutation	Solution-coding change	Memristance change
Crossover	Addition of individual	Addition of memristor

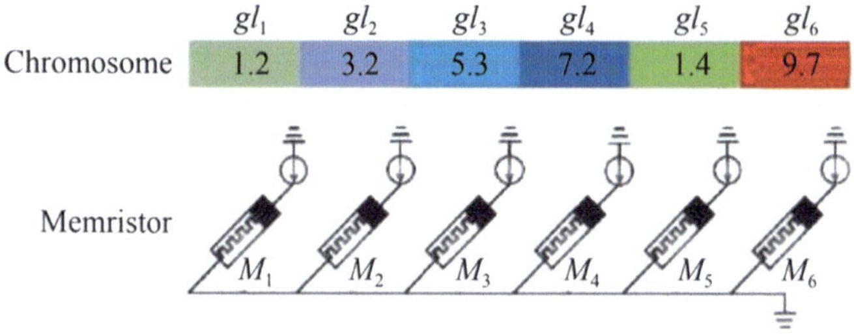

Fig. 6.19 Sample for implementing chromosome containing six genes by six memristors

6.2.4 Edge Detection Based on MNGA

Detection of significant changes in an image is called edge detection. In this section, an MNGA-based edge detection algorithm is presented. Before introducing this algorithm, two design obstacles need to be considered. Firstly, in our early attempts, we find that the detection result is ineffective because of the noise in images. Secondly, it is difficult to design the mapping between an individual's fitness and an image's edge information. For the first obstacle, the filtering process which can suppress the white Gaussian noise is adopted. Furthermore, gray images are chosen for simulation to verify the feasibility of the algorithm. For the second obstacle, we design a fitness function to quantitatively analyse every pixel, which means that the more similar one pixel is to the edge pixel, the higher level fitness this pixel has. After solving these two obstacles, the MNGA-based edge detection algorithm can be proposed.

The MNGA for image edge detection contains three steps. The first step is filtering. In order to properly detect the edges, there is a need to cancel local random variations caused by noise. The next step is called preprocessing which consists of computing the fitness of all pixels. The final step is edge detection where the new algorithm performs. In edge detection, the image is treated as the population and every pixel has its fitness which is calculated by $FF(x, y)$. Furthermore, an MN is used to express whether the corresponding pixel is an edge or not. With the GA performing, the low-level fitness pixels are eliminated while the corresponding memristor's memristance is installed to 0. Finally, at the end of the GA, the MN describes the edge information.

6.2.4.1 Filtering

To remove the impact caused by noise and promote the property of image edge detection, filtering is adopted.

6.2.4.2 GA Operator Based on Memristive Network

In this step, several preprocessing approaches are done. Firstly, the 468×468 image Lena is presented (see Fig. 6.20), and the same sized MN needs to be completed where all memristors are two-value type memristors (for simplicity, memristance = 0 or 1) to map the corresponding pixel. If memristance is 1, this pixel is an edge pixel, otherwise, it is not an edge pixel. Then all memristor's memristance is initialized to 1 which shows the corresponding pixel is an edge, and the fitness of pixel is presented as follows:

$$FF(x, y) = \sum_{\substack{i=-1,0,1 \\ j=-1,0,1}} |P(x, y) - P(x+i, y+j)| \tag{6.16}$$

where $P(x, y)$ denotes the xth row and yth column pixel value in the image. $FF(x, y)$ denotes the fitness of pixel (x, y). We use (6.16) which, is simple to compute and has high performance. From (6.16), we know that the more different a pixel (x, y) is with its nearby pixels, the more higher $FF(x, y)$ is. A C_c ording to this rule, it is easy to distinguish the edge pixels with the others.

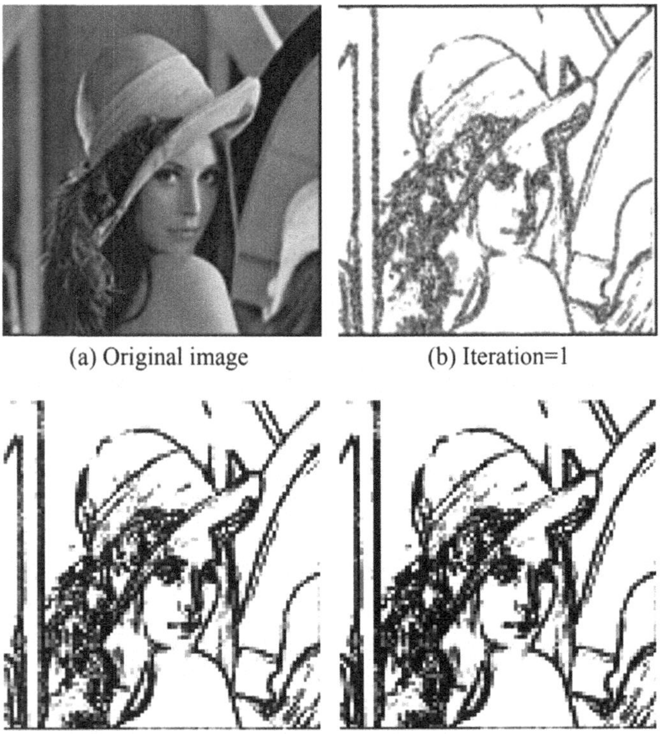

Fig. 6.20 Comparison of the quality of the edges detected from the Lena image

Remark Equation (6.16) calculates the difference in pixel values between a pixel and its neighbors. In general, the greater the difference is, the more likely it is to be an edge pixel. Spikes and noise are often isolated points eliminated by mutation during population iteration. In this case, more and more pixels remaining in the population are edges.

6.2.4.3 Edge Detection

After the previous two processes, the model of the MNGA-based edge detection model can be presented as follows. The indispensable parameters and their meanings are shown in Table 6.4. The parameters need to be adjusted in Table 6.4 including R_c, R_b, F_m, C_c. All parameters are adjusted several times to select the optimal ones. Furthermore, the selection processes of F_m and C_c are illustrated in Figs. 6.23 and 6.24 respectively.

Step 1 Load the $m \times n$ pixels image to a matrix $\mathbf{P}(x, y)$, the pixel value is from 0 to 255, and prepare a same sized memristive network $MN(x, y)$ where memristors are two-valued and $MN(x, y) = 0$ or 1. x and y denote the coordinate information and their ranges are $[1, m]$ and $[1, n]$.

Step 2 Randomly choose 50% memristors of the MN and set their memristance as 1 and the others as 0. Meanwhile the image is processed by the Gaussian-filter whose σ is 1.6.

Step 3 Count all pixels' fitness $FF(x, y)$. Set up parameters as shown in Table 6.2.

Step 4 Perform the selection process $S_r(b)$, where b is an elect pixel which will be eliminated if its fitness is lower than F_m.

Step 5 Perform the crossover processing $C_r(b_1, b_2)$, where b_1 and b_2 are two elect pixels and a new pixel will generate if the condition R_c is satisfied.

Step 6 Perform the biomutation processing $B_r(b)$, where b is an elect pixel.

Table 6.4 Parameters of MNGA-based edge detection

Symbol	Value	Meaning
R_c	0.4s	Rate of crossover
R_b	0.05	Rate of biomutation
F_m	Mean value	Threshold value of fitness
$w\dot{h}$	468 × 468	Resolution of image
$MN(x, y)$	0 or 1	Memristance
$P(x, y)$	0–255	Pixel value
$F(x, y)$	Calculated by fitness function	Fitness of $P(x, y)$ and $MN(x, y)$
C_c	4	The neighborhood size of crossover processing
Mean	33.6678	Mean value of fitness

Step 7 Perform Steps 4–6, if the number of generations by Step 3 is larger than the number of eliminations by Step 6.

Step 8 The MN becomes the edge information of the image, with 1 denoting the edge.

6.2.5 *Simulation Results*

6.2.5.1 Experiment and Analysis

By initializing parameters shown in Table 6.4, the simulation results are presented in Figs. 6.20 and 6.21. The edge information is picked out gradually. In this case, the algorithm is convergent at the 39th generation, as shown in Fig. 6.21.

In order to properly detect the edges, there is a need to cancel local random variations caused by noise. The detection result without median filtering is presented in Fig. 6.22c where there is much residual noise compared with Fig. 6.22d. By comparing Fig. 6.22c and d, we can easily know that the filtering plays an important role in the detection algorithm at the same experimental conditions.

In this case, the time complexity $T(n)$ of our detection algorithm can be obtained as follows:

$$T(n) = L \times O\left(T_{S_r}(n) + T_{C_r}(n) + T_{B_r}(n)\right) \tag{6.17}$$

where n denotes the number of pixels, $T_{S_r} - (n)$ denotes the time complexity of $S_r(b)$, $T_{C_r}(n)$ denotes the time complexity of $C_r(n)$, $T_{B_r}(n)$ denotes the time complexity of $B_r(b)$, and L is iteration which is independent of n. Firstly, $S_r(b)$ will iterate over the population. Then, the crossover process is determined by R_c. Finally, R_b affects $T_{B_r}(n)$. Based on Table 6.4, the calculation of $T_{S_r}(n)$, $T_{C_r}(n)$, and $T_{B_r}(n)$ is provided as follows:

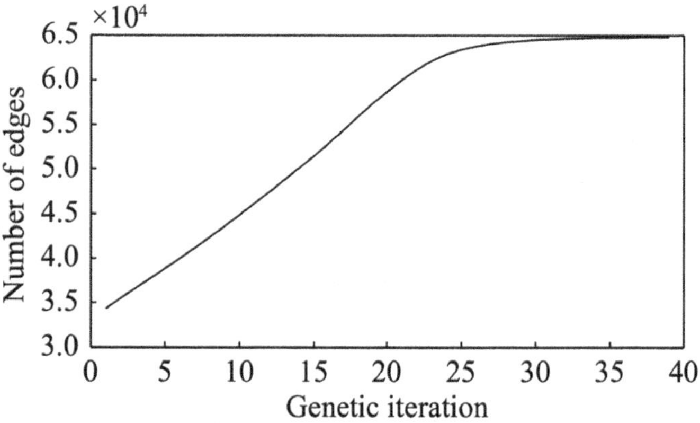

Fig. 6.21 Convergence rate of GA

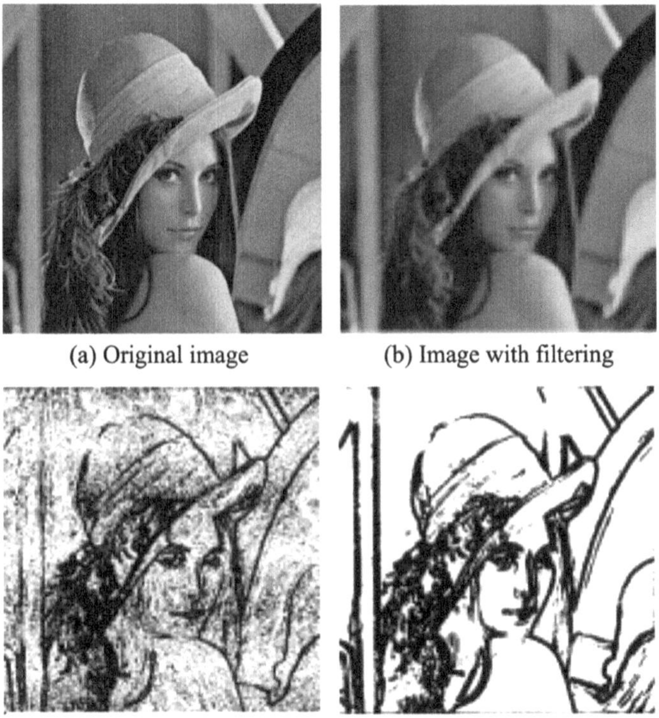

Fig. 6.22 Comparison of the edge detection quality with and without filtering conditions

$$\begin{cases} T_{S_r}(n) = O(n) \\ T_{C_r}(n) = O\left[R_c \cdot (15n)\right] = O\left[0.4 \cdot (15n)\right] = O(n) \\ T_{B_r}(n) = O(R_b \cdot n) = O(0.05 \cdot n) = O(n) \end{cases} \tag{6.18}$$

Consider (6.17) and (6.18) together and we get the time complexity $T(n)$ is equal to $O(n)$. The space complexity $S(n)$ consists of the storage of image, the storage of fitness, and the storage of MN. After introducing the MNGA and the edge detection algorithm, we know that $S(n)$ is $O(n)$.

To compare the quality with different F_m, Fig. 6.23a and b show the experimental results which are compared with the results as shown in Fig. 6.20d. In Fig. 6.23a, the value of F_m is initialized to 0.5 mean, which retains more edges than Fig. 6.23b. With F_m increasing, fewer edges are selected but more noise filtered. Meanwhile, the detailed relationship between F_m and the number of edges is presented in Fig. 6.24. We know that the result in Fig. 6.20d is a compromising choice to balance the discrete degree of edges and the number of details, which means our MNGA edge detection can perform various effects as differently required. To demonstrate this point, Figs. 6.24 and 6.25 vividly present the changes of the number of edges with different parameters. Three different values of F_m are employed to reflect the influ-

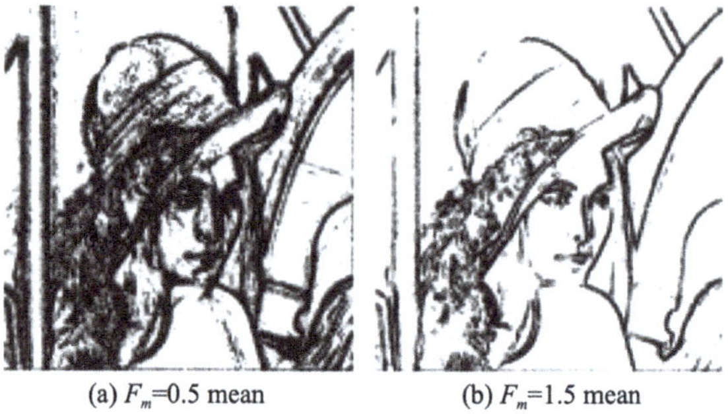

Fig. 6.23 Comparison of the quality of the edges detected from Lena image shown in Fig. 6.19a with different parameters

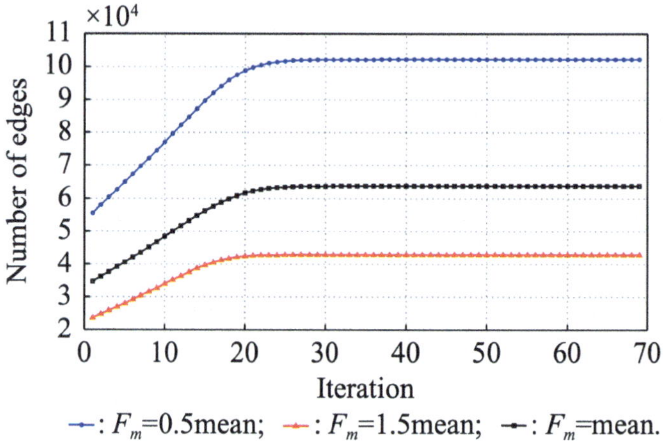

Fig. 6.24 Impact of F_m

ence of F_m. The number of edges is proportional to F_m. Three different values of C_c are employed to reflect the influence of C_c. The number of edges is proportional to C_c. The other key parameter is C_c which defines a square window to limit the range of crossover between two individuals. If C_c is equal to 4, two individuals can reproduce a new individual only when they are in the window with a diameter of 4. Suitable size of C_c can effectively eliminate noise and retain serial edges, which can be proved in Fig. 6.25. With C_c increasing from 2 to 10, the number of edges is maximal when C_c is 4.

6.2.5.2 Contrastive Analysis

In this subsection, a contrastive analysis is presented. Figures 6.26 and 6.27 show the edge detection results, where Figs. 6.26a and 6.27a are the original images. Based

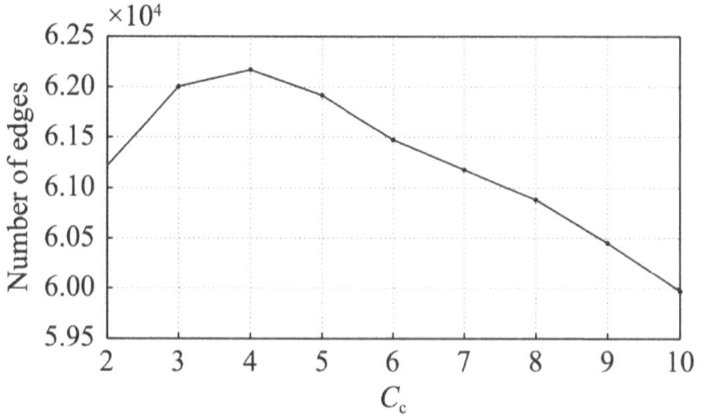

Fig. 6.25 Impact of C_c

Fig. 6.26 Comparison of the Lena image edge detection with that in [15]

on those two comparisons, we find that our MNGA can obtain more details than the algorithm in [15]. Meanwhile, in [15], there are many discrete edges which should be continuous originally. In our result, more continuous edges are retained than those in [15].

Figure 6.28 shows the FoM [53, 54] based on the correlation between results obtained by the edge detector and the results obtained by the Canny algorithm. After only one iteration, our result is proved to be higher than the best result from [15].

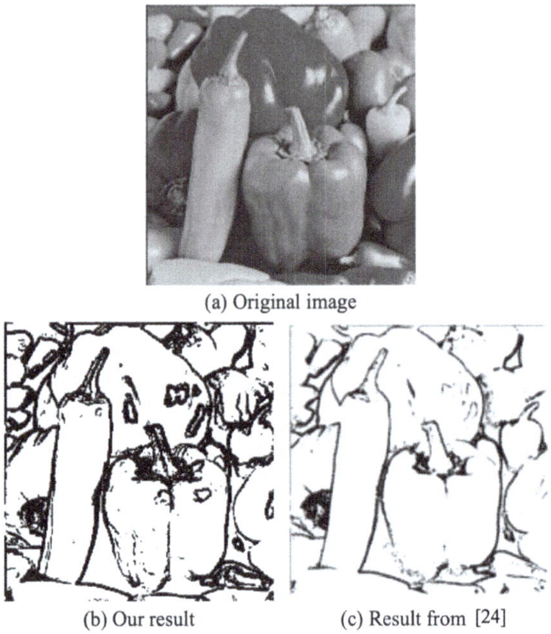

Fig. 6.27 Comparison of the Pepper image edge detection with that in [15]

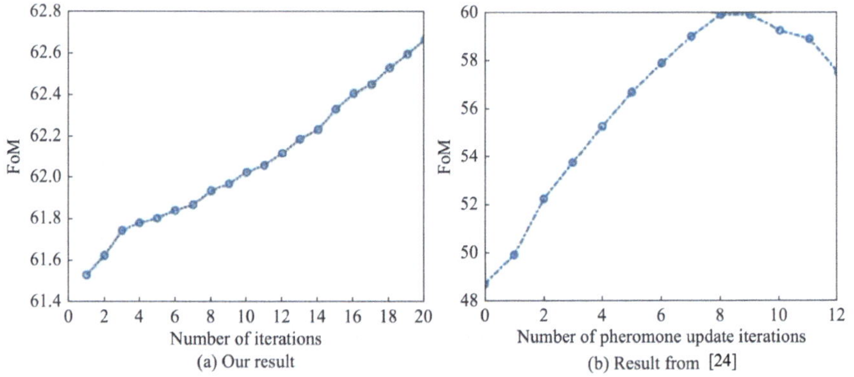

Fig. 6.28 Comparison of the FoM on Pepper image with that in [15]

Figure 6.29 shows the change of the mean and the standard deviation of fitness with iteration. Firstly, it can be found that the mean and the standard deviation of edge pixels are both higher than that of the non-edge pixels. This shows that the fitness of the edge pixels is often higher than that of the non-edge pixels. However, because the standard deviation is large enough, the edge pixels are not concentrated on the high-fitness process.

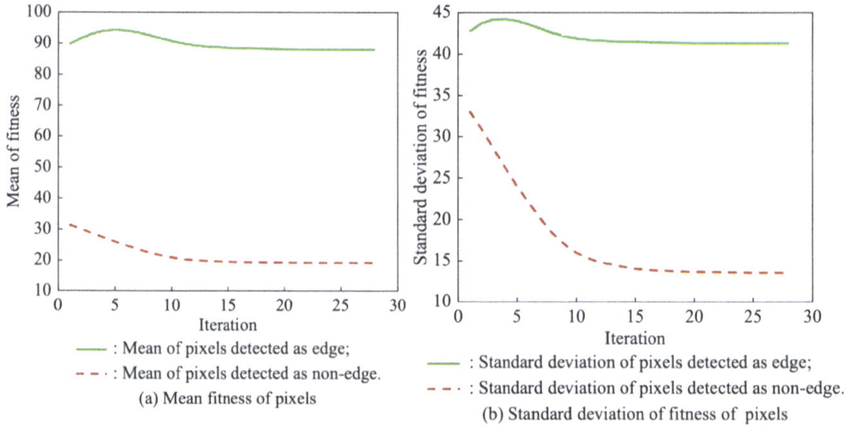

Fig. 6.29 Change of mean and standard deviation of fitness with iteration

6.2.6 Conclusions

In this chapter, a new MNGA is proposed. In light of the computing and memory features of the MN, the relationship between the MN and the GA is built. After comparing the MN and the GA system structures, self-feedback, and storage of unit information, we find that the GA processes high suitability level with MN. With this mapping implemented, an edge detection algorithm is designed for images, where the pixels of images are treated as individuals of a population (also the units of the MN). Experimental results show the high quality of our algorithm compared with a similar computing model. Furthermore, we discuss some parameters of the MNGA which can impact the detection result, thus more experiments to confirm the most suitable parameters are required. This design is a novel point to explore the possibility of MN-based biological intelligence algorithms. There are other algorithms like GA that can be implemented in the similar way.

These preliminary results advocate further investigation of the proposed design method in the future. For the next step, we would like to compare more other well-known related algorithms regarding detection quality. Then, the adaptation and self-adaptation of those parameters studied in this chapter should be worthy of study. Also, more other MN-based nature-inspired meta-heuristic algorithms can be designed to form a unified system.

6.3 Memristive Crossbar Array-Based Artificial Fish Swarm Algorithm

6.3.1 Introduction

This section explores the relationship between swarm intelligence, specifically the Artificial Fish Swarm Algorithm (AFSA), and memristive circuits, a novel concept in the field of neural networks. Building upon previous research that has established connections between Particle Swarm Optimization (PSO) and memristive circuits [55], this investigation aims to uncover and analyze the potential synergies between AFSA and memristive networks.

AFSA is a nature-inspired optimization algorithm that simulates the social behaviors of fish schools. It encompasses four primary behaviors: prey, follow, swarm, and move [56]. These behaviors collectively enable the algorithm to navigate complex solution spaces and converge on optimal solutions. The core insight driving this research is the recognition that memristors, with their unique ability to store and process information, can be utilized to model and implement the fundamental mechanisms of AFSA. Specifically, memristors can be employed to record and update the positions of artificial fish within the algorithm, providing a hardware-based approach to swarm intelligence optimization.

This section delves into the intricacies of both AFSA and memristive circuits, establishing a clear mapping between the algorithmic components of AFSA and the electrical characteristics of memristive networks. By doing so, it aims to demonstrate how the adaptive and memory properties of memristors can be harnessed to create efficient, hardware-implemented versions of swarm intelligence algorithms. The investigation will cover the design and analysis of memristive circuits that correspond to each of the four AFSA behaviors. Particular attention will be given to how these circuits can replicate the decision-making processes and movement patterns of artificial fish in the algorithm.

Furthermore, this section explores the potential advantages of implementing AFSA using memristive networks, including increased processing speed, reduced power consumption, and the possibility of parallel computation. These benefits could make memristive AFSA implementations particularly suitable for optimization problems in real-time systems and resource-constrained environments.

By bridging the gap between swarm intelligence algorithms and neuromorphic computing, this research contributes to the broader field of bio-inspired computing and opens up new avenues for the application of memristive technologies in optimization and decision-making systems.

6.3.2 *Artificial Fish Swarm Algorithm*

Artificial fish swarm algorithm is a kind of swarm intelligence that simulates the fundamental behaviors of artificial fish. The concept of the algorithm is that in the area of the water, the place where the greatest number of fish has the most nutrients distributed [55–57]. Based on this concept, the basic model of AFSA can be constructed, different behaviors can be realized, and optimization problems can be solved by using this algorithm. For the basic model, Let X be the current state of artificial fish, record the location of the number of n fishes as $(x_1, x_2, \ldots, x_n)$, let visual be the radius of the visual domain of each single fish, and X_υ be the set of target points of every fish within the visual domain, recorded as $(x_{1\upsilon}, x_{2\upsilon}, \ldots, x_{n\upsilon})$, let X_{next} be the next state of each member of fish. If the target situation is better than the current situation, the fish will step forward to the target location for a random distance (Fig. 6.30). The basic model can be expressed in the following formulas:

$$X_\upsilon = X + visual * rand() \tag{6.19}$$

$$X_{next} = X + \frac{X_\upsilon - X}{|X_\upsilon - X|} * step * rand() \tag{6.20}$$

Equation (6.19) above expresses the target location of each fish, $rand()$ is the random number between 0 and 1. The expression can control the target point within vision domain. Equation (6.20) shows the formula of the next state of artificial fish, step means the distance of the motion of artificial fish every time. Based on the basic model of AFSA, the behavior of artificial fish can be divided into four kinds of behavior, they are prey behavior, follow behavior, swarm behavior, and move behavior. These four behaviors obey the basic model, but the variables have different meanings under different scenarios.

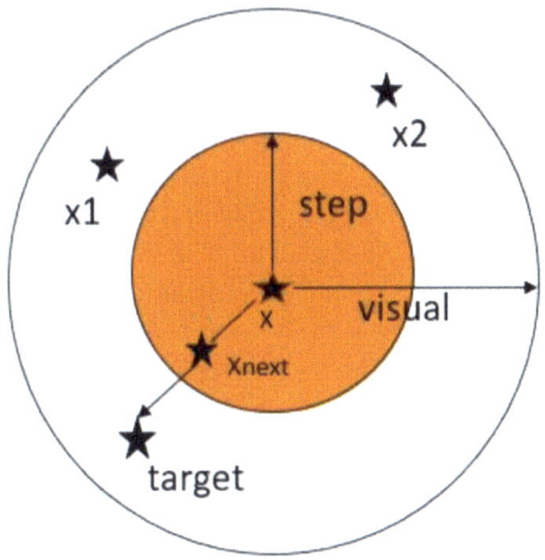

Fig. 6.30 The basic model of AFSA

Prey behavior is a kind of food-seeking behavior where each artificial fish in the group selects the target point. They know the food concentration of the current location, and assess the food concentration of the target point, if the food concentration of the target point is higher than that of the current location, it will move forward to the target location in a certain distance, otherwise, they choose the target point again. Let Y_c be the concentration of the current location, Y_t be the concentration of target location. The algorithm can be expressed in the following [56]:

$$\text{If } Y_t > Y_c$$
$$X_i^{t+1} = X^t{}_i + \frac{X_{iv} - X^{t+1}{}_i}{|X_{iv} - X^{t+1}{}_i|} * step * rand() \tag{6.21}$$

The equation above shows the motion of a single artificial fish under the condition that the concentration of the target location is higher, i is the number between 0 and n. If $Y_t < Y_c$, the fish will search for the target again. Typically, there is a limit on the number of choosing the target location. If the trying number exceeds the limit, the fish will choose the point randomly and move forward at a certain distance (move behavior).

Swarm behavior is the action of swarming to guarantee the colony and avoid danger. Like prey behavior, the current state of single artificial fish is X_i, the X_c is the center of companions of fish within a visual domain (target point), δ is the crowd factor of fish. If the concentration of the target point is greater than the current location, and the target companions are not crowded, the fish will step forward to the center of companions. Otherwise, the fish will continue searching for target points (prey behavior). The algorithm can be shown below [56]:

$$\text{If } Y_t > Y_c \text{ and } \frac{nf}{n} < \delta, then$$
$$X^{t+1}{}_i = X^t{}_i + \frac{X_c - X^{t+1}{}_i}{|X_c - X^{t+1}{}_i|} * step * rand() \tag{6.22}$$

The swarm behavior is similar to prey behavior, the main difference is the meaning of the target point, the target point of prey behavior is the single location, and the target point of swarming behavior is the center of the companion. Fishes also consider the scale of the group companion. The crowd factor can limit the size of the group and make fish aggregate in optimal locations in the whole area.

Following the behavior of artificial fish is the action of AF to follow the companion which has a higher food concentration. The computation of follow behavior is nearly the same as that of swarming behavior, both considering the concentration of the target point and crowding factor. The only difference is the objective of the behavior, the swarm behavior is to aggregate together to avoid dander, and the target point is typically stationary, however, the objective of the following behavior is to be optimal and ensure the target point is continuously moving.

Move behavior is a kind of random behavior of artificial fish. In some conditions, the artificial fish must conduct moving behavior. For example, when the fish conduct prey behavior, the number of finding target points exceeds the limit, and the fish will move randomly within the visual domain, the equation of move behavior is shown below.

$$X^{t+1}{}_i = X^t{}_i + visual * rand() \tag{6.23}$$

6.3.3 Memristive Circuits of Afsa

After introducing and explaining different behaviors of AFSA, let us move on to the memristive network. The memristive circuit contains electrical components with memory such as a memristor. In the chapter, we focus on four behaviors mentioned in the last section of AFSA, and the prey behavior of artificial fish will be discussed. Based on the circuit of prey behavior, the other three behaviors can be simulated by modifying the circuit of prey behavior. First, we explain why the memristive circuit is used to establish the AFSA model, then elaborate on the memristive circuit of prey behavior, and show the mapping relationship between the circuit and prey behavior. Next, we discuss how to achieve the other three behaviors by the modifying circuit of prey behavior. Finally, we perform a simulation in MATLAB Simulink in the next section.

Let's first move on to the characteristic of the memristor. The memristor is an electrical component describing the relationship between flux and charge and the value of memristance is continuous and time-varying which makes high precision of storage. On the other hand, because the internal construction of the memristor is changed (solid-state TiO_2 thin films), it won't lose their electrics easily compared with traditional electrical components. What's more, a memristor can considered as a kind of normal electrical component, which means it can be applied in a normal circuit with another traditional electrical component such as a resistor and capacitor.

For the memristive circuit, which is shown in Fig. 6.31, a single layer of the memristive neural circuit and the comparative circuit is proposed. The assumption of this circuit is that the food concentration of the area has the function of a linear relationship ($C = ax + by$). The memristive neural circuit is based on a memristor array and compounded with an inverting operational amplifier comparator and feedback circuit. This single-layer memristive neural circuit is to simulate the current state (location and food concentration). The comparative circuit is used to simulate the state of the target point, and the output of 2 circuits passes through the comparator and op-amp, which is shown in Fig. 6.32. Design in Figs. 6.31 and 6.32 connect to form the final design. The memristor array is distributed in 2 rows and 3 columns in Fig. 6.31. The number of columns is the number of artificial fish, it depends on the sample space of artificial fish. The number of rows is 2 and shows x-axis and y-axis locations respectively.

The value of the memristor is directly related to the current location of each fish. The operational amplifiers above and below the memristor array are used to control

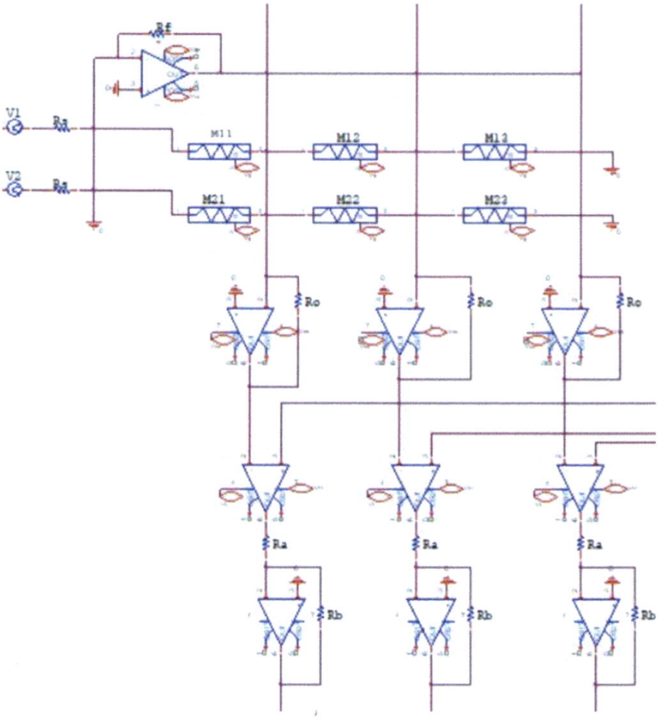

Fig. 6.31 Memristive neural circuit used to simulate AFSA

the voltage. Let the input voltage of the first row be V_{1i} and the second row be V_{2i}. Both input voltages are the square wave. K is the nth iteration (nth trial of each fish). The output voltage gets from each column. Let W be a matrix of the transfer function of all memristors in this circuit, for the circuit in Fig. 6.31, since memristors are arrayed in 2 rows and 3 columns, W is 2×3 matrix. The basic equation of the output voltage at j_{th} column can be expressed as following

$$V_{oj}{}^{(k)} = W_{1j}{}^{(k)} V_1{}^{(k)} + W_{2j}{}^{(k)} V_2{}^{(k)} \tag{6.24}$$

Let's consider the circuit specifically. According to Kirchhoff's law, the voltage V_f can be expressed in (6.23), the jth output voltage can be expressed in (6.24). Let G be the conductance of the resistor and memristor. Equation (6.25) of the output voltage can be obtained and each element in matrix W can be expressed in (6.26).

$$V_f = -\frac{R_f}{R_s}(V_1 + V_2) \tag{6.25}$$

$$V_{oj} = -\left(\frac{R_o}{R_{1j}} V_1 + \frac{R_o}{R_{2j}} V_2 + \frac{R_o}{R_f} V_f\right) \tag{6.26}$$

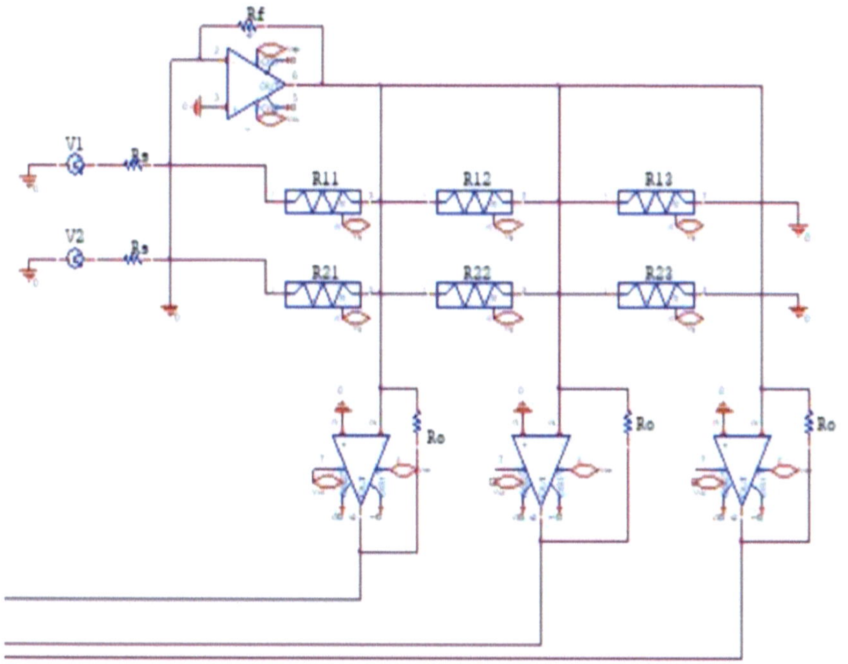

Fig. 6.32 Comparative circuit used to simulate AFSA

$$V_{oj} = R_o(G_s - G_{1j})V_1 + R_o(G_s - G_{1j})V_2 \tag{6.27}$$

$$W_{ji} = R_o(G_s - G_{ij}) \tag{6.28}$$

As previously mentioned, we assume that the food concentration of area is $C = ax + by$, so according to (6.22), the value of W can be thought as the location of fish in ASFA, for example, W_{1j} is the x-axis value of the jth fish, W_{2j} is the y-axis value of the jth fish. What's more, V_{oj} is the value of food concentration of the location of jth fish, on the other hand, the value of two input voltages can be thought as a and b respectively. One important step is to compare the food concentration of the current location and target location, the comparators are proposed to simulate this step.

According to Fig. 6.31, the comparators are in the second layer of the set of amplifiers, and two inputs of the comparator are output voltages of the memristive circuit and comparative circuit. The structure of the comparative circuit is the same as that of the memristive circuit, the only difference is that the comparative circuit uses a variable resistor rather than a memristor. The range of variable resistor is set from R_{on} to R_{off} value of memristor. For the characteristic of the comparator, if the positive input voltage is greater than the input voltage at the negative side, the output voltage will be a higher level voltage, otherwise, it will be a lower level voltage. The output voltage from the comparator will be amplified in the third layer and fed to the

Table 6.5 Mapping relationship between AFSA and memristive circuit

AFSA	Memristive circuit
Artificial fish	Each column of memristor
Current state of fish	Memristive circuit part
Food concentration of current state	Output voltage of memristive circuit
Food concentration of target point	Output voltage of comparative circuit
Location of artificial fish	Value of W
Prey behavior of artificial fish	The change of the value of memristor

input side to change the value of W, so the location of the artificial fish is changed. The specific relation between AFSA prey behavior and memristive neural network circuit is shown in Table 6.5.

After showing the circuit and explaining the mapping relationship of prey behavior of artificial fish, let us move on to the other three behaviors (swarm behavior, follow behavior, and move behavior). The basic mapping relationship of the other three behaviors are same as that of prey behavior, the output voltage is the concentration of food at a specific point, and the value of W determines the position of artificial fish. It is important to specify the difference between prey behavior and other behaviors and modify the corresponding part of a circuit that makes other behaviors different.

For swarm behavior of artificial fish, there are two main differences from prayer behavior. The first difference is that both target and crowding factors determine whether the artificial fish move to the target point. The second difference is that the target point is stationary. Based on these differences, we can modify the circuit in Fig. 6.31. To be specific, one more layer of the comparator can be added between the comparator in layer 2 and the amplifier in layer 3, the negative input voltage side connects to the output of the second layer of the comparator, and the positive input side connects to output voltage comparator which compare crowding factor and constant voltage. On the other hand, because the target position is stationary, we replace the variable resistor by using a traditional resistor in the comparative circuit. For the "Follow" behavior, the only difference is that the target point is not stationary and is the position of a fish, a reasonable way to achieve this behavior is to replace the variable resistor in a comparative circuit by using a memristor but assigned in a different order. For the "Move" behavior, in this chapter, we assume that it is in the same circuit as Fig. 6.31, but in the condition that the output voltage is less than that of the comparative circuit. The modification of the circuit is summarized in Table 6.6.

Table 6.6 Modification of circuit of other behaviors

Swarm behavior	One more layer of comparator, traditional resistor used
Follow behavior	Memristor in comparative circuit used
Move behavior	Same as Fig. 6.31

6.3.4 Simulation

In this chapter, the Pspice is used to construct circuit, and MATLAB Simulink is used to simulate the motion of artificial fish. We simulate the memristor used in circuit, get the value of the memristor at different times and $I - V$ relationships, then explore the motion of single fish, and then simulate 4 fishes and observe their location change. Notice that the value of V_H is 0 V and the value of V_L is −0.9 V, that means that when the voltage output of memristive circuit is greater, there is no voltage feedback to input and the value of input will not change.

6.3.4.1 Simulation of Memristor

To simulate memristor, the simulation of memristor is based on several formulas below [27, 58]

$$R_m = R_{on}x + R_{off}(1 - x) \tag{6.29}$$

$$\frac{dx}{dt} = kf(x)i(t), k = \frac{\mu_v + R_{on}}{D^2} \tag{6.30}$$

$$f(x) = 1 - (2x - 1)^{2p} \tag{6.31}$$

R_on and R_off are bond value of memristor, and $f(x)$ is window function to find internal value x. Setting parameter as following: $R_on = 100$ ohm, $R_off = 16000$ ohm, $p = 10$, $D = 10$ nm, $u_1 = 10^{-14}\,\text{m}^2\text{s}^{-1}\text{V}^{-1}$. First, the sinusoidal signal is added in the simulation of memristor, $V_0 = 1$ V, $f = 1$ Hz, and no phase shift in this wave. As a result, the Lissajous figure of relationship between $v(t)$ and $i(t)$ is obtained, and the result shows a pinched hysteresis loop relationship. In this experiment, the input voltage signal is pulse wave instead of sinusoidal signal. The amplitude of the pulse is 0.9 V and period is 0.01 s, and the width of pulse occupies 50% of the period. The value of the memristor in first 2 s are shown in Fig. 6.33.

6.3.4.2 Simulation of Single Fish

For the memritive circuit, $R_o = 201$ ohm, $R_f = 200$ ohm, $R_S = 199$ ohm. $R_a = 1000$ ohm, $R_b = 1010$ ohm. For the comparator, $VH = 0$ V, $VL = -0.9$ V. The parameters of the circuit of single fish and 4 fish are same. According to (6.26), the range of W is $[-1, 1]$ under this parameter setting. That means artificial fish move in the square domain from $(-1, -1)$ to $(1, 1)$. When the value of the output voltage of the memristive circuit is lower than that of the comparative circuit, a higher level voltage will be output, which means no voltage change in the feedback circuit, otherwise the lower voltage of −0.9 V will be output and pass through the amplifier, resulting in negative voltage change to memristor, and the both value of memristor

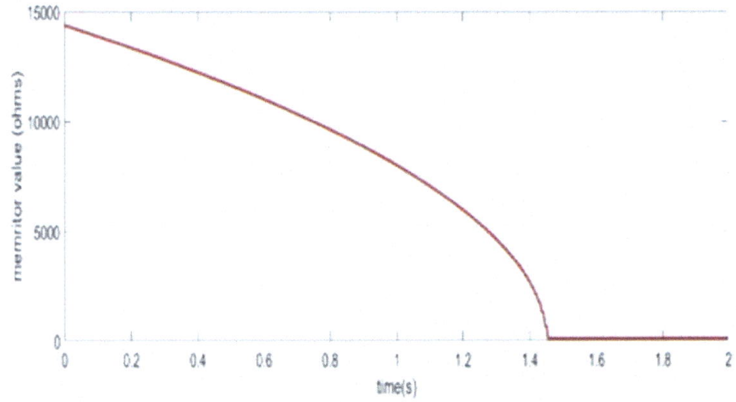

Fig. 6.33 The value of memristor in first 2 s

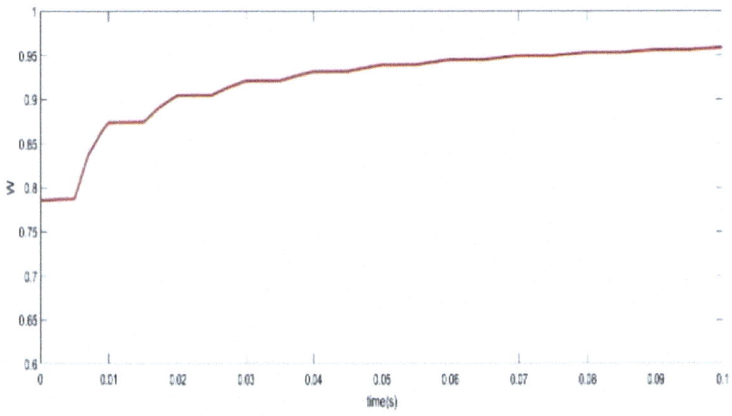

Fig. 6.34 W value of first row of memristor in first 0.1 s

and W are increased. The Value of W of the first and second row in each column and time are shown in Figs. 6.34 and 6.35, respectively.

The value of W in the first 0.1 s are obtained based on the mapping table in the third section. It also represents the changes of the location of fish in the x and y component. Figures 6.34 and 6.35 show the value of W keep increasing and finally approaching 1, which means that the location of fish is moving toward the location of $(1, 1)$, the location with greatest food concentration. Moreover, the value of output change in the first 0.5 s is shown in Fig. 6.36. The figure shows that the output voltage increases and approaches to 1.8 V. That means the food concentration of the final position of fish approaches to 1.8, according to the food concentration $C = ax + by$, the position which has the largest food concentration is $(1, 1)$, so

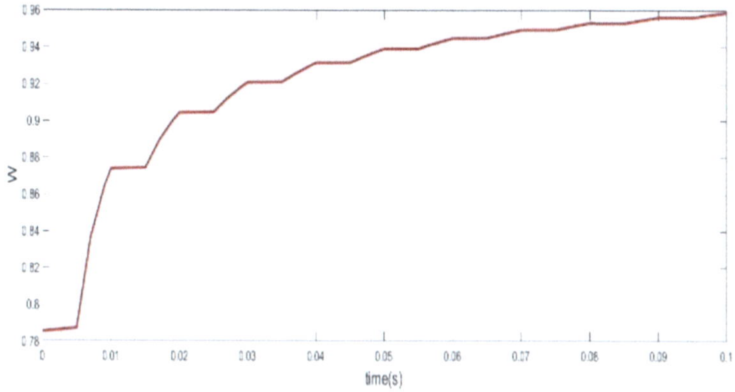

Fig. 6.35 *W* value of second row of memristor in first 0.1 s

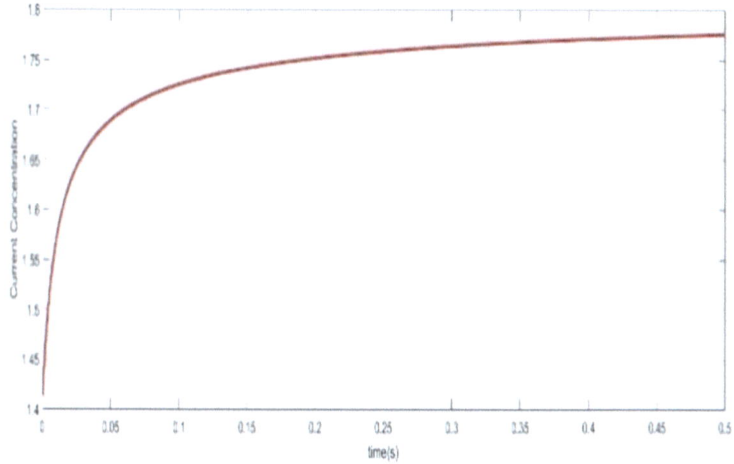

Fig. 6.36 Food concentration in first 0.5 s

$C = a + b$ is the largest concentration. In this simulation experiment, a and b are the input voltage which are 0.9 V, so the largest concentration is 1.8, which makes the result of simulation reasonable.

6.3.4.3 Simulation of 4 Fishes

For simulation of prey behavior of 4 fishes, the parameter of each resistor is similar to that of a single fish. In multiple fish simulation, the starting point is depended on the initial value of each memristor, and the initial value of memristor can be set in the subsystem of memristor in MATLAB Simulink. The result on simulation of the x and y components are shown in Figs. 6.37 and 6.38. In these two figures, the position

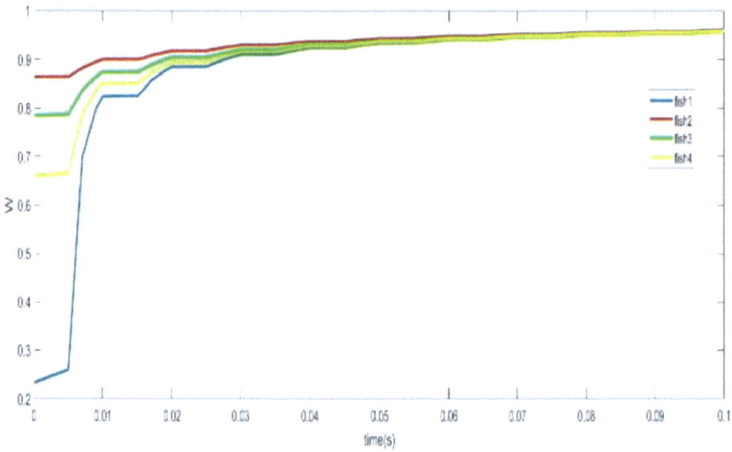

Fig. 6.37 X-axis position of 4 fishes in the first 0.1 s

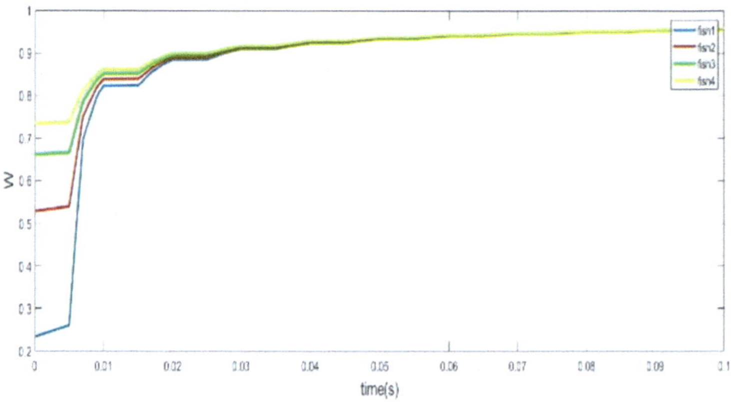

Fig. 6.38 Y-axis position of 4 fishes in the first 0.1 s

of 4 fish approaches the point (1, 1) and aggregate at that point, which simulates the behavior of prey behavior of artificial fish swarm, and the optimal solution is found. Comparing with the prey behavior of AFSA, the trend of artificial fish and the value of W are similar, both approaching to the position which has the largest food concentration. With this method, the number of fish can be increased, and the result will be the same.

6.3.4.4 Simulation of Larger Sample Space

In previous part, the position change of 4 fishes was successfully simulated. What if the sample space is larger? To make this mapping method of memristive network

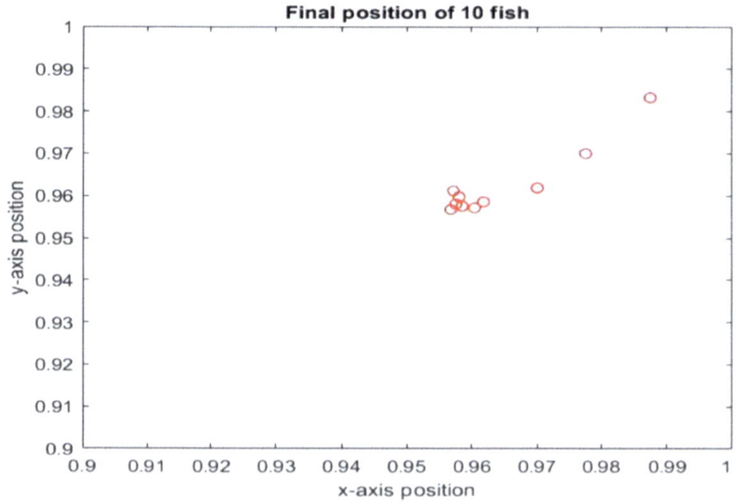

Fig. 6.39 Final position of 10 artificial fish

and AFSA more convincing. Sample spaces of 10 and 50 fishes are used in the simulation. Unlike showing the position change of each fish in the previous part, I will show the final position of each fish after 0.1 s simulation period. The function of food concentration is same as previous part, which means the point (1, 1) has the largest food concentration theoretically. The final positions of 10 and 50 fishes are shown in Figs. 6.39 and 6.40 respectively. According to these two figures, the final position of these two figures aggregates from 0.95 to 0.99 in both x and y axis directions, which means the final position of all sample fish approaches the theoretical maximum of food concentration.

6.3.5 *Application*

In the last part, AFSA is successfully simulated by using the memristive network. In this part, some possible real-life applications will be discussed using AFSA and memristive networks. The basic function of AFSA is to find the global maximum or minimum in one specific area, on the other hand, the important characteristic memristive network is to store memory of previous trends or information. One of the possible applications of this method is a taxi dispatching system.

In real life, the taxi is an important transportation in cities. However, the passengers are not evenly distributed in cities, for example, the Central Business District (CBD) has more passengers in other areas, so the AFSA and memristive network are helpful to taxi companies and taxi drivers to aggregate in places that have more passengers. To be specific, a big city can be divided into several parts, and each part has the place

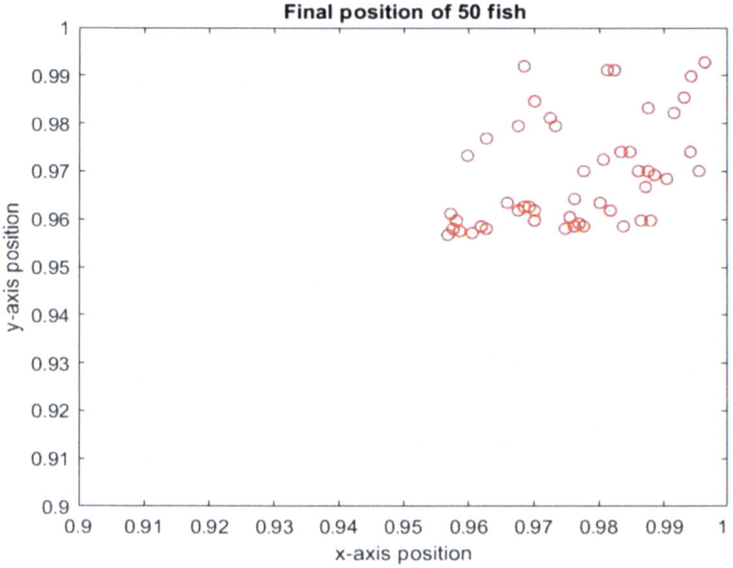

Fig. 6.40 Final position of 50 artificial fish

with the most passengers, which can be mapped to the characteristics of AFSA, the place with the most passengers is like the place with the largest food concentration in AFSA and each taxi is like artificial fish, on the other hand, memristive network and help to store information to determine which place has the most passenger based on historical data, because the value of memristance is time-varing based on trend of voltage. In the last part, we know that the value of the memristor determines the position change of artificial fish. The place of most passengers may change, for example on weekdays CBD area has the most passengers, but on weekends, the entertainment center has the most passengers. The memristive network can help taxi drivers head to the place with the most passengers based on data close to that day. The mapping relationship of the taxi dispatching system can be generalized in Table 6.7.

Table 6.7 Mapping the relationship between AFSA/memristive circuit and taxi dispatching system

AFSA/Memristive network	Taxi dispatching system
Artificial Fish	Taxi
Area	One part of city
Food concentration	Number of passengers
Output voltage of comparative circuit	Number of passengers in target place
The change of the value of memristor	Find the place with most passenger
Value of W	Position of taxi

6.3.6 Conclusion

The results of the simulation show that all the values of W increase as time goes on, which means the x-axis and y-axis components increase. Because we set the pulse value to be positive, the food concentration will increase as the x position and y position increase, which means artificial fish move concentrate to the location that has higher food concentration. The result of the experiment is reasonable in simulating the prey behavior of AFSA. However, there are still some restrictions for this circuit, for example, what if the food concentration is not linear? How to simulate the other three behaviors of AFSA? Further exploration is needed to improve the circuit in order to make it more usable in more scenarios.

6.4 Memristive Crossbar Array-Based Matrix-Friendly Genetic Algorithm for Parallel Computing

6.4.1 Introduction

This section explores an innovative approach to implementing genetic algorithms (GAs) using memristive crossbar arrays, with a focus on enhancing parallel computing capabilities. The research builds upon the growing trend of parallel GA implementations, which have been realized on various platforms such as GPUs [59–61] and FPGAs [62–64]. Recent advancements in cloud and distributed computing have also led to the development of parallel GAs based on platforms like Hadoop [65–67].

In response to the need for higher parallelism and hardware-friendly implementations, this section introduces a matrix-friendly GA (MGA). The MGA approach reformulates GA operators as matrix operations, offering two key advantages. First, it decouples the core operations of GA operators from their strategic implementations, providing greater flexibility. Second, it aligns with the efficient parallel computation capabilities of hardware when processing matrices [68, 69], thus offering a hardware-friendly parallelism.

The proposed MGA is designed to leverage the unique properties of memristive crossbar arrays, which have shown promise in neuromorphic computing and parallel processing applications. By mapping the GA operations onto memristive circuits, this approach aims to achieve higher computational efficiency and parallelism compared to traditional software implementations. To demonstrate the effectiveness of the MGA, the section presents a case study on feature selection using the Sonar dataset. The results show that the MGA-based approach improves the average accuracy of logistic regression by 11.9% while using 76.7% fewer features, highlighting its potential in optimization problems.

The section details the mathematical formulation of the MGA, its implementation on memristive crossbar arrays, and comparative analyses with traditional GA approaches. By bridging the gap between evolutionary algorithms and emerging memristive technologies, this research contributes to the development of more efficient, scalable, and hardware-implementable optimization techniques.

6.4.2 Matrix-Friendly Genetic Algorithm

6.4.2.1 Baseline Genetic Algorithm

The key data structure of GA is the population, which contains chromosomes with many genes and every chromosome is an encoded solution to the target problem. Besides, binary and float are two common encoding value types of the gene. Here, the population of t iteration is represented by a matrix P_t with $N_c \times N_g$ (N_c is the number of chromosomes and N_g is the number of genes). Typically, there are three operators, which are core steps of GA, including the selection operator, crossover operator, and mutation operator. The selection operator is used to select chromosomes for generating the next generation. Crossover operation generates next-generation chromosomes from parent chromosomes. The mutation operator generates a richer solution by mutating genes. In addition, a function named the fitness function is to judge the quality of chromosomes and the fitness determines the possibility of chromosome selection.

Figure 6.41 reveals the process of GA. First, the population matrix P_0 and the number of iterations t are initialized. Then, the fitness function is calculated, three operators are executed in each iteration, and the population of next-generation P_{t+1} is generated from P_t. In addition, the solution with the best fitness is saved. Finally, the process is terminated when t reaches the maximum number of iterations N_e.

6.4.2.2 Proposed Model

MGA, which transforms the three operators into matrix operations, is a matrix-friendly GA. The model can be described as:

$$P_t = \left(S_t P_{t-1} \bigodot C_t + S_t' P_{t-1} \bigodot C_t'\right) \bigoplus M_t \tag{6.32}$$

$$P_t = \left(S_t P_{t-1} \bigodot C_t + S_t' P_{t-1} \bigodot C_t'\right) + M_t \tag{6.33}$$

where t is the number of iterations, P_t is the population matrix, S_t and S_t' are matrices for selection operator, C_t and C_t are matrices for crossover operator, M_t is the matrix for the mutation operator, $\bigodot$ is the elementwise multiple operation, and $\bigoplus$ is the elementwise XOR operation. The difference between (6.32) and (6.33) is the mutation operation. XOR operation and add operation are used for the binary chromosome and float chromosome, respectively.

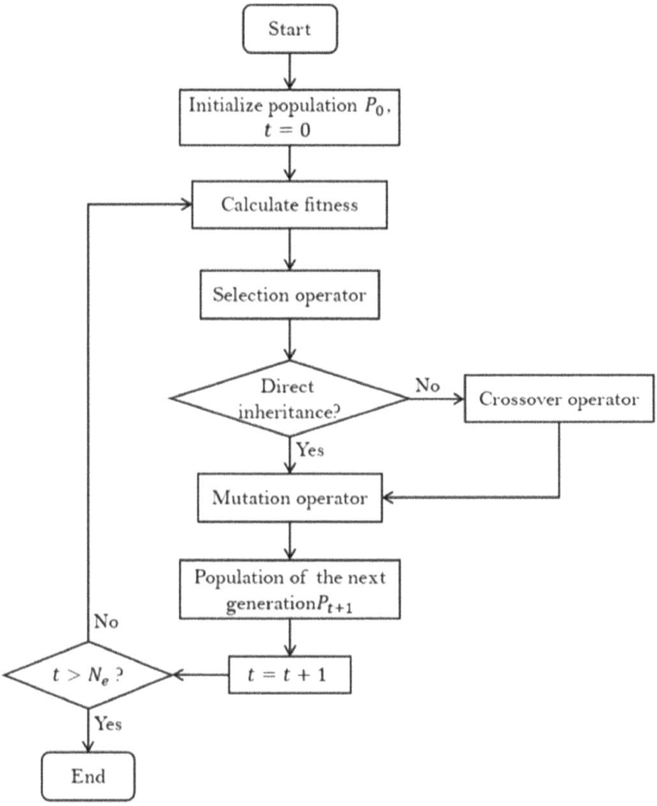

Fig. 6.41 Flowchart of GA. Fitness, used to evaluate each chromosome, is calculated by the fitness function. In order to generate the next generation of population, three operators are executed in turn

The population of the proposed model is represented by a matrix and each row is a chromosome. The population matrix is described as:

$$
\begin{aligned}
P_t &= \begin{pmatrix} \overrightarrow{p_1} & \overrightarrow{p_2} & \cdots & \overrightarrow{p_{N_c}} \end{pmatrix}^{\mathrm{T}} \\
&= \begin{bmatrix} p_{1,1} & p_{1,2} & \cdots & p_{1,N_g} \\ p_{2,1} & p_{2,2} & \cdots & p_{1,N_g} \\ \vdots & \vdots & \ddots & \vdots \\ p_{N_c,1} & p_{N_c,2} & \cdots & p_{N_c,N_g} \end{bmatrix}
\end{aligned}
\tag{6.34}
$$

where $p_i(i = 1, 2, \cdots, N_c)$ is a row vector of Pt and represents a chromosome, $p_{i,j}(i = 1, 2 \cdots, N_c; j = 1, 2, \cdots, N_g)$ is a gene which is binary value for binary chromosome and float value for float chromosome.

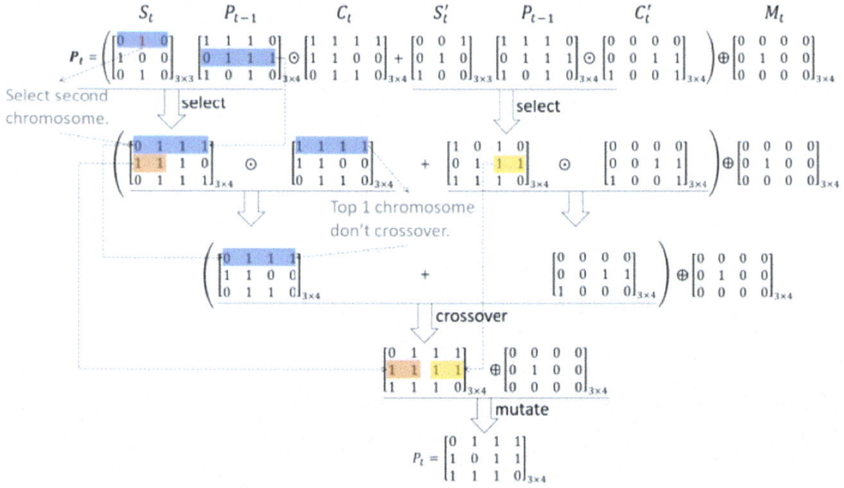

Fig. 6.42 Example of (6.32). In this example, the population has three chromosomes, each with four genes. The three operators of GA are realized by the matrix operation

The selection matrix is described as:

$$S_t = \left(\vec{s_1} \quad \vec{s_2} \quad \cdots \quad \vec{s_{N_c}}\right)^{\mathbf{T}} labeleqsps7sps4.2sps4 \tag{6.35}$$

$$S_t' = \left(\vec{s_1'} \quad \vec{s_2'} \quad \cdots \quad \vec{s_{N_c}'}\right)^{\mathbf{T}} \tag{6.36}$$

where S_t and S_t' are both selection matrices because the crossover operator needs parent chromosomes, $\vec{s_i}$ $(i = 1, 2, \ldots, N_c)$ is one-hot vector for selecting one-row vector in Pt, so is $\vec{s_i'}$ $(i = 1, 2, \ldots, N_c)$.

The crossover matrix is described as:

$$C_t = \left(\vec{c_1} \quad \vec{c_2} \quad \cdots \quad \vec{c_{N_c}}\right)^{\mathbf{T}} = \begin{bmatrix} c_{1,1} & c_{1,2} & \cdots & c_{1,N_g} \\ c_{2,1} & c_{2,2} & \cdots & c_{1,N_g} \\ \vdots & \vdots & \ddots & \vdots \\ c_{N_c,1} & c_{N_c,2} & \cdots & c_{N_c,N_g} \end{bmatrix} \tag{6.37}$$

$$C_t' = \left(\vec{c_1'} \quad \vec{c_2'} \quad \cdots \quad \vec{c_{N_c}'}\right)^{\mathbf{T}} = I - C_t \tag{6.38}$$

where C_t and C_t' are both crossover matrices, $\vec{c_i}$ $(i = 1, 2, \ldots, N_c)$ is a binary vector, $c_i = (1, 1, \ldots, 1)$ when $i = 1, 2, \ldots, N_c \cdot Ps$ makes the top $N_c \cdot P_s$ selected chromosomes not crossover, and $c_{i,j}$ $(i = 1, 2, \ldots, N_c;\ \mathrm{j} = 1, 2, \ldots, N_g)$ determines whether the gene on position (i, j) is retained (retained if $c_{i,j} = 1$, discarded otherwise).

The mutation matrix is described as

$$M_t = \left(\overrightarrow{m_1}\ \overrightarrow{m_2}\ \cdots\ \overrightarrow{m_{N_c}}\right)^{\mathrm{T}} = \begin{bmatrix} m_{1,1} & m_{1,2} & \cdots & m_{1,N_g} \\ m_{2,1} & m_{2,2} & \cdots & m_{1,N_g} \\ \vdots & \vdots & \ddots & \vdots \\ m_{N_c,1} & m_{N_c,2} & \cdots & m_{N_c,N_g} \end{bmatrix} \tag{6.39}$$

where $m_{i,j}$ is the mutation value for the gene on position (i, j), and $m_{i,j}$ is the binary value for binary chromosome and float value for float chromosome.

An example of (6.32) is shown in Fig. 6.42. The number of chromosomes of population N_c is 3. The number of genes of a chromosome N_g is 4. The probability of selection P_s is 1/3 which means that only the top 1 ($N_c \cdot P_s = 1$) selected chromosome does not need to crossover.

As shown in Fig. 6.43, in MGA, the generation of the selection matrix, crossover matrix, and mutation matrix can run in parallel, while in GA, the selection operator, crossover operator, and mutation operator are generated in turn, demonstrating that the process of MGA is simpler than GA. Like GA, P_0 and t are also initialized first in MGA. In each iteration, the fitness function is calculated, and S_t, S_t', C_t, C_t', and M_t are generated according to fitness value. Then, P_{t+1} is gotten by (6.32) or (6.33). Finally, the process is terminated when t reaches the maximum number of iterations N_e.

6.4.2.3 Solving the Maximum Value of Binary Function

With the aim of comparing the efficiency of GA and MGA, a binary fitness function is utilized to solve the maximum value. The function is shown below:

$$F(x, y) = 3(1-x)^2 e^{-x^2-(y+1)^2} - 10\left(\frac{x}{5} - x^3 - y^5\right)e^{-x^2-y^2} - \frac{1}{3}e^{-(x+1)^2-y^2} \tag{6.40}$$

where $F(x, y)$ is the binary fitness function, x and y are inputs. Surface plots of $(\partial F/\partial x)$ and $(\partial F/\partial y)$ are shown in Fig. 6.44.

From Fig. 6.44, in the range of [–3, 3], there exists a significant fluctuation. Then, a contour map of this fitness function is shown in Fig. 6.45. An approximate maximum value is obtained when $x \approx 0$ and $y \approx 1.5$. The maximum value is more than 7.

In order to solve this optimization problem with GA and MGA, some algorithm strategies need to be declared in advance. First, the most parameters of both GA and MGA are shown in Table 6.8. The search range of x and y is both [–3, 3]. Second, the binary encoding is used for embedding x and y to a chromosome and each of x and y needs $N_g/2$ genes. The accuracy of the binary encoding is $(3-(-3))/2^{N_g/2} \approx 5.72 \times 10^{-6}$. Finally, roulette selection, multipoint crossover, and multipoint mutation are adopted.

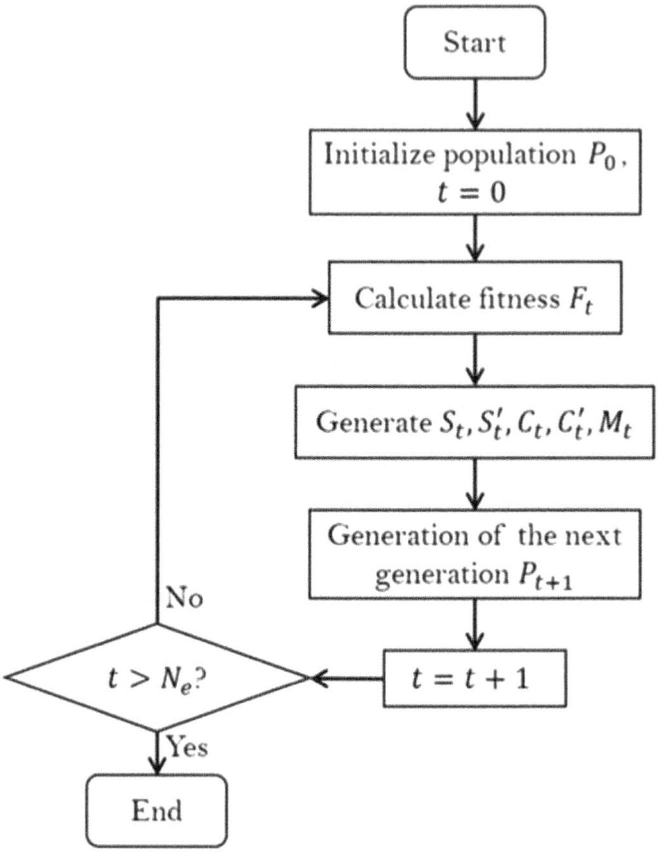

Fig. 6.43 Flowchart of MGA. Like GA, fitness needs to be calculated first. The difference between GA and MGA is that MGA only needs to generate five matrices, and then the next-generation population can be obtained by calculating these matrices and population matrix

The simulations are completed with Python language and NumPy library. Baseline GA, MGA, and elitist GA (the elite ratio is set to 0.01, other parameters are shown in Table 6.8) [70] run N_e times, respectively, and we saved all the best fitness values of N_i iterations of N_e epochs. The fitness curves are obtained as shown in Fig. 6.46.

In Fig. 6.46, the convergence rate and final best fitness of MGA are both better than GA and elitist GA. Specifically, the average final best fitness in GA is about 6.7636, which is about 6.0054 in elitist GA and 6.9636 in MGA. The total running time of GA, elitist GA, and MGA are 113, 163, and 44 s, respectively. Since MGA uses more matrix operations, MGA executes more efficiently than GA and elitist GA when using the NumPy library.

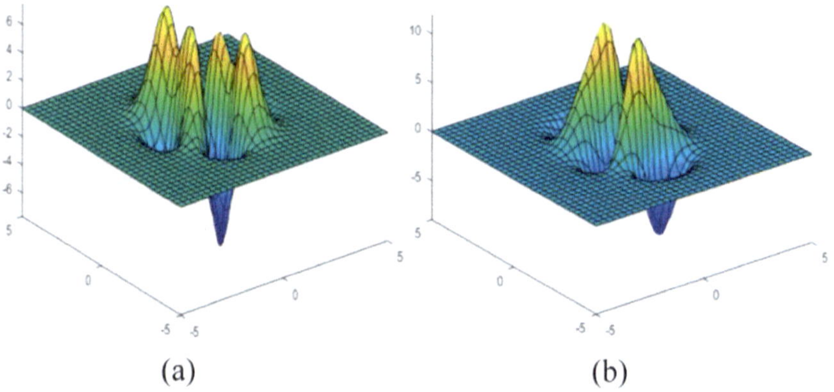

Fig. 6.44 Partial derivation of (6.40). The partial derivative value is close to 0 when the range is in $(-\infty, -3) \cup (3, +\infty)$. **a** Surface plot of $(\partial F/\partial x)$. **b** Surface plot of $(\partial F/\partial y)$

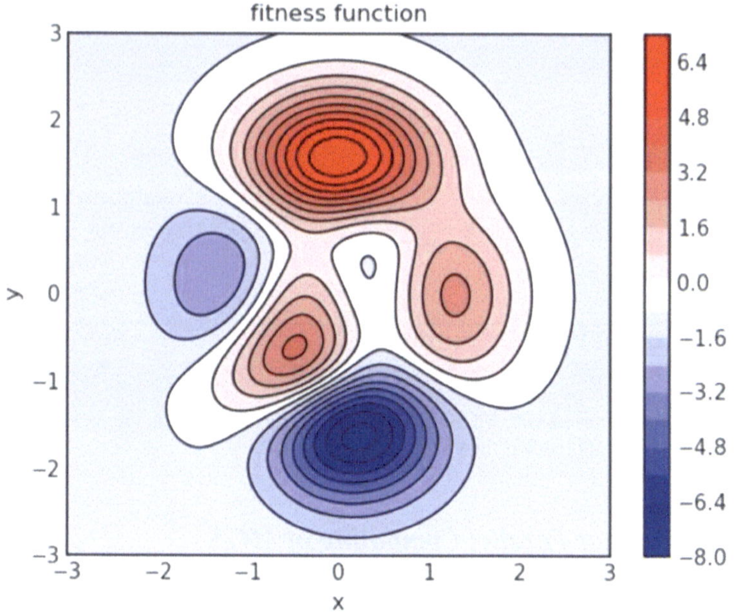

Fig. 6.45 Contour map of fitness function. Red represents a positive value and blue represents a negative value. The darker the color, the greater the absolute value

Table 6.8 Parameters for solving the maximum value of binary function

Parameter	Description	Value
N_e	Number of epochs	100
N_i	Number of iterations	80
N_c	Number of chromosomes	100
N_g	Number of genes	40
P_s	Probability of select operator	0.1
P_c	Probability of crossover	0.2
P_m	Probability of mutation	0.01

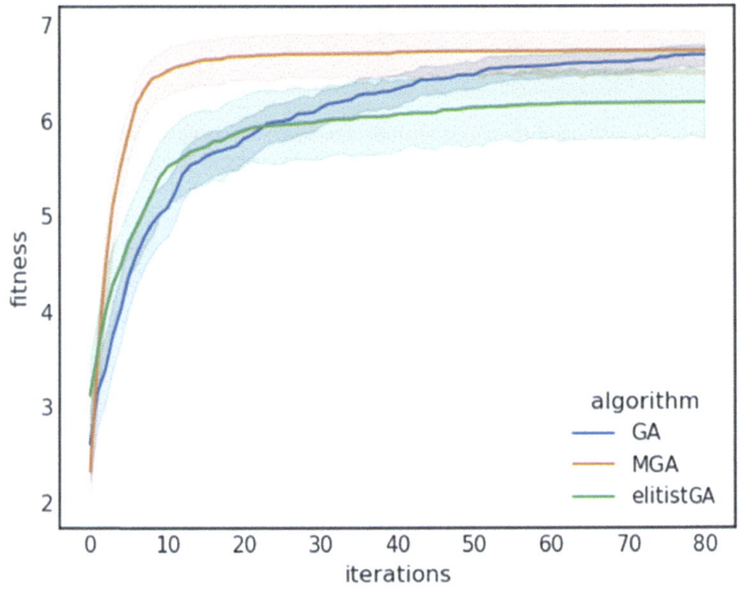

Fig. 6.46 Fitness curves of GA, MGA, and elitist GA. The solid line is the mean value of Ne experiments and the light shade is the 95% confidence interval

6.4.3 *Memristor Parallel Computing of MGA*

The pipeline of memristor circuits deployment of MGA is shown in Fig. 6.47. In this pipeline, the input is a column vector of P_t and the output is the corresponding column vector of P_{t+1}. The selection matrix S_t and S_t' only need to write to memristor array once in each iteration while C_t, C_t', and M_t need to write to memristor array column by column in each calculation.

There are three advantages of inputting one column of Pt at each step: (1) the memristor circuit of vector-matrix multiplication is easier to design; (2) N_g is fixed for some problems, so the time complexity of each iteration is a constant; and (3) the

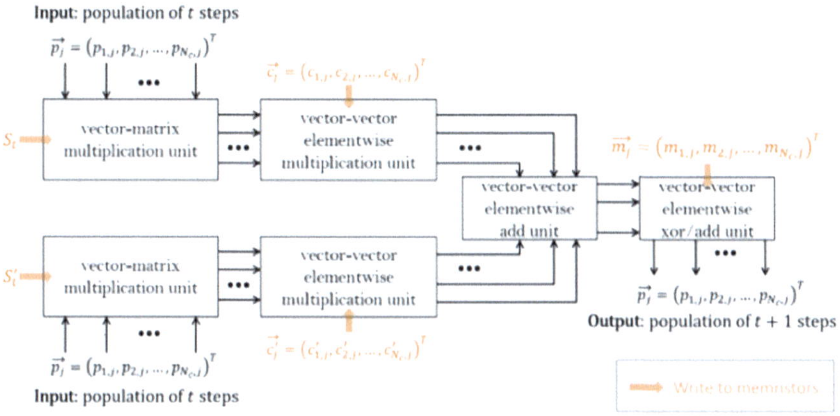

Fig. 6.47 Pipeline of the calculation process. The four circuits need to be designed, including the vector-matrix multiplication unit, vector-vector elementwise multiplication unit, vector-vector elementwise add unit, and vector-vector elementwise XOR/add unit

performance is always related to N_c, so different performance levels can be extended by integrating memristor circuits with different chromosome numbers.

The HP memristor model [71] is used for simulation, and this model is described as

$$M = M_{on}\xi + M_{off}(1-\xi), 0 \leq x \leq 1 \tag{6.41}$$

$$\frac{d\xi}{dt} = I\frac{W(\xi)}{Q_0} \tag{6.42}$$

where M is the resistance of the memristor, M_{on} and M_{off} are the resistances when the entire memristor is in state ON and OFF, respectively, ξ is the fraction of the memristor that is in state on, Q_0 is total charge required for full state transition and I is the current from the positive to the negative terminal. $W(\xi)$ is a window function which is shown as

$$W(\xi) = 1 - (2\xi - 1)^{2p_w} \tag{6.43}$$

where p_w is a positive integer. This function is modified when ξ is close to either 0 or 1, to improve numerical stability.

Moreover, the distributions of M_{on} and M_{off} follow the lognormal probability density function [72, 73], which is

$$f_x(x;\mu,\sigma) = \frac{1}{x\sigma\sqrt{2\pi}} exp\left(-\frac{(\ln x/\mu)^2}{2\sigma^2}\right) \tag{6.44}$$

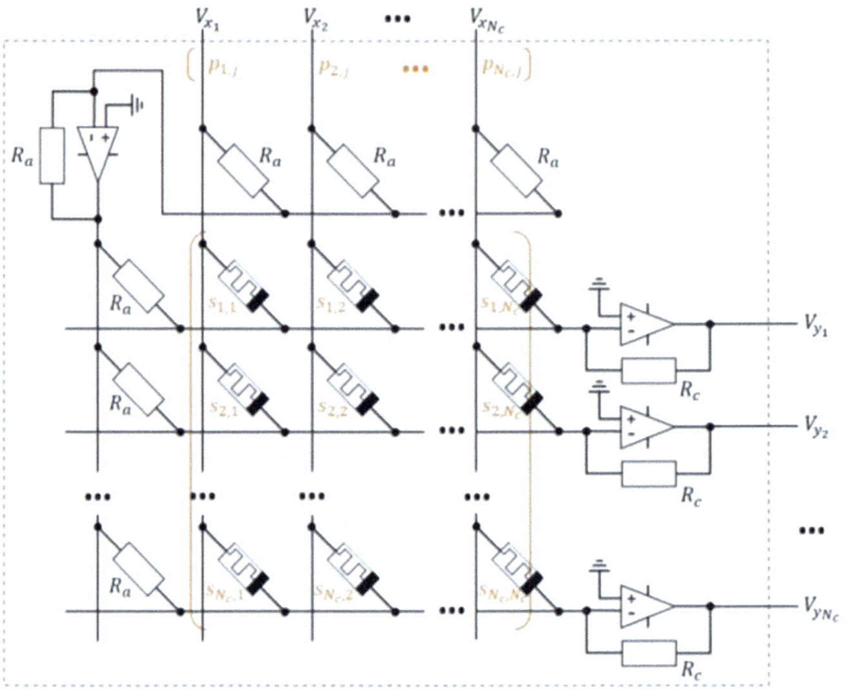

Fig. 6.48 Memristor circuit of vector-matrix multiplication. The memristor array stores the selection matrix. The input is the voltage value corresponding to a column of genes of the population, and the output is a column of the selected population matrix

where μ is the normal mean and σ is the standard deviation. Therefore, the sampling data of M_{on} and M_{off} can be generated by M_{on} = Lognormal(μ_{on}, σ_{on}) and M_{off} = Lognormal(μ_{off}, σ_{off}), respectively.

The parameters of the memristor model used in the simulation are set as $Q_0 = 0.01C$, $p_w = 2$, $\mu_{on} = 10^5\Omega$, $\mu_{off} = 108$, $\sigma_{on} = 0.1$, and $\sigma_{off} = 1$. All simulation experiments are repeated 1000 times.

As shown in Fig. 6.48, a memristor crossbar array is used to realize vector-matrix multiplication. Fig. 6.49 shows the simulation of vector-matrix multiplication based on MATLAB Simulink. For the first output, the output value in binary should be "0," which means that the output voltage is correspondingly 0 V. Although the mean of the output voltage is about 20 mV and the max is about 30 mV, the output voltage is close to 0 V and has little impact on the subsequent circuit. According to the second and third waveforms, the errors are negligible, which means that the variation of the memristor has little effect on non-zero values.

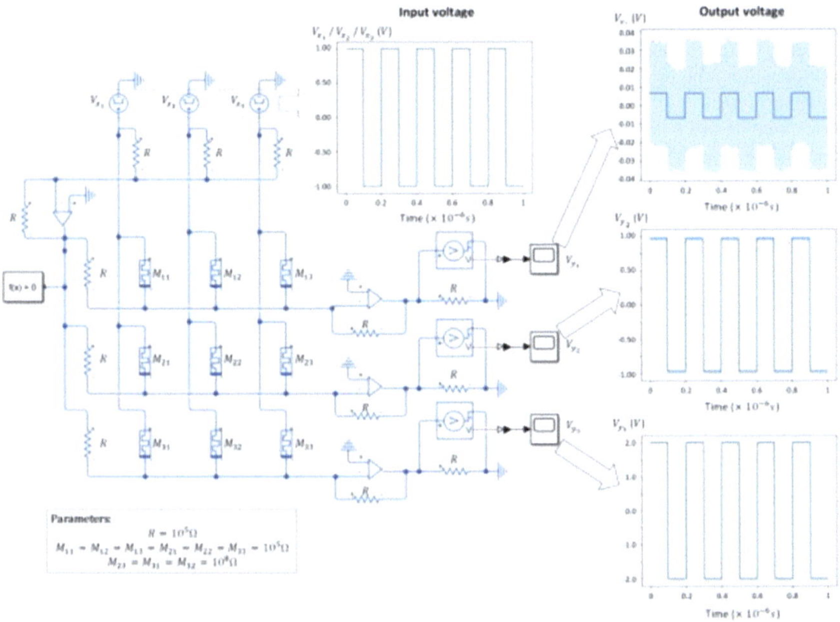

Fig. 6.49 Simulation of vector-matrix multiplication based on MATLAB Simulink. The input gene values are all 1, and the selection matrix of the memristor array is [0, 0, 0; 0, 0, 1; 1, 1, 0]. The output is [0, 1, 2], and the result of the vector multiplication matrix is correct. The shaded part represents the 95% confidence interval of 1000 experiments

The relationship between inputs and outputs of circuits in Fig. 6.48 can be described as:

$$V_{y_i} = \sum_{j=1}^{N_c} V_{x_i} \cdot \left(\frac{1}{R_a} - \frac{1}{M_{i,j}} \right) \cdot R_c \tag{6.45}$$

$$s_{i,j} = \left(\frac{1}{R_a} - \frac{1}{M_{i,j}} \right) \cdot R_c \tag{6.46}$$

where V_{x_i} and V_{y_i} (i = 1, 2, $\cdots$, N_c) are the ith input and output voltages, namely, R_a and R_c are resistors, $M_{i,j}$ is the resistance of the corresponding memristor and is preset by $s_{i,j}$, and the relationship is shown in (6.46).

As shown in Fig. 6.50, a vector-vector elementwise multiplication is realized by memristors and an operational amplifier. Figure 6.51 shows the simulation of vector-vector elementwise multiplication based on MATLAB Simulink. According to the first waveform, the output voltage is negligible (10.3V). The error brought by the variation of the memristor is also negligible according to the second waveform.

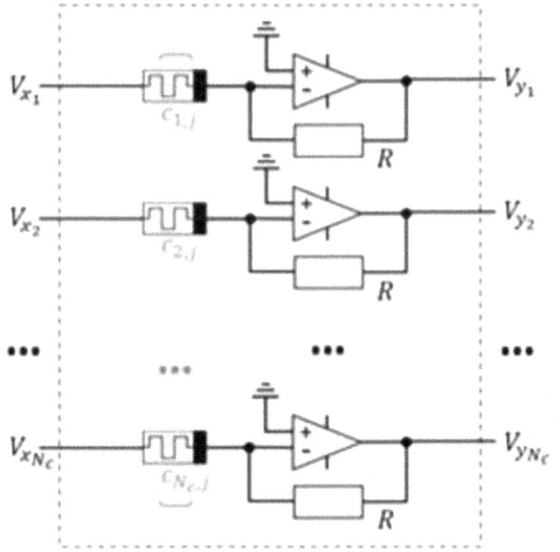

Fig. 6.50 Memristor circuit of vector-vector elementwise multiplication. The input is the voltage value corresponding to a column of genes of the selected population, and the memristor array stores a column of the crossover matrix

The circuit can be described as:

$$V_{y_i} = V_{x_i} \cdot \frac{R}{M_i} \tag{6.47}$$

$$c_{i,j} = \frac{R}{M_i} \tag{6.48}$$

where V_{x_i} and V_{y_i} (i = 1, 2, ..., N_c) are the ith input and output voltages, respectively, R is the resistance of the resistor, M_i is the resistance of the corresponding memristor and is preset by $c_{i,j}$ and the relationship is shown in (6.48).

As shown in Fig. 6.52, a vector-vector elementwise add unit is designed. Figure 6.53 shows the simulation of vector-vector elementwise add based on MATLAB Simulink. This circuit can be described as:

$$V_{y_i} = V_{x_i} + V_{x_i'} \tag{6.49}$$

As shown in Fig. 6.54, a vector-vector elementwise XOR/add unit is designed. Figure 6.55 shows the simulation of vector-vector elementwise XOR based on MATLAB Simulink. For the first and fourth outputs, the output value in binary should be "0" which means that the output voltages are 0 V. According to the first and fourth waveforms, the output voltages are close to 0 V and have a little impact on the subsequent circuit. The errors brought by the variation of the memristor are also negligible according to the second and third waveforms.

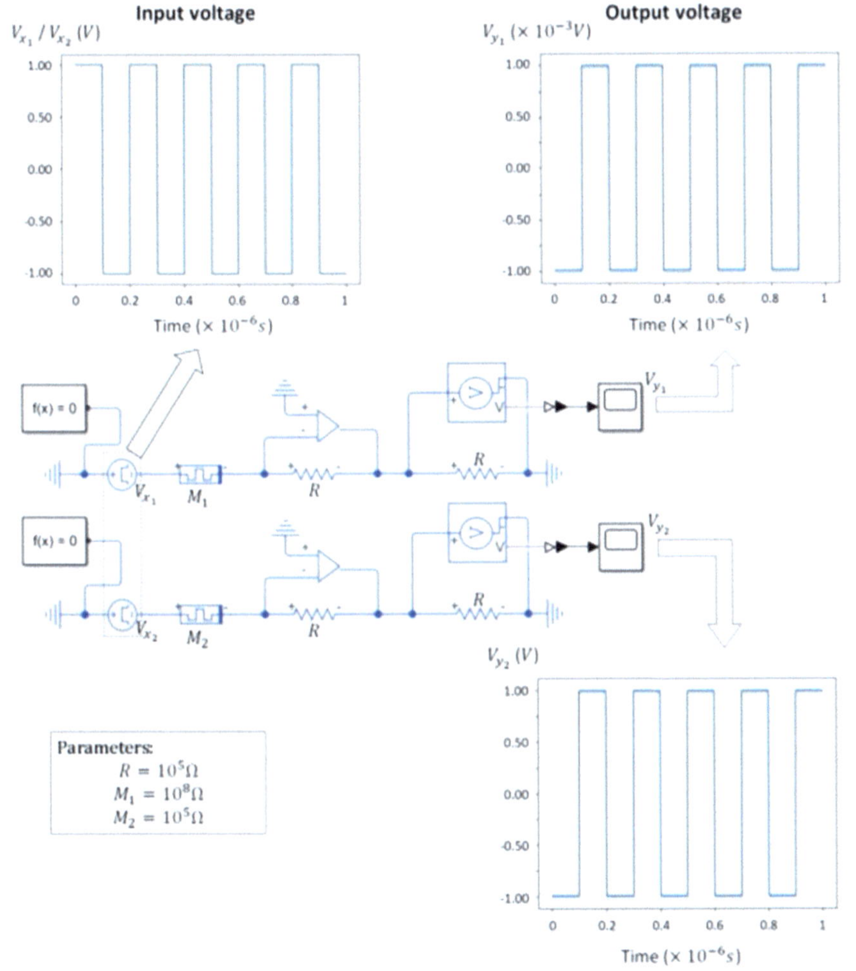

Fig. 6.51 Simulation of vector-vector elementwise multiplication based on MATLAB Simulink. The column of the crossover matrix of the memristor array is [0, 1]. The input gene values are all 1 and the output is [0, 1], which is consistent with the calculation results. The shaded part represents the 95% confidence interval of 1000 experiments

The unit is different between the vector-vector elementwise add unit because it needs to preset half of the addends. This circuit can be described as:

$$V_{y_i} = V_{x_i} + V_1 \frac{R}{M_i} \tag{6.50}$$

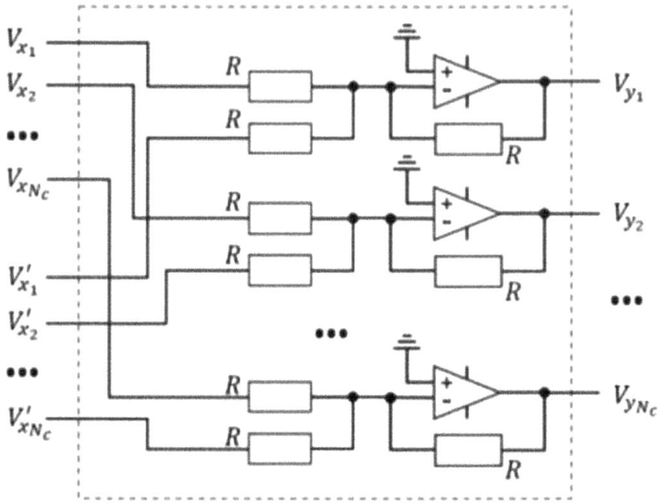

Fig. 6.52 Circuit of vector-vector elementwise add

where V_{x_i} and V_{y_i} (i = 1, 2, ..., N_c) are the ith input and output voltages, respectively, V_1 is the fixed value related to the encoding strategy, R is the resistance of the resistor, and M_i is the resistance of the corresponding memristor.

When float encoding is used, the add operation rather than the XOR operation should be realized. Thus, the relationship between $m_{i,j}$ and M_i can be obtained as:

$$m_{i,j} = V_1 \frac{R}{M_i} \tag{6.51}$$

where $m_{i,j}$ is the mutation value for the gene on position (i, j).

In the practical use of a memristor, the input must be high-frequency AC voltage for keeping the resistance of the memristor, so values 1 and 0 cannot be represented by positive and negative voltage. When binary encoding is used, the value is 0 if the voltage is equal to 0 and 1 otherwise. To realize the XOR operation, V_1 should be of value -1 (it means voltage which is reverse to voltage of value 1). In this way, a truth table, which is the same as the truth table of the XOR operation, is obtained as Table 6.9. The relationship between $m_{i,j}$ and M_i can be obtained as follows:

$$m_{i,j} = \begin{cases} 1, \ if M_i = R \\ 0, \ if M_i \approx +\infty \end{cases} \tag{6.52}$$

where M_i is the resistance of the corresponding memristor, and $M_i \gg R$ when $M_i = M_{off}$. In this case, M_i can be assumed as $+\infty$. Although, theoretically, the population size of the proposed algorithm can be expanded arbitrarily, the accuracy of the memristor will be severely limited in the device with the increase of the scale of

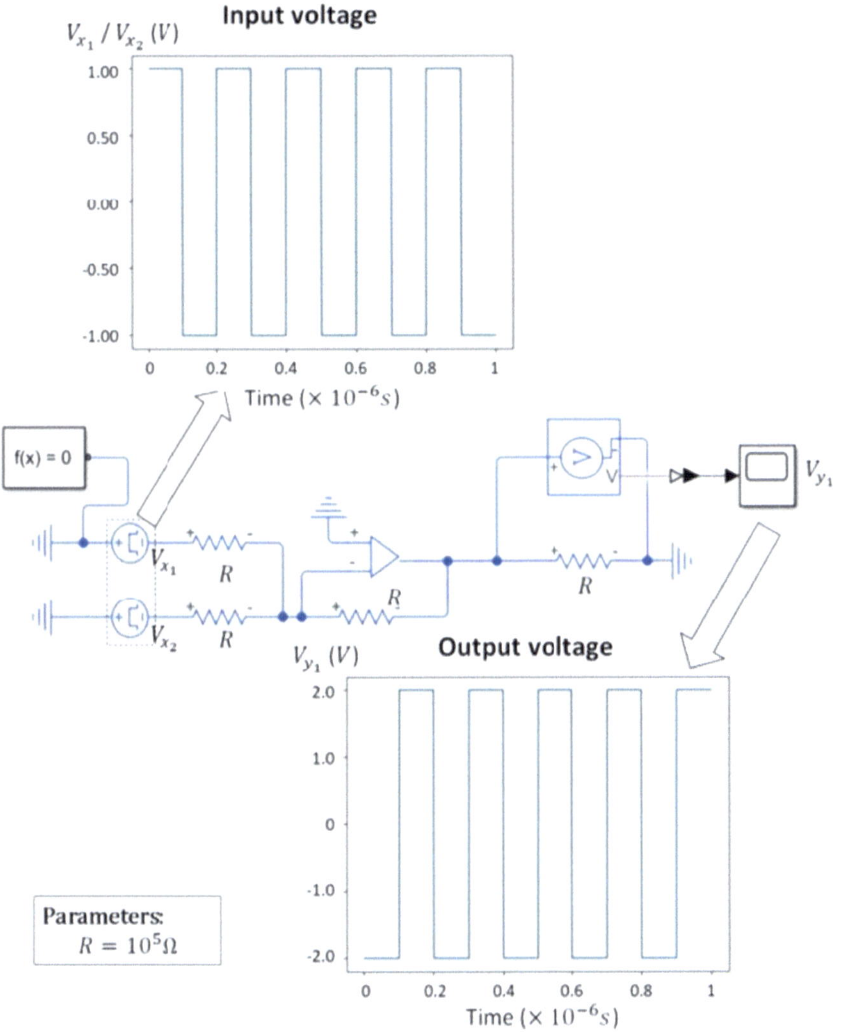

Fig. 6.53 Simulation of vector-vector elementwise add based on MATLAB Simulink

the memristor array [74]. Therefore, in practical application, it is necessary to block the large population and divide it into multiple memristor arrays for calculation.

The MGA circuit diagram is shown in Fig. 6.56. In MGA, each row of the selection matrix is one-hot coding. Correspondingly, the outputs of the multiplication of digital input signals from a memristor-based matrix are digital signals so that the outputs can be directly transferred to the next unit. However, for ADD and XOR units, their outputs should be converted to be binary cause the output voltage amplitude might be high. So, the voltage limiter circuits are added behind ADD and XOR circuits to

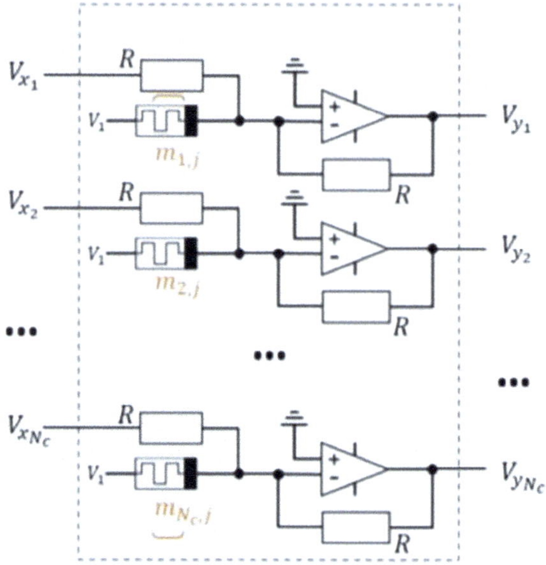

Fig. 6.54 Memristor circuit of vector-vector elementwise XOR/add. The input is the voltage value corresponding to a column of genes of the population obtained after selection and crossover, and a column of the mutation matrix is stored in the memristor array

obtain binary outputs. When all inputs are setting to 1, the outputs of two matrices are both [1, 1, 1]. Next, the outputs of vector-vector elementwise multiplication are [1, 0, 1] and [0, 1, 0], respectively. Then, the output of addition is [1, 1, 1]. The final output is [1, 1, 0], and the corresponding waveforms are shown in Fig. 6.56.

For different kinds of memristor window sizes, a survey has been made to illustrate their design. In [75], M_{on} ranging from 100 KΩ to 1 MΩ and M_{off} ranging from 100 to 500 MΩ are used in the experiment. Small window size memristors are implemented in [11]. The [76] tests different ON/OFF ratios between the maximum conductance and minimum conductance, specifically, 10, 100, and 1000 are all included, showing the feasibility of these memristor window sizes.

This section also explores the impact of the window size of memristor (ratio between M_{off} and M_{on}) on the accuracy of output voltages. In light of (6.45), when all memristances are M_{on}, the expected output having little relevance with window size is 0 V; otherwise, the expected output is related with window size. An experiment with different outputs of vector-matrix multiplication unit obtained in different window sizes is proposed and the result is shown in Fig. 6.57. Figure 6.57a shows that the window size of the memristor has almost no effect on output voltage corresponding to binary value "0." Specifically, the outputs in Fig. 6.57b and c get closer to the expected voltages (1 and 2 V, respectively) with the increase in the window size. Therefore, a larger window size of the memristor can increase the accuracy of output voltage. At the same time, the reduction of the memristor window will induce a lower accuracy.

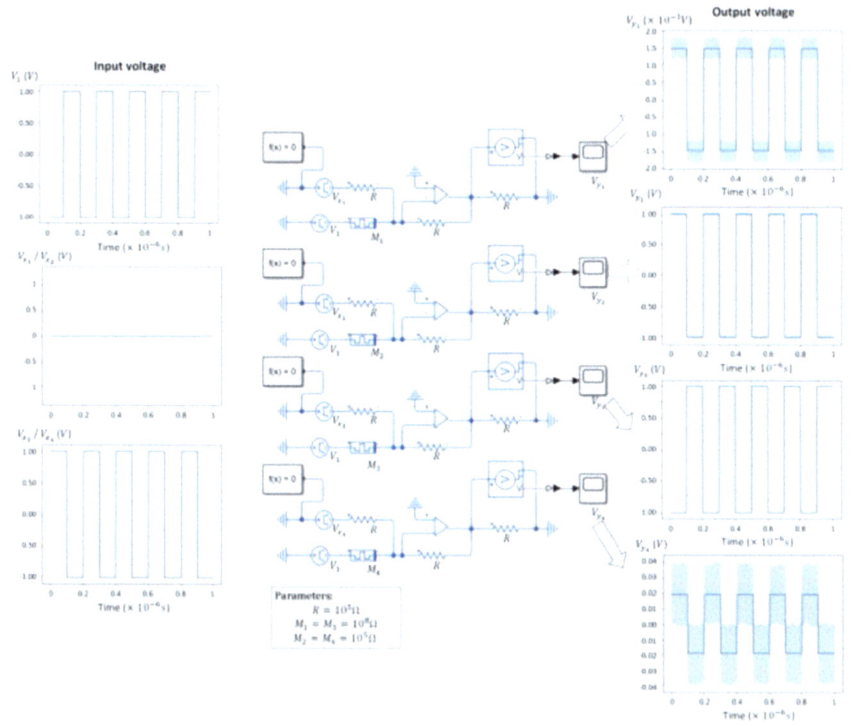

Fig. 6.55 Simulation of vector-vector elementwise XOR based on MATLAB Simulink. The four circuits correspond to 0 XOR 0, 0 XOR 1, 1 XOR 0, and 1 XOR 1, respectively. The shaded part represents the 95% confidence interval of 1000 experiments

Table 6.9 Truth table of vector-vector elementwise XOR/Add unit when binary encoding is used

Input	V_x	$m_{i,j}$	M_i	Output	V_y
0	0	0	M_{off}	0	0
0	0	1	R	1	$-V_1$
1	$-V_1$	0	M_{off}	1	V_1
1	$-V_1$	1	R	0	0

It is important to note that the problem caused by the memristor with the small window can be addressed by adding auxiliary components in practice to switch output. In other words, the design of auxiliary components is to optimize output by restraining low voltage and stabilizing high voltage.

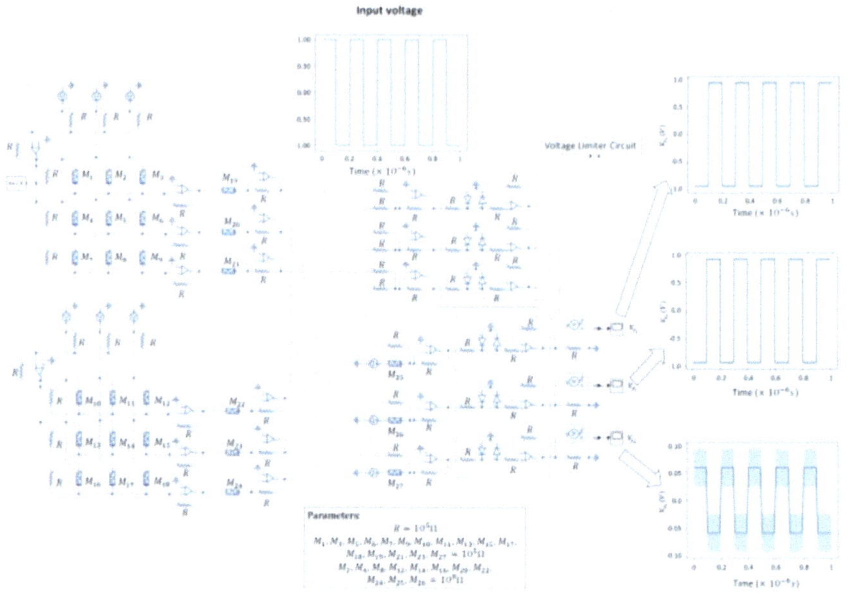

Fig. 6.56 Simulation of memristor parallel computing of MGA based on MATLAB Simulink. The top-left selection matrix of memristor is [0, 1, 0; 1, 0, 1; 0, 1, 0], and the bottom-left is [0, 0, 1; 0, 1, 0; 1, 0, 0]. The column of the crossover matrix of memristor array is [1, 0, 1] and [0, 1, 0], respectively. The three XOR calculations correspond to XOR 0, XOR 0, and XOR 1, respectively. With all inputs being 1, the final output is [1, 1, 0]. The shaded part represents the 95% confidence interval of 1000 experiments

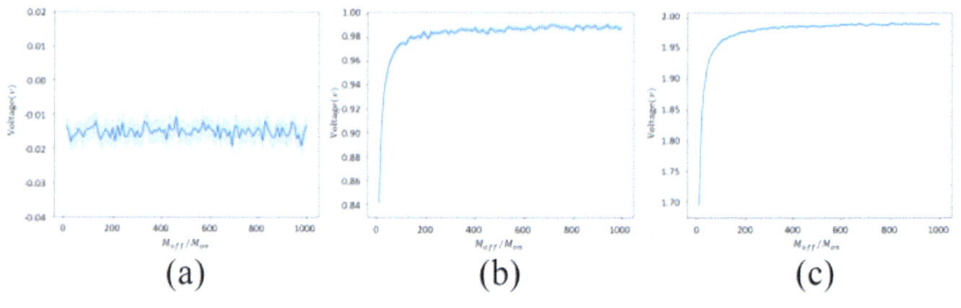

Fig. 6.57 (a)–(c) plot the effect of window size of memristor on output voltage. The shaded part represents the 95% confidence interval of 1000 experiments

6.4.4 Memristor-Based MGA for Feature Selection

GA can be used for feature selection in machine learning. We simulate the method proposed in Section III and use it for feature selection on the Sonar dataset and LR. The Sonar dataset, described in Table 6.10, involves predicting the target object as a rock or mine according to the strength returned by a given sonar from different

Table 6.10 Description of the sonar dataset

Mine samples	Rock samples	Features	Feature range
111	97	60	[0,1]

Table 6.11 Result comparisons between LR without feature selection and LR with feature selection. The result obtained by fixing random states of LR to 0 and the accuracy of memristor-based MGA is the average of 100 experiments

Method	No. of features	Accuracy (%)
Without feature selection	60	63.95
Variance threshold	15	57.76
Chi square test	15	73.07
Memristor-based MGA	**14.12**	**74.05**

angles. The main reason this dataset is chosen is that it has many features but only two classes. Therefore, there is a high probability of redundancy.

There are some algorithm strategies that need to be modified for complete feature selection. First, N_g must be fixed to the number of features, and the value of each gene is binary (select if 1, do not select if 0). Second, a new parameter named P_p, representing the probability of the initial gene value is equal to 1, is added, and we set it to 0.2. Finally, the primary fitness is the average accuracy of 10-fold cross-validation of LR on the Sonar dataset, and the secondary fitness is the negativity of the number of features. Therefore, a chromosome is better when its primary fitness is higher or when primary fitness is equal but secondary fitness is higher. All other algorithm strategies are the same as in Section II. Other parameters are set to the values shown in Table 6.8.

As shown in Table 6.11, the accuracy is increased by 11.9%, and the number of features is reduced by about 46. Then, the analysis of the influence of different parameters on the fitness (the average accuracy of 10-fold cross-validation of LR on the Sonar dataset) is shown in Fig. 6.58. Except for a parameter that needs to be changed, the other parameters of all these subfigures in Figs. 6.58 and 6.59 are fixed to the corresponding values shown in Table 6.11. The conclusions drawn from these figures include: (1) the greater number of chromosomes N_c, the better results can be searched; (2) there is no significant difference in the results when P_s is small, but the results become worse when P_s equal to 0.8; (3) P_c has no significant effect on the results; (4) smaller P_m can get better results; and (5) all these parameters have little effect on the convergence rate.

Furthermore, we analyze the influence of the probabilities of gene values equal to 1 in the initial population on the results. As shown in Fig. 6.59, the fitness is higher, and the number of features is less with less value of 1 in the initial population. The reason is the less initial number of features has more potential to search for the

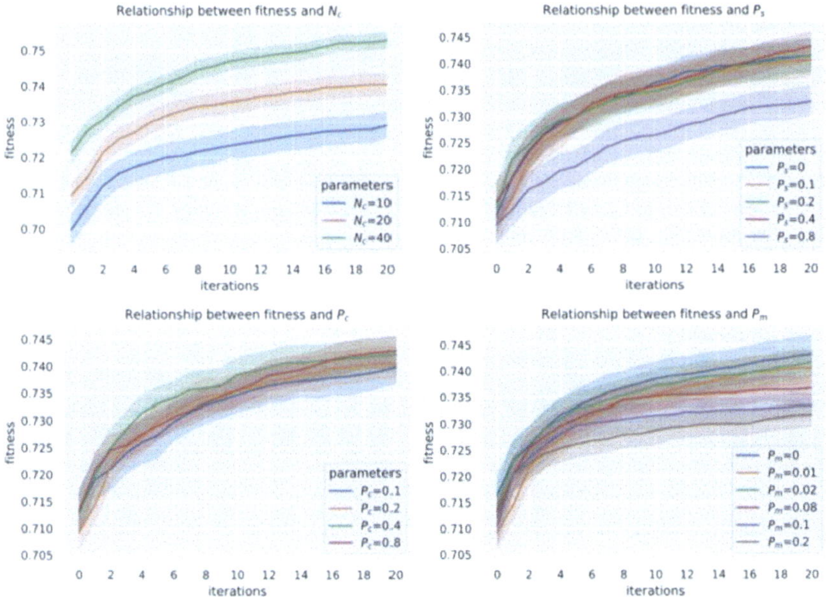

Fig. 6.58 Relationships of fitness and different parameters. All the solid lines represent the average results of 100 experiments and all the bars represent the 95% confidence interval of 100 experiments

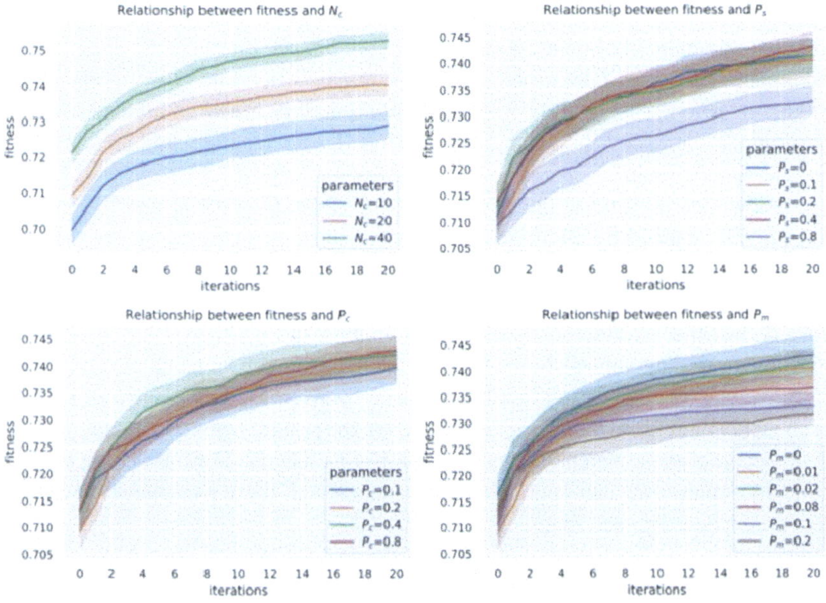

Fig. 6.59 Influence of P_p on the fitness and the number of features. All the solid lines represent the average results of 100 experiments and all the bars represent the 95% confidence interval of 100 experiments

optimal solution with few features. However, it should be noted that the final number of features does not change compared to the initial value, which may lead to local optimization.

6.4.5 Conclusion

In this article, MGA was proposed and implemented in memristor circuits. The motivation of MGA is to transform GA to matrix form for better parallelism. The advantages of the proposed method include: (1) using matrix calculation, which is easy to parallelize; (2) the three operators are decoupled and the operator matrices can be generated in parallel; and (3) the algorithm strategy of the operator is transformed into the generation strategy of the operator matrix. However, MGA may limit the usage of some algorithm strategies because of the limitation of the matrix operation. In the feature selection experiment, the simulation of the memristor circuit implementing MGA can effectively extract features to improve the accuracy of LR.

Considering that the precision of floating-point coding is low, which cannot support the GA of floating-point coding, we will explore the implementation of the high-precision memristor circuit in the future. Besides, the future work will also focus on the manufacturing of memristor, where the practical algorithms such as GA will be implemented by hardware-in-the-loop simulation. At that time, power, delay, detailed data mapping relationship, overhead, and other factors will continuously be key points of our research for the development of the memristor and its wider application.

6.5 Memristive Crossbar Array-Based Genetic Algorithm for Restoring Images

6.5.1 Introduction

In recent years, image restoration has become important in many fields, such as medical imaging and graphic printing. Digital images are often corrupted by degradation [77]. The reason for image blur is complicated, the restoration cannot be processed by the regular linear function. Hence, it is important to use GA for image restoration.

Because of the advantage of memristors in storage, the memristor-based array is proposed, which also can be called the memristive network. This array is capable of solving more complex problems compared with the single memristor unit. Besides, it has a similarity to the image gray value matrix, and its structure is similar to chromosomes and genes. In the memristor-based array, every pixel of the image has a corresponding memristor unit to store its value, making it possible to realize storage for binary, grayscale, and color images. In this chapter, our goals are to verify the feasibility of deploying GA with memristors, explore mapping rules between GA and the memristor-based array, and deploy GA with memristors through the mapping rules in image restoration.

6.5.2 Computing Model

6.5.2.1 GA

GA can be divided into three parts: (1) The first part is initialization. Engineering problems are mapped into population evolution problems in the biological world. Coding and generating population and feasible solutions to problems can be defined by chromosomes and genes. (2) The second part is to design the fitness function, that is, choosing the appropriate standard to quantify the individual's adaptation to the living environment. Fitness is an important standard to evaluate the individual and the basis of other operations. Because the requirement of the fitness function is not particularly strict and it is not forced to be differentiable, derivable, and so on, GA is suitable for dealing with nonlinear problems. (3) The third part is the core operations of GA, including selection, crossover, and mutation. Selection is to select individuals with good characters in the current population, which is the basis of crossover and mutation operations. Crossover is the exchange of the segment of parents' genes, that is, the process of reproduction in biology, which is an important part of GA. The genes of parents can be continued because of crossover. Mutation is the idea of mutation in biology, sometimes, gene mutation occurs during the process that offspring inherit the parent gene, and mutation is beneficial for the population to jump out of the local optimal solution. The specific process of GA is shown in Fig. 6.60.

6.5.2.2 Memristor-Based Array

Limited by the computing ability of a single memristor, the memristor-based array was designed for more complex problems. The memristor-based array has the structure similar to a two-dimensional matrix, which is composed of multiple memristors. Its structure is shown in Fig. 6.61. Each memristor can be added to this array through two connecting wires, and the deletion of the memristor only needs to disconnect the two connecting wires between the memristor and array [78].

There have been applications of using the memristive network to implement other swarm intelligence algorithms. For example, Pershin and Ventra [14] compared the memory characteristics of memristors with the pheromones left by ants on the path, proposed a mapping relationship, and used memristors to deploy the ant colony optimization, which inspires us to explore memristor-based GA.

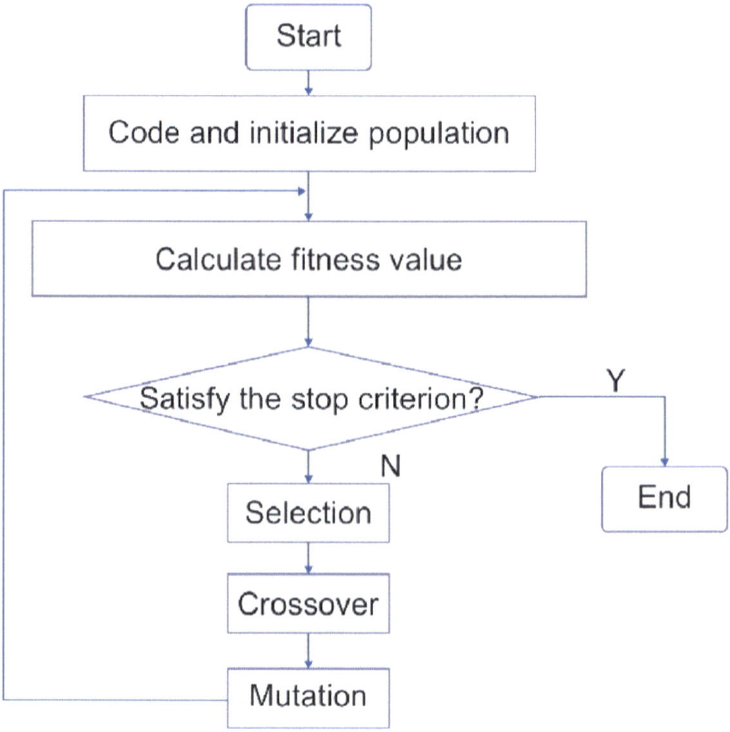

Fig. 6.60 Process of GA

6.5.2.3 Memristor-Based GA

To solve optimization problems, the analogy and mapping between GA and the memristor array is then conducted. The structure of the memristor-based array is similar to that of the chromosome, so a single memristor in the array can be abstracted into a gene unit. Next, the operation in GA is mapped to the change of the resistance value of the memristor in the memristor-based array. The mapping relationship is shown in Table 6.12. The memristor is the basic unit of the memristor network and the gene is the basic unit of the chromosome; the sum of information of each pixel constitutes an image, so the memristor is mapped to genes in GA, which corresponds to the gray value information of a pixel in an image. Multiple genes form a chromosome or individual, so the chromosome corresponds to a set of memristors, that is, a memristor network and a set of gray value information of pixels in the memristor network.

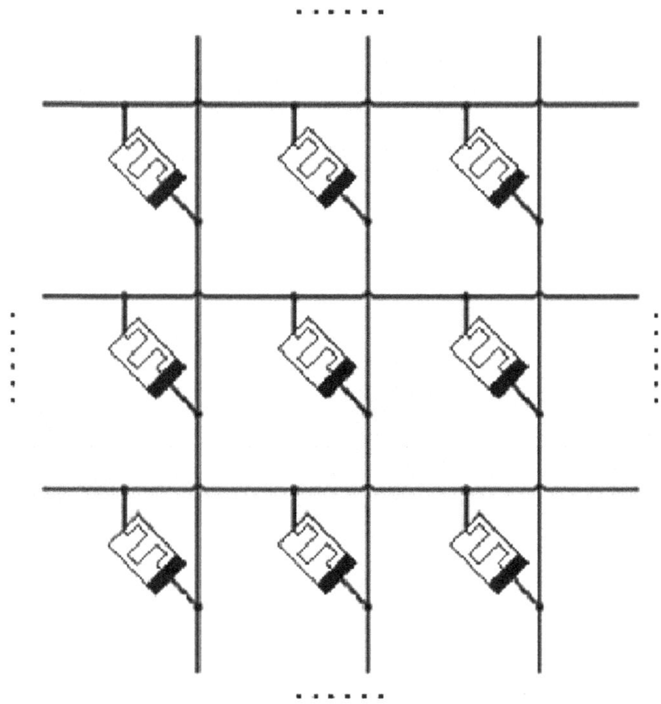

Fig. 6.61 Structure of memristor-based array

6.5.3 Image Restoration Based on Memristor-Based GA

This section introduces the theory of image restoration, and the detailed design of image restoration based on GA, including the way of coding and generating the initial population, the selection of the fitness function, and the selection of the basic operation of GA.

6.5.3.1 Theory of Image Restoration

In the process of image generation, transportation, and processing, the phenomenon of image blur caused by man-made and natural factors, such as camera jitter, inappropriate light, and haze, is called image degradation. Except for image blur, the degradation is often manifested as distortion, additional noise, etc., resulting in a decline in image quality. In the image degradation model, image degradation is modeled by using the degradation function and adding the external noise. The model is shown in Fig. 6.62, where x and y denote coordinates in the spatial domain, $f(x, y)$ represents the original clear image in the spatial domain, $h(x, y)$ is the degradation function, $n(x, y)$ represents the noise, $g(x, y)$ denotes the degraded image.

Table 6.12 Mapping rules among nature, GA, memristive network, and image restoration

Nature	GA	Memristive network	Image restoration
Gene	Feasible solution partial coding	Single memristor	Single pixel gray value
Chromosome	One feasible solution	Memristor network subnet	A set of pixel gray values
Population	Multiple feasible solutions	Multiple memristor network subnets	Multiple sets of pixel gray values
Selection	Selection function	Select memristor network subnets	Select gray values
Crossover	Crossover function	Exchange memristor resistance	Exchange gray values
Mutation	Mutation function	Change memristor resistance	Change gray values

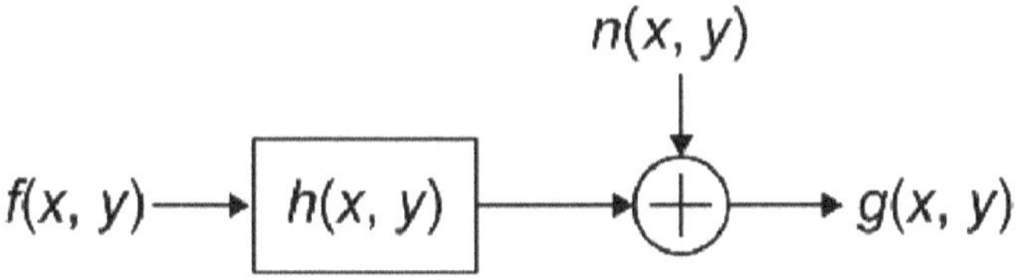

Fig. 6.62 Image degradation model

The mathematical expression of the degenerate model can be expressed as [79]:

$$g(x, y) = f(x, y) * h(x, y) + n(x, y) \tag{6.53}$$

where $*$ denotes the convolution operation. After the Fourier transform, (6.53) in the frequency domain is represented as

$$G(u, v) = F(u, v)H(u, v) + N(u, v) \tag{6.54}$$

It can be inferred that the restored images in the frequency domain are given as:

$$F'(u, v) = G(u, v)H^{-1}(u, v) \tag{6.55}$$

$G(u, v)$ is defined in (6.54), then substitute it into (6.55),

$$F'(u, v) = F(u, v) + N(u, v)H^{-1}(u, v) \tag{6.56}$$

The restored image is obtained by using the inverse Fourier transform:

$$f'(x, y) = F^{-1}(N(u, v)H^{-1}(u, v))) \tag{6.57}$$

where $f'(x, y)$ denotes the restored image. So, it is possible to restore a blurred image without noise.

Before the start of GA, the significance and design of image segmentation will be introduced. During the process of image restoration, the number of pixels in a picture is very big, so the computing burden will be heavy. Suppose that the size of the image to be restored is 100×30 pixels, and the image is not divided into blocks. At this time, an image is an individual has $100 \times 30 = 3000$ genes, and the range of each gene is an integer of 0–255. For one graph, there are 256^{3000} different feasible solutions, which is a very big number. Usually, the size of the initial population is about 20–120. In this case, the size of the randomly generated initial population is only about $100/256^{3000}$, and the result of this ratio is close to zero. Therefore, the probability of getting individuals with good characters is very small, that is, the fitness of individuals is very low, and the use of improved selection, crossover, and mutation algorithms cannot get good results afterward. The speed of convergence is very slow because the individual does not perform well from the beginning. Furthermore, in order to increase the probability of producing good initial individuals, we have to choose to increase the size of the initial population. However, increasing the size of the initial population means that more space is needed to store individual information, and each additional individual of the initial population needs more than 3000 memories to store individual information, so the space cost is very huge. The most critical problem is that even if the size of the initial population is increased to thousands, tens of thousands, or even hundreds of thousands, it is still a very small number compared with 256^{3000}. Moreover, when the size of the picture increases slightly, this number will grow exponentially, and the increasing population size will lead to greatly increasing the amount of memristors needed and the space cost. The effect of improving the efficiency of the algorithm is negligible. Above all, we put forward to process the image by blocks.

6.5.3.2 Image Division into Small Blocks

In this situation, the size of the image block is set as 3×3 pixels, and the image block is moved according to the rules from left to right and from top to bottom. The 3×3-image block first traverses from left to right, moves one pixel down after traversing a whole line and continues to traverse a whole row until all pixels have been traversed.

Thus, an individual is not a whole image, but the size of a block, and its structure is like that a chromosome has nine genes. To simplify the operation, the data matrix of a 3×3 block will be converted into a matrix of 1×9. Assuming that the number of the initial population is 20, then the size of the initial population is the two-dimensional matrix of 20×9. Besides, each individual also needs a memory to store the value of fitness, so the size of the matrix is extended to 20×10. This matrix is called the operation matrix, in which the first nine genes in each row form an individual. The number of rows represents the size of the population, and the last one stores the individual fitness value. So two memristor-based arrays are required, one is to store the blurred image (called MNblur instead), and another is to store the restored image (called MNres instead). In addition, an operation-matrix memristor-based array is also needed, GA operations all happen in the operation-matrix memristor-based array.

6.5.3.3 GA Operation for Image Restoration

Coding and Initializing Population

Two-dimensional integer coding is adopted by this chapter because the picture can be seen as a two-dimensional matrix storing pixel information. The gray value of the image is expressed as the genetic value directly, which not only retains the similarity between the data and the image but also avoids the complex coding and decoding operations. It is helpful to improve the operation speed, and then stochastically generate the initial population.

Fitness Function

The fitness function is a function to evaluate the fitness of an individual, which is generally required to be nonnegative. The purpose of image restoration is to obtain the best original image, but the restoration quality of an image can only be known by comparing it with the original image and restored image, especially, if the original image is unknown. However, we discussed the image degradation model in Sect. 6.5.3.1, and the degraded image $g(x, y)$ is degraded from the original image $f(x, y)$. If the restored image $f'(x, y)$ can also produce another degraded image $g'(x, y)$ after the same degradation function, then both $g(x, y)$ and $g'(x, y)$ are known and can be compared. If the final $g(x, y)$ is closer to $g(x, y)$, $f(x, y)$ will be more similar to $f(x, y)$, so the fitness function is designed by [80].

$$E(f_i'(x, y)) = \frac{1}{\left|g(x, y) - h(x, y) * f_i'(x, y)\right|^2} \tag{6.58}$$

where $f_i'(x, y)$ denotes the restored image of individual i, $g(x, y)$ represents the degraded image that is to be restored, and $h(x, y)$ is the degradation function.

Selection Function

The selection function in this research adopts the roulette selection: The probability of each individual being selected is defined as:

$$p(x_i) = \frac{f(x_i)}{\sum_{n}^{i=1} f(x_i)} \tag{6.59}$$

where x_i represents individual, and $f(x_i)$ denotes the fitness value of individual x_i.

Crossover Function

As discussed before, we designed image block processing, and one image block is an individual. The number of genes is very small, so instead of the complex way of crossover, one-point crossover is used. The crossing point is selected by a random function. According to the position of the selected crossing point, a chromosome is divided into two parts before and after the crossing point, where the matching parents exchange the front or back gene values of the crossing point.

Mutation Function

The mutation function imitates the mutation of biological genes in nature. First, a gene bit on the chromosome is selected randomly, and the decimal gray value on the gene position is converted into the binary code of 8 bits, then one bit is randomly selected to change. Finally, the changed binary value is converted into the decimal gray value.

6.5.4 Simulation Results

6.5.4.1 Data of Experiment

The experimental data is the captured picture shown in Fig. 6.63, the size is 101 × 26 pixels. The degraded image is obtained from the original image by a 4 × 4-point spread function, as shown in Fig. 6.64.

Fig. 6.63 Test image

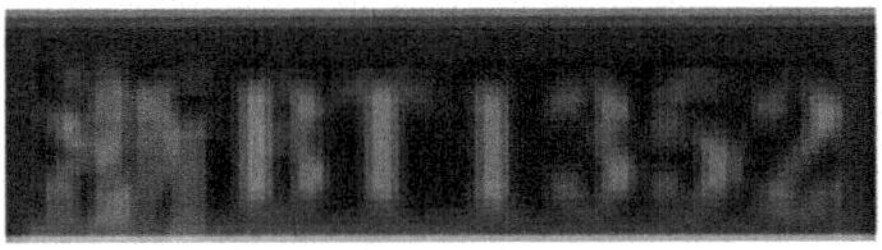

Fig. 6.64 Degraded image

6.5.4.2 Experiment Results

We discussed the design of dividing the image into blocks, so in Experiment 1, the results of dividing the image into blocks or not are compared. The set of parameters is shown in Table 6.13, and the results are shown in Fig. 6.65.

Figure 6.65a and b are the image restoration results of Case 1 and Case 2, designed in Table 6.13, respectively. They are set with the same value of the initial population and iteration, the only difference is the operation of dividing the image into blocks or not. Obviously, as shown in Fig. 6.65b, by dividing the image into blocks, Case 2 has a better restoration result. Because of the large number of pixels and the wide range of pixels in the image processed without blocks, it is difficult to get good individuals by randomly generating the initial population. As a result, the convergence speed is very slow in the later stage, and the individual performance is still very poor after the same number of iterations. It is less likely to obtain more information through the restored image in Case 1.

Then in Experiment 2, different values of the population size and iterations are set in Table 6.14. For all cases in Table 6.14, the dividing operation with the block size of 3×3 pixels is adopted.

Then, we get different results and their running time, as shown in Fig. 6.66.

Table 6.13 Case design

ID	Divide image to blocks	Initial population size	Iteration	Size of block
Case 1	No	120	150	/
Case 2	Yes	120	150	3×3 pixels

Fig. 6.65 Image restoration results of case design in Table 1.13: **a** result of Case 1 and **b** result of Case 2

Table 6.14 Case design

ID	Population size	Iteration
Case 2	120	150
Case 3	30	300
Case 4	30	120
Case 5	20	120

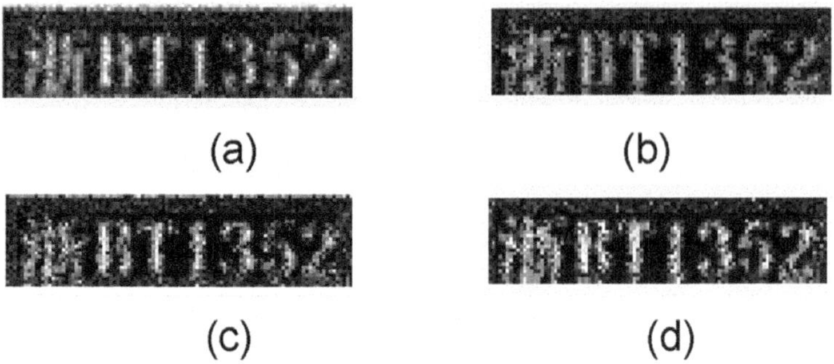

Fig. 6.66 Experiment 2 results: **a** $ID = 2$ and $t = 50$ min, **b** $ID = 3$ and $t = 22$ min, **c** $ID = 4$ and $t = 10.5$ min, and **d** $ID = 5$ and $t = 7$ min

6.5.4.3 Compare Study

The contrastive study is shown in Table 6.15, and experimental results are given in Fig. 6.67.

Figure 6.67a and b are the results of applying the Gaussian and median filters, respectively, and Fig. 6.67c is the results of applying the proposed model, which is better than the other two. From both Table 6.15 and Fig. 6.67, it can be seen that image restoration results are not good by directly performing the filtering operation on the image, and our proposed model performs better, which verifies the improvement of our model.

Compared with Case 4, Case 3 has a larger number of iterations, leading to a longer running time and a better result. Compared with Case 5, Case 4 has a larger population size, leading to a longer running time and a clearer result. Cases 1 to 5 demonstrate the feasibility of deploying GA with memristors in image restoration. Besides, when the initial population and the number of iterations are too small (as shown in Fig. 6.66d), the restoration result of the image is the worst: The words and

Table 6.15 Case design

Method	Algorithm	Application
Gaussian-filter	/	Image restoration
Median filter	/	Image restoration
Our proposed model	GA	Image restoration

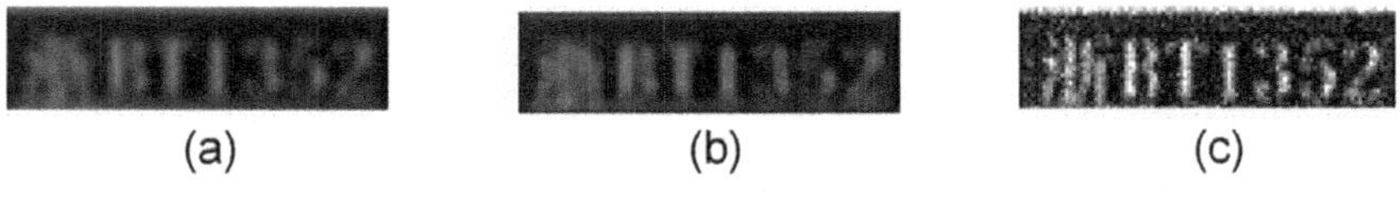

Fig. 6.67 Comparative study results: **a** Gaussian-filter, **b** median filter, and **c** our proposed model

numbers are vague. After increasing the size of the initial population and the number of iterations, the restoration results are improved, but it is not that: The larger the size of the population and the number of iterations are, the better the results. Therefore, the appropriate population size and iteration times should be chosen according to the actual needs, without blindly pursuing the large size of the initial population and iterations. Although the restoration result could be improved, the time cost is very large. Hence, the improvement of restoration and time cost should be taken into consideration.

6.5.5 Conclusion

In this chapter, memristor-based GA for image restoration is proposed. The mapping rules between the memristor and GA are shown. According to the rules, memristor-based GA is designed and carried out successfully. The comparative study shows that the proposed method has better performance. It is demonstrated that using memristors to implement GA is possible and memristor-based GA for image restoration is feasible.

Memristor, GA, and image restoration all have wide applications in different fields. There are still some other algorithms, like the ant colony optimization, which can be implemented in a similar way. However, the existing methods are not completely unified yet. Once unified, more different algorithms can be implemented with the memristor-based array, which is helpful to increase the use of memristors. Thus, more efforts are needed to design a unified system and enrich the application cases.

6.6 Comments

Swarm intelligence algorithms have been widely applied in fields such as signal processing, image processing, production scheduling, and mechanical design due to their robustness, self-organization, and flexibility. In recent years, algorithms such as Bee Colony Optimization, Elephant Herding Optimization, and Harris Hawks Optimization (refer to Fig. 6.68) have made significant breakthroughs in their respective domains, demonstrating strong problem-solving capabilities. However, the integration of these algorithms with memristors is still in its early stages, holding vast exploration space and research potential.

Currently, introducing memristors into swarm intelligence algorithms is a major research direction that may break through the inherent limitations of traditional algorithms and provide a theoretical basis for parameter selection. Parameter selection in swarm intelligence algorithms largely relies on researchers' prior knowledge and experimental results, lacking a rigorous theoretical foundation. This dependence not only limits the generality of the algorithms but also increases the difficulty and time cost of algorithm tuning. The nonlinear characteristics and intrinsic memory

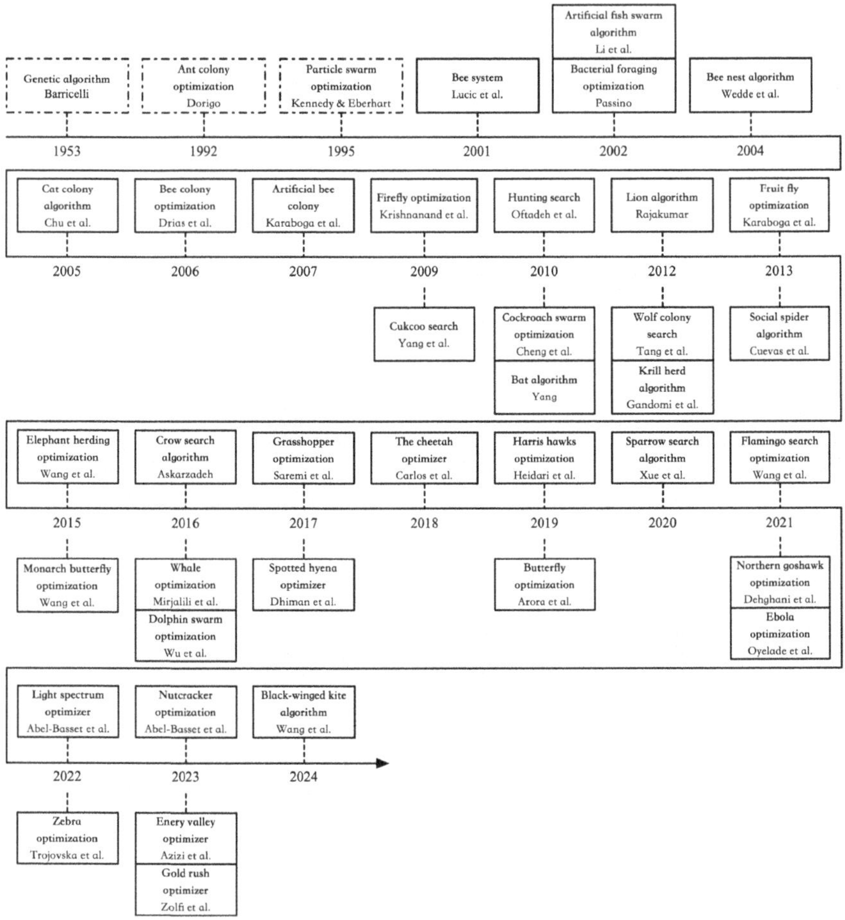

Fig. 6.68 Development trends of swarm intelligence algorithms

functions of memristors may provide a new paradigm for real-time dynamic adjustment of algorithm parameters, potentially solving the problem of parameter selection being overly dependent on experience and trial-and-error. Memristors can serve as dynamic weight storage units, adjusting parameters in real time based on problem characteristics and optimization processes, improving the adaptability and optimization efficiency of the algorithms.

Exploring the impact mechanisms of memristor characteristics on the working principles, mathematical foundations, and dynamic characteristics of swarm intelligence algorithms is a key research direction. The nonlinear voltage-current characteristics of memristors may provide new mathematical models for swarm intelligence algorithms, changing the traditional iterative update methods. This new mathematical description may lead to a reassessment of the convergence and stability of the algorithms, opening up new avenues for theoretical research on swarm intelligence algorithms. In terms of dynamic characteristics, the hysteresis properties of memristors may introduce new oscillation and bifurcation behaviors, enriching the dynamic features of swarm intelligence algorithms and potentially enhancing their ability to escape local optima.

Memristor networks may simulate complex group interaction functions such as competition, cooperation, and evolution, achieving more efficient information transmission and decision-making processes. For example, the multi-value storage characteristics of memristors can be used to record the historical competition and cooperation states of individuals, enhancing the long-term memory capability of the group and improving the adaptability of the algorithms in dynamic environments. With memristors as the cornerstone, exploring new modes of information sharing, collaborative cooperation, and learning optimization among groups and establishing swarm intelligence collaborative cognitive mechanisms are urgently needed. The continuous state change characteristics of memristors may support more fine-grained evolutionary processes, avoiding the problem of information loss in traditional discrete evolutionary algorithms and improving the exploration ability and convergence speed of the algorithms.

At the practical application level, most optimization problems are essentially discrete, such as node load balancing and resource scheduling. Utilizing the unique characteristics of memristors to improve the modeling methods and parameter settings of swarm intelligence algorithms for these problems, in order to obtain more efficient and robust solutions, will be an important breakthrough point. The multi-value characteristics of memristors may provide more natural representation methods for discrete problems, and their nonvolatile storage capabilities may support incremental solutions for large-scale optimization problems. Moreover, integrating memristor-based swarm intelligence algorithms with advanced methods such as deep learning and reinforcement learning may give birth to entirely new hybrid algorithm paradigms. For example, memristor networks can be used to simultaneously realize the forward propagation of neural networks and the optimization process of swarm intelligence algorithms, creating new intelligent systems with both learning and optimization capabilities.

The deep integration of memristors and swarm intelligence algorithms is expected to achieve qualitative leaps in algorithm performance, application fields, and theoretical foundations. This innovative cross-disciplinary research may not only significantly enhance the performance limits of existing algorithms but also give rise to revolutionary new algorithm paradigms. By combining the hardware advantages of memristors and the software flexibility of swarm intelligence algorithms, we hope to develop optimization systems with higher energy efficiency, larger scale, and stronger adaptability.

References

1. Pereira, F.B., Tavares, J.: Bio-Inspired Algorithms for the Vehicle Routing Problem, vol. 161. Springer (2008)
2. Das, T.K., Venayagamoorthy, G.K., Aliyu, U.O.: Bio-inspired algorithms for the design of multiple optimal power system stabilizers: Sppso and bfa. IEEE Trans. Ind. Appl. **44**(5), 1445–1457 (2008)
3. Dorigo, M., Blum, C.: Ant colony optimization theory: a survey. Theoret. Comput. Sci. **344**(2–3), 243–278 (2005)
4. Tian, J., Yu, W., Xie, S.: An ant colony optimization algorithm for image edge detection. In: 2008 IEEE Congress on Evolutionary Computation (IEEE World Congress on Computational Intelligence), pp. 751–756. IEEE (2008)
5. Baterina, A.V., Oppus, C.: Image edge detection using ant colony optimization. Wseas Trans. Signal Proc. **6**(2), 58–67 (2010)
6. Tian, J., Yu, W., Chen, L., Ma, L.: Image edge detection using variation-adaptive ant colony optimization. Transactions on Computational Collective Intelligence V, pp. 27–40 (2011)
7. Xia, Q., Yang, J.J., Wu, W., Li, X., Williams, R.S.: Self-aligned memristor cross-point arrays fabricated with one nanoimprint lithography step. Nano Lett. **10**(8), 2909–2914 (2010)
8. Zidan, M., Omran, H., Naous, R., Sultan, A., Fahmy, H., Lu, W., Salama, K.N.: Single-readout high-density memristor crossbar. Sci. Rep. **6**(1), 18863 (2016)
9. Milo, V., Pedretti, G., Carboni, R., Calderoni, A., Ramaswamy, N., Ambrogio, S., Ielmini, D.: Demonstration of hybrid cmos/rram neural networks with spike time/rate-dependent plasticity. In: 2016 IEEE International Electron Devices Meeting (IEDM), pp. 16–8. IEEE (2016)
10. Alibart, F., Zamanidoost, E., Strukov, D.B.: Pattern classification by memristive crossbar circuits using ex situ and in situ training. Nat. Commun. **4**(1), 2072 (2013)
11. Prezioso, M., Merrikh-Bayat, F., Hoskins, B.D., Adam, G.C., Likharev, K.K., Strukov, D.B.: Training and operation of an integrated neuromorphic network based on metal-oxide memristors. Nature **521**(7550), 61–64 (2015)
12. Sheridan, P.M., Cai, F., Du, C., Ma, W., Zhang, Z., Lu, W.D.: Sparse coding with memristor networks. Nat. Nanotechnol. **12**(8), 784–789 (2017)
13. Choi, S., Shin, J.H., Lee, J., Sheridan, P., Lu, W.D.: Experimental demonstration of feature extraction and dimensionality reduction using memristor networks. Nano Lett. **17**(5), 3113–3118 (2017)
14. Pershin, Y.V., Di Ventra, M.: Memcomputing implementation of ant colony optimization. Neural Process. Lett. **44**, 265–277 (2016)
15. Pajouhi, Z., Roy, K.: Image edge detection based on swarm intelligence using memristive networks. IEEE Trans. Comput. Aided Des. Integr. Circuits Syst. **37**(9), 1774–1787 (2017)
16. Yu, J., Tan, M., Zhang, H., Rui, Y., Tao, D.: Hierarchical deep click feature prediction for fine-grained image recognition. IEEE Trans. Pattern Anal. Mach. Intell. **44**(2), 563–578 (2019)
17. Keysers, D., Deselaers, T., Gollan, C., Ney, H.: Deformation models for image recognition. IEEE Trans. Pattern Anal. Mach. Intell. **29**(8), 1422–1435 (2007)
18. Yu, J., Zhu, C., Zhang, J., Huang, Q., Tao, D.: Spatial pyramid-enhanced netvlad with weighted triplet loss for place recognition. IEEE Trans. Neural Netw. Learn. Syst. **31**(2), 661–674 (2019)
19. Redmon, J., Divvala, S., Girshick, R., Farhadi, A.: You only look once: unified, real-time object detection. In: Proceedings of the IEEE Conference on Computer Vision and Pattern Recognition, pp. 779–788 (2016)
20. Liu, W., Anguelov, D., Erhan, D., Szegedy, C., Reed, S., Fu, C.Y., Berg, A.C.: Ssd: single shot multibox detector. In: Computer Vision–ECCV 2016: 14th European Conference, Amsterdam, The Netherlands, October 11–14, 2016, Proceedings, Part I 14, pp. 21–37. Springer (2016)
21. Hong, C., Yu, J., Zhang, J., Jin, X., Lee, K.H.: Multimodal face-pose estimation with multitask manifold deep learning. IEEE Trans. Industr. Inf. **15**(7), 3952–3961 (2018)
22. Hong, C., Yu, J., Wan, J., Tao, D., Wang, M.: Multimodal deep autoencoder for human pose recovery. IEEE Trans. Image Process. **24**(12), 5659–5670 (2015)

23. Yu, J., Zhang, B., Kuang, Z., Lin, D., Fan, J.: iprivacy: image privacy protection by identifying sensitive objects via deep multi-task learning. IEEE Trans. Inf. Forensics Secur. **12**(5), 1005–1016 (2016)
24. Yu, J., Tao, D., Wang, M., Rui, Y.: Learning to rank using user clicks and visual features for image retrieval. IEEE Trans. Cybern. **45**(4), 767–779 (2014)
25. Zhang, J., Yu, J., Tao, D.: Local deep-feature alignment for unsupervised dimension reduction. IEEE Trans. Image Process. **27**(5), 2420–2432 (2018)
26. Ding, L., Goshtasby, A.: On the canny edge detector. Pattern Recogn. **34**(3), 721–725 (2001)
27. Chua, L.: Memristor-the missing circuit element. IEEE Trans. Circuit Theory **18**(5), 507–519 (1971)
28. Chua, L.O.: The fourth element. Proc. IEEE **100**(6), 1920–1927 (2012)
29. Chua, L.O.: How we predicted the memristor. Nature Electronics **1**(5), 322–322 (2018)
30. Gao, W., Zhang, X., Yang, L., Liu, H.: An improved sobel edge detection. In: 2010 3rd International Conference on Computer Science and Information Technology, vol. 5, pp. 67–71. IEEE (2010)
31. Xie, S., Tu, Z.: Holistically-nested edge detection. In: Proceedings of the IEEE International Conference on Computer Vision, pp. 1395–1403 (2015)
32. Pande, S., Bhadouria, V.S., Ghoshal, D.: A study on edge marking scheme of various standard edge detectors. Int. J. Comput. Appl. **44**(9), 33–37 (2012)
33. Du, C., Cai, F., Zidan, M.A., Ma, W., Lee, S.H., Lu, W.D.: Reservoir computing using dynamic memristors for temporal information processing. Nat. Commun. **8**(1), 2204 (2017)
34. Fister, I., Jr., Perc, M., Kamal, S.M., Fister, I.: A review of chaos-based firefly algorithms: perspectives and research challenges. Appl. Math. Comput. **252**, 155–165 (2015)
35. Fister, I., Jr., Perc, M., Ljubič, K., Kamal, S.M., Iglesias, A., Fister, I.: Particle swarm optimization for automatic creation of complex graphic characters. Chaos, Solitons & Fractals **73**, 29–35 (2015)
36. Mirjalili, S., Lewis, A.: The whale optimization algorithm. Adv. Eng. Softw. **95**, 51–67 (2016)
37. Holland, J.H.: Adaptation in Natural and Artificial Systems: An Introductory Analysis with Applications to Biology, Control, and Artificial Intelligence. MIT press (1992)
38. Deb, K., Pratap, A., Agarwal, S., Meyarivan, T.: A fast and elitist multiobjective genetic algorithm: Nsga-ii. IEEE Trans. Evol. Comput. **6**(2), 182–197 (2002)
39. Liu, J., Tang, X.: Evolutionary search for faces from line drawings. IEEE Trans. Pattern Anal. & Mach. Intell. **27**(06), 861–872 (2005)
40. Das, S., Suganthan, P.N.: Differential evolution: a survey of the state-of-the-art. IEEE Trans. Evol. Comput. **15**(1), 4–31 (2010)
41. Ghamisi, P., Benediktsson, J.A.: Feature selection based on hybridization of genetic algorithm and particle swarm optimization. IEEE Geosci. Remote Sens. Lett. **12**(2), 309–313 (2014)
42. Xue, B., Zhang, M., Browne, W.N., Yao, X.: A survey on evolutionary computation approaches to feature selection. IEEE Trans. Evol. Comput. **20**(4), 606–626 (2015)
43. Bianco, S., Ciocca, G., Schettini, R.: Combination of video change detection algorithms by genetic programming. IEEE Trans. Evol. Comput. **21**(6), 914–928 (2017)
44. Faliszewski, P., Sawicki, J., Schaefer, R., Smolka, M.: Multiwinner voting in genetic algorithms. IEEE Intell. Syst. **32**(1), 40–48 (2017)
45. Nagamani, A., Anuktha, S., Nanditha, N., Agrawal, V.K.: A genetic algorithm-based heuristic method for test set generation in reversible circuits. IEEE Trans. Comput. Aided Des. Integr. Circuits Syst. **37**(2), 324–336 (2017)
46. Gong, D., Sun, J., Miao, Z.: A set-based genetic algorithm for interval many-objective optimization problems. IEEE Trans. Evol. Comput. **22**(1), 47–60 (2016)
47. Gao, B., Li, X., Woo, W.L., yun Tian, G.: Physics-based image segmentation using first order statistical properties and genetic algorithm for inductive thermography imaging. IEEE Trans. Image Proc. **27**(5), 2160–2175 (2017)
48. Fister, I., Suganthan, P.N., Fister, I., Kamal, S.M., Al-Marzouki, F.M., Perc, M., Strnad, D.: Artificial neural network regression as a local search heuristic for ensemble strategies in differential evolution. Nonlinear Dyn. **84**, 895–914 (2016)

49. Di Ventra, M., Pershin, Y.V.: The parallel approach. Nat. Phys. **9**(4), 200–202 (2013)
50. Gonzales, R.C., Wintz, P.: Digital Image Processing. Addison-Wesley Longman Publishing Co. Inc (1987)
51. Gudmundsson, M., El-Kwae, E.A., Kabuka, M.R.: Edge detection in medical images using a genetic algorithm. IEEE Trans. Med. Imaging **17**(3), 469–474 (1998)
52. Saenthon, A., Kaitwanidvilai, S.: Development of new edge-detection filter based on genetic algorithm: an application to a soldering joint inspection. Int. J. Adv. Manuf. Technol. **46**, 1009–1019 (2010)
53. Pratt, W.K.: Digital Image Processing. A Wiley-Interscience Publication by John Wiley & Sons, New York (1978)
54. Yi, S., Labate, D., Easley, G.R., Krim, H.: A shearlet approach to edge analysis and detection. IEEE Trans. Image Process. **18**(5), 929–941 (2009)
55. Pershin, Y.V., Di Ventra, M.: Memcomputing and swarm intelligence (2014). arXiv:1408.6741
56. Neshat, M., Adeli, A., Sepidnam, G., Sargolzaei, M., Toosi, A.N.: A review of artificial fish swarm optimization methods and applications. Int. J. Smart Sens. Intell. Syst. **5**(1), 107–148 (2012)
57. Kennedy, J.: Swarm intelligence. In: Handbook of Nature-Inspired and Innovative Computing: Integrating Classical Models with Emerging Technologies, pp. 187–219. Springer (2006)
58. Joglekar, Y.N., Wolf, S.J.: The elusive memristor: properties of basic electrical circuits. Eur. J. Phys. **30**(4), 661 (2009)
59. Pospichal, P., Jaros, J., Schwarz, J.: Parallel genetic algorithm on the cuda architecture. In: EvoApplications (2010)
60. Yu, Q., Chen, C., Pan, Z.: Parallel genetic algorithms on programmable graphics hardware. In: International Conference on Computing, Networking and Communications (2005)
61. Li, J.M., Wang, X., He, R.S., Chi, Z.X.: An efficient fine-grained parallel genetic algorithm based on gpu-accelerated. 2007 IFIP International Conference on Network and Parallel Computing Workshops (NPC 2007) pp. 855–862 (2007)
62. Guo, L., Funie, A.I., Thomas, D.B., Fu, H., Luk, W.W.C.: Parallel genetic algorithms on multiple fpgas. ACM SIGARCH Comput. Archit. News **43**, 86–93 (2016)
63. Vavouras, M., Papademetriou, K., Papaefstathiou, I.: High-speed fpga-based implementations of a genetic algorithm. In: 2009 International Symposium on Systems, Architectures, Modeling, and Simulation, pp. 9–16 (2009)
64. Vega-Rodríguez, M.A., Gutierrez-Gil, R., Avila-Roman, J.M., Sánchez-Pérez, J.M., Pulido, J.A.G.: Genetic algorithms using parallelism and fpgas: the tsp as case study. In: 2005 International Conference on Parallel Processing Workshops (ICPPW'05), pp. 573–579 (2005)
65. Ferrucci, F., Salza, P., Sarro, F.: Using hadoop mapreduce for parallel genetic algorithms: a comparison of the global, grid and island models. Evol. Comput. **26**, 535–567 (2018)
66. Ferrucci, F., Salza, P., Kechadi, M.T., Sarro, F.: A parallel genetic algorithms framework based on hadoop mapreduce. In: Proceedings of the 30th Annual ACM Symposium on Applied Computing (2015)
67. Verma, A., Llorà, X., Goldberg, D.E., Campbell, R.H.: Scaling genetic algorithms using mapreduce. In: 2009 Ninth International Conference on Intelligent Systems Design and Applications, pp. 13–18 (2009)
68. Haron, A., Yu, J., Nane, R., Taouil, M., Hamdioui, S., Bertels, K.: Parallel matrix multiplication on memristor-based computation-in-memory architecture. In: 2016 International Conference on High Performance Computing & Simulation (HPCS), pp. 759–766 (2016)
69. Mahmoodi, M.R., Vincent, A.F., Nili, H., Strukov, D.B.: Intrinsic bounds for computing precision in memristor-based vector-by-matrix multipliers. IEEE Trans. Nanotechnol. **19**, 429–435 (2020)
70. Baluja, S., Caruana, R.: Removing the genetics from the standard genetic algorithm. In: International Conference on Machine Learning (1995)
71. Strukov, D.B., Snider, G.S., Stewart, D.R., Williams, R.S.: The missing memristor found. Nature **453**(7191), 80–83 (2008)

72. Hu, M., Wang, Y., Qiu, Q., Chen, Y., Li, H.H.: The stochastic modeling of tio2 memristor and its usage in neuromorphic system design. In: 2014 19th Asia and South Pacific Design Automation Conference (ASP-DAC), pp. 831–836 (2014)
73. Wang, Y., Wen, W., Li, H.H., Hu, M.: A novel true random number generator design leveraging emerging memristor technology. In: Proceedings of the 25th Edition on Great Lakes Symposium on VLSI (2015)
74. Xia, L., Gu, P., Li, B., Tang, T., Yin, X., Huangfu, W., Yu, S., Cao, Y., Wang, Y., Yang, H.: Technological exploration of rram crossbar array for matrix-vector multiplication. J. Comput. Sci. Technol. **31**, 3–19 (2015)
75. Kim, K.H., Gaba, S., Wheeler, D.C., Cruz-Albrecht, J.M., Hussain, T., Srinivasa, N., Lu, W.D.: A functional hybrid memristor crossbar-array/cmos system for data storage and neuromorphic applications. Nano Lett. **12**, 389–95 (2012)
76. Kwon, S., Kim, T.W., Jang, S., Lee, J.H., Kim, N.D., Ji, Y., Lee, C.H., Tour, J.M., Wang, G.: Structurally engineered nanoporous ta2o5-x selector-less memristor for high uniformity and low power consumption. ACS Appl. Mater. & Interf. **9**, 34015–34023 (2017)
77. eddine Joudar, N., Harchli, F., Abdelatif, E.S., Ettaouil, M.: A genetic algorithm for solving a new image restoration model based on selective filtering. In: 2015 10th International Conference on Intelligent Systems: Theories and Applications (SITA), pp. 1–5 (2015)
78. Yang, C.Y.: Research on memcomputing model and its application on image edge detection. M.S. Thesis, pp. 41–47 (2019)
79. Sheta, A.F.: Restoration of Medical Images Using Genetic Algorithms. In: 2017 IEEE Applied Imagery Pattern Recognition Workshop (AIPR), pp. 1–8 (2017)
80. Sun, Y., Xu, C., Li, G., Xu, W.F., Kong, J., Jiang, D., Tao, B., Disi, C.: Intelligent human computer interaction based on non redundant emg signal. Alex. Eng. J. **59**, 1149–1157 (2020)

Chapter 7
Dynamic Analysis of Memristive Neural Network

Memristive neural networks (MNNs) have emerged as a promising paradigm for neuromorphic computing, combining the unique properties of memristors with the powerful computational capabilities of neural networks. This chapter presents a comprehensive analysis of the dynamic behavior and stability of MNNs under various conditions, including stochastic disturbances, time-varying delays, fuzzy switching, and connection faults. This chapter explores novel control strategies and analytical methods to address the challenges associated with these conditions and ensure the reliable operation of MNNs.

In Sect. 7.1, the robust exponential stability of fuzzy-switched memristive inertial neural networks (FSMINNs) with time-varying delays is investigated under a mode-dependent destabilizing impulsive control protocol. The memristive model is treated as a switched system, and hybrid mode-dependent destabilizing impulsive and adaptive feedback controllers are employed to optimize the robust exponentially stable process and reduce the time cost. The new model considers multiple impulsive effects between switched modes and multiple switched effects between impulsive instants. Using switched analysis techniques, the Takagi-Sugeno (T-S) fuzzy method, and the average dwell time, extended robust exponential stability conditions are derived and validated through simulations.

Section 7.2 focuses on the exponential synchronization of stochastic delayed memristive neural networks (SDMNNs) using a hybrid control (HC) approach. The proposed HC scheme employs a state-dependent trigger condition to determine the impulsive instants while incorporating switching and quantification strategies to handle interval parameters, impulses, stochastic disturbances, and time-varying delays. The designed controller not only reduces control costs but also saves communication channels and bandwidths. Novel Lyapunov functions and analytical methods are developed to achieve exponential synchronization of SDMNNs via HC, with numerical simulations validating the theoretical results.

Y. Yu et al., *Memristive Computing*,
https://doi.org/10.1007/978-981-96-0332-9_7

Section 7.3 focuses on the exponential synchronization of coupled memristive neural networks (CMNNs) with multiple mismatched parameters, using a topology-based probability impulsive mechanism (TPIM) on time scales. The model addresses mismatched dimensions, connection weights, and time-varying delays, employing auxiliary state variables to synchronize CMNNs either within the system or with an external node. The TPIM enhances robustness against non-ideal factors compared to traditional methods. Exponential synchronization conditions for both continuous and discrete-time systems are derived using Lyapunov-Krasovskii functionals and robust analysis, with numerical examples validating the model's effectiveness.

The research presented in this chapter significantly advances the understanding of the dynamic behavior and stability of memristive neural networks under various challenging conditions. By developing novel control strategies, analytical methods, and stability conditions, this chapter paves the way for the reliable and efficient implementation of MNNs in real-world applications. The insights gained from the analysis of stochastic disturbances, time-varying delays, fuzzy switching, and connection faults contribute to the development of robust and fault-tolerant MNNs, enabling their deployment in complex and dynamic environments.

7.1 Robust Exponential Stability of Fuzzy Switched Memristive Inertial Neural Networks

7.1.1 Introduction

With Recurrent Neural Networks, the connections between their nodes form a directed graph along a temporal sequence. They have received extensive attention [1–8] because RNNs can process sequences of inputs by using their internal states. If the resistors are replaced with the memristors in RNNs, then a new model can be built, which means that their connection weights can be changed according to the states of the system. These types of networks are called memristive recurrent neural networks (MRNNs) and simplified into memristive neural networks (MNNs), which are state-dependent switched networks. The dynamic characteristic of MNNs has drawn plenty of attention, such as exponential stability [9–11], finite-time stability [12, 13], global mittag-leffler stabilization [14], finite-time synchronization [15], input-to-state stability [16], and so forth. The system consisting of these types of networks is a nonlinear system, for the purpose of researching the nonlinear dynamical system more effectively, some types of fuzzy models are elaborated such as Mamdani models [17, 18], Takagi-Sugeno (T-S) models [19] and Neuro-fuzzy models [20], which the aim of these models is to achieve greater accuracy, and dimensionality and also desire to simplify the structure of nonlinear systems. Specifically, the T-S fuzzy logic has aroused extensive attention [10, 21, 21–27], because it can approximate nonlinear smooth functions with arbitrary accuracy linear functions.

The evolution trend of the MNNs system not only depends on the current state of the system but also relates to the state of the past moment or several moments, so time-varying delays are inevitable in MNNs. Moreover, in the hardware implementation of MNNs, time delays, in particular time-varying delays among the networks, are very common because of the finite switched speed of amplifiers, the finite information transmission (the network bandwidth restricts the rapid transmission of information), and processing speed among the units, which may degrade the system performance, such as oscillatory behavior, divergence, and system instability. Meanwhile, the spatial extent is ubiquitous in neural networks because of the presence of an amount of parallel pathways of varying axon size and lengths, which may cause distributed time delays in the networks. Thus, dynamic analysis for MNNs with time-varying delays [13, 28–30] or distributed time delays [31–33] has widely aroused the enthusiasm of various scholars in the past few years. If the circuit model contains inductance elements, the inertial items will be obtained [11, 34]. The inertia terms are considered to be an effective method for generating chaos and complex bifurcation behaviors, such as homoclinic and heteroclinic orbits, and limit cycles, which can be interpreted as stored memories [35–38]. Therefore, the stability analysis of MNNs with time-varying delays and inertial items is an important research subject.

In many practical dynamic systems, uncertain parameters are inevitably introduced in the process of parameters design because the weights of neurons essentially depend on certain resistance and capacitance values. In order to ensure that the system can maintain its stability under parameter perturbation, many researchers consider robustness of the system [9, 39–43]. Moreover, the state of neural networks often subjects to instantaneous interference in some moments due to the interference of internal structure or external environment, and the phenomenon that can be considered as a kind of impulse perturbation [44–47]. In [44], the impulsive instants are determined by certain state-dependent triggering conditions, and the authors study the synchronization of MNNs. The globally exponential stability of MNNs with time-varying delays and impulsive effects has been investigated in [45]. In [46], authors study the exponential synchronization of coupled stochastic MNNs with probabilistic time-varying delay coupling and time-varying impulsive delay. This chapter considers the asymptotic synchronization problem for a class of uncertain complex networks with hybrid switched and impulsive effects, and the switches and impulses occur by following a time sequence which is characterized by dwell time constraint [47]. In addition, although although recent works have studied the switched impulsive neural networks and have obtained some good results [16, 47–50], none of these results have involved the MNNs especially the inertial MNNs.

In this section, $\mathbb{N}$ and $\mathbb{N}^+$ denote the set of nonnegative integers and the set of positive integers, respectively; $\mathbb{R}^n$ and $\mathbb{R}^{n\times m}$ denote the set of the n-dimensional euclidean space and $n \times m$ real matrix space, $C([-\tau, 0], \mathbb{R}^n)$ represents a Banach space of all continuous functions. $\|x\|$ represents the Euclidean norm of the vector x. I_n represents a identity matrix of $n \times n$ dimension. The symmetric term in a symmetric matrix is denoted by $*$, and $\mathrm{Sym}(H) = H + H^T$.

7.1.2 Problem Description and Preliminaries

$$
\begin{cases}
\ddot{z}_i(t) = & -(a_i + \Delta a_i(t))\dot{z}_i(t) - (b_i + \Delta b_i(t))z_i(t) \\
& + \sum_{j=1}^{n} (c_{ij}(z_i(t)) + \Delta c_{ij}(t))g_j(z_j(t)) \\
& + \sum_{j=1}^{n} (d_{ij}(z_i(t)) + \Delta d_{ij}(t))g_j(z_j(t - \tau(t))) \\
& +\bar{u}_i(t, z_i(t), \dot{z}_i(t)) + \tilde{u}_i(t, z_i(t), \dot{z}_i(t)) \\
z_i(s) = & \psi_i(s), \quad s \in [-\tau, 0] \\
\dot{z}_i(s) = & \varphi_i(s), \quad s \in [-\tau, 0]
\end{cases}
\tag{7.1}
$$

where the second derivative is called an inertial term of (7.1); $\tau(t)$ corresponds to the time-varying delays; $z_i(t)$ and $g_j(\cdot)$ represent the state of the ith neuron and active function of the jth neuron, respectively; a_i and b_i are known positive constants, and b_i denotes the rate with which the ith neuron will reset its potential to the resetting state in isolation when disconnected from the network and external inputs; $\bar{u}_i(t, z_i(t), \dot{z}_i(t))$ and $\tilde{u}_i(t, z_i(t), \dot{z}_i(t))$ represent impulsive controllers and adaptive feedback controllers, respectively; $\psi_i(s)$ and $\varphi_i(s)$ represent the initial values of $z_i(t)$ and $\dot{z}_i(t)$, respectively. $c_{ij}(z_i(t))$ and $d_{ij}(z_i(t))$ represent connection weights. According to [51], the state-dependent parameters in (7.1) have the following form:

$$
\begin{aligned}
c_{ij}(z_i(t)) &= \begin{cases} \hat{c}_{ij}, & z_i(s) \searrow or \rightarrow, s \in (t - \Delta, t], \\ \check{c}_{ij}, & z_i(s) \nearrow, s \in (t - \Delta, t], \end{cases} \\
d_{ij}(z_i(t)) &= \begin{cases} \hat{d}_{ij}, & z_i(s) \searrow or \rightarrow, s \in (t - \Delta, t], \\ \check{d}_{ij}, & z_i(s) \nearrow, s \in (t - \Delta, t], \end{cases} ,
\end{aligned}
\tag{7.2}
$$

where $\hat{c}_{ij}, \check{c}_{ij}, \hat{d}_{ij}$ and $\check{d}_{ij}$ are known constants. $\searrow$, $\nearrow$ and $\rightarrow$ represent "decrease", "increase", and "unchange", respectively; Δ is a sufficiently small constant with $\Delta > 0$.

Assumption 7.1 For $j = 1, \ldots, n$, the neuron activation function $g_j(\cdot)$ satisfies the following condition:

$$
l_j^- \leq \frac{g_j(y_j) - g_j(x_j)}{y_j - x_j} \leq l_j^+, \quad (y_j \neq x_j)
\tag{7.3}
$$

where l_j^- and l_j^+ are some known constants.

Assumption 7.2 There exists a positive constant τ satisfying $0 \leq \tau(t) \leq \tau$.

For the sake of representation, the vector form of (7.1) can be described as follows:

$$\begin{cases} \ddot{z}(t) = & -(A+\Delta A(t))\dot{z}(t) - (B+\Delta B(t))z(t) + (C(z(t)) \\ & +\Delta C(t))g(z(t)) + (D(z(t)) + \Delta D(t))g(z(t-\tau(t))) \\ & +\bar{u}(t, z(t), \dot{z}(t)) + \tilde{u}(t, z(t), \dot{z}(t)) \\ z(s) = & \psi(s), \quad s \in [-\tau, 0] \\ \dot{z}(s) = & \varphi(s), \quad s \in [-\tau, 0] \end{cases} \tag{7.4}$$

where $z(t) = [z_1(t), \ldots, z_n(t)]^T$, $g(z(\cdot)) = [g_1(z_1(\cdot)), \ldots, g_n(z_n(\cdot))]^T$, $\bar{u}(t, z(t), \dot{z}(t)) = [\bar{u}_1(t, z_1(t), \dot{z}_1(t)), \ldots, \bar{u}_n(t, z_n(t), \dot{z}_n(t))]^T$, $\tilde{u}(t, z(t), \dot{z}(t)) = [\tilde{u}_1(t, z_1(t), \dot{z}_1(t)), \ldots, \tilde{u}_n(t, z_n(t), \dot{z}_n(t))]^T$, $A = \text{diag}(a_1, \ldots, a_n)$, $B = \text{diag}(b_1, \ldots, b_n)$, $C(z(t)) = [c_{ij}(z_i(t))]_{n\times n}$, $D(z(t)) = [d_{ij}(z_i(t))]_{n\times n}$, $\Delta A(t) = \text{diag}(\Delta a_1(t), \ldots, \Delta a_n(t))$, $\Delta B(t) = \text{diag}(\Delta b_1(t), \ldots, \Delta b_n(t))$, $\Delta C(t) = [\Delta c_{ij}(t)]_{n\times n}$, $\Delta D(t) = [\Delta d_{ij}(t)]_{n\times n}$.

According to the feature of (7.2), we observe that $c_{ij}(z_i(t))$ and $d_{ij}(z_i(t))$ have two cases, that is to say, $c_{ij}(z_i(t))$ and $d_{ij}(z_i(t))$ may be $\hat{c}_{ij}$ or $\check{c}_{ij}$ and $\hat{d}_{ij}$ or $\check{d}_{ij}$ at any fixed t, respectively. Then the number of possible combinations of matrices $C(x(t))$ and $D(x(t))$ is 2^n. The 2^n cases can be designed as the following ordered form:

$$(C^1, D^1), \ldots, (C^{\wp}, D^{\wp}) \tag{7.5}$$

where $\wp = 2^n$.

At fixed time t, it can be seen that $C(x(t))$ and $D(x(t))$ must be one of the above $\wp$ cases, which can be described as the fact that there exists the signal $\jmath \in \mathbb{I}_{\wp} = \{1, \ldots, \wp\}$ such that $C(x(t)) = C^{\jmath}$ and $D(x(t)) = D^{\jmath}$. Therefore, (7.4) can be reconstituted into the following form:

$$\begin{cases} \ddot{z}(t) = & -(A+\Delta A(t))\dot{z}(t) - (B+\Delta B(t))z(t) + (C^{\jmath} + \Delta C(t))g(z(t)) \\ & + (D^{\jmath} + \Delta D(t))g(z(t-\tau(t))) + \bar{u}(t, z(t), \dot{z}(t)) + \tilde{u}(t, z(t), \dot{z}(t)) \\ z(s) = & \psi(s), \quad s \in [-\tau, 0] \\ \dot{z}(s) = & \varphi(s), \quad s \in [-\tau, 0] \end{cases} \tag{7.6}$$

For any $\jmath \in \mathbb{I}_{\wp}$, the characteristic function of $C(x(t))$ and $D(x(t))$ is defined as follows:

$$o_{\jmath}(t) = \begin{cases} 1, & C(x(t)) = C^{\jmath} \quad \text{and} \quad D(x(t)) = D^{\jmath} \\ 0, & \text{otherwise} \end{cases} \tag{7.7}$$

It can be easily concluded that $\sum_{\jmath=1}^{\wp} o_{\jmath}(t) = 1$. Hence, (7.6) can be translated into the following:

$$\begin{cases} \ddot{z}(t) = -(A + \Delta A(t))\dot{z}(t) - (B + \Delta B(t))z(t) + (C(t) + \Delta C(t))g(z(t)) \\ \qquad + (\mathcal{D}(t) + \Delta D(t))g(z(t - \tau(t))) + \bar{u}(t, z(t), \dot{z}(t)) + \tilde{u}(t, z(t), \dot{z}(t)) \\ z(s) = \psi(s), \quad s \in [-\tau, 0] \\ \dot{z}(s) = \varphi(s), \quad s \in [-\tau, 0] \end{cases} \tag{7.8}$$

where $\mathcal{D}(t) = \sum_{\jmath=1}^{\wp} o_\jmath(t) D^\jmath$, $C(t) = \sum_{\jmath=1}^{\wp} o_\jmath(t) C^\jmath$.

Remark 7.1 From (7.2), we observe that (7.4) is a discontinuous system. In order to obtain the solution of the system in Filippov's sense, the theory of differential inclusion and set-value map have been widely used, and some important results have been presented [13, 14, 28, 33, 36, 45]. Since the system consisting of MNNs is a state-dependent switched system, and its connection weights depend on the networks states, we can use the switched method to process model (7.4).

Inspired by [12, 16, 52], the switched method can be introduced into (7.8). Assume that the switched signal $\sigma(t) : [0, \infty) \to \Lambda_\sigma$, which is a right continuous piecewise constant function, $\Lambda_\sigma = \{1, \ldots, m\}$; $\sigma(t) = \imath$ represents the information of the $\imath$th subsystem is adopted for $t \in [t_k, t_{k+1})$. Hence, (7.8) can be further described as the following switched system:

$$\begin{cases} \ddot{z}(t) = -(A + \Delta A(t))\dot{z}(t) - (B + \Delta B(t))z(t) + (C_{\sigma(t)}(t) \\ \qquad + \Delta C(t))g(z(t)) + (\mathcal{D}_{\sigma(t)}(t) + \Delta D(t))g(z(t - \tau(t))) \\ \qquad + \bar{u}(t, z(t), \dot{z}(t)) + \tilde{u}(t, z(t), \dot{z}(t)) \\ z(s) = \psi(s), \quad s \in [-\tau, 0] \\ \dot{z}(s) = \varphi(s), \quad s \in [-\tau, 0] \end{cases} \tag{7.9}$$

where, for any $\sigma(t) = \imath$, $(C_\imath(t), \mathcal{D}_\imath(t)) \in Co\daleth_\imath$, $\daleth_\imath = \{(C_\imath^\jmath, D_\imath^\jmath) | \jmath \in \mathbb{I}_\wp\}$. The definition means that the connection weight matrixes $C(x(t))$ and $D(x(t))$ of the system depend on signal $\sigma(t)$.

For the convenience of writing and the beauty of form, $C_{\sigma(t)}(t)$ is abbreviated as $C_{\sigma(t)}$, $\mathcal{D}_{\sigma(t)}(t)$ is abbreviated as $\mathcal{D}_{\sigma(t)}$. Now, we introduce the method of optimizing fuzzy model, and consider the following T-S fuzzy system [53]:

Model Rule r:

If $\aleph_1(t)$ is Γ_1^r, ..., $\aleph_\varsigma(t)$ is Γ_ς^r

Then

$$\begin{cases} \ddot{z}(t) = -(A_r + \Delta A_r(t))\dot{z}(t) - (B_r + \Delta B_r(t))z(t) + (C_{\sigma(t)} \\ \qquad + \Delta C_r(t))g(z(t)) + (\mathcal{D}_{\sigma(t)} + \Delta D_r(t))g(z(t - \tau(t))) \\ \qquad + \bar{u}(t, z(t), \dot{z}(t)) + \tilde{u}(t, z(t), \dot{z}(t)) \\ z(s) = \psi(s), \quad s \in [-\tau, 0] \\ \dot{z}(s) = \varphi(s), \quad s \in [-\tau, 0] \end{cases} \tag{7.10}$$

where $r \in \mathbb{I}_{\bar{\varsigma}} = \{1, ..., \bar{\varsigma}\}$, and $\bar{\varsigma}$ is the number of IF-THEN; $\aleph_i(t)(i \in \mathbb{I}_{\tilde{\varsigma}} = \{1, ..., \tilde{\varsigma}\})$ are premise variables, which are some measurable variables of the system, for example, the state variables; $\Gamma_i^r(i \in \mathbb{I}_{\tilde{\varsigma}})$ represent the fuzzy sets of rule r.

For the fuzzy system, the state vector $z(t)$ of (7.10) depends on a weighted average of all local models [54]. By using a standard fuzzy system inference method, including singleton fuzzier, product fuzzy inference, and center average defuzzifier, we can get the following UFSMINNs model.

$$\begin{cases} \ddot{z}(t) = -(\mathcal{A}(t) + \Delta\mathcal{A}(t))\dot{z}(t) - (\mathcal{B}(t) + \Delta\mathcal{B}(t))z(t) + (C_{\sigma(t)} + \Delta C(t))g(z(t)) \\ \qquad + (\mathcal{D}_{\sigma(t)} + \Delta\mathcal{D}(t))g(z(t-\tau(t))) + \bar{u}(t, z(t), \dot{z}(t)) + \tilde{u}(t, z(t), \dot{z}(t)) \\ z(s) = \psi(s), \quad s \in [-\tau, 0] \\ \dot{z}(s) = \varphi(s), \quad s \in [-\tau, 0] \end{cases} \tag{7.11}$$

where $\mathcal{A}(t) = \sum_{r=1}^{\bar{\varsigma}} \varpi_r(\aleph(t))A_r(t)$, $\mathcal{B}(t) = \sum_{r=1}^{\bar{\varsigma}} \varpi_r(\aleph(t))B_r(t)$,

$\Delta\mathcal{A}(t) = \sum_{r=1}^{\bar{\varsigma}} \varpi_r(\aleph(t))\Delta A_r(t)$, $\Delta\mathcal{B}(t) = \sum_{r=1}^{\bar{\varsigma}} \varpi_r(\aleph(t))\Delta B_r(t)$,

$\Delta C(t) = \sum_{r=1}^{\bar{\varsigma}} \varpi_r(\aleph(t))\Delta C_r(t)$, $\Delta\mathcal{D}(t) = \sum_{r=1}^{\bar{\varsigma}} \varpi_r(\aleph(t))\Delta D_r(t)$,

and the normalized membership function $\varpi_r(\aleph(t)) \geq 0$ satisfies the following conditions:

$$\varpi_r(\aleph(t)) = \frac{\prod_{i=1}^{\tilde{\varsigma}} \Gamma_i^r(\aleph_i(t))}{\sum_{r=1}^{\bar{\varsigma}} \prod_{i=1}^{\tilde{\varsigma}} \Gamma_i^r(\aleph_i(t))}, \quad \sum_{r=1}^{\bar{\varsigma}} \varpi_r(\aleph(t)) = 1.$$

where $\Gamma_i^r(\aleph_i(t))$ is the grade of membership of $\aleph_i(t)$ in the fuzzy sets Γ_i^r.

According to the method of variable transformation, let $x(t) = \dot{z}(t) + \Lambda z(t)$, (7.11) can be transformed into the form of first-order system, which can be described as follows:

$$\begin{cases} \dot{z}(t) = -\Lambda z(t) + x(t) \\ \dot{x}(t) = -(-\mathcal{A}(t)\Lambda - \Delta\mathcal{A}(t)\Lambda + \mathcal{B}(t) + \Delta\mathcal{B}(t) + \Lambda\Lambda)z(t) \\ \qquad - (-\Lambda + \mathcal{A}(t) + \Delta\mathcal{A}(t))x(t) + (C_{\sigma(t)} + \Delta C(t))g(z(t)) \\ \qquad + (\mathcal{D}_{\sigma(t)} + \Delta\mathcal{D}(t))g(z(t-\tau(t))) \\ \qquad + \bar{u}(t, z(t), \dot{z}(t)) + \tilde{u}(t, z(t), \dot{z}(t)) \\ z(s) = \psi(s), \quad s \in [-\tau, 0] \\ x(s) = \Lambda\psi(s) + \varphi(s), \quad s \in [-\tau, 0] \end{cases} \tag{7.12}$$

Next, we introduce the mathematical model of an evolving process with impulsive effects.

Without loss of generality, $\bar{u}(t, z(t), \dot{z}(t))$ and $\tilde{u}(t, z(t), \dot{z}(t))$ can be designed as follows:

$$\begin{cases} \bar{u}(t, z(t), \dot{z}(t)) = \sum_{l_k=1}^{d_k} \left(\sum_{k=0}^{+\infty} H_{k,l_k}(\dot{z}(t) + \Lambda z(t)) - (\dot{z}(t) + \Lambda z(t)) \right) \delta(t - t_{k,l_k}^-) \\ \tilde{u}(t, z(t), \dot{z}(t)) = \mathcal{K}_{\sigma(t)}(\dot{z}(t) + \Lambda z(t)) \end{cases} \tag{7.13}$$

where $x(t) = \dot{z}(t) + \Lambda z(t)$. Similar to (7.9), for the sake of simplicity, the controller (7.13) can be further described as follows:

$$\begin{cases} \bar{u}(t, x(t)) = \sum_{l_k=1}^{d_k} \left(\sum_{k=0}^{+\infty} H_{k,l_k}(x(t)) - x(t) \right) \delta(t - t_{k,l_k}^-) \\ \tilde{u}(t, x(t)) = \mathcal{K}_{\sigma(t)}(x(t)) \end{cases} \tag{7.14}$$

where $H_{k,l_k}(x(t)) = H_{k,l_k}(\dot{z}(t) + \Lambda z(t)), \mathcal{K}_{\sigma(t)}(x(t)) = \mathcal{K}_{\sigma(t)}(\dot{z}(t) + \Lambda z(t))$. The functions $H_{k,l_k}(\cdot)$ and $\mathcal{K}_{\sigma(t)}(\cdot)$ are called impulsive functions and adaptive feedback functions, respectively. For any $\sigma(t) = \iota$ with $t \in [t_k, t_{k+1})$, the time sequences $\{t_{k,l_k}\}(k \in \mathbb{N})$ are strictly increased. $l_k \in \{0, 1, 2, \ldots, d_k\}$, and d_k represents the maximum number of impulses in the interval $t \in [t_k, t_{k+1})$, if $d_k = 0$, there are no impulses. $\delta(\cdot)$ is the Dirac delta function. In this section, we assume that $x(t)$ is right continuous at $t = t_{k,l_k}$.

Remark 7.2 Most papers implicitly assume that the impulsive effects depend on switched instants of the system, and the impulsive effects occur at switched instants t_k or there only has an impulse in the interval $[t_k, t_{k+1})$ [16, 47, 49, 55–59]. The cases can be explained as follows:

- **Case 1**: If the state of the system is not switched, the impulsive effects will not occur. That is, when $t \in [t_k, t_{k+1})$, the impulsive effects and the switched signal $\sigma(t) = \iota$ will occur simultaneously at $t = t_k$ with $t_{k,l_k} = t_k$, and $t_{k+1,l_{k+1}} = t_{k+1}$. For the uniform symbology specification and the simplicity of description, in this case, we assume that the impulsive instants t_{k,l_k} satisfy $l_k \in \{1\}$, for $t > 0$, the constraint relationship between impulsive instance sets $\{t_{k,l_k}\}$ and switched instance sets $\{t_k\}$ is explained as $0 \leq t_0 = t_{0,1} < t_1 = t_{1,1} < \cdots < t_k = t_{k,1} < t_{k+1} = t_{k+1,1} < \cdots$ (See Fig. 7.1). It is obvious that t_{k,l_k} is limited.

To give a more general and practical impulse control, this section assumes that the impulsive effects do not depend on switched instants of the system, and the impulsive effects occur in the interval $[t_k, t_{k+1})$. To better describe the impulsive effects, we provide the following explanations:

- **Case 2**: If the state of the system is not switched, the impulsive effects perhaps will occur. That is, when $t \in [t_k, t_{k+1})$, the switched signal $\sigma(t) = \iota$ will occur at $t = t_k$, and there will be multiple impulsive effects with $t_k \leq t_{k,1} < t_{k,2} \ldots < t_{k,d_k} < t_{k+1}$

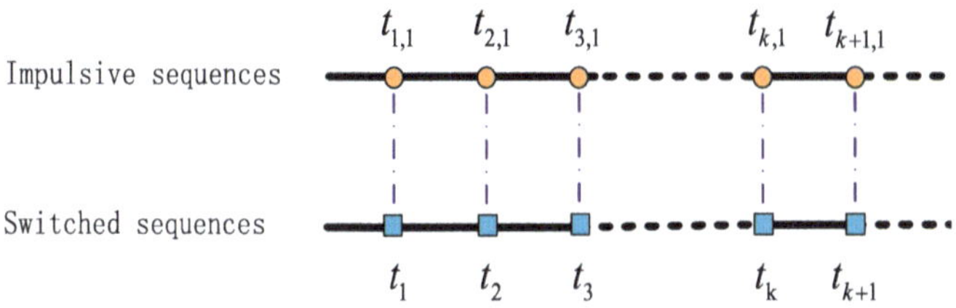

Fig. 7.1 Explanation of Case 1

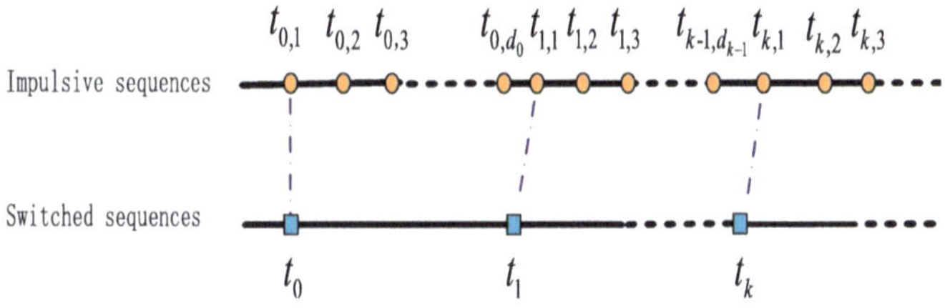

Fig. 7.2 Explanation of Case 2

or no impulsive effects with $t_k < t_{k+1}$. For the uniform symbology specification and the simplicity of description, in this case, for $t > 0$, the constraint relationship between impulsive instance sets $\{t_{k,l_k}\}$ and switched instance sets $\{t_k\}$ is explained as $0 \leq t_0 \leq t_{0,1} < t_{0,2} < \cdots < t_{0,d_0} < t_1 \leq t_{1,1} < t_{1,2} < \cdots < t_{1,d_1} < t_2 < \cdots < t_k \leq t_{k,1} < t_{k,2} < \cdots < t_{k,d_k} < t_{k+1} < \cdots$ (See Fig. 7.2).

In Case 2, according to the define of d_k, it is obvious that if $d_k = 0$, there is no impulses in the interval $[t_k, t_{k+1})$, otherwise there is at least one impulse in the interval $[t_k, t_{k+1})$.

Remark 7.3 As shown in Case 2, the impulsive effects can be divided into three types in terms of switched instant. It is worth to point out that many existing literature such as switched systems or switched systems with impulsive effects can be regarded as special cases of Case 2. More specifically, the specific sub-cases of Case 2 can be further described as follows:

- Case A: For any $\sigma(t) = \imath$ with $t \in [t_k, t_{k+1})$, if each $d_k = 0$, there are no impulsive effects in each interval $[t_k, t_{k+1})$. Thus, we assume that the system only contains switched effects. For the uniform symbology specification and the simplicity of the description, in this case, for $t > 0$, the switched instances set $\{t_k\}$ is explained as $0 < t_0 < t_1 < t_2 < \cdots < t_k < \cdots$.
- Case B: For any $\sigma(t) = \imath$ with $t \in [t_k, t_{k+1})$, if there are finite numbers of $d_k = 0$, then multiple switched effects can occur between two impulsive effects instants. For the uniform symbology specification and the simplicity of the description, in this case, for $t > 0$, the constraint relationship between impulsive instances set $\{t_{k,l_k}\}$ and switched instances set $\{t_k\}$ is explained as $0 < t_0 \leq t_{0,1} < t_{0,2} < \cdots < t_{0,d_0} < t_1 \leq t_{1,1} < t_{1,2} < \cdots < t_{1,d_1} < t_2 < \cdots < t_{k-1} \leq t_{k-1,1} < t_{k-1,2} < \cdots <$

$t_{k-1,d_{k-1}} < t_k < t_{k+1} < \cdots < t_{k+\ell} < t_{k+\ell+1} \leq t_{k+\ell+1,1} < t_{k+\ell+1,2} < \cdots < t_{k+\ell+1,d_{k+\ell+1}} < t_{k+\ell+2} < \cdots$, and $k+\ell \in \{0, 1, \ldots\}$, $d_{k+\ell} = 0$.

- Case C: For any $\sigma(t) = \iota$ with $t \in [t_k, t_{k+1})$, if there are finite numbers of $d_k \neq 0$, then multiple impulsive effects can occur between two switched instants.

Remark 7.4 Case 1 is the comparative ideal situation, and we can say that the impulsive signals and the switched signals are designed to be synchronously updated in this case. However, in the real case, the system or circuit will always be subject to external interference or internal disturbance, thus the switched signals and impulsive signals cannot occur at the same time. In other words, the node clock of the switched device and the pulse generator fails to keep consistent, and transmission and reception of signals are not synchronized. At this point, we think the impulsive generator may fall out of step with the switched device.

Remark 7.5 In impulse control systems, most impulse generators and switched devices are digital devices with their own sampling time, and these devices can be characterized as clock-driven samplers with very high-precision clocks. Therefore, if we choose these devices with very high-precision clocks, then Case 1 is very easy to achieve in the ideal case, but the cost will be high, and there is no guarantee that it will be synchronized from time to time. Thus, we need to consider Case 2, in which the situation is easily implemented and constrained even if the system suffers interference from the external environment or internal components of the circuit, because Case 2 includes both cases of synchronization and out of synchronization.

Combining (7.12) and (7.14), integrating both sides of (7.12) from $t_{k,l_k} - s$ to t_{k,l_k}, and $s \to 0^+$, one can get

$$\begin{aligned} \Delta z(t_{k,l_k}) &= z(t_{k,l_k}^+) - z(t_{k,l_k}^-) = z(t_{k,l_k}) - z(t_{k,l_k}^-) \\ &= \lim_{s\to 0^+} \int_{t_{k,l_k}-s}^{t_{k,l_k}} dz(t) = 0 \end{aligned} \tag{7.15}$$

$$\begin{aligned} \Delta x(t_{k,l_k}) &= x(t_{k,l_k}^+) - x(t_{k,l_k}^-) = x(t_{k,l_k}) - x(t_{k,l_k}^-) \\ &= \lim_{s\to 0^+} \int_{t_{k,l_k}-s}^{t_{k,l_k}} \bar{u}(t, x(t))dt \\ &= H_{k,l_k}(x(t_{k,l_k}^-)) - x(t_{k,l_k}^-) \end{aligned} \tag{7.16}$$

where $z(t_{k,l_k}) = z(t_{k,l_k}^+) = \lim_{h\to 0^+} z(t_{k,l_k} + h)$, $z(t_{k,l_k}^-) = \lim_{h\to 0^-} z(t_{k,l_k} + h)$, $x(t_{k,l_k})$ and $x(t_{k,l_k}^-)$ are defined similarly to $z(t_{k,l_k})$ and $z(t_{k,l_k}^-)$.

Assumption 7.3 For any $\sigma(t) = \iota$ with $t \in [t_k, t_{k+1})$, the function $\mathcal{K}_\iota(\cdot)$ and $H_{k,l_k}(\cdot)$ satisfy the Lipschitz condition with $\mathcal{K}_\iota(0) = 0$ and $H_{k,l_k}(0) = 0$, and

$$
\begin{aligned}
|\mathcal{K}_\iota(y(t)) - \mathcal{K}_\iota(x(t))| &\le S_{1\iota}|y(t) - x(t)| \\
|H_{k,l_k}(y(t)) - H_{k,l_k}(x(t))| &\le S_{2k,l_k}|y(t) - x(t)|
\end{aligned}
\tag{7.17}
$$

where $S_{1\iota}$ and S_{2k,l_k} are positive definite diagonal matrices.

Remark 7.6 Similar to the method of [45], the specific mathematical definition of destabilizing impulsive effects can be described as: when $|H_{k,\sigma(t)}(x(t_{k,l_k}^-))| > |x(t_{k,l_k}^-)|$, the impulsive effects may be considered as impulsive disturbances. Additionally, when the mode-dependent impulses suppress the stability of the system, we expect the impulse interval to be as large as possible [48].

Based on the above introduction, the controlled UFSMINNs can be described as follows:

$$
\begin{cases}
\dot{z}(t) = -\Lambda z(t) + x(t) \\
\dot{x}(t) = -(-\mathcal{A}(t)\Lambda - \Delta\mathcal{A}(t)\Lambda + \Lambda\Lambda + \mathcal{B}(t) + \Delta\mathcal{B}(t))z(t) \\
\qquad - (-\Lambda + \mathcal{A}(t) + \Delta\mathcal{A}(t))x(t) + (C_{\sigma(t)} + \Delta C(t))g(z(t)) \\
\qquad + (\mathcal{D}_{\sigma(t)} + \Delta\mathcal{D}(t))g(z(t - \tau(t))) + \mathcal{K}_{\sigma(t)}(x(t)), t \ne t_{k,l_k} \\
x(t_{k,l_k}) = H_{k,l_k}(x(t_{k,l_k}^-)), \qquad t = t_{k,l_k} \\
z(s) = \psi(s), \quad s \in [-\tau, 0] \\
x(s) = \Lambda\psi(s) + \varphi(s), \quad s \in [-\tau, 0]
\end{cases}
\tag{7.18}
$$

Assumption 7.4 The parametric uncertainties $\Delta\mathcal{A}(t)$, $\Delta\mathcal{B}(t)$, $\Delta C(t)$ and $\Delta\mathcal{D}(t)$ are norm bounded, and $[\Delta\mathcal{A}(t), \Delta\mathcal{B}(t), \Delta C\ (t), \Delta\mathcal{D}(t)] = \sum_{r=1}^{\bar{\varsigma}} \varpi_r(\aleph(t))[\Delta A(t),$ $\Delta B(t), \Delta C(t), \Delta D(t))$, the following condition is satisfied:

$$
\left[\, \Delta\mathcal{A}(t)\ \Delta\mathcal{B}(t)\ \Delta C(t)\ \Delta\mathcal{D}(t) \,\right] = \sum_{r=1}^{\bar{\varsigma}} \varpi_r(\aleph(t)) M_r F_r(t) \left[\, N_{1r}\ N_{2r}\ N_{3r}\ N_{4r} \,\right]
$$

where M_r, N_{1r}, N_{2r}, N_{3r} and N_{4r} are some given real constant matrices with appropriate dimensions, $F_r(t)$ is an unknown matrix function with Lebesgue measures elements, which representing the parameter perturbations, and satisfies the following condition:

$$
F_r{}^T(t) F_r(t) \le I \tag{7.19}
$$

To promote the stability of the system, we have the following definitions and lemmas.

Definition 7.1 ([48]) For any switched signal $\sigma(t)$, $N_\sigma(T, t)$ represents the number of switching in the time interval (t, T), $N_0 > 0$ and $\Delta_t > 0$ are called the chatter bound and the average dwell time, one can have

$$
\frac{T - t}{\Delta_t} + N_0 \ge N_\sigma(T, t), \forall\ \ 0 \le t \le T
$$

Definition 7.2 If there are constants $\eta_0 > 0$, $\eta_1 > 0$ and $\eta_2 > 0$, and the following condition holds:

$$(\|x(t)\|^2 + \|z(t)\|^2) \leq (\eta_1 \sup_{-\tau \leq s \leq 0} ||z(s)||^2 + \eta_2 \sup_{-\tau \leq s \leq 0} ||x(s)||^2)e^{-\eta_0 t}$$

then the trivial solution of (7.18) is said to be exponentially stable. $\eta_0 > 0$ is called the global exponential stable estimate of the rate index, $z(s)$ and $x(s)$ represent the initial states of the system.

Lemma 7.1 (S-procedure [60]) *Given real matrices $\Omega = \Omega^T$, Λ_1 and Λ_2 with appropriate dimensions, for all $F(t)$ satisfying $F^T(t)F(t) < I$, then we can have $\Omega + \Lambda_1 F(t)\Lambda_2 + \Lambda_2^T F(t)\Lambda_1^T < 0$ if and only if there exists a positive scalar $\beta > 0$, such that*

$$\Omega + \beta^{-1}\Lambda_1\Lambda_1^T + \beta\Lambda_2^T\Lambda_2 < 0 \tag{7.20}$$

Lemma 7.2 *Suppose that $a_1 > a_2 \geq 0$, $v(t)$ is continuous in $[-\tau, 0]$, and $v(t)$ satisfies the following differential inequality*

$$\begin{cases} D^+v(t) \leq -a_1 v(t) + a_2 v_\tau(t) \\ v_\tau(t) = \sup_{-\tau \leq s \leq 0} v(t+s) \end{cases} \tag{7.21}$$

Then

$$v(t) \leq v_\tau(t_0)e^{-\alpha(t-t_0)} \tag{7.22}$$

where $\alpha > 0$ is the unique solution of the equation $\alpha - a_1 + a_2 e^{\alpha\tau} = 0$.

Proof For any $\varepsilon > 0$, we provide a reasonable proof to verify that the following inequality holds by using the reduction to absurdity

$$v(t) < v_\tau(t_0)e^{-\alpha(t-t_0)} + \varepsilon \tag{7.23}$$

Obviously, $v(t_0) \leq v_\tau(t_0) < v_\tau(t_0) + \varepsilon$.

On one hand, if the inequality (7.22) is not true, there exists at least one solution t^* such that

$$v(t^*) = v_\tau(t_0)e^{-\alpha(t^*-t_0)} + \varepsilon \tag{7.24}$$

and for any $t \in [t_0, t^*)$, one has

$$v(t) < v_\tau(t_0)e^{-\alpha(t-t_0)} + \varepsilon \tag{7.25}$$

Then, we obtain the following:

$$D^+v(t^*) \geq -\alpha v_\tau(t_0)e^{-\alpha(t^*-t_0)} \tag{7.26}$$

On the other hand, we obtain the following:

$$\begin{aligned} D^+v(t^*) \le & -a_1 v(t^*) + a_2 v_\tau(t^*) \\ \le & -a_1(v_\tau(t_0)e^{-\alpha(t^*-t_0)} + \varepsilon) \\ & + a_2(v_\tau(t_0)e^{-\alpha(t^*-t_0)}e^{\alpha\tau} + \varepsilon) \\ = & (-a_1 + a_2 e^{\alpha\tau})v_\tau(t_0)e^{-\alpha(t^*-t_0)} - (a-b)\varepsilon \\ < & -\alpha v_\tau(t_0)e^{-\alpha(t^*-t_0)} \end{aligned} \tag{7.27}$$

The result of (7.27) in contradiction to (7.26). Consequently, (7.22) holds.
This completes the proof. □

Theorem 7.1 *Assume that Assumptions 7.1–7.4 hold. Then (7.18) can achieve globally exponentially stable in (7.14), if there exist positive scalars* $\alpha, \tilde{\alpha}_{1k}, \tilde{\alpha}_{2k}, \beta_1, \beta_2$ *and* $\tilde{\beta}$, *and any symmetric positive definite matrices* P_1 *and* P_2, *positive diagonal matrices* T_1, T_2 *and* T_3 *with appropriate dimensions such that*

$$\begin{bmatrix} \tilde{\tilde{\Theta}} & \hat{\Theta}_0^T & \hat{\Theta}_1^T \\ * & -\beta_1 I_n & 0 \\ * & * & -\beta_2 I_n \end{bmatrix} < 0 \tag{7.28}$$

$$\tilde{\alpha}_1 > \tilde{\alpha}_2 > 0, \frac{1}{k}\sum_{j=0}^{k} \ln \tilde{\vartheta}_j^* \le \tilde{\beta}, \Delta_t > \frac{\tilde{\beta}}{\alpha} \tag{7.29}$$

where $\alpha > 0$ *is the unique solution of the equation* $\alpha - \tilde{\alpha}_1 + \tilde{\alpha}_2 e^{\alpha\tau} = 0$, $\tilde{\alpha}_1 = \min_{k\in\mathbb{N}}\{\tilde{\alpha}_{1k}\}$, $\tilde{\alpha}_2 = \max_{k\in\mathbb{N}}\{\tilde{\alpha}_{2k}\}$,

$\tilde{\tilde{\Theta}} = \tilde{\Theta} + \beta_1 \check{\Theta}_0^T \check{\Theta}_0 + \beta_2 \check{\Theta}_1^T \check{\Theta}_1$,
$\hat{\Theta}_0 = [M_r^T P_2 \;\; 0 \;\; 0 \;\; 0 \;\; 0 \;\; 0 \;\; 0 \;\; 0]$,
$\check{\Theta}_0 = [-N_{1r} \;\; -N_{2r} \;\; 0 \;\; 0 \;\; N_{3r} \;\; N_{4r} \;\; 0 \;\; 0]$,
$\hat{\Theta}_1 = [M_r^T \Lambda^T P_2 \;\; 0 \;\; 0 \;\; 0 \;\; 0 \;\; 0 \;\; 0 \;\; 0]$,
$\check{\Theta}_1 = [0 \;\; N_{1r} \;\; 0 \;\; 0 \;\; 0 \;\; 0 \;\; 0 \;\; 0]$,

$$\tilde{\Theta} = \begin{bmatrix} \tilde{\Theta}_{11} & \tilde{\Theta}_{12} & 0 & 0 & \tilde{\Theta}_{15} & \tilde{\Theta}_{16} & 0 & \tilde{\Theta}_{18} \\ * & \tilde{\Theta}_{22} & 0 & 0 & \tilde{\Theta}_{25} & 0 & 0 & 0 \\ * & * & \tilde{\Theta}_{33} & 0 & 0 & \tilde{\Theta}_{36} & 0 & 0 \\ * & * & * & \tilde{\Theta}_{44} & 0 & 0 & \tilde{\Theta}_{47} & 0 \\ * & * & * & * & \tilde{\Theta}_{55} & 0 & 0 & 0 \\ * & * & * & * & * & \tilde{\Theta}_{66} & 0 & 0 \\ * & * & * & * & * & * & \tilde{\Theta}_{77} & 0 \\ * & * & * & * & * & * & * & \tilde{\Theta}_{88} \end{bmatrix}$$

and $\tilde{\Theta}_{11} = (\alpha + \tilde{\alpha}_1)P_2 + \mathrm{Sym}(P_2\Lambda) - \mathrm{Sym}(P_2A_r) + S_{1\iota}T_4S_{1\iota}$, $\tilde{\Theta}_{12} = P_1 + P_2A_r\Lambda - P_2\Lambda\Lambda - P_2B_r$, $\tilde{\Theta}_{15} = P_2C_\iota^J$, $\tilde{\Theta}_{16} = P_2D_\iota^J$, $\Theta_{18} = P_2$, $\tilde{\Theta}_{22} = (\alpha + \tilde{\alpha}_1) P_1 - \mathrm{Sym}(P_1\Lambda) - 2L^-T_1L^+$, $\tilde{\Theta}_{25} = T_1(L^+ + L^-)$, $\tilde{\Theta}_{33} = -2L^-T_2L^+ - \tilde{\alpha}_2P_1e^{-\alpha\tau}$, $\tilde{\Theta}_{36} = T_2(L^+ + L^-)$, $\tilde{\Theta}_{44} = -\tilde{\alpha}_2e^{-\alpha\tau}P_2 - 2L^-T_3L^+$, $\tilde{\Theta}_{47} = T_3(L^+ + L^-)$, $\tilde{\Theta}_{55} = -2T_1$, $\tilde{\Theta}_{66} = -2T_2$, $\tilde{\Theta}_{77} = -2T_3$, $\tilde{\Theta}_{88} = -T_4$.

Remark 7.7 The impulsive effects can be divided into three types in terms of impulsive strength, which are called destabilizing impulsive effects, inactive impulsive effects, and stabilizing impulsive effects [61, 62], respectively. Most scholars focus on researching stabilizing impulsive effects, which can optimize the exponentially stable process, improve the speed of convergence, and reduce the cost of time. However, destabilizing impulsive effects may be considered as impulsive disturbances, since the transient state of the system may jump up along trajectories of the system at impulsive instants t_{k,l_k}, and these impulses will potentially destroy the stability of the (7.12). How to make the system stable under the influence of a series of destabilizing impulses has always been a difficult and important point of research.

Remark 7.8 In this section, the result is different from existing research results [48], which the adaptive controller is not taken into consideration. Moreover, for any interval $[t_k, t_k + 1)$, it is worth pointing out that the impulsive instances are implicitly assumed to have at least one impulse in [48], which can be regarded as a special case of this section.

Remark 7.9 On one hand, since the adaptive feedback controllers can minimize the cost function of the system and optimize the exponentially stable process [45, 63], they have been widely researched [11, 13, 33, 64–66]. However, it should be noted that all these presented results do not take the impulsive controllers. On the other hand, impulsive controllers also attract the attention of many researchers [16, 57, 67, 68], since they can reduce data transmission, improve data security and improve system robustness. However, these researches, particularly the results of MINNs, reckon without adaptive controllers. In this section, hybrid impulsive and adaptive feedback controllers are simultaneously exerted on MINNs.

Remark 7.10 Recently, MINNs have been extensively researched, and some great results of MINNs have been obtained in [11, 32, 33, 68, 69]. Especially, [11, 33] directly study the stability of MINNs without the reduced-order method. Also, the connection weights of MINNs are disposed of by using the theory of differential inclusion and set-value map. Different from the method used in those previous works, we use the switched method to process the model (7.1) in this section. Therefore, the result presented here extends the existing research.

In some results, the theorem is also suitable for the situations, in which only one controller is considered. Several important corollaries will be derived as follows:

Corollary 7.1 *Assume that Assumptions 7.1–7.4 hold. For any signal* $\sigma(t) = \iota$ *and the rule r, and the impulsive strengths are exactly same in any interval* $[t_k, t_k + 1)$. *(7.18) can achieve globally exponentially stable in the (7.14), if there exist positive* *scalars* $\alpha, \tilde{\alpha}_{1k}, \tilde{\alpha}_{2k}, £_{1k}, \beta_1, \beta_2$ *and* $\tilde{\vartheta}_k$, *and any symmetric positive definite matrices* P_1 *and* P_2, *positive diagonal matrices* T_1, T_2 *and* T_3 *with appropriate dimensions such that*

$$\begin{bmatrix} \tilde{\check{\Theta}} & \hat{\Theta}_0^T & \hat{\Theta}_1^T \\ * & -\beta_1 I_n & 0 \\ * & * & -\beta_2 I_n \end{bmatrix} < 0$$

$$\tilde{\alpha}_1 > \tilde{\alpha}_2 > 0, \frac{d_k + 1}{k} \sum_{j=0}^{k} \ln \tilde{\vartheta}_j^* \leq \tilde{\beta}, \Delta_\iota > \frac{\tilde{\beta}}{\alpha}$$

where the other elements have been defined in Theorem 1.

Corollary 7.2 *For any signal* $\sigma(t) = \iota$ *and the rule* r, *let* $\mathcal{K}_\iota(x(t)) = 0$ *and* $H_{k,l_k}(x(t)) = \nu_{k,l_k} x(t)$. *Assume that Assumptions 7.1–7.4 hold. (7.18) can achieve globally exponentially stable in (7.14), if there exist positive scalars* $\alpha, \tilde{\alpha}_{1k}, \tilde{\alpha}_{2k}, £_{1k}, \beta_1, \beta_2$ *and* $\tilde{\vartheta}_k$, *and any symmetric positive definite matrices* P_1 *and* P_2, *positive diagonal matrices* T_1, T_2 *and* T_3 *with appropriate dimensions such that*

$$\begin{bmatrix} \tilde{\check{\Theta}} & \hat{\Theta}_0^T & \hat{\Theta}_1^T \\ * & -\beta_1 I_n & 0 \\ * & * & -\beta_2 I_n \end{bmatrix} < 0$$

$$\tilde{\alpha}_1 > \tilde{\alpha}_2 > 0, \frac{1}{k} \sum_{j=0}^{k} \ln \tilde{\vartheta}_j^* \leq \tilde{\beta}, \Delta_\iota > \frac{\tilde{\beta}}{\alpha}$$

where $\tilde{\check{\Theta}} = \check{\Theta} + \beta_1 \check{\Theta}_0^T \check{\Theta}_0 + \beta_2 \check{\Theta}_1^T \check{\Theta}_1$,

$$\check{\Theta} = \begin{bmatrix} \tilde{\Theta}_{11} & \tilde{\Theta}_{12} & 0 & 0 & \tilde{\Theta}_{15} & \tilde{\Theta}_{16} & 0 & S_{1\iota} \\ * & \tilde{\Theta}_{22} & 0 & 0 & \tilde{\Theta}_{25} & 0 & 0 & 0 \\ * & * & \tilde{\Theta}_{33} & 0 & 0 & \tilde{\Theta}_{36} & 0 & 0 \\ * & * & * & \tilde{\Theta}_{44} & 0 & 0 & \tilde{\Theta}_{47} & 0 \\ * & * & * & * & \tilde{\Theta}_{55} & 0 & 0 & 0 \\ * & * & * & * & * & \tilde{\Theta}_{66} & 0 & 0 \\ * & * & * & * & * & * & \tilde{\Theta}_{77} & 0 \\ * & * & * & * & * & * & * & -T_4 \end{bmatrix}$$

$\tilde{\vartheta}_{k,l_k} = \max\{\vartheta_{k,l_k}, e^{\alpha\tau}\}$ *with* $\vartheta_{k,l_k} = \nu_{k,l_k}^2$, *and the other elements have been defined in Theorem 1.*

Remark 7.11 In most of the previous literature, a series of exponential stability results of impulsive system are derived by using the Halanay-type inequality [70, 71] and Comparison theorem [16, 72, 73], and the stabilizing impulsive effects are implicitly considered to obtain these results. In this section, an inductive method is adopted to establish the stability condition of FSMINNs. Moreover, in this corollary, both stabilizing and destabilizing impulsive effects make the system stable.

Corollary 7.3 *For any signal $\sigma(t) = \iota$ and the rule r, let $\mathcal{K}_\iota(x(t)) = F_\iota x(t)$ with $F_\iota = \mathrm{diag}(F_{1\iota}, \ldots, F_{n\iota})$, and $H_{k,l_k}(x(t)) = 0$. Assume that Assumptions 7.1–7.4 hold. The (7.18) can achieve globally exponentially stable in the (7.14), if there exist positive scalars $\alpha, \tilde{\alpha}_{1k}, \tilde{\alpha}_{2k}, £_{1k}, \beta_1, \beta_2$ and $\tilde{\vartheta}_k$, and any symmetric positive definite matrices P_1 and P_2, positive diagonal matrices T_1, T_2 and T_3 with appropriate dimensions such that*

$$\begin{bmatrix} \tilde{\tilde{\Theta}} & \hat{\Theta}_0^T & \hat{\Theta}_1^T \\ * & -\beta_1 I_n & 0 \\ * & * & -\beta_2 I_n \end{bmatrix} < 0$$

where $\tilde{\tilde{\Theta}} = \hat{\Theta} + \beta_1 \check{\Theta}_0^T \check{\Theta}_0 + \beta_2 \check{\Theta}_1^T \check{\Theta}_1$,

$$\hat{\Theta} = \begin{bmatrix} \hat{\Theta}_{11} & \tilde{\Theta}_{12} & 0 & 0 & \tilde{\Theta}_{15} & \tilde{\Theta}_{16} & 0 & S_{1\iota} \\ * & \tilde{\Theta}_{22} & 0 & 0 & \tilde{\Theta}_{25} & 0 & 0 & 0 \\ * & * & \tilde{\Theta}_{33} & 0 & 0 & \tilde{\Theta}_{36} & 0 & 0 \\ * & * & * & \tilde{\Theta}_{44} & 0 & 0 & \tilde{\Theta}_{47} & 0 \\ * & * & * & * & \tilde{\Theta}_{55} & 0 & 0 & 0 \\ * & * & * & * & * & \tilde{\Theta}_{66} & 0 & 0 \\ * & * & * & * & * & * & \tilde{\Theta}_{77} & 0 \\ * & * & * & * & * & * & * & -T_4 \end{bmatrix}$$

and $\hat{\Theta}_{11} = (\alpha + \tilde{\alpha}_1)P_2 + \mathrm{Sym}(P_2\Lambda) - \mathrm{Sym}(P_2 A_r) + \tilde{F}_\iota$, $F_\iota = P_2 \tilde{F}_\iota$, *the others elements has been defined in Theorem 1.*

Corollary 7.4 *For any signal $\sigma(t) = \iota$ and the rule r, if the system does not contain uncertainties, assume that Assumptions 7.1–7.4 hold. (7.18) can achieve globally exponentially stable in the variable hybrid switched-dependent impulsive effects, if there exist positive scalars $\alpha, \tilde{\alpha}_{1k}, \tilde{\alpha}_{2k}, £_{1k}$ and $\tilde{\vartheta}_k$, and any symmetric positive definite matrices P_1 and P_2, positive diagonal matrices T_1, T_2 and T_3 with appropriate dimensions such that*

$$\tilde{\Theta} < 0, \tilde{\alpha}_1 > \tilde{\alpha}_2 > 0, \frac{1}{k}\sum_{j=0}^{k} \ln \tilde{\vartheta}_j^* \leq \tilde{\beta}, \Delta_t > \frac{\tilde{\beta}}{\alpha}$$

where the other elements have been defined in Theorem 1.

7.1.3 Numerical Simulation

A numerical example with simulation results is given to show the effectiveness of the proposed theoretical results.

Example 7.1 Consider the two-dimensional FSMINNs with the following two fuzzy rules and two subsystems [53]:

Rule 1: IF $z_1(t)$ is Γ_1^1, Then

$$\begin{aligned}\ddot{z}(t) = &-(A_1 + \Delta A_1(t))\dot{z}(t) - (B_1 + \Delta B_1(t))z(t) + (C_{\sigma(t)} \\ &+ \Delta C_1(t))g(z(t)) + (\mathcal{D}_{\sigma(t)} + \Delta D_1(t))g(z(t - \tau(t))) \\ &+ \bar{u}(t, z(t), \dot{z}(t)) + \tilde{u}(t, z(t), \dot{z}(t))\end{aligned} \tag{7.30}$$

Rule 2: IF $z_1(t)$ is Γ_1^2, Then

$$\begin{aligned}\ddot{z}(t) = &-(A_2 + \Delta A_2(t))\dot{z}(t) - (B_2 + \Delta B_2(t))z(t) + (C_{\sigma(t)} \\ &+ \Delta C_2(t))g(z(t)) + (\mathcal{D}_{\sigma(t)} + \Delta D_2(t))g(z(t - \tau(t))) \\ &+ \bar{u}(t, z(t), \dot{z}(t)) + \tilde{u}(t, z(t), \dot{z}(t))\end{aligned} \tag{7.31}$$

where Γ_1^1 is $z_1(t) \le 0$ and Γ_1^2 is $z_1(t) > 0$, and $\hat{c}_{11} = \check{c}_{11} = 0.2$, $\hat{c}_{12} = \check{c}_{12} = 0.3$, $\hat{c}_{21} = \check{c}_{21} = -0.1$, $\hat{c}_{22} = -0.2$, $\check{c}_{22} = -0.4$, $\hat{d}_{11} = \check{d}_{11} = -0.2$, $\hat{d}_{12} = \check{d}_{12} = 0.6$, $\hat{d}_{21} = \check{d}_{21} = 0.5$, $\hat{d}_{22} = 0.8$, $\check{c}_{22} = 0.5$,

$$A_1 = \begin{bmatrix} 2.8 & 0 \\ 0 & 2.5 \end{bmatrix}, \quad A_2 = \begin{bmatrix} 3.3 & 0 \\ 0 & 3.1 \end{bmatrix}, \Lambda = \begin{bmatrix} 0.02 & 0 \\ 0 & 0.02 \end{bmatrix},$$

$$B_1 = \begin{bmatrix} 2.2 & 0 \\ 0 & 2.7 \end{bmatrix}, \quad B_2 = \begin{bmatrix} 2.6 & 0 \\ 0 & 2.2 \end{bmatrix},$$

$$\Delta A_1(t) = \Delta A_2(t) = \begin{bmatrix} 0.002\sin(t) & 0 \\ 0 & 0.002\sin(t) \end{bmatrix},$$

$$\Delta B_1(t) = \Delta B_2(t) = \begin{bmatrix} 0.002\sin(2t) & 0 \\ 0 & 0.002\sin(2t) \end{bmatrix},$$

$$\Delta C_1(t) = \Delta C_2(t) = \begin{bmatrix} 0.002\sin(3t) & 0 \\ 0 & 0.002\sin(3t) \end{bmatrix},$$

$$\Delta D_1(t) = \Delta D_2(t) = \begin{bmatrix} 0.002\sin(2t) & 0 \\ 0 & 0.002\sin(2t) \end{bmatrix}.$$

The time-varying delay is taken as $\tau(t) = 0.5 + 0.3\sin(t)$. The activation function is taken as $g(x(t)) = 0.5\tanh(x(t))$, it can be verified that the activation function satisfies Assumption 7.1 with $l_i^+ = 0.5, l_i^- = -0.5$. The switched signal $\sigma(t) \in \{1, 2\}$, and $\mathcal{K}_{\sigma(t)}(x(t)) = 1.4x(t)$, $H_{k,l_k}(x(t)) = 1.4x(t)$, $M_r = I_2$, $N_{1r} = N_{2r} = N_{3r} = N_{4r} = 0.002I_2$ with $r = 1, 2$.

By solving the conditions in Theorem 7.1 by MATLAB LMI Control Toolbox, we can get the following feasible solutions:

$$P_1 = \begin{bmatrix} 3.7387 & -0.2731 \\ -0.2731 & 2.4566 \end{bmatrix}, \quad P_2 = \begin{bmatrix} 0.4491 & -0.0604 \\ -0.0604 & 0.2673 \end{bmatrix},$$

$$T_1 = 10^3 \begin{bmatrix} 3.7131 & 0 \\ 0 & 3.7131 \end{bmatrix}, \quad T_2 = \begin{bmatrix} 0.7475 & 0 \\ 0 & 0.7475 \end{bmatrix},$$

$$T_3 = \begin{bmatrix} 0.1274 & 0 \\ 0 & 0.1274 \end{bmatrix}, \quad T_4 = \begin{bmatrix} 0.2237 & 0 \\ 0 & 0.2237 \end{bmatrix}.$$

According to the condition in Theorem 1, if we choose the initial values $z(0) = [-1.3\ 1.6]^T$, $x(0) = [0.8\ \ -1.3]^T$, then we can get the following simulation diagrams. On one hand, Fig. 7.3 depicts the multiple impulsive effects that exist between two switched modes. Under the control case of Figs. 7.3, 7.4 and 7.5 respectively

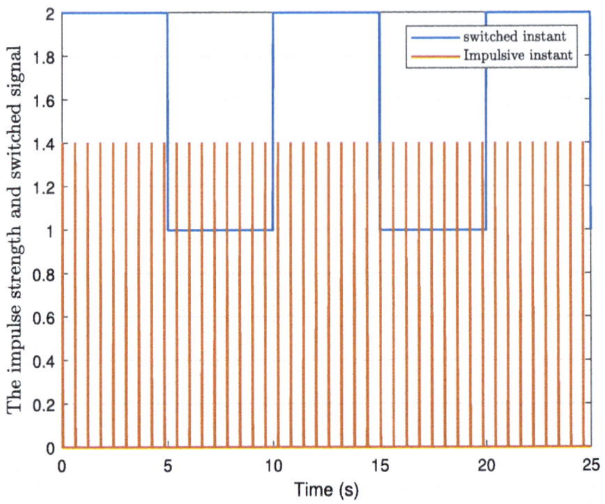

Fig. 7.3 The relationship between impulsive signal and switched signal

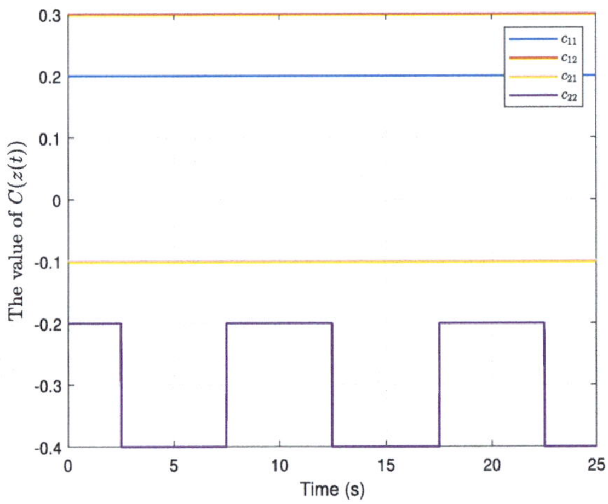

Fig. 7.4 The change in weights c_{11}, c_{12}, c_{21} and c_{22} as the switched modes change

describe the values of weights $c_{11}, c_{12}, c_{21}, c_{22}$ and $d_{11}, d_{12}, d_{21}, d_{22}$ under different switched modes, the trajectories of state $x_1(t)$, $x_2(t)$ and $z_1(t)$, $z_2(t)$ are shown in Figs. 7.6 and 7.7. Figure 7.8 depicts the multiple switched effects that occur between two impulsive effect instants. Under the control case of Figs. 7.8, 7.9 and 7.10 respectively describe the values of weights $c_{11}, c_{12}, c_{21}, c_{22}$ and $d_{11}, d_{12}, d_{21}, d_{22}$ under different switched modes, Fig. 7.11 describes the trajectories of state $x_1(t)$ and $x_2(t)$. Hence, the simulation results confirm the effectiveness of Theorem 1 for fuzzy-switched memristive inertial recurrent neural networks with time-varying delays on mode-dependent destabilizing impulsive control protocol.

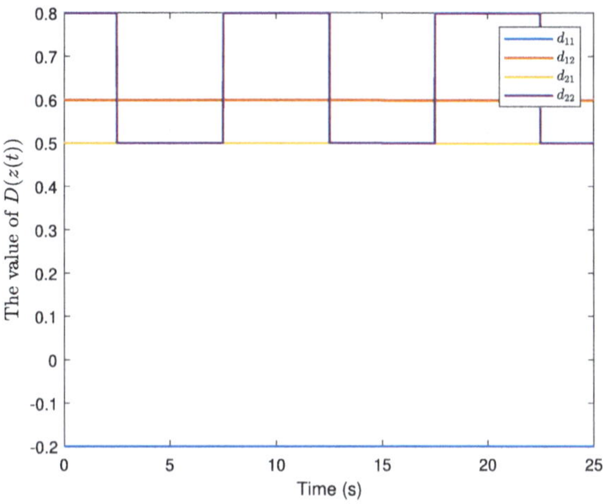

Fig. 7.5 The change in weights d_{11}, d_{12}, d_{21} and d_{22} as the switched modes change

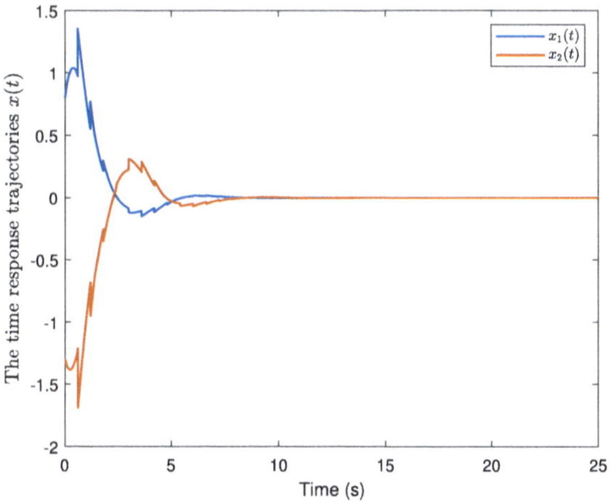

Fig. 7.6 Trajectories of state $x_1(t)$ and $x_2(t)$ on multiple impulsive effects

7.1.4 Conclusion

In this section, the original second-order system is transformed into a first-order differential system by using variable transmission, and the memristive model is described as a switched system rather than employing the theory of differential inclusion and set-value map. Hybrid mode-dependent destabilizing impulsive and adaptive feedback controllers are simultaneously applied to study the model. In the FSMINNs model, multiple impulsive effects can exist between two switched modes, and the multiple switched effects may also occur between two impulsive instants. Based on the Lyapunov method, T-S fuzzy method and the average dwell time, and so on, the extended robust exponential stability condition of FSMINNs is derived. Finally, a simulation example is given to illustrate the effectiveness of the main results.

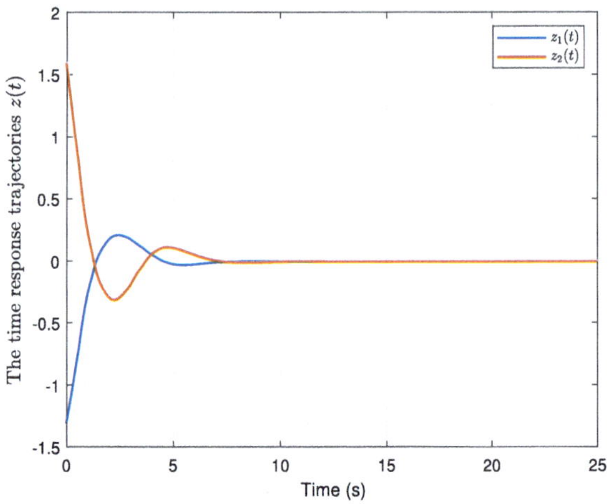

Fig. 7.7 Trajectories of state $z_1(t)$ and $z_2(t)$ on multiple impulsive effects

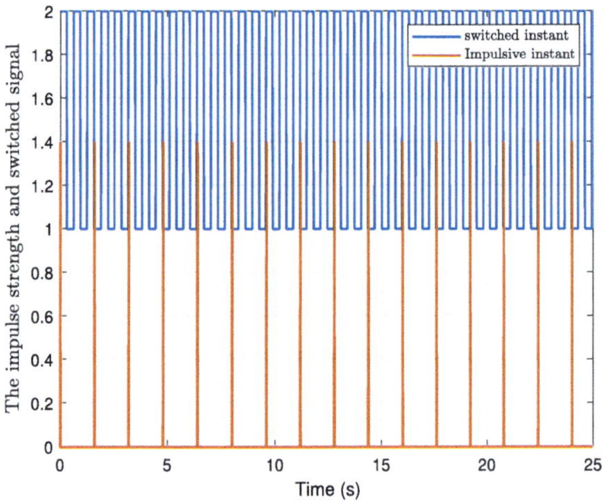

Fig. 7.8 The relationship between impulsive signal and switched signal

7.1.5 *Proof of the Theorem 1*

Proof The Lyapunov-Krasovskii functional is constructed as follows:

$$V(t, x(t), z(t)) = V_1(t, x(t), z(t)) + V_2(t, x(t), z(t)), t \in [t_k, t_{k+1}) \tag{7.32}$$

where

$$V_1(t, x(t), z(t)) = e^{\alpha t} z^T(t) P_1 z(t),$$
$$V_2(t, x(t), z(t)) = e^{\alpha t} x^T(t) P_2 x(t)$$

For $t \in [t_{k,l_k}, t_{k,l_{k+1}})$, the time derivative of $V(t, x(t), z(t))$ along the trajectory of (7.18) can be derived as follows:

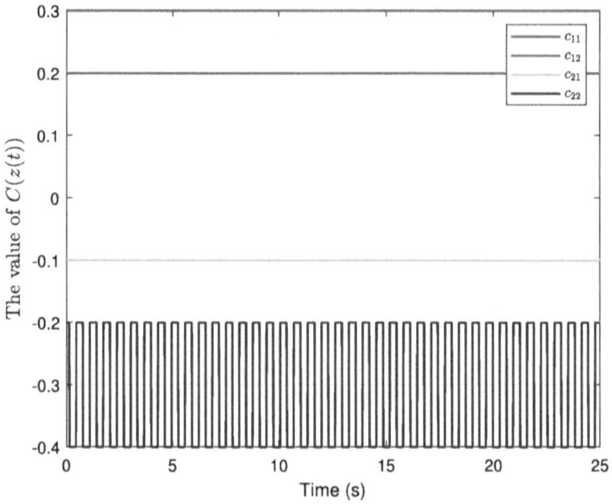

Fig. 7.9 The change in weights c_{11}, c_{12}, c_{21} and c_{22} as the switched modes change

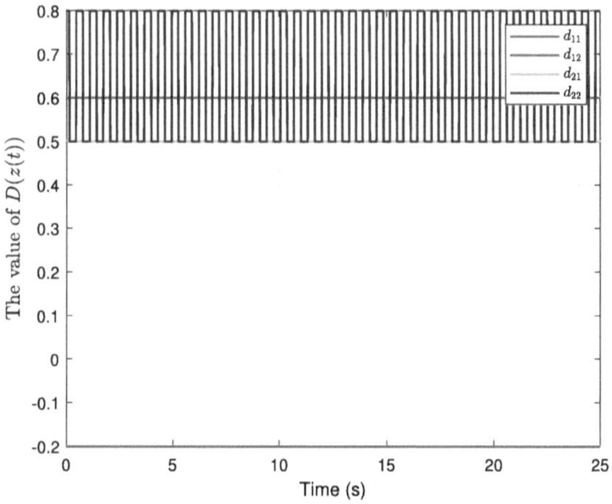

Fig. 7.10 The change in weights d_{11}, d_{12}, d_{21} and d_{22} as the switched modes change

$$D^+V(t, x(t), z(t)) = D^+V_1(t, x(t), z(t)) + D^+V_2(t, x(t), z(t)) \tag{7.33}$$

For any given very small constants $\tilde{\alpha}_{1k} > 0$ and $\tilde{\alpha}_{2k} > 0$, we can obtain the following:

$$\begin{aligned}
& D^+V_1(t, x(t), z(t)) + \tilde{\alpha}_{1k} V_1(t, x(t), z(t)) \\
& - \tilde{\alpha}_{2k} V_1(t - \tau(t), x(t - \tau(t)), z(t - \tau(t))) \\
= & \alpha e^{\alpha t} z^T(t) P_1 z(t) + 2e^{\alpha t} z^T(t) P_1 \dot{z}(t) + \tilde{\alpha}_{1k} e^{\alpha t} z^T(t) P_1 z(t) \\
& - \tilde{\alpha}_{2k} e^{\alpha(t-\tau(t))} z^T(t - \tau(t)) P_1 z(t - \tau(t)) \\
\leq & (\alpha + \tilde{\alpha}_{1k}) e^{\alpha t} z^T(t) P_1 z(t) + 2e^{\alpha t} z^T(t) P_1 (-\Lambda z(t) + x(t)) \\
& - \tilde{\alpha}_{2k} e^{-\alpha \tau} e^{\alpha t} z^T(t - \tau(t)) P_1 z(t - \tau(t))
\end{aligned} \tag{7.34}$$

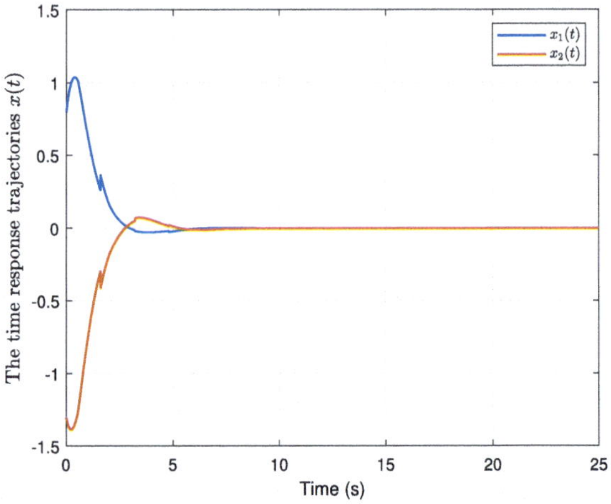

Fig. 7.11 Trajectories of state $x_1(t)$ and $x_2(t)$ on multiple switched effects

The $D^+V_2(t, x(t), z(t))$ can be derived as follows:

$$
\begin{aligned}
& D^+V_2(t, x(t), z(t)) + \tilde{\alpha}_{1k}V_2(t, x(t), z(t)) - \tilde{\alpha}_{2k}V_2(t - \tau(t), \\
& x(t - \tau(t)), z(t - \tau(t))) \\
&= \alpha e^{\alpha t}x^T(t)P_2x(t) + 2e^{\alpha t}x^T(t)P_2\dot{x}(t) + \tilde{\alpha}_{1k}e^{\alpha t}x^T(t)P_2x(t) \\
&\quad - \tilde{\alpha}_{2k}e^{\alpha(t-\tau(t))}x^T(t - \tau(t))P_2x(t - \tau(t)) \\
&= \alpha e^{\alpha t}x^T(t)P_2x(t) + 2e^{\alpha t}x^T(t)P_2(-(-\mathcal{A}(t)\Lambda - \Delta\mathcal{A}(t)\Lambda \\
&\quad + \Lambda\Lambda + \mathcal{B}(t) + \Delta\mathcal{B}(t))z(t) - (-\Lambda + \mathcal{A}(t) + \Delta\mathcal{A}(t))x(t) \\
&\quad + (C_{\sigma(t)} + \Delta C(t))g(z(t)) + (\mathcal{D}_{\sigma(t)} + \Delta\mathcal{D}(t))g(z(t - \tau(t))) \\
&\quad + \mathcal{K}_{\sigma(t)}(x(t))) + \tilde{\alpha}_{1k}e^{\alpha t}x^T(t)P_2x(t) \\
&\quad - \tilde{\alpha}_{2k}e^{\alpha(t-\tau(t))}x^T(t - \tau(t))P_2x(t - \tau(t)) \\
&\le (\alpha + \tilde{\alpha}_{1k})e^{\alpha t}x^T(t)P_2x(t) + 2e^{\alpha t}x^T(t)P_2(\mathcal{A}(t)\Lambda \\
&\quad + \Delta\mathcal{A}(t)\Lambda - \Lambda\Lambda - \mathcal{B}(t) - \Delta\mathcal{B}(t))z(t) \\
&\quad + 2e^{\alpha t}x^T(t)P_2(\Lambda - \mathcal{A}(t) - \Delta\mathcal{A}(t))x(t) \\
&\quad + 2e^{\alpha t}x^T(t)P_2(C_{\sigma(t)} + \Delta C(t))g(z(t)) \\
&\quad + 2e^{\alpha t}x^T(t)P_2(\mathcal{D}_{\sigma(t)} + \Delta\mathcal{D}(t))g(z(t - \tau(t))) \\
&\quad + 2e^{\alpha t}x^T(t)P_2\mathcal{K}_{\sigma(t)}(x(t)) \\
&\quad - \tilde{\alpha}_{2k}e^{-\alpha\tau}e^{\alpha t}x^T(t - \tau(t))P_2x(t - \tau(t))
\end{aligned}
\tag{7.35}
$$

According to the Assumption 7.1, for any positive define diagonal matrices T_1, T_2 and T_3, one can have

$$\begin{aligned} e^{\alpha t}\chi_1^T(t)\begin{bmatrix} -2L^-T_1L^+ & T_1(L^++L^-) \\ * & -2T_1 \end{bmatrix}\chi_1(t) \geq 0 \\ e^{\alpha t}\chi_2^T(t-\tau(t))\begin{bmatrix} -2L^-T_2L^+ & T_2(L^++L^-) \\ * & -2T_2 \end{bmatrix}\chi_2(t-\tau(t)) \geq 0 \\ e^{\alpha t}\chi_3^T(t-\tau(t))\begin{bmatrix} -2L^-T_3L^+ & T_3(L^++L^-) \\ * & -2T_3 \end{bmatrix}\chi_3(t-\tau(t)) \geq 0 \end{aligned} \tag{7.36}$$

where $\chi_1^T(t) = [z^T(t)\ g^T(z(t))]$, $\chi_2^T(t-\tau(t)) = [z^T(t-\tau(t))\ g(z^T(t-\tau(t)))]$, $\chi_3^T(t-\tau(t)) = [x^T(t-\tau(t))\ g(x^T(t-\tau(t)))]$, $L^- = \text{diag}(l_1^-, l_2^-, \ldots, l_n^-)$, $L^+ = \text{diag}(l_1^+, l_2^+, \ldots, l_n^+)$.

According to Assumption 7.3, for any positive definite diagonal matrix T_4, one has

$$e^{\alpha t}\left(x(t)S_{1\sigma(t)}^T T_4 S_{1\sigma(t)} x(t) - \mathcal{K}_{\sigma(t)}^T(x(t)) T_4 \mathcal{K}_{\sigma(t)}(x(t))\right) \geq 0 \tag{7.37}$$

Considering the following augmented vector
$\digamma^T(t) = [x^T(t)\ z^T(t)\ z^T(t-\tau(t))\ x^T(t-\tau(t))\ g^T(z(t))\ g^T(z(t-\tau(t)))\ g^T(x(t-\tau(t)))\ \mathcal{K}_{\sigma(t)}(x(t))]$.

Combing (7.33)–(7.37), one has

$$\begin{aligned} & D^+V(t,x(t),z(t)) + \tilde{\alpha}_{1k}V(t,x(t),z(t)) - \tilde{\alpha}_{2k}V(t-\tau(t),x(t-\tau(t)),z(t-\tau(t))) \\ = & D^+V_1(t,x(t),z(t)) + D^+V_2(t,x(t),z(t)) \\ & + \tilde{\alpha}_{1k}V_1(t,x(t),z(t)) + \tilde{\alpha}_{1k}V_2(t,x(t),z(t)) \\ & - \tilde{\alpha}_{2k}V_1(t-\tau(t),x(t-\tau(t)),z(t-\tau(t))) \\ & - \tilde{\alpha}_{2k}V_2(t-\tau(t),x(t-\tau(t)),z(t-\tau(t))) \\ \leq & e^{\alpha t}\digamma^T(t)\Theta\digamma(t) \end{aligned} \tag{7.38}$$

where

$$\Theta = \begin{bmatrix} \Theta_{11} & \Theta_{12} & 0 & 0 & \Theta_{15} & \Theta_{16} & 0 & \Theta_{18} \\ * & \Theta_{22} & 0 & 0 & \Theta_{25} & 0 & 0 & 0 \\ * & * & \Theta_{33} & 0 & 0 & \Theta_{36} & 0 & 0 \\ * & * & * & \Theta_{44} & 0 & 0 & \Theta_{47} & 0 \\ * & * & * & * & \Theta_{55} & 0 & 0 & 0 \\ * & * & * & * & * & \Theta_{66} & 0 & 0 \\ * & * & * & * & * & * & \Theta_{77} & 0 \\ * & * & * & * & * & * & * & \Theta_{88} \end{bmatrix}$$

$\Theta_{11} = (\alpha + \tilde{\alpha}_{1k})P_2 + \text{Sym}(P_2\Lambda) - \text{Sym}(P_2\mathcal{A}(t)) - \text{Sym}(P_2\Delta\mathcal{A}(t)) + S_{1\sigma(t)}T_4 S_{1\sigma(t)}$, $\Theta_{12} = P_1 + P_2\Lambda\mathcal{A}(t) + P_2\Lambda\Delta\mathcal{A}(t) - P_2\Lambda\Lambda - P_2\mathcal{B}(t) - P_2\Delta\mathcal{B}(t)$, $\Theta_{15} = P_2C_{\sigma(t)} + P_2\Delta C(t)$, $\Theta_{16} = P_2\mathcal{D}_{\sigma(t)} + P_2\Delta\mathcal{D}(t)$, $\Theta_{18} = P_2$, $\Theta_{22} = (\alpha + \tilde{\alpha}_{1k})P_1 - \text{Sym}(P_1\Lambda) - 2L^-T_1L^+$, $\Theta_{25} = T_1(L^+ + L^-)$, $\Theta_{33} = -2L^-T_2L^+ - \tilde{\alpha}_{2k}P_1e^{-\alpha\tau}$,

$\Theta_{36} = T_2(L^+ + L^-)$, $\Theta_{44} = -\tilde{\alpha}_{2k}e^{-\alpha\tau}P_2 - 2L^-T_3L^+$, $\Theta_{47} = T_3(L^+ + L^-)$, $\Theta_{55} = -2T_1, \Theta_{66} = -2T_2, \Theta_{77} = -2T_3, \Theta_{88} = -T_4$.

For any signal $\sigma(t) = \iota$ and the rule r, it is obvious that the matrix Θ can be rewritten as:

$$\Theta = \bar{\Theta} + \text{Sym}(\hat{\Theta}_0^T F_r(t)\check{\Theta}_0) + \text{Sym}(\hat{\Theta}_1^T F_r(t)\check{\Theta}_1) \tag{7.39}$$

where

$$\bar{\Theta} = \begin{bmatrix} \bar{\Theta}_{11} & \bar{\Theta}_{12} & 0 & 0 & \bar{\Theta}_{15} & \bar{\Theta}_{16} & 0 & \bar{\Theta}_{18} \\ * & \bar{\Theta}_{22} & 0 & 0 & \Theta_{25} & 0 & 0 & 0 \\ * & * & \bar{\Theta}_{33} & 0 & 0 & \Theta_{36} & 0 & 0 \\ * & * & * & \bar{\Theta}_{44} & 0 & 0 & \Theta_{47} & 0 \\ * & * & * & * & \Theta_{55} & 0 & 0 & 0 \\ * & * & * & * & * & \Theta_{66} & 0 & 0 \\ * & * & * & * & * & * & \Theta_{77} & 0 \\ * & * & * & * & * & * & * & \Theta_{88} \end{bmatrix}$$

$\bar{\Theta}_{11} = (\alpha + \tilde{\alpha}_{1k})P_2 + \text{Sym}(P_2\Lambda) - \text{Sym}(P_2 A_r) + S_{1\iota}T_4S_{1\iota}$, $\bar{\Theta}_{12} = P_1 + P_2\Lambda A_r - P_2\Lambda\Lambda - P_2B_r$, $\bar{\Theta}_{15} = P_2C_\iota^J$, $\bar{\Theta}_{16} = P_2D_\iota^J$, $\bar{\Theta}_{18} = P_2$, $\bar{\Theta}_{22} = (\alpha + \tilde{\alpha}_{1k})P_1 - \text{Sym}(P_1\Lambda) - 2L^-T_1L^+$, $\bar{\Theta}_{33} = -2L^-T_2L^+ - \tilde{\alpha}_{2k}P_1e^{-\alpha\tau}$, $\bar{\Theta}_{44} = -\tilde{\alpha}_{2k}e^{-\alpha\tau}P_2 - 2L^-T_3L^+$, $\hat{\Theta}_0 = [M_r^T P_2 \ \ 0 \ \ 0 \ \ 0 \ \ 0 \ \ 0 \ \ 0 \ \ 0]$ $\check{\Theta}_0 = [-N_{1r} \ \ -N_{2r} \ \ 0 \ \ 0 \ \ N_{3r} \ \ N_{4r} \ \ 0 \ \ 0]$ $\hat{\Theta}_1 = [M_r^T\Lambda^T P_2 \ \ 0 \ \ 0 \ \ 0 \ \ 0 \ \ 0 \ \ 0 \ \ 0]$ $\check{\Theta}_1 = [0 \ \ N_{1r} \ \ 0 \ \ 0 \ \ 0 \ \ 0 \ \ 0 \ \ 0]$.

Applying Lemma 7.2, one can get

$$\begin{aligned} \Theta &= \bar{\Theta} + \text{Sym}(\hat{\Theta}_0^T F_r(t)\check{\Theta}_0) + \text{Sym}(\hat{\Theta}_1^T F_r(t)\check{\Theta}_1) \\ &\leq \bar{\Theta} + \beta_1^{-1}\hat{\Theta}_0\hat{\Theta}_0^T + \beta_1\check{\Theta}_0^T\check{\Theta}_0 + \beta_2^{-1}\hat{\Theta}_1\hat{\Theta}_1^T + \beta_2\check{\Theta}_1^T\check{\Theta}_1 \end{aligned} \tag{7.40}$$

By using Schur lemma and condition (7.28), one has $\Theta < 0$.

Based on the above conditions and derivations, for $t \in [t_{k,l_k}, t_{k,l_{k+1}})$, we can know

$$\begin{aligned} D^+V(t, x(t), z(t)) \leq & -\tilde{\alpha}_{1k}V(t, x(t), z(t)) \\ & + \tilde{\alpha}_{2k}V(t - \tau(t), x(t - \tau(t)), z(t - \tau(t))) \end{aligned} \tag{7.41}$$

For the sake of representation, $V(t, x(t), z(t))$ is restated as $\tilde{V}(t)$. It is obvious that

$$D^+\tilde{V}(t) \leq -\tilde{\alpha}_1\tilde{V}(t) + \tilde{\alpha}_2\tilde{V}(t - \tau(t)) \tag{7.42}$$

where $\tilde{\alpha}_1 = \min_{k\in\mathbb{N}}\{\tilde{\alpha}_{1k}\}$, $\tilde{\alpha}_2 = \max_{k\in\mathbb{N}}\{\tilde{\alpha}_{2k}\}$.

If $\tilde{\alpha}_1 > \tilde{\alpha}_2$, then one has

$$\tilde{V}(t) \leq \tilde{V}_\tau(t_0)e^{-\lambda(t-t_0)} \tag{7.43}$$

where $\tilde{V}_\tau(t) = \sup_{-\tau\leq s\leq 0}\tilde{V}(t + s)$.

When $t = t_{k,l_k}$, according to (7.18) and (7.32), one has

$$\begin{aligned}\tilde{V}(t_{k,l_k}) &= e^{\alpha t^+_{k,l_k}} x^T(t^+_{k,l_k}) P_2 x(t^+_{k,l_k}) \\ &= e^{\alpha t^+_{k,l_k}} H^T_{k,l_k}(x(t^-_{k,l_k})) P_2 H_{k,l_k}(x(t^-_{k,l_k})) \\ &\leq \lambda_{\max}(P_2) e^{\alpha t^+_{k,l_k}} x^T(t^-_{k,l_k}) S^T_{2k,l_k} S_{2k,l_k} x(t^-_{k,l_k}) \\ &\leq \frac{\lambda_{\max}(P_2)\lambda_{\max}(S_{2k,l_k} S_{2k,l_k})}{\lambda_{\min}(P_2)} \tilde{V}_2(t^-_{k,l_k}) \\ &= \vartheta_{k,l_k} \tilde{V}(t^-_{k,l_k})\end{aligned} \tag{7.44}$$

where $\vartheta_{k,l_k} = \frac{\lambda_{\max}(P_2)\lambda_{\max}(S_{2k,l_k} S_{2k,l_k})}{\lambda_{\min}(P_2)}$, implies that

$$\tilde{V}(t_k, l_k) = \tilde{V}(t^+_{k,l_k}) \leq \vartheta_{k,l_k} \tilde{V}(t^-_{k,l_k}) \tag{7.45}$$

For any $t \in [t_{k,l_k}, t_{k+1,l_{k+1}}]$, according to the Lemma (7.2), one has

$$\tilde{V}(t) \leq \tilde{V}_\tau(t_{k,l_k}) e^{-\alpha(t - t_{k,l_k})} \tag{7.46}$$

Next, for any $t \in [t_{k,l_k-1}, t_{k,l_k})$, we will give the proof to verify that the following inequality is true.

$$\tilde{V}_\tau(t_{k,l_k}) \leq \tilde{\vartheta}_{k,l_k} \tilde{V}_\tau(t_{k,l_k-1}) e^{-\alpha(t_{k,l_k} - t_{k,l_k-1})} \tag{7.47}$$

where $\tilde{\vartheta}_{k,l_k} = \max\{\vartheta_{k,l_k}, e^{\alpha\tau}\}$.

Obviously, there exists $t^* \in [t_{k,l_k-\tau}, t_{k,l_k})$ that satisfies

$$\tilde{V}(t^*) = \tilde{V}_\tau(t_{k,l_k}) \tag{7.48}$$

Now, for any k, we assume that (7.47) holds. Then it can be proven by considering three situations.

- (a): If $t^* = t_{k,l_k}$, from (7.45) and (7.46), it is obvious that

$$\tilde{V}(t^*) = \tilde{V}(t_{k,l_k}) \leq \vartheta_{k,l_k} \tilde{V}(t^-_{k,l_k}) \tag{7.49}$$

According to (7.48), one derives

$$\begin{aligned}\tilde{V}_\tau(t_{k,l_k}) = \tilde{V}(t_{k,l_k}) &\leq \vartheta_{k,l_k} \tilde{V}(t^-_{k,l_k}) \\ &\leq \vartheta_{k,l_k} \tilde{V}_\tau(t_{k,l_k-1}) e^{-\alpha(t_{k,l_k} - t_{k,l_k-1})} \\ &\leq \tilde{\vartheta}_{k,l_k} \tilde{V}_\tau(t_{k,l_k-1}) e^{-\alpha(t_{k,l_k} - t_{k,l_k-1})}\end{aligned} \tag{7.50}$$

- (b): If $t^* \in [t_{k,l_k-1}, t_{k,l_k})$, one can have

$$\begin{aligned}\tilde{V}(t^*) &= \tilde{V}_\tau(t_{k,l_k}) \\ &\leq \tilde{V}_\tau(t_{k,l_k-1})e^{-\alpha(t^*-t_{k,l_k-1})} \\ &= e^{\alpha(t_{k,l_k}-t^*)}\tilde{V}_\tau(t_{k,l_k-1})e^{-\alpha(t_{k,l_k}-t_{k,l_k-1})} \\ &\leq e^{\alpha\tau}\tilde{V}_\tau(t_{k,l_k-1})e^{-\alpha(t_{k,l_k}-t_{k,l_k-1})} \\ &\leq \tilde{\vartheta}_{k,l_k}\tilde{V}_\tau(t_{k,l_k-1})e^{-\alpha(t_{k,l_k}-t_{k,l_k-1})}\end{aligned} \tag{7.51}$$

- (c): If $t^* \in [t_{k,l_k-\tau}, t_{k,l_k-1})$, one can have

$$\begin{aligned}\tilde{V}(t^*) &= \tilde{V}_\tau(t_{k,l_k}) \leq \tilde{V}_\tau(t_{k,l_k-1}) \\ &\leq e^{\alpha(t_{k,l_k}-t_{k,l_k-1})}\tilde{V}_\tau(t_{k,l_k-1})e^{-\alpha(t_{k,l_k}-t_{k,l_k-1})} \\ &\leq e^{\lambda\tau}\tilde{V}_\tau(t_{k,l_k-1})e^{-\alpha(t_{k,l_k}-t_{k,l_k-1})} \\ &\leq \tilde{\vartheta}_{k,l_k}\tilde{V}_\tau(t_{k,l_k-1})e^{-\alpha(t_{k,l_k}-t_{k,l_k-1})}\end{aligned} \tag{7.52}$$

Therefore, from (7.50), (7.51) and (7.52), we can easily derive the inequality (7.47) by reasonable induction.

Now, for any $t > t_k$, one can obtain

$$\begin{aligned}\tilde{V}(t) &\leq \tilde{\vartheta}_{k,l_k}\tilde{V}_\tau(t_{k,l_k-1})e^{-\alpha(t-t_{k,l_k-1})} \\ &\leq \tilde{\vartheta}_{k,l_k}\tilde{\vartheta}_{k,l_k-1}\tilde{V}_\tau(t_{k,l_k-2})e^{-\alpha(t-t_{k,l_k-2})} \\ &\dots \\ &\leq \tilde{\vartheta}_k^*\tilde{V}_\tau(t_k)e^{-\alpha(t-t_k)}\end{aligned} \tag{7.53}$$

where $\tilde{\vartheta}_k^* = \tilde{\vartheta}_{k,l_k}\tilde{\vartheta}_{k,l_k-1}\dots\tilde{\vartheta}_{k,1}$.

When $d_k = 0$, there is no impulses in the interval $[t_{k-1}, t_k)$, and $\tilde{\vartheta}_{k,0} = e^{\alpha\tau}$, and we can obtain

$$\tilde{V}_\tau(t_k) \leq \tilde{\vartheta}_{k,0}\tilde{V}_\tau(t_{k-1,d_{k-1}})e^{-\alpha(t_k-t_{k-1,d_{k-1}})} \tag{7.54}$$

Similarly, we can derive

$$\begin{aligned}&\tilde{V}_\tau(t_{k-1,d_{k-1}}) \\ &\leq \tilde{\vartheta}_{k-1,d_{k-1}}\tilde{\vartheta}_{k-1,d_{k-1}-1}\dots\tilde{\vartheta}_{k-1,1}V_\tau(t_{k-1})e^{-\alpha(t_k-t_{k-1})}\end{aligned} \tag{7.55}$$

From (7.54) and (7.55), one has

$$\tilde{V}_\tau(t_k) \leq \tilde{\vartheta}_{k-1}^* V_\tau(t_{k-1})e^{-\alpha(t_k-t_{k-1})} \tag{7.56}$$

where $\tilde{\vartheta}_{k-1}^* = \tilde{\vartheta}_{k,0}\tilde{\vartheta}_{k-1,d_{k-1}}\tilde{\vartheta}_{k-1,d_{k-1}-1}\dots\tilde{\vartheta}_{k-1,1}$.

Further, for any $t > t_0$, we can get

$$\tilde{V}(t) \leq \tilde{\vartheta}_k^*\tilde{\vartheta}_{k-1}^*\tilde{\vartheta}_{k-2}^*\dots\tilde{\vartheta}_0^* V_\tau(t_0)e^{-\alpha(t-t_0)} \tag{7.57}$$

Let $\tilde{\vartheta}_k^{**} = \tilde{\vartheta}_k^{*}\tilde{\vartheta}_{k-1}^{*}\tilde{\vartheta}_{k-2}^{*}\ldots\tilde{\vartheta}_0^{*}$, then $\tilde{\vartheta}_k^{**}$ can be described as:

$$\begin{aligned}\tilde{\vartheta}_k^{**} &= e^{\ln\tilde{\vartheta}_k^{**}} = e^{\ln(\tilde{\vartheta}_k^{*}\tilde{\vartheta}_{k-1}^{*}\tilde{\vartheta}_{k-2}^{*}\ldots\tilde{\vartheta}_0^{*})} \\ &= e^{\sum_{j=0}^{k}\ln\tilde{\vartheta}_j^{*}} \leq e^{\frac{1}{k}\sum_{j=0}^{k}\ln\tilde{\vartheta}_j^{*}\left(\frac{t-t_0}{\Delta_t}+N_0\right)}\end{aligned} \tag{7.58}$$

By using the condition (7.29), it is obvious that

$$\tilde{\vartheta}_k^{**} \leq e^{\tilde{\beta}\left(\frac{t-t_0}{\Delta_t}+N_0\right)} = e^{\tilde{\beta}N_0}e^{\frac{\tilde{\beta}}{\Delta_t}(t-t_0)} \tag{7.59}$$

Combing (7.59) and (7.57), one can get

$$\tilde{V}(t) \leq V_\tau(t_0)e^{\tilde{\beta}N_0}e^{-(\alpha-\frac{\tilde{\beta}}{\Delta_t})(t-t_0)} \tag{7.60}$$

Noting that $\min\{V(t,x(t),z(t))\} \leq V(t,x(t),z(t)) \leq \max\{V(t,x(t),z(t))\}$, for each $\sigma(t)=\iota$, one has

$$\begin{aligned}&V(t,x(t),z(t)) \\ &= e^{\alpha t}z^T(t)P_1z(t) + e^{\alpha t}x^T(t)P_2x(t) \\ &\geq e^{\alpha t}\lambda_{\min}(P_1)z^T(t)z(t) + e^{\alpha t}\lambda_{\min}(P_2)x^T(t)x(t) \\ &\geq e^{\alpha t}\lambda_{\min}(P_{12})(||z(t)||^2 + ||x(t)||^2)\end{aligned} \tag{7.61}$$

where $\lambda_{\min}(P_{12}) = \min\{\lambda_{\min}(P_1), \lambda_{\min}(P_2)\}$.

Noting $t_0 = 0$, and combine (7.60) and (7.61), one can obtain

$$\begin{aligned}&e^{\alpha t}\lambda_{\min}(P_{12})(||z(t)||^2 + ||x(t)||^2) \\ &\leq V_\tau(t_0)e^{\tilde{\beta}N_0}e^{-(\alpha-\frac{\tilde{\beta}}{\Delta_t})(t-t_0)} \\ &\leq e^{\tilde{\beta}N_0}(\sup_{-\tau\leq s\leq 0}||z(s)||^2 + \sup_{-\tau\leq s\leq 0}||x(s)||^2)e^{-(\alpha-\frac{\tilde{\beta}}{\Delta_t})t}\end{aligned} \tag{7.62}$$

Thus, we obtain the following result:

$$\begin{aligned}&(||z(t)||^2 + ||x(t)||^2) \\ &\leq \frac{e^{\tilde{\beta}N_0}}{\lambda_{\min}(P_{12})}(\sup_{-\tau\leq s\leq 0}||z(s)||^2 + \sup_{-\tau\leq s\leq 0}||x(s)||^2)e^{-(2\alpha-\frac{\tilde{\beta}}{\Delta_t})t} \\ &\leq (\eta_1\sup_{-\tau\leq s\leq 0}||z(s)||^2 + \eta_2\sup_{-\tau\leq s\leq 0}||x(s)||^2)e^{-\eta_0 t}\end{aligned} \tag{7.63}$$

where $\eta_1 = \eta_2 = \frac{e^{\tilde{\beta}N_0}}{\lambda_{\min}(P_{12})}$, $\eta_0 = 2\alpha - \frac{\tilde{\beta}}{\Delta_t}$. Hence, we assert that the trivial solution of system (7.18) can achieve exponentially stable in Definition 7.2 and Case 2.

This completes the proof. □

7.2 Exponential Synchronization of Stochastic Delayed Memristive Neural Networks

7.2.1 Introduction

In the implementation of MNNs, time delay and noise are inevitable and should be considered. Time-varying delay is a hot topic recently and has been applied in synchronization of NNs [74–78]. In a neural network (NN), synaptic transmission can be seen as a noise process because of random fluctuations that are caused by the release of neuron transmitters and other probabilities [79]. Thus, it is of great practical significance to study the dynamics of stochastic memristive neural networks (SMNNs) with time delays.

In recent years, a lot of efforts have been devoted to study the dynamics of NNs [16, 22, 80–84]. The synchronization of MNNs plays an important role, due to its application in secure communication, image encryption, information science, etc. [85, 86]. There are various types of synchronization, such as anticipation synchronization [87], lag synchronization [88], projective synchronization [89], etc. Among these types, projective synchronization has great engineering significance due to its proportional projective factor between master and slave systems. For example, the proportional feature of projective synchronization can be utilized to generalize binary to M-nary ($M > 2$) for fast digital communication. Projective synchronization is a special case of WPS. In addition, WPS also establishes a bridge between stability and synchronization. Uniformly bounded, quasi-synchronization and quasi-anti-synchronization are special cases of the WPS. Therefore, the WPS is of great importance to be considered. Paper [90] researched fixed-time projective synchronization of complex dynamical networks. Paper [91] investigated projective synchronization of nonidentical fractional-order MNNs by using sliding mode control. Paper [92] proposed the projective synchronization of complex NNs. However, the research of the WPS of SMNNs is inadequate. Hence, it is significant to analyze the WPS of the SMNNs.

The controller plays a significant role in achieving synchronization for most MNNs. There are many controllers which are designed in the nonlinear dynamics, such as intermittent controller [93], nonfragile asynchronous controller [94], impulsive controller [95, 96], sampled-data controller [97–100], and so on. It is worth mentioning that signal transmission is usually limited by the bandwidth and capacity of the communication channels in practical applications. To reduce the information redundancy in the transmitted signal and increase robustness against disturbances and not easily being attacked, impulsive control should be considered. Recently, many research articles about impulsive controllers have been published. For example, paper [44] utilized an event-based impulsive control scheme to investigate the exponential stabilization and synchronization problem of continuous-time dynamic systems. Paper [101] promoted the synchronization control problem by structuring hybrid impulsive and feedback controllers. Paper [102] studied the synchronization and passivity in multiple weighted coupled NNs with impulsive control. In this research, the impulsive control is used to reduce control costs.

In a practical situation, parameter mismatch is inevitable in the realization of synchronization between the master system (MS) and the slave system (SS). It has been proved that parameter mismatch may have effects on synchronization quality: in the case of small parameter mismatch, synchronization error will not decay to zero with time, but it is very small near zero, and non-zero means small fluctuation. If the parameter mismatch is large, it may occur out of step [103]. Hence, it is necessary to research the impacts of parameter mismatch for MNNs synchronization.

In this section, $\mathbf{R}$ stands for the set of real numbers. $\mathbf{R}^n$ is n-dimensional Euclidean space. diag$(\cdot)$ denotes a block-diagonal matrix. P^T is the transpose of matrix P. I and 0 denote the identity and zero matrices with appropriate dimensions, respectively. $\|\cdot\|_q$ represents the standard q norm of a vector or a matrix. Moreover, let $(\Omega, \mathcal{F}, \{\mathcal{F}_t\}_{t\geqslant 0}, P)$ be a complete probability space with filtration $\{\mathcal{F}_t\}_{t\geqslant 0}$ satisfying the usual conditions. Denote by $\mathcal{L}^2_{\mathcal{F}_t}([t_0-\bar{\tau}, t_0]; \mathbf{R}^n)$ the family of all $\mathcal{F}_t$ measurable $C([t_0-\bar{\tau}, t_0]; \mathbf{R}^n)$ -valued stochastic variables $\xi = \{\xi(\theta) : t_0 - \bar{\tau} \leq \theta \leq t_0\}$ such that $\sup_{t_0-\tau\leq\theta\leq t_0} \mathbb{E}\|\xi(\theta)\|^2 < \infty$, where $\mathbb{E}$ stands for mathematical expectation operator. Let $a \vee b$ $(a \wedge b)$ be the maximum (minimum) value of a and b. $\mathfrak{F}([t_0-\zeta, t_0]; \mathbf{R}_+)$ denotes the class of piecewise continuous functions mapping $[t_0-\zeta, t_0]$ to $\mathbf{R}_+$. $\mathbf{R}_+$ denotes the set of positive constants. $\lambda_{\max}$ denotes the maximum eigenvalue. $co\{\cdot,\cdot\}$ is the closure of the convex hull. D^+ denotes Dini derivatives. $|\cdot|$ represents the absolute value.

7.2.2 Problem Description and Preliminaries

Consider the following time-delay SMNNs as the MS:

$$\begin{aligned} \mathrm{d}x_i(t) = \Big(&- c_i x_i(t) + \sum_{j=1}^{n} a_{ij}(x_i(t)) f_j(x_j(t)) + \sum_{j=1}^{n} b_{ij}(x_i(t)) f_j(x_j(t-\tau(t))) \\ &+ I_i(t)\Big)\mathrm{d}t + g_i(t, x_i(t), x_i(t-\tau(t)))\,\mathrm{d}W_i(t),\ i = 1, \ldots, n \end{aligned} \tag{7.64}$$

with

$$a_{ij}(x_i(t)) = \begin{cases} \hat{a}_{ij}, & \text{for } |x_i(t)| \leq \nu_i, \\ \check{a}_{ij}, & \text{for } |x_i(t)| > \nu_i, \end{cases} \qquad b_{ij}(x_i(t)) = \begin{cases} \hat{b}_{ij}, & \text{for } |x_i(t)| \leq \nu_i, \\ \check{b}_{ij}, & \text{for } |x_i(t)| > \nu_i, \end{cases}$$

where $\nu_i > 0, c_i > 0, \hat{a}_{ij}, \check{a}_{ij}, \hat{b}_{ij}, \check{b}_{ij}(i, j = 1, \ldots, n)$ are all constant numbers. $a_{ij}(x_i(t))$, $b_{ij}(x_i(t))$ represent connection weights. $f_j : \mathbf{R} \to \mathbf{R}$ is bounded continuous activation function. $f_j(\cdot)$ satisfies the Lipschitz condition. $I_i(t)$ is external input. $g_i(t, x_i(t), x_i(t-\tau(t)))$ denotes the noise intensity. $W_i(t)$ is a Brownian movement defined on a complete probability space $(\Omega, \mathcal{F}, \{\mathcal{F}_t\}_{t\geqslant 0}, P)$, $W_i(t)$ and $W_j(t)$ are independent process of one another for $j \neq i$. $\tau(t)$ is a continuous function satisfying $0 \leq \tau(t) \leq \bar{\tau}$. Suppose the initial condition of (7.64) is $x_i(s) = \varphi_i(s)$, $s \in [t_0 - \bar{\tau}, t_0]$, in which $\varphi_i(\cdot) \in \mathcal{L}^2_{\mathcal{F}_t}([t_0 - \bar{\tau}, t_0]; \mathbf{R}^n)$.

Denoting the following notations:
$x(t) = (x_1(t), \dots, x_n(t))^T$,
$\overline{a}_{ij} = \max\{\hat{a}_{ij}, \check{a}_{ij}\}$, $\underline{a}_{ij} = \min\{\hat{a}_{ij}, \check{a}_{ij}\}$,
$\overline{b}_{ij} = \max\{\hat{b}_{ij}, \check{b}_{ij}\}$, $\underline{b}_{ij} = \min\{\hat{b}_{ij}, \check{b}_{ij}\}$,
$\hat{A} = (\hat{a}_{ij})_{n\times n}$, $\check{A} = (\check{a}_{ij})_{n\times n}$, $\hat{B} = (\hat{b}_{ij})_{n\times n}$, $\check{B} = (\check{b}_{ij})_{n\times n}$,
$\overline{A} = (\overline{a}_{ij})_{n\times n}$, $\underline{A} = (\underline{a}_{ij})_{n\times n}$, $\overline{B} = (\overline{b}_{ij})_{n\times n}$, $\underline{B} = (\underline{b}_{ij})_{n\times n}$,
$C = \mathrm{diag}(c_1, \dots, c_n)$, $A(x(t)) = (a_{ij}(x_i(t)))_{n\times n}$,
$B(x(t)) = (b_{ij}(x_i(t)))_{n\times n}$, $f(x(t)) = (f_1(x_1(t)), \dots, f_n(x_n(t)))^T$,
$f(x(t-\tau(t))) = (f_1(x_1(t-\tau(t))), \dots, f_n(x_n(t-\tau(t))))^T$,
$G(t, x(t), x(t-\tau(t))) = \mathrm{diag}(g_1(t, x_1(t), x_1(t-\tau(t))), \dots, g_n(t, x_n(t), x_n(t-\tau(t))))^T$,
$I(t) = (I_1(t), \dots, I_n(t))^T$.

Then, using the theories of set-valued maps and differential inclusions, there exist $co\{\hat{A}, \check{A}\} = [\underline{A}, \overline{A}]$, $co\{\hat{B}, \check{B}\} = [\underline{B}, \overline{B}]$, $A_x \in [\underline{A}, \overline{A}]$, $B_x \in [\underline{B}, \overline{B}]$ such that (7.64) is equivalently written in the following form

$$\begin{aligned} \mathrm{d}x(t) =& (-Cx(t) + A_x f(x(t)) + B_x f(x(t-\tau(t))) + I(t))\mathrm{d}t \\ &+ G(t, x(t), x(t-\tau(t)))\mathrm{d}W(t). \end{aligned} \tag{7.65}$$

Then the impulse controlled SS for (7.65) can be constructed as follows:

$$\begin{aligned} \mathrm{d}y(t) =& (-\check{C}y(t) + A_y f(y(t)) + B_y f(y(t-\tau(t))) + J(t) + U(t))\mathrm{d}t \\ &+ G(t, y(t), y(t-\tau(t)))\mathrm{d}W(t) \end{aligned} \tag{7.66}$$

where $y(t) = (y_1(t), \dots, y_n(t))^T \in \mathbf{R}^n$ is the state vector and $A_y \in [\underline{A}, \overline{A}]$, $B_y \in [\underline{B}, \overline{B}]$. $J(t)$ is external input. $\check{C} > 0$ represents the reset rate matrix of each neuron cell. Suppose the initial condition is $\phi(s)$. $U(t)$ is the impulsive controller, which will be constructed later.

Remark 7.12 The parameters between the MS and the SS have three different cases: perfectly matched, partially mismatched, and completely mismatched. Without loss of generality, we consider the parameters to be completely mismatched in this section. Furthermore, the systems with perfectly matched parameters and the systems with partially mismatched parameters are special cases of the presented method.

Consider the parameters between MS and SS mismatched, i.e., $C \neq \check{C}$, $A_x \neq A_y$, $B_x \neq B_y$. Then define the ES as $e(t) = y(t) - \alpha x(t)$, where α is a projective factor. The error system which subtract (7.65) from (7.66) can be written as follows:

$$
\begin{aligned}
\mathrm{d}e(t) =& (-\check{C}e(t) + \alpha(-\check{C} + C)x(t) + A_y\bar{f}(e(t)) + A_y f(\alpha x(t)) \\
& - \alpha A_x f(x(t)) + B_y\bar{f}(e(t-\tau(t))) \\
& + B_y f(\alpha x(t-\tau(t))) - \alpha B_x f(x(t-\tau(t))) + J(t) \\
& - \alpha I(t) + U(t))\mathrm{d}t + \bar{G}(t, e(t), e(t-\tau(t)))\mathrm{d}W(t)
\end{aligned} \tag{7.67}
$$

where $\bar{f}(e(t)) = f(y(t)) - f(\alpha x(t))$, $\bar{f}(e(t-\tau(t))) = f(y(t-\tau(t))) - f(\alpha x(t-\tau(t)))$, $\bar{G}(t, e(t), e(t-\tau(t))) = G(t, y(t), y(t-\tau(t))) - G(t, \alpha x(t), \alpha x(t-\tau(t)))$.

Then, the impulsive controller can be designed as follows:

$$
U(t) = -\Gamma e(t) + \sum_{k=1}^{\infty} \mu e(t)\delta(t-t_k), \tag{7.68}
$$

where $\Gamma = \mathrm{diag}(\gamma_1, \ldots, \gamma_n) \geq 0$ is feedback gain matrix, and μ is the impulsive gain. $\{t_k\}$ denotes the time sequence satisfying $0 \leq t_1 < t_2 < \cdots < t_k < \ldots, \lim_{k\to\infty} t_k = \infty, k \in \overline{N}, \overline{N} = \{0, 1, 2, \ldots\}$. $\delta(\cdot)$ is the Dirac function.

Therefore, (7.67) can be rewritten as:

$$
\begin{aligned}
\mathrm{d}e(t) =& (-(\check{C} + \Gamma)e(t) + \alpha(-\check{C} + C)x(t) + A_y\bar{f}(e(t)) \\
& + A_y f(\alpha x(t)) - \alpha A_x f(x(t)) + B_y\bar{f}(e(t-\tau(t))) \\
& + B_y f(\alpha x(t-\tau(t))) - \alpha B_x f(x(t-\tau(t))) + J(t) - \alpha I(t))\mathrm{d}t \\
& + \bar{G}(t, e(t), e(t-\tau(t)))\mathrm{d}W(t), \quad t \neq t_k, \\
\Delta e(t_k) =& e(t_k^+) - e(t_k^-) = \mu e(t_k^-), \quad t = t_k.
\end{aligned} \tag{7.69}
$$

It is assumed that $e(t)$ is right continuous at t_k, i.e., $e(t_k) = e(t_k^+)$. In addition, the initial condition of (7.69) satisfies $e(t) = \phi(t) - \alpha\varphi(t)$.

By setting $\varepsilon(t) = [x^T(t) \;\; e^T(t)]^T$, the following augmented system can be obtained by combining (7.65) and (7.69):

$$
\begin{aligned}
\mathrm{d}\varepsilon(t) =& (-\tilde{C}\varepsilon(t) + \tilde{A}\tilde{f}(\varepsilon(t)) + \tilde{\tilde{A}}\tilde{f}(H\varepsilon(t)) + \tilde{B}\tilde{f}(\varepsilon(t-\tau(t))) \\
& + \tilde{\tilde{B}}\tilde{f}(H\varepsilon(t-\tau(t))) + \tilde{I}(t))\mathrm{d}t + \tilde{G}(t, \varepsilon(t), \varepsilon(t-\tau(t)))\mathrm{d}W(t), \quad t \neq t_k, \\
\Delta\varepsilon(t_k) =& \varepsilon(t_k^+) - \varepsilon(t_k^-) = \tilde{\mu}\varepsilon(t_k^-), \quad t = t_k, \quad k = 1, 2, 3\ldots
\end{aligned} \tag{7.70}
$$

where $\tilde{C} = \begin{bmatrix} C & 0 \\ \alpha(\check{C} - C) & \check{C} + \Gamma \end{bmatrix}$, $\tilde{f}(\varepsilon(t)) = \begin{bmatrix} f(x(t)) \\ \bar{f}(e(t)) \end{bmatrix}$, $\tilde{A} = \begin{bmatrix} A_x & 0 \\ -\alpha A_x & A_y \end{bmatrix}$, $\tilde{B} = \begin{bmatrix} B_x & 0 \\ -\alpha B_x & B_y \end{bmatrix}$, $\tilde{f}(\varepsilon(t-\tau(t))) = \begin{bmatrix} f(x(t-\tau(t))) \\ \bar{f}(e(t-\tau(t))) \end{bmatrix}$, $\tilde{\tilde{A}} = \begin{bmatrix} 0 & 0 \\ A_y & 0 \end{bmatrix}$, $\tilde{\tilde{B}} = \begin{bmatrix} 0 & 0 \\ B_y & 0 \end{bmatrix}$,

$H = \begin{bmatrix} \alpha I & 0 \\ 0 & 0 \end{bmatrix}$, $\tilde{I}(t) = \begin{bmatrix} I(t) \\ J(t) - \alpha I(t) \end{bmatrix}$, $\tilde{\mu} = \begin{bmatrix} 0 & 0 \\ 0 & \mu I \end{bmatrix}$, $\tilde{G}(t, \varepsilon(t), \varepsilon(t-\tau(t))) = \begin{bmatrix} G(t, x(t), x(t-\tau(t))) \\ \bar{G}(t, e(t), e(t-\tau(t))) \end{bmatrix}$.

Assumption 7.5 (**H1**) For each j, $f_j : \mathbf{R} \to \mathbf{R}$ is bounded, and satisfying Lipschitz condition, i.e., there exists a constant $L_j > 0$ such that $|f_j(z_1) - f_j(z_2)| \le L_j|z_1 - z_2|$, where $z_1, z_2 \in \mathbf{R}$, $j = 1, \ldots, n$.
(**H2**) For $\forall \varepsilon(t) \in \mathbf{R}^{2n}$, there exist positive diagonal matrices $\bar{L}_1$, $\bar{L}_2$ with appropriate dimensions, such that $Tr[\tilde{G}^T(t, \varepsilon(t), \varepsilon(t-\tau(t)))\tilde{G}(t, \varepsilon(t), \varepsilon(t-\tau(t)))] \le \varepsilon^T(t)\bar{L}_1^2\varepsilon(t) + \varepsilon^T(t-\tau(t))\bar{L}_2^2\varepsilon(t-\tau(t))$.
(**H3**) It is assumed that $\tilde{I}(t)$ is bounded.

Remark 7.13 Comparing constraint conditions with [104] which need to ensure the solution of the MS and external input are bounded, this section only needs to assume the external input $\tilde{I}(t)$ is bounded because the ES is processed by using the augmented matrix form. It can not only utilize more information on the system but also reduce the conservatism.

Remark 7.14 The circuit of memristive neural network has been proposed in paper [105]. In most cases, the parameters of MNNs are mismatched. On this basis, we consider the external destabilization factors. Signal disturbance of electronic devices, interference of external environment, limitation of network bandwidth, and delay of equipment communication are inevitable in hardware implementation. In this section, stochastic noise and time delay are used to represent these disturbances. Impulse control is a kind of simple and efficient control strategy because of its fast response, simple operation, and strong anti-interference ability. In order to improve the efficiency of the controller under the energy constraint, we used the impulsive control in system design. Therefore, the implementation of the model design is feasible.

Remark 7.15 It is of practical significance to consider MNNs with stochastic noise and time delays. The time delays could also product in the impulses. Thus, it's necessary to deal with impulses with time delays. The time-delayed impulsive controller will be designed as (7.115). Meanwhile, parameter mismatch is an important characteristic of MNNs. Complete parameter mismatch, stochastic disturbance and delay impulse make the system more practical and more complex. Therefore, the system augmentation is used to reduce conservatism.

Definition 7.3 ([104]) (7.67) and (7.68) are said to be weak projective synchronized, if there exist positive constants $\mathfrak{M} > 0$, $\bar{\bar{\gamma}} > 0$, such that the error NNs (7.70) satisfying that

$$\|\varepsilon(t)\|_q \le e^{-\bar{\bar{\gamma}}(t-t_0)}\mathfrak{M} + \bar{\varepsilon} \tag{7.71}$$

Then, $\|\varepsilon(t)\|_q$ is exponentially convergent to the compact domain $\bar{\Delta} = \{\varepsilon(t) \in \mathbf{R}^{2n} | \|\varepsilon(t)\|_q \le \bar{\varepsilon}\}$ for $t \to \infty$, where $\bar{\varepsilon}$ is an error bound.

Definition 7.4 ([61]) T is said to be the average impulsive interval for the impulsive sequence $\eta = \{t_1, t_2, t_3, \ldots\}$ if there exists a positive integer N_0 and impulsive number $N_\eta(s, t)$ over the interval (s, t) such that:

$$\frac{t-s}{T} - N_0 \leq N_\eta(s,t) \leq \frac{t-s}{T} + N_0, \quad \forall t \geq s > 0. \tag{7.72}$$

Lemma 7.3 ([106]) *Suppose that $V(t)$ and $v(t)$ satisfy*

$$\begin{cases} D^+V(t) \leq F(t, V(t), V(t-\tau(t))), & t \neq t_k, \\ V(t_k) \leq I_k\left(V\left(t_k^-\right)\right), & t = t_k, \end{cases} \tag{7.73}$$

and

$$\begin{cases} D^+v(t) > F(t, v(t), v(t-\tau(t))), & t \neq t_k, \\ v(t_k) \geq I_k\left(v\left(t_k^-\right)\right), & t = t_k, \end{cases} \tag{7.74}$$

where $k \in \overline{N}$, $F(t, V(t), V(t-\tau(t))) : \mathbf{R}_+ \times \mathbf{R} \times \mathbf{R} \to \mathbf{R}$ is nondecreasing in $V(t-\tau(t))$ for any fixed $(t, V(t))$ and $I_k(V(t)) : \mathbf{R} \to \mathbf{R}$ is nondecreasing in $V(t)$. Then $V(t) \leq v(t)$, $\forall t \in [t_0 - \bar{\tau}, t_0]$ implies that $V(t) \leq v(t)$ for $t > t_0$.

Lemma 7.4 ([107]) *For $0 \leq \tau(t) \leq \bar{\tau}$, $\grave{\tau} \leq \tau_k \leq \acute{\tau}$, $\zeta = \hat{\tau} \vee \acute{\tau}$, suppose that $V(t)$ and $v(t)$ satisfy*

$$\begin{cases} D^+V(t) \leq F(t, V(t), V(t-\tau(t))), & t \neq t_k, \\ V(t_k) \leq I_k\left(V\left(t_k^- - \tau_k\right)\right), & t = t_k, \end{cases} \tag{7.75}$$

and

$$\begin{cases} D^+v(t) > F(t, v(t), v(t-\tau(t))), & t \neq t_k, \\ v(t_k) \geq I_k\left(v\left(t_k^- - \tau_k\right)\right), & t = t_k, \end{cases} \tag{7.76}$$

where $k \in \overline{N}$, $F(t, V(t), V(t-\tau(t))) : \mathbf{R}_+ \times \mathbf{R} \times \mathbf{R} \to \mathbf{R}$ is nondecreasing in $V(t-\tau(t))$ for any fixed $(t, V(t))$ and $I_k(V(t)) : \mathbf{R} \to \mathbf{R}$ is nondecreasing in $V(t)$ and $\tau_k > 0$. Then $V(t) \leq v(t)$, $\forall t \in [t_0 - \zeta, t_0]$ implies that $V(t) \leq v(t)$ for $t > t_0$.

Lemma 7.5 ([107]) *Let function $u(t) \in \mathfrak{F}([t_0 - \zeta, +\infty]; \mathbf{R}_+)$. If there exist positive constants Ω_1, Ω_2 and $\breve{\mu}$ such that for $\forall t \geq t_0$,*

$$\begin{cases} \dot{u}(t) = \Omega_1 u(t) + \Omega_2 u(t-\tau(t)), & t \neq t_k, \\ u(t_k^+) = \breve{\mu} u(t_k^- - \tau_k), & t = t_k, \end{cases}$$

then the following inequality holds:

$$
\begin{aligned}
u(t) <& \breve{\mu}^{N_\eta(t_0,t)} e^{(-\Omega_1(t_0-t+\sum_{k=1}^{N_\eta(t_0,t)}\tau_k))} u(t_0) \\
&+ \int_{t_0}^{t} \breve{\mu}^{N_\eta(t,s)} e^{(-\Omega_1(s-t+\sum_{k=N_\eta t(t_0,s)+1}^{N_\eta(t_0,t)}\tau_k))} \Omega_2 u(s-\tau(s))ds, \quad \forall t \geq t_0.
\end{aligned}
\tag{7.77}
$$

7.2.3 Impulsive Controller Without Time Delays

Theorem 7.2 *Suppose that Assumption 7.5 holds. If there exist positive constants* $a_\iota(\iota = 1, 2, 3, 4, 5)$ *and constant* $\breve{I}$, *symmetric positive definite matric* P, *positive definite matrices* $R_\jmath(\jmath = 1, 2, 3, 4, 5)$, *positive diagonal matrices* L, $\bar{L}_1$, $\bar{L}_2$, *such that for* $\breve{\mu} > 0$,

$$
\Omega_1 + \frac{\ln \breve{\mu}}{T} + \max\left\{\breve{\mu}^{-N_0}, \breve{\mu}^{N_0}\right\}\Omega_2 < 0,
\tag{7.78}
$$

where $\Omega_1 = \frac{\lambda_{\max}(Q_1)}{\lambda_{\min}(P)}$, $\Omega_2 = \frac{\lambda_{\max}(Q_2)}{\lambda_{\min}(P)}$,

$$
Q_1 = \begin{bmatrix}
\bar{Q} & P\tilde{A} & P\tilde{\tilde{A}} & P\tilde{B} & P\tilde{\tilde{B}} & P \\
* & -a_1R_1 & 0 & 0 & 0 & 0 \\
* & * & -a_2R_2 & 0 & 0 & 0 \\
* & * & * & -a_3R_3 & 0 & 0 \\
* & * & * & * & -a_4R_4 & 0 \\
* & * & * & * & * & -a_5R_5
\end{bmatrix},
\tag{7.79}
$$

$\bar{Q} = P(-\tilde{C}) + (-\tilde{C})^T P + a_1LR_1L + a_2HLR_2LH + \lambda_{\max}(P)\bar{L}_1^2$, $Q_2 = a_3LR_3L + a_4HLR_4LH + \lambda_{\max}(P)\bar{L}_2^2$, $\breve{\mu} = \|(I + \tilde{\mu})\|_q^2$, $L = \text{diag}(L_1, \ldots, L_n)$. *Then, when* $t \to \infty$, *the ES (7.70) exponentially converges into compact domain*

$$
\tilde{\Delta} = \left\{\varepsilon(t) \in \mathbf{R}^{2n} : \|\varepsilon(t)\|_q \leq \bar{\varepsilon} = \left(\frac{\breve{I}}{\lambda_{min}(P) * \breve{M}}\right)^{\frac{1}{2}}\right\}
\tag{7.80}
$$

with convergence rate $\breve{\gamma}/2$, *in which* $\breve{I} \geq a_5\tilde{I}^T(t)R_5\tilde{I}(t)$, $\breve{M} = -(\Omega_1 + \frac{\ln \breve{\mu}}{T})$ $\min\{\breve{\mu}^{-N_0}, \breve{\mu}^{N_0}\} - \Omega_2$ *and* $\breve{M} > 0$, $\breve{\gamma}$ *is the unique solution of* $\psi(\breve{\gamma}) = \breve{\gamma} + \Omega_1 + \frac{\ln \breve{\mu}}{T} + \max\left\{\breve{\mu}^{-N_0}, \breve{\mu}^{N_0}\right\}\Omega_2 e^{\breve{\gamma}\bar{\tau}} = 0$. *Hence, (7.65) and (7.66) are weak projective synchronized.*

Proof Consider the Lyapunov function candidate

$$
V(t) = \varepsilon^T(t)P\varepsilon(t).
$$

Along with the trajectories of the ES (7.70), we can obtain the following differential system by the $It\hat{o}$ formula:

$$\mathrm{d}V(t) = \mathscr{L}V(t)dt + \mathscr{H}V(t)dW(t), \quad t \in \left[t_k, t_{k+1}\right), \tag{7.81}$$

where

$$\mathscr{L}V(t) = 2\varepsilon^T(t)P(-\tilde{C}\varepsilon(t) + \tilde{A}\tilde{f}(\varepsilon(t)) + \tilde{\tilde{A}}\tilde{f}(H\varepsilon(t)) + \tilde{B}\tilde{f}(\varepsilon(t-\tau(t))) \tag{7.82}$$

$$+ \tilde{\tilde{B}}\tilde{f}(H\varepsilon(t-\tau(t))) + \tilde{I}(t))$$
$$+ Tr(G^T(t, \varepsilon(t), \varepsilon(t-\tau(t)))PG(t, \varepsilon(t), \varepsilon(t-\tau(t)))), \tag{7.83}$$

$$\mathscr{H}V(t) = 2\varepsilon^T(t)PG(t, \varepsilon(t), \varepsilon(t-\tau(t))). \tag{7.84}$$

It is achieved that

$$\begin{aligned} 2\varepsilon^T(t)P\tilde{A}\tilde{f}(\varepsilon(t)) \leq& a_1^{-1}\varepsilon^T(t)P\tilde{A}R_1^{-1}\tilde{A}^T P\varepsilon(t) + a_1\tilde{f}^T(\varepsilon(t))R_1\tilde{f}(\varepsilon(t)) \\ \leq& \varepsilon^T(t)(a_1^{-1}P\tilde{A}R_1^{-1}\tilde{A}^T P)\varepsilon(t) + a_1\varepsilon^T(t)LR_1L\varepsilon(t), \end{aligned} \tag{7.85}$$

$$\begin{aligned} 2\varepsilon^T(t)P\tilde{\tilde{A}}\tilde{f}(H\varepsilon(t)) \leq& a_2^{-1}\varepsilon^T(t)P\tilde{\tilde{A}}R_2^{-1}\tilde{\tilde{A}}^T P\varepsilon(t) \\ &+ a_2\tilde{f}^T(H\varepsilon(t))R_2\tilde{f}(H\varepsilon(t)) \\ \leq& \varepsilon^T(t)(a_2^{-1}P\tilde{\tilde{A}}R_2^{-1}\tilde{\tilde{A}}^T P)\varepsilon(t) \\ &+ a_2\varepsilon^T(t)HLR_2LH\varepsilon(t), \end{aligned} \tag{7.86}$$

$$\begin{aligned} 2\varepsilon^T(t)P\tilde{B}\tilde{f}(\varepsilon(t-\tau(t))) \leq& \varepsilon^T(t)(a_3^{-1}P\tilde{B}R_3^{-1}\tilde{B}^T P)\varepsilon(t) \\ &+ a_3\varepsilon^T(t-\tau(t))LR_3L\varepsilon(t-\tau(t)), \end{aligned} \tag{7.87}$$

$$\begin{aligned} 2\varepsilon^T(t)P\tilde{\tilde{B}}\tilde{f}(H\varepsilon(t-\tau(t))) \leq& a_4^{-1}\varepsilon^T(t)P\tilde{\tilde{B}}R_4^{-1}\tilde{\tilde{B}}^T P\varepsilon(t) \\ &+ a_4\tilde{f}^T(H\varepsilon(t-\tau(t)))R_4\tilde{f}(H\varepsilon(t-\tau(t))) \\ \leq& \varepsilon^T(t)(a_4^{-1}P\tilde{\tilde{B}}R_4^{-1}\tilde{\tilde{B}}^T P)\varepsilon(t) \\ &+ a_4\varepsilon^T(t-\tau(t))HLR_4LH\varepsilon(t-\tau(t)), \end{aligned} \tag{7.88}$$

$$2\varepsilon^T(t)P\tilde{I}(t) \leq a_5^{-1}\varepsilon^T(t)PR_5^{-1}P\varepsilon(t) + a_5\tilde{I}^T(t)R_5\tilde{I}(t), \tag{7.89}$$

$$\begin{aligned} &Tr(\tilde{G}^T(t, \varepsilon(t), \varepsilon(t-\tau(t)))P\tilde{G}(t, \varepsilon(t), \varepsilon(t-\tau(t)))) \\ &\leq \lambda_{\max}(P)Tr(\tilde{G}^T(t, \varepsilon(t), \varepsilon(t-\tau(t)))\tilde{G}(t, \varepsilon(t), \varepsilon(t-\tau(t)))) \\ &\leq \lambda_{\max}(P)(\varepsilon^T(t)\bar{L}_1^2\varepsilon(t) + \varepsilon^T(t-\tau(t))\bar{L}_2^2\varepsilon(t-\tau(t))). \end{aligned} \tag{7.90}$$

Substituting (7.85)–(7.90) into (7.82) implies that

$$\begin{aligned}\mathscr{L}V(t) \leq & \varepsilon^T(t)Q_1\varepsilon(t) + \varepsilon^T(t-\tau(t))Q_2\varepsilon(t-\tau(t)) + \breve{I} \\ \leq & \frac{\lambda_{\max}(Q_1)}{\lambda_{\min}(P)}V(t) + \frac{\lambda_{\max}(Q_2)}{\lambda_{\min}(P)}V(t-\tau(t)) + \breve{I} \\ = & \Omega_1 V(t) + \Omega_2 V(t-\tau(t)) + \breve{I},\end{aligned} \tag{7.91}$$

where $\breve{I} \geq a_5\tilde{I}^T(t)R_5\tilde{I}(t)$ and $\breve{I}$ is a constant because $\tilde{I}(t)$ is bounded form Assumption (H3). Q_1 and Q_2 are defined in (7.78).

There exists a small enough constant $\epsilon > 0$ such that $t + \epsilon \in [t_k, t_{k+1})$, and then integrating (7.81) from t to $t+\epsilon$, and taking expectations on the both sides, it can be obtained that

$$\mathbb{E}V(t+\epsilon) = \mathbb{E}V(t) + \mathbb{E}\int_t^{t+\epsilon} \mathscr{L}V(s)ds + \mathbb{E}\int_t^{t+\epsilon} \mathscr{H}V(s)dW(s), \tag{7.92}$$

$$D^+\mathbb{E}V(t) \leq \Omega_1\mathbb{E}V(t) + \Omega_2\mathbb{E}V(t-\tau(t)) + \breve{I}, \quad t \in [t_k, t_{k+1}). \tag{7.93}$$

On the other hand, when $t = t_k$,

$$V(t_k) = V(t_k^+) = \varepsilon^T(t_k^+)P\varepsilon(t_k^+) \leq \|(I+\tilde{\mu})\|_q^2\varepsilon^T(t_k^-)P\varepsilon(t_k^-) = \breve{\mu}V(t_k^-). \tag{7.94}$$

According to the comparison principle, the following system can be constructed:

$$\begin{cases} \dot{v}(t) = \Omega_1 v(t) + \Omega_2 v(t-\tau(t)) + \breve{I} + \varrho, & t \neq t_k, \\ v(t_k^+) = \breve{\mu}v(t_k^-), & t = t_k, \\ v(t) = \lambda_{\max}(P)\|\phi(t) - \alpha\varphi(t)\|_q^2, & \forall t \in [t_0 - \bar{\tau}, t_0], \end{cases} \tag{7.95}$$

where $v(t)$ is a unique solution and $\varrho > 0$. By Lemma 7.3, we can obtain that $V(t) \leq v(t)$. Moreover, $v(t)$ can be written as

$$v(t) = Z(t,0)v(0) + \int_0^t Z(t,s)[\Omega_2 v(s-\tau(t)) + \breve{I} + \varrho]ds, \tag{7.96}$$

where $Z(t,s) = e^{\Omega_1(t-s)}\Pi_{s\leq t_k \leq t}\breve{\mu}$ can be derived by following linear system:

$$\begin{cases} \dot{v}(t) = \Omega_1 v(t), & t \neq t_k, \\ v(t_k^+) = \breve{\mu}v(t_k^-), & t = t_k. \end{cases} \tag{7.97}$$

Then, $Z(t,s)$ can be calculated through the following two cases.

Case 7.1 When $0 < \breve{\mu} \leq 1$, $N_\eta(s,t) \geq (t-s)/T - N_0$, it follows that

$$\begin{aligned} Z(t,s) &= e^{\Omega_1(t-s)} \prod_{s \leq t_k \leq t} \breve{\mu} = \quad e^{\Omega_1(t-s)} \breve{\mu}^{N_\eta(t,s)} \\ &\leq e^{\Omega_1(t-s)} \breve{\mu}^{\frac{t-s}{T} - N_0} \\ &= \breve{\mu}^{-N_0} e^{\left(\Omega_1 + \frac{\ln \breve{\mu}}{T}\right)(t-s)}. \end{aligned} \tag{7.98}$$

Then, let $\kappa = \breve{\mu}^{-N_0} \lambda_{\max}(P) \sup_{t_0 - \bar{\tau} < t < t_0} \|\phi(t) - \alpha\varphi(t)\|_q^2$. We can obtain that

$$v(t) \leq \kappa e^{\left(\Omega_1 + \frac{\ln \breve{\mu}}{T}\right)(t-t_0)} + \int_{t_0}^{t} \breve{\mu}^{-N_0} e^{\left(\Omega_1 + \frac{\ln \breve{\mu}}{T}\right)(t-s)} (\Omega_2 v(s - \tau(t)) + \breve{I} + \varrho) ds. \tag{7.99}$$

Suppose $\psi(\delta) = \delta + \Omega_1 + \frac{\ln \breve{\mu}}{T} + \breve{\mu}^{-N_0} \Omega_2 e^{\delta \bar{\tau}}$. From (7.78), it can be derived that $\psi(0) < 0$ and $\psi(+\infty) = +\infty$. Meanwhile, $\psi'(\delta) = 1 + \delta \breve{\mu}^{-N_0} \Omega_2 e^{\delta \bar{\tau}} > 0$. Hence, there exists a constant $\gamma > 0$ such that $\psi(\gamma) = \gamma + \Omega_1 + \frac{\ln \breve{\mu}}{T} + \breve{\mu}^{-N_0}$ $\Omega_2 e^{\gamma \bar{\tau}} = 0$. Let $M = -(\Omega_1 + \frac{\ln \breve{\mu}}{T}) \breve{\mu}^{N_0} - \Omega_2$, we can get that if $t_0 - \bar{\tau} < t < t_0$,

$$v(t) = \lambda_{\max}(P) \|\phi(t) - \alpha\varphi(t)\|_q^2 \leq \kappa \leq \kappa e^{-\gamma(t-t_0)} + \frac{\breve{I} + \varrho}{M}. \tag{7.100}$$

Next, it can be seen that when $t > t_0$, $v(t) \leq \kappa e^{-\gamma(t-t_0)} + \frac{\breve{I}+\varrho}{M}$. This can be proofed by contradiction, which means that there exists t^* such that

$$v(t^*) > \kappa e^{-\gamma(t-t_0)} + \frac{\breve{I} + \varrho}{M}. \tag{7.101}$$

Meanwhile, $v(t) \leq \kappa e^{-\gamma(t-t_0)} + \frac{\breve{I}+\varrho}{M}$ for $t_0 < t < t^*$. Then, $v(t^*)$ can be derived as follows:

$$\begin{aligned} v(t^*) &\leq \kappa e^{\left(\Omega_1 + \frac{\ln \breve{\mu}}{T}\right)(t^*-t_0)} + \int_{t_0}^{t^*} \breve{\mu}^{-N_0} e^{\left(\Omega_1 + \frac{\ln \breve{\mu}}{T}\right)(t^*-s)} [\Omega_2 v(s - \tau(t)) + \breve{I} + \varrho] ds \\ &\leq \kappa e^{\left(\Omega_1 + \frac{\ln \breve{\mu}}{T}\right)(t^*-t_0)} + \int_{t_0}^{t^*} \breve{\mu}^{-N_0} e^{\left(\Omega_1 + \frac{\ln \breve{\mu}}{T}\right)(t^*-s)} \Omega_2 (\kappa e^{-\gamma(s-\bar{\tau})} + \frac{\breve{I} + \varrho}{M}) \end{aligned} \tag{7.102}$$

$$+ \breve{I} + \varrho] ds$$

$$
\begin{aligned}
&\leq \kappa e^{\left(\Omega_1+\frac{\ln \breve{\mu}}{T}\right)(t^*-t_0)} + \breve{\mu}^{-N_0} e^{\left(\Omega_1+\frac{\ln \breve{\mu}}{T}\right)(t^*-t_0)} \int_{t_0}^{t^*} e^{-\left(\gamma+\Omega_1+\frac{\ln \breve{\mu}}{T}\right)s} \Omega_2 \kappa e^{\gamma \bar{\tau}} ds \\
&+ (\Omega_2 \frac{\breve{I}+\varrho}{M} + \breve{I} + \varrho) \int_{t_0}^{t^*} \breve{\mu}^{-N_0} e^{\left(\Omega_1+\frac{\ln \breve{\mu}}{T}\right)(t^*-s)} ds \\
&\leq \kappa e^{\left(\Omega_1+\frac{\ln \breve{\mu}}{T}\right)(t^*-t_0)} + \kappa e^{\left(\Omega_1+\frac{\ln \breve{\mu}}{T}\right)(t^*-t_0)} \int_{t_0}^{t^*} e^{-\left(\gamma+\Omega_1+\frac{\ln \breve{\mu}}{T}\right)s} (-\left(\gamma+\Omega_1+\frac{\ln \breve{\mu}}{T}\right)) ds \\
&+ \frac{(\Omega_2+M)(\breve{I}+\varrho)}{M} \int_{t_0}^{t^*} \breve{\mu}^{-N_0} e^{(\Omega_1+\frac{\ln \breve{\mu}}{T})(t^*-s)} ds \\
&\leq \kappa e^{-\gamma(t^*-t_0)} + \frac{\breve{I}+\varrho}{M}.
\end{aligned}
\tag{7.103}
$$

It can be obtained that (7.103) is contrary to the assumption (7.101). Therefore, (7.103) holds for $t > t_0$. Then, let $\varrho \to 0$, one derives that

$$
\lambda_{min}(P)\|\varepsilon(t)\|_q^2 \leq V(t) \leq v(t) \leq \kappa e^{-\gamma(t-t_0)} + \frac{\breve{I}}{M}. \tag{7.104}
$$

Then, (7.104) can be written as follows.

$$
\|\varepsilon(t)\|_q \leq \left(\frac{\kappa}{\lambda_{min}(P)}\right)^{\frac{1}{2}} e^{-\frac{\gamma}{2}(t-t_0)} + \left(\frac{\breve{I}}{\lambda_{min}(P) * M}\right)^{\frac{1}{2}}. \tag{7.105}
$$

Hence, it is obvious that (7.70) is exponentially convergent with convergence rate $\gamma/2$, where γ is the unique solution of $\psi(\gamma) = \gamma + \Omega_1 + \frac{\ln \breve{\mu}}{T} + \breve{\mu}^{-N_0}\Omega_2 e^{\gamma\bar{\tau}} = 0$. Meanwhile, when $t \to \infty$, (7.70) converges into compact domain

$$
\bar{\Delta} = \left\{ \varepsilon(t) \in \mathbf{R}^{2n} : \|\varepsilon(t)\|_q \leq \bar{\varepsilon} = \left(\frac{\breve{I}}{\lambda_{min}(P) * M}\right)^{\frac{1}{2}} \right\}, \tag{7.106}
$$

The first case is completed.

Case 7.2 When $\breve{\mu} > 1$, $N_\eta(s,t) \leq (t-s)/T + N_0$, it follows that

$$
\begin{aligned}
Z(t,s) &= e^{\Omega_1(t-s)} \prod_{s \leq t_k \leq t} \breve{\mu} \\
&= e^{\Omega_1(t-s)} \breve{\mu}^{N_\eta(t,s)} \\
&\leq e^{\Omega_1(t-s)} \breve{\mu}^{\frac{t-s}{T}+N_0} \\
&= \breve{\mu}^{N_0} e^{\left(\Omega_1+\frac{\ln \breve{\mu}}{T}\right)(t-s)}.
\end{aligned}
\tag{7.107}
$$

Then, let $\bar{\kappa} = \breve{\mu}^{N_0}\lambda_{\max}(P)\sup_{t_0-\bar{\tau}<t<t_0} \|\phi(t) - \alpha\varphi(t)\|_q^2$. We can obtain that

$$v(t) \leq \bar{\kappa}e^{\left(\Omega_1+\frac{\ln\breve{\mu}}{T}\right)(t-t_0)} + \int_{t_0}^{t} \breve{\mu}^{N_0}e^{\left(\Omega_1+\frac{\ln\breve{\mu}}{T}\right)(t-s)}(\Omega_2 v(s-\tau(s)) + \breve{I} + \varrho)ds \tag{7.108}$$

for all $t > t_0$. Calculating as Case 1, it can be obtained that

$$v(t) \leq \bar{\kappa}e^{-\gamma'(t-t_0)} + \frac{\breve{I}}{\bar{M}}, \qquad t > t_0, \tag{7.109}$$

where $\bar{M} = -(\Omega_1 + \frac{\ln\breve{\mu}}{T})\breve{\mu}^{-N_0} - \Omega_2$.

Then, one derives that

$$\|\varepsilon(t)\|_q \leq \left(\frac{\bar{\kappa}}{\lambda_{min}(P)}\right)^{\frac{1}{2}} e^{-\frac{\gamma'}{2}(t-t_0)} + \left(\frac{\breve{I}}{\lambda_{min}(P) * \bar{M}}\right)^{\frac{1}{2}}. \tag{7.110}$$

Hence, it is obvious that (7.70) is exponentially convergent with convergence rate $\gamma'/2$, where γ' is the unique solution of $\psi(\gamma') = \gamma' + \Omega_1 + \frac{\ln\breve{\mu}}{T} + \breve{\mu}^{N_0}\Omega_2 e^{\gamma'\bar{\tau}} = 0$. Meanwhile, when $t \to \infty$, (7.70) converges into compact domain

$$\bar{\Delta} = \left\{\varepsilon(t) \in \mathbf{R}^{2n} : \|\varepsilon(t)\|_q \leq \tilde{\varepsilon} = \left(\frac{\breve{I}}{\lambda_{min}(P) * \bar{M}}\right)^{\frac{1}{2}}\right\}. \tag{7.111}$$

This completes the second case. □

Corollary 7.5 *Consider (7.65) and (7.66) without the noise $G(t, x(t), x(t-\tau(t))$ and $G(t, y(t), y(t-\tau(t))$. Suppose that Assumption 7.5 holds. If there exist positive constants a_ι ($\iota = 1, 2, 3, 4, 5$) and constant $\breve{I}$, symmetric positive definite matric P, positive definite matrices $R_\jmath$ ($\jmath = 1, 2, 3, 4, 5$), positive diagonal matrices L, $\bar{L}_1$, $\bar{L}_2$, such that for $\breve{\mu} > 0$,*

$$\Omega_1 + \frac{\ln\breve{\mu}}{T} + \max\left\{\breve{\mu}^{-N_0}, \breve{\mu}^{N_0}\right\}\Omega_2 < 0 \tag{7.112}$$

where $\Omega_1 = \frac{\lambda_{\max}(\tilde{Q}_1)}{\lambda_{\min}(P)}$, $\Omega_2 = \frac{\lambda_{\max}(\tilde{Q}_2)}{\lambda_{\min}(P)}$,

$$\tilde{Q}_1 = \begin{bmatrix} \bar{Q}_1 & P\tilde{A} & P\tilde{\tilde{A}} & P\tilde{B} & P\tilde{\tilde{B}} & P \\ * & -a_1R_1 & 0 & 0 & 0 & 0 \\ * & * & -a_2R_2 & 0 & 0 & 0 \\ * & * & * & -a_3R_3 & 0 & 0 \\ * & * & * & * & -a_4R_4 & 0 \\ * & * & * & * & * & -a_5R_5 \end{bmatrix}, \tag{7.113}$$

$\tilde{Q}_1 = P(-\tilde{C}) + (-\tilde{C})^T P + a_1 L R_1 L + a_2 H L R_2 L H$, $\tilde{Q}_2 = a_3 L R_3 L + a_4 H L R_4 L H$, $\breve{\mu} = \|(I + \tilde{\mu})\|_q^2$, $L = \mathrm{diag}(L_1, \ldots, L_n)$. *Then, it can be obtained that the ES exponentially converges into the compact domain*

$$\tilde{\Delta} = \left\{ \varepsilon(t) \in \mathbf{R}^{2n} : \|\varepsilon(t)\|_q \leq \bar{\varepsilon} = \left(\frac{\breve{I}}{\lambda_{min}(P) * \breve{M}} \right)^{\frac{1}{2}} \right\} \tag{7.114}$$

with convergence rate $\breve{\gamma}/2$ *when* $t \to \infty$, *where* $\breve{\gamma}$ *is the unique solution of* $\psi(\breve{\gamma}) = \breve{\gamma} + \Omega_1 + \frac{\ln \breve{\mu}}{T} + \max\left\{\breve{\mu}^{-N_0}, \breve{\mu}^{N_0}\right\} \Omega_2 e^{\breve{\gamma}\bar{\tau}} = 0$, $\breve{I} \geq a_5 \tilde{I}^T(t) R_5 \tilde{I}(t)$, $\breve{M} = -(\Omega_1 + \frac{\ln \breve{\mu}}{T}) - \min\left\{\breve{\mu}^{-N_0}, \breve{\mu}^{N_0}\right\} \Omega_2$ *and* $\breve{M} > 0$. *Hence, MS and SS are weak projective synchronized.*

Remark 7.16 If $\alpha = 1$ for the ES $e(t) = y(t) - \alpha x(t)$, the WPS of the time-delay SMNNs (7.65) and SS (7.66) can be extended to quasi-synchronization. Furthermore, when $\alpha = -1$, the MS (7.65) and SS (7.66) are quasi-antisynchronized.

Remark 7.17 Consider MS (7.65) and SS (7.66) with the same external input, which means $I(t) = J(t)$. It can be obtained that $\breve{I}(t) = 0$. Case 1. For $0 < \|(I + \tilde{\mu})\|_q^2 \leq 1$, if the condition $\Omega_1 + \frac{\ln \breve{\mu}}{T} + \breve{\mu}^{-N_0}\Omega_2 < 0$ is satisfied, it can be obtained that, when $t \to \infty$, the ES is convergent into compact domain $\tilde{\Delta} = \left\{\varepsilon(t) \in \mathbf{R}^{2n} : \|\varepsilon(t)\|_q \leq 0\right\}$ with convergence rate $\gamma/2$, where γ is the unique solution of $\psi(\gamma) = \gamma + \Omega_1 + \frac{\ln \breve{\mu}}{T} + \breve{\mu}^{-N_0}\Omega_2 e^{\gamma\bar{\tau}} = 0$.

Case 2. For $\|(I + \tilde{\mu})\|_q^2 > 1$, if the condition $\Omega_1 + \frac{\ln \breve{\mu}}{T} + \breve{\mu}^{N_0}\Omega_2 < 0$ is satisfied, when $t \to \infty$, the ES exponentially converges into compact domain $\bar{\Delta} = \left\{\varepsilon(t) \in \mathbf{R}^{2n} : \|\varepsilon(t)\|_q \leq \tilde{\varepsilon} = 0\right\}$ with convergence rate $\gamma'/2$, where γ' is the unique solution of $\psi(\gamma') = \gamma' + \Omega_1 + \frac{\ln \breve{\mu}}{T} + \breve{\mu}^{N_0}\Omega_2 e^{\gamma'\bar{\tau}} = 0$. Hence, in both cases, the MS and the SS are complete exponential synchronization.

Remark 7.18 Compared with [108], which considered the projective synchronization of MNNs with feedback controller, this section studied the MNNs with impulsive control and stochastic noise. This section's research is more practical and less control signal transmission. Compared with [95], which only studied the exponential synchronization of SMNNs, we analyzed the WPS in this section. Exponential synchronization is a special case of WPS. Therefore, we expanded the dynamic characteristics of SMNNs.

7.2.4 Impulsive Controller with Time Delays

Consider the impulsive controller with time delays as follows:

$$U(t) = -\Gamma e(t) + \sum_{k=1}^{\infty} \mu e(t - \tau_k)\delta(t - t_k), \tag{7.115}$$

where for any given $\beta > 0$, $t_k - t_{k-1} \leq \beta$. τ_k denotes the delay in impulse satisfying $\grave{\tau} \leq \tau_k \leq \acute{\tau}$, where $\acute{\tau} \in (0, \beta)$ and $\grave{\tau} \geq 0$ are two constants.

Then, the ES can be rewritten as:

$$\begin{aligned} de(t) =& (-(\breve{C} + \Gamma)e(t) + \alpha(-\breve{C} + C)x(t) + A_y \bar{f}(e(t)) + A_y f(\alpha x(t)) \\ & - \alpha A_x f(x(t)) + B_y \bar{f}(e(t - \tau(t))) + B_y f(\alpha y(t - \tau(t))) \\ & - \alpha B_x f(x(t - \tau(t))) + J(t) - \alpha I(t))dt \\ & + \bar{G}(t, e(t), e(t - \tau(t)))dW(t), \quad t \neq t_k, \\ \Delta e(t_k) =& \mu e(t_k^- - \tau_k), \quad t = t_k. \end{aligned} \tag{7.116}$$

Therefore, the following augmented system can be obtained from (7.65) and (7.116):

$$\begin{aligned} d\varepsilon(t) =& (-\tilde{C}\varepsilon(t) + \tilde{A}\tilde{f}(\varepsilon(t)) + \tilde{\tilde{A}}\tilde{f}(H\varepsilon(t)) + \tilde{I}(t) + \tilde{B}\tilde{f}(\varepsilon(t - \tau(t))) \\ & + \tilde{\tilde{B}}\tilde{f}(H\varepsilon(t - \tau(t))))dt \\ & + \tilde{G}(t, \varepsilon(t), \varepsilon(t - \tau(t)))dW(t), \quad t \neq t_k, \\ \Delta\varepsilon(t_k) =& \tilde{\mu}\varepsilon(t_k^- - \tau_k), \quad t = t_k, \quad k = 1, 2, 3 \ldots \end{aligned}$$

Theorem 7.3 *Suppose that Assumption 7.5 holds. If there exist positive constants* $a_\iota (\iota = 1, 2, 3, 4, 5)$ *and constant* $\breve{I}$, $\ell > e^{(\bar{\tau}/\beta+1)\Omega_1\acute{\tau}}$, *symmetric positive definite matric* P, *positive definite matrices* $R_\jmath (\jmath = 1, 2, 3, 4, 5)$, *positive diagonal matrices* L, $\bar{L}_1$, $\bar{L}_2$, *such that for* $\breve{\mu} > 0$,

$$\Omega_1 + \frac{\ln \breve{\mu}}{T} + \ell \max\{\breve{\mu}^{-N_0}, \breve{\mu}^{N_0}\}\Omega_2 < 0 \tag{7.117}$$

where $\Omega_1 = \frac{\lambda_{\max}(Q_1)}{\lambda_{\min}(P)}$, $\Omega_2 = \frac{\lambda_{\max}(Q_2)}{\lambda_{\min}(P)}$,

$$Q_1 = \begin{bmatrix} \bar{Q} & P\tilde{A} & P\tilde{\tilde{A}} & P\tilde{B} & P\tilde{\tilde{B}} & P \\ * & -a_1R_1 & 0 & 0 & 0 & 0 \\ * & * & -a_2R_2 & 0 & 0 & 0 \\ * & * & * & -a_3R_3 & 0 & 0 \\ * & * & * & * & -a_4R_4 & 0 \\ * & * & * & * & * & -a_5R_5 \end{bmatrix}, \tag{7.118}$$

$\bar{Q} = P(-\tilde{C}) + (-\tilde{C})^T P + a_1 L R_1 L + a_2 H L R_2 L H + \lambda_{\max}(P)\bar{L}_1^2$, $Q_2 = a_3 L R_3 L + a_4 H L R_4 L H + \lambda_{\max}(P)\bar{L}_2^2$, $\breve{\mu} = \|(I + \tilde{\mu})\|_q^2$, $L = \mathrm{diag}(L_1, \ldots, L_n)$. *Then, it can be obtained that when* $t \to \infty$, *the ES (7.117) exponentially converges into compact domain*

$$\tilde{\Delta} = \left\{ \varepsilon(t) \in \mathbf{R}^{2n} : \|\varepsilon(t)\|_q \leq \bar{\varepsilon} = \left(\frac{\breve{I}}{\lambda_{min}(P) * \breve{M}} \right)^{\frac{1}{2}} \right\} \tag{7.119}$$

with convergence rate $\frac{\breve{\wp}}{2} + \frac{\Omega_1 \bar{\tau}}{2T}$, *where* $\breve{\wp}$ *is the unique solution of* $\bar{\psi}(\breve{\wp}) = \breve{\wp} + \Omega_1 + \frac{\ln \breve{\mu}}{T} + \ell \max\left\{\breve{\mu}^{-N_0}, \breve{\mu}^{N_0}\right\} \Omega_2 e^{\breve{\wp}\bar{\tau}} = 0$, $\breve{I} \geq a_5 \tilde{I}^T(t) R_5 \tilde{I}(t)$, $\breve{M} = -(\Omega_1 + \frac{\ln \breve{\mu}}{T}) \min\left\{\breve{\mu}^{-N_0}, \breve{\mu}^{N_0}\right\} - \Omega_2$ *and* $\breve{M} > 0$. *Hence, the MS and the SS are weak projective synchronized.*

Proof Consider the following Lyapunov function:

$$V(t) = \varepsilon^T(t) P \varepsilon(t). \tag{7.120}$$

Then, similarly as in Theorem 1, we can get

$$D^+ \mathbb{E}V(t) \leq \Omega_1 \mathbb{E}V(t) + \Omega_2 \mathbb{E}V(t - \tau(t)) + \breve{I}, \quad t \in \left[t_k, t_{k+1}\right). \tag{7.121}$$

On the other hand, when $t = t_k$,

$$\begin{aligned} V(t_k) = V(t_k^+) &= \varepsilon^T(t_k^+) P \varepsilon(t_k^+) \\ &\leq \|(I + \tilde{\mu})\|_q^2 \varepsilon^T(t_k^- \tau_k) P \varepsilon(t_k^- \tau_k) \\ &= \breve{\mu} V(t_k^- - \tau_k). \end{aligned} \tag{7.122}$$

According to the comparison principle, the following system can be constructed:

$$\begin{cases} \dot{u}(t) = \Omega_1 v(t) + \Omega_2 u(t - \tau(t)) + \breve{I} + \varrho, & t \neq t_k, \\ u(t_k^+) = \breve{\mu} u(t_k^- - \tau_k), & t = t_k, \\ u(t) = \lambda_{\max}(P) \|\phi(t) - \alpha\varphi(t)\|_q^2, & \forall t \in [t_0 - \bar{\tau}, t_0], \end{cases} \tag{7.123}$$

where $u(t)$ is a unique solution. By Lemma 7.4, we can obtain that $V(t) \leq u(t)$.

In addition, by Lemma 7.5, we obtain that $\forall t \geq t_0$,

$$\begin{aligned} u(t) <& \breve{\mu}^{N_\eta(t_0,t)} e^{(-\Omega_1(t_0 - t + \sum_{k=1}^{N_\eta(t_0,t)} \tau_k))} u(t_0) \\ &+ \int_{t_0}^{t} \breve{\mu}^{N_\eta(t,s)} e^{(-\Omega_1(s - t + \sum_{k=N_\eta(t_0,s)+1}^{N_\eta(t_0,t)} \tau_k))} \\ &\times (\Omega_2 u(s - \tau(s)) + \breve{I} + \varrho) ds. \end{aligned} \tag{7.124}$$

Case 1. When $0 < \breve{\mu} \leq 1$, $N_\eta(s, t) \geq (t - s)/T - N_0$, we can get $\forall t \geq s \geq t_0$,

$$e^{\Omega_1(t-s)} \breve{\mu}^{N_\eta(t,s)} \leq e^{\Omega_1(t-s)} \breve{\mu}^{\frac{t-s}{T} - N_0} = \breve{\mu}^{-N_0} e^{\left(\Omega_1 + \frac{\ln \breve{\mu}}{T}\right)(t-s)}. \tag{7.125}$$

Then, it can be obtained that

$$u(t) \leq \kappa e^{\left(\Omega_1+\frac{\ln\breve{\mu}}{T}\right)(t-t_0)-\Omega_1\sum_{k=1}^{N_\eta(t_0,t)}\tau_k}$$
$$+\int_{t_0}^{t}\breve{\mu}^{-N_0}e^{\left(\Omega_1+\frac{\ln\breve{\mu}}{T}\right)(t-s)-\Omega_1\sum_{k=N_\eta(t_0,s)+1}^{N_\eta(t_0,t)}\tau_k}(\Omega_2 u(s-\tau(s))+\breve{I}+\varrho)ds. \tag{7.126}$$

Suppose $\ell > e^{(\bar{\tau}/\beta+1)\Omega_1\acute{\tau}} > e^{\Omega_1\sum_{k=N_\eta(t-\bar{\tau}(t),t_0)}^{N_\eta(t,t_0)+1}\tau_k}$, $\bar{\psi}(\sigma)=\sigma+\Omega_1+\frac{\ln\breve{\mu}}{T}+\ell\breve{\mu}^{-N_0}\Omega_2 e^{\sigma\bar{\tau}}$. From the condition (7.117), it can be derived that $\psi(0)<0$ and $\bar{\psi}(+\infty)=+\infty$. Meanwhile, $\bar{\psi}'(\sigma)=1+\ell\sigma\breve{\mu}^{-N_0}\Omega_2 e^{\sigma\bar{\tau}}>0$. Hence, there exists a constant $\gamma>0$ such that $\bar{\psi}(\wp)=\wp+\Omega_1+\frac{\ln\breve{\mu}}{T}+\ell\breve{\mu}^{-N_0}\Omega_2 e^{\gamma\bar{\tau}}=0$. Let $M=-(\Omega_1+\frac{\ln\breve{\mu}}{T})\breve{\mu}^{N_0}-\Omega_2$, we can get that if $t_0-\bar{\tau}<t<t_0$,

$$u(t)=\lambda_{\max}(P)\|\phi(t)-\alpha\varphi(t)\|_q^2\leq\kappa\leq\kappa e^{-\wp(t-t_0)-\Omega_1\sum_{k=1}^{N_\eta(t_0,t)}\tau_k}+\frac{\breve{I}+\varrho}{M}. \tag{7.127}$$

Next, similar to the method of calculation in Theorem 1, we can get that

$$u(t)\leq\kappa e^{-\wp(t-t_0)-\Omega_1\sum_{k=1}^{N_\eta(t_0,t)}\tau_k}+\frac{\breve{I}+\varrho}{M},\qquad t>t_0. \tag{7.128}$$

Then, let $\varrho\to 0$, one derives that

$$\lambda_{min}(P)\|\varepsilon(t)\|_q^2\leq V(t)\leq u(t)\leq\kappa e^{-\wp(t-t_0)-\Omega_1\sum_{k=1}^{N_\eta(t_0,t)}\tau_k}+\frac{\breve{I}}{M}. \tag{7.129}$$

According to Definition 2 and $\grave{\tau}\leq\tau_k\leq\acute{\tau}$, one can derive that $-\Omega_1\sum_{k=1}^{N_\eta(t_0,t)}\tau_k\leq -\Omega_1 N_\eta(t_0,t)\,\grave{\tau}\leq-\Omega_1\grave{\tau}\frac{t-t_0}{T}+\Omega_1 N_0\grave{\tau}$. Then, (7.129) can be written as follows.

$$\|\varepsilon(t)\|_q\leq\left(\frac{\hat{\kappa}}{\lambda_{min}(P)}\right)^{\frac{1}{2}}e^{-\left(\frac{\wp}{2}+\frac{\Omega_1\grave{\tau}}{2T}\right)(t-t_0)}+\left(\frac{\breve{I}}{\lambda_{min}(P)*M}\right)^{\frac{1}{2}}, \tag{7.130}$$

where $\hat{\kappa}=\kappa e^{\Omega_1 N_0\grave{\tau}}$. Hence, it is obvious that the ES (7.117) exponentially converges with convergence rate $\frac{\wp}{2}+\frac{\Omega_1\grave{\tau}}{2T}$, where $\wp$ is the unique solution of $\bar{\psi}(\wp)=\wp+\Omega_1+\frac{\ln\breve{\mu}}{T}+\ell\breve{\mu}^{-N_0}\Omega_2 e^{\gamma\bar{\tau}}=0$. Meanwhile, when $t\to\infty$, the ES (7.117) converges into compact domain

$$\bar{\Delta} = \left\{ \varepsilon(t) \in \mathbf{R}^{2n} : \|\varepsilon(t)\|_q \leq \bar{\varepsilon} = \left(\frac{\breve{I}}{\lambda_{min}(P) * M} \right)^{\frac{1}{2}} \right\}. \tag{7.131}$$

The first case is completed.

Case 2. When $\breve{\mu} > 1$, $N_\eta(s,t) \leq (t-s)/T + N_0$,

$$e^{\Omega_1(t-s)} \breve{\mu}^{N_\eta(t,s)} \leq e^{\Omega_1(t-s)} \breve{\mu}^{\frac{t-s}{T} + N_0} = \breve{\mu}^{N_0} e^{\left(\Omega_1 + \frac{\ln \breve{\mu}}{T}\right)(t-s)}. \tag{7.132}$$

Then, let $\bar{\kappa} = \breve{\mu}^{N_0} \lambda_{\max}(P) \sup_{t_0 - \bar{\tau} < t < t_0} \|\phi(t) - \alpha\varphi(t)\|_q^2$. We can obtain that

$$\begin{aligned} u(t) \leq & \bar{\kappa} e^{\left(\Omega_1 + \frac{\ln \breve{\mu}}{T}\right)(t - t_0) - \Omega_1 \sum_{k=1}^{N_\eta(t_0,t)} \tau_k} + \int_{t_0}^{t} \breve{\mu}^{N_0} e^{\left(\Omega_1 + \frac{\ln \breve{\mu}}{T}\right)(t-s) - \Omega_1 \sum_{k=N_\eta(t_0,s)+1}^{N_\eta(t_0,t)} \tau_k} \\ & \times (\Omega_2 u(s - \tau(s)) + \breve{I} + \varrho) ds \end{aligned} \tag{7.133}$$

is hold for all $t > t_0$. Calculating as Case 1, it can be obtained that

$$u(t) \leq \bar{\kappa} e^{-\wp'(t-t_0) - \Omega_1 \sum_{k=1}^{N_\eta(t_0,t)} \tau_k} + \frac{\breve{I}}{\bar{M}}, \qquad t > t_0, \tag{7.134}$$

where $\bar{M} = -(\Omega_1 + \frac{\ln \breve{\mu}}{T})\breve{\mu}^{-N_0} - \Omega_2$. Then, one derives that

$$\|\varepsilon(t)\|_q \leq \left(\frac{\breve{\kappa}}{\lambda_{min}(P)} \right)^{\frac{1}{2}} e^{-\left(\frac{\wp'}{2} + \frac{\Omega_1 \dot{\tau}}{2T}\right)(t - t_0)} + \left(\frac{\breve{I}}{\lambda_{min}(P) * M} \right)^{\frac{1}{2}}. \tag{7.135}$$

Hence, it is obvious that the ES (7.117) is exponentially convergent with convergence rate $\frac{\wp'}{2} + \frac{\Omega_1 \dot{\tau}}{2T}$, where $\wp'$ is the unique solution of $\bar{\psi}(\wp') = \wp' + \Omega_1 + \frac{\ln \breve{\mu}}{T} + \ell \breve{\mu}^{N_0} \Omega_2 e^{\wp' \bar{\tau}} = 0$. Meanwhile, when $t \to \infty$, the ES (7.117) converges into compact domain

$$\bar{\Delta} = \left\{ \varepsilon(t) \in \mathbf{R}^{2n} : \|\varepsilon(t)\|_q \leq \tilde{\varepsilon} = \left(\frac{\breve{I}}{\lambda_{min}(P) * \bar{M}} \right)^{\frac{1}{2}} \right\}. \tag{7.136}$$

This completes the second case. □

7.2.5 Simulation

As an application of the results obtained previously, a numerical example is given to show that our WPS criteria are effective. Moreover, the anti-synchronization, a special case of the proposed WPS approach, is used to realize a secure communication application.

7.2.6 Numerical Example

In this section, a numerical example will be shown. The simulation platform is MATLAB 2020a on 64-bit Windows 10. Consider the ES (7.67) with the following parameters:

$C = \begin{bmatrix} 6 & 0 \\ 0 & 6 \end{bmatrix}, \breve{C} = \begin{bmatrix} 5 & 0 \\ 0 & 3 \end{bmatrix},$

$$a_{11}(x_1(t)) = \begin{cases} 1.8, & |x_1(t)| \le 0.03, \\ 2.2, & |x_1(t)| > 0.03, \end{cases} \quad a_{12}(x_1(t)) = \begin{cases} -0.05, & |x_1(t)| \le 0.03, \\ -0.15, & |x_1(t)| > 0.03, \end{cases}$$

$$a_{21}(x_2(t)) = \begin{cases} -4, & |x_2(t)| \le 0.05, \\ -6, & |x_2(t)| > 0.05, \end{cases} \quad a_{22}(x_2(t)) = \begin{cases} 4, & |x_2(t)| \le 0.05, \\ 5, & |x_2(t)| > 0.05, \end{cases}$$

$$b_{11}(x_1(t)) = \begin{cases} -1.3, & |x_1(t)| \le 0.03, \\ -1.7, & |x_1(t)| > 0.03, \end{cases} \quad b_{12}(x_1(t)) = \begin{cases} -0.15, & |x_1(t)| \le 0.03, \\ -0.05, & |x_1(t)| > 0.03, \end{cases}$$

$$b_{21}(x_2(t)) = \begin{cases} -0.18, & |x_2(t)| \le 0.05, \\ -0.22, & |x_2(t)| > 0.05, \end{cases} \quad b_{22}(x_2(t)) = \begin{cases} -3, & |x_2(t)| \le 0.05, \\ -5, & |x_2(t)| > 0.05, \end{cases}$$

$$a_{11}(y_1(t)) = \begin{cases} 1.6, & |y_1(t)| \le 0.03, \\ 2, & |y_1(t)| > 0.03, \end{cases} \quad a_{12}(y_1(t)) = \begin{cases} -0.2, & |y_1(t)| \le 0.03, \\ -0.1, & |y_1(t)| > 0.03, \end{cases}$$

$$a_{21}(y_2(t)) = \begin{cases} -5.4, & |y_2(t)| \le 0.05, \\ -5, & |y_2(t)| > 0.05, \end{cases} \quad a_{22}(y_2(t)) = \begin{cases} 3, & |y_2(t)| \le 0.05, \\ 4, & |y_2(t)| > 0.05, \end{cases}$$

$$b_{11}(y_1(t)) = \begin{cases} -1.9, & |y_1(t)| \le 0.03, \\ -1.5, & |y_1(t)| > 0.03, \end{cases} \quad b_{12}(y_1(t)) = \begin{cases} -0.14, & |y_1(t)| \le 0.03, \\ -0.1, & |y_1(t)| > 0.03, \end{cases}$$

$$b_{21}(y_2(t)) = \begin{cases} -0.29, & |y_2(t)| \le 0.05, \\ -0.23, & |y_2(t)| > 0.05, \end{cases} \quad b_{22}(y_2(t)) = \begin{cases} -3, & |y_2(t)| \le 0.05, \\ -2, & |y_2(t)| > 0.05. \end{cases}$$

The noise intensities are taken as follows: $g_1(t, x_1(t), x_1(t-\tau(t))) = 0.3x_1(t) - 0.004x_1(t-\tau(t))$ and $g_2(t, x_2(t), x_2(t-\tau(t))) = 0.6x_2(t) - 0.002x_2(t-\tau(t))$. The nonlinear functions $f(\cdot) = 0.01\tanh(\cdot)$. The external inputs $I = [0.07; 0.014]$, $J = [0.015; 0.021]$. The Γ of the impulsive controller is $\Gamma = [30, 0; 0, 30]$. The Lipschitz matrices of augmented systems are shown as follows:

$$L = \begin{bmatrix} 0.01 & 0 & 0 & 0 \\ 0 & 0.01 & 0 & 0 \\ 0 & 0 & 0.01 & 0 \\ 0 & 0 & 0 & 0.01 \end{bmatrix}, \bar{L}_2 = \begin{bmatrix} 1\,0\,0\,0 \\ 0\,1\,0\,0 \\ 0\,0\,1\,0 \\ 0\,0\,0\,1 \end{bmatrix}, \bar{L}_1 = \begin{bmatrix} 0.001 & 0 & 0 & 0 \\ 0 & 0.01 & 0 & 0 \\ 0 & 0 & 0.01 & 0 \\ 0 & 0 & 0 & 0.001 \end{bmatrix}.$$

Select $a_\iota(\iota = 1, \ldots, 5) = 1$, $T = 0.9$, $N_0 = 1$, $\bar{\tau} = 0.6$, $\varphi(s) = [1, 0.1]$, $\phi(s) = [1, 0.1]$, $s \in [t_0 - \bar{\tau}, t_0]$. For Theorem 1 Case 1, the remaining parameters are chosen as $\alpha = -0.9$, $\mu = -0.01$ or $\breve{\mu} = 1$. By using the above-specified parameter values and solving the LMIs (7.78) with the aid of MATLAB LMI toolbox, the feasibility can be obtained and we can get $\Omega_1 = -20.1647$, $\Omega_2 = 4.4410$, $\Omega_1 + \frac{\ln \breve{\mu}}{T} + \breve{\mu}^{-N_0}\Omega_2 = -15.7237 < 0$. The ES (7.70) converges into compact domain $\tilde{\Delta} = \left\{\varepsilon(t) \in \mathbf{R}^{2n} : \|\varepsilon(t)\|_q \le \bar{\varepsilon} = 0.2344\right\}$ with the convergence rate of 1.1591. From Fig. 7.12a and b, it is obvious that the MS and the SS have mismatched parameters. Figure 7.13a shows the MS (7.65) and the SS (7.66) are weak projective synchronized. Additionally, the error bound is 0.12.

For Theorem 1 Case 2, we select $\alpha = -0.2$, $\mu = 0.9$ or $\breve{\mu} = 3.61 > 1$. After calculating, it can be obtained that $\Omega_1 = -23.6805$, $\Omega_2 = 5.9407$, $\Omega_1 + \frac{\ln \breve{\mu}}{T} + \breve{\mu}^{N_0}\Omega_2 = -0.8082 < 0$. The ES (7.70) converges into compact domain $\tilde{\Delta} = \left\{\varepsilon(t) \in \mathbf{R}^{2n} : \|\varepsilon(t)\|_q \le \bar{\varepsilon} = 1.3825\right\}$ with the convergence rate of 0.0287. Figure 7.12c and d show that the MS and the SS have mismatched parameters in Case 2. Figure 7.13b shows the MS (7.65) and the SS (7.66) are weak projective synchronizated. Additionally, the error bound is 0.09.

For Theorem 2, select $\grave{\tau} = 0.2$, $\acute{\tau} = 1.2$, $\beta = 1.3$, $\tau_k = 1$. For Case 1, parameters are same as Theorem 1 Case 1. After calculating, $\Omega_1 + \frac{\ln \breve{\mu}}{T} + \ell\breve{\mu}^{-N_0}\Omega_2 = -20.1647 < 0$. The ES (7.117) converges into compact domain $\tilde{\Delta} = \{\varepsilon(t) \in \mathbf{R}^{2n} : \|\varepsilon(t)\|_q \le \bar{\varepsilon} = 0.2344\}$ with the convergence rate of 7.8418. From Fig. 7.14a and b, it is obvious that the MS and the SS have mismatched parameters. Figure 7.15a shows WPS of the MS and the SS via delayed impulsive control. For Case 2, $\Omega_1 + \frac{\ln \breve{\mu}}{T} + \ell\breve{\mu}^{N_0}\Omega_2 = -22.2542 < 0$ and the convergence rate is 8.4959. Figure 7.14c and d show the MS and the SS have mismatched parameters. Figure 7.15b shows WPS of the MS and the SS via delayed impulsive control. Comparing Figs. 7.15a with 7.13a, it can be seen that due to the influence of time delay, the larger fluctuation in Fig. 7.15a lags behind that in Fig. 7.13. For example, the first big fluctuation is about at 0.9s in Fig. 7.13, while about at 1s in Fig. 7.15. Therefore, it can be proved that the controller is time-delay efficient.

Remark 7.19 In the simulation, the feasible solution and related parameters of the system can be obtained by MATLAB. The system and control conditions are constrained according to the parameters. In addition, the validity of the theorem can be verified by simulation in Simulink. Therefore, the computation is feasible and the conditions we obtained provide ideas for the selection of practical application parameters.

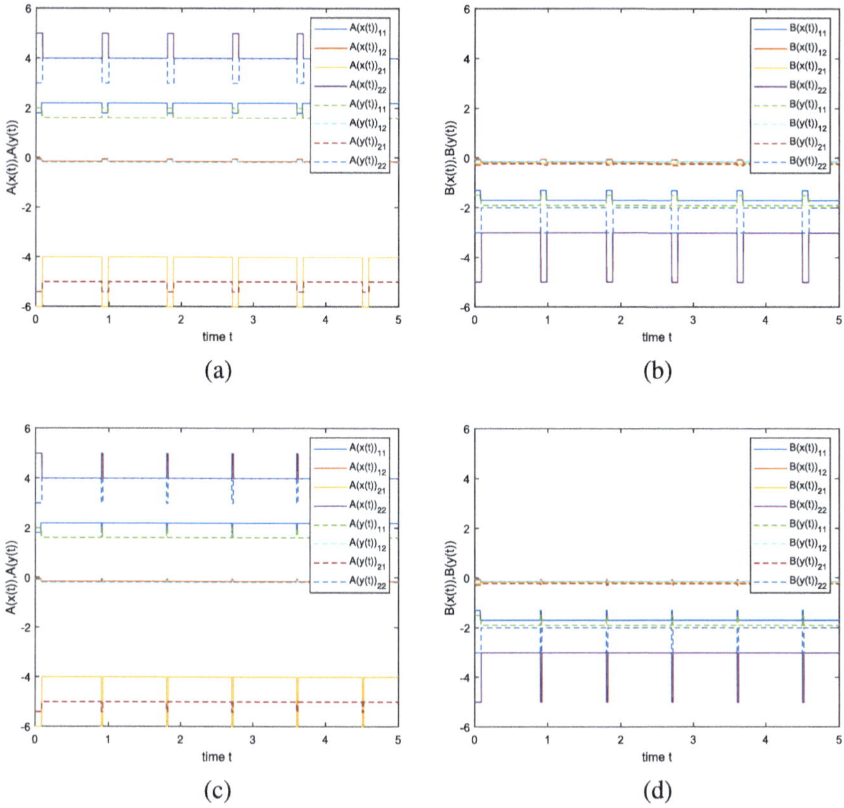

Fig. 7.12 Mismatched parameters. **a** $A(x(t))$ and $A(y(t))$ are mismatched in Theorem 1 Case 1. **b** $B(x(t))$ and $B(y(t))$ are mismatched in Theorem 1 Case 1. **c** $A(x(t))$ and $A(y(t))$ are mismatched in Theorem 1 Case 2. **d** $B(x(t))$ and $B(y(t))$ are mismatched in Theorem 1 Case 2

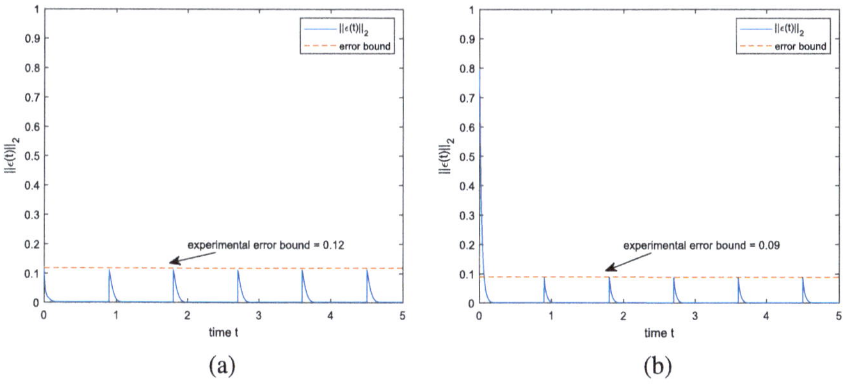

Fig. 7.13 Weak projective synchronization error. **a** WPS error of in Theorem 1 Case 1. **b** WPS error of in Theorem 1 Case 2

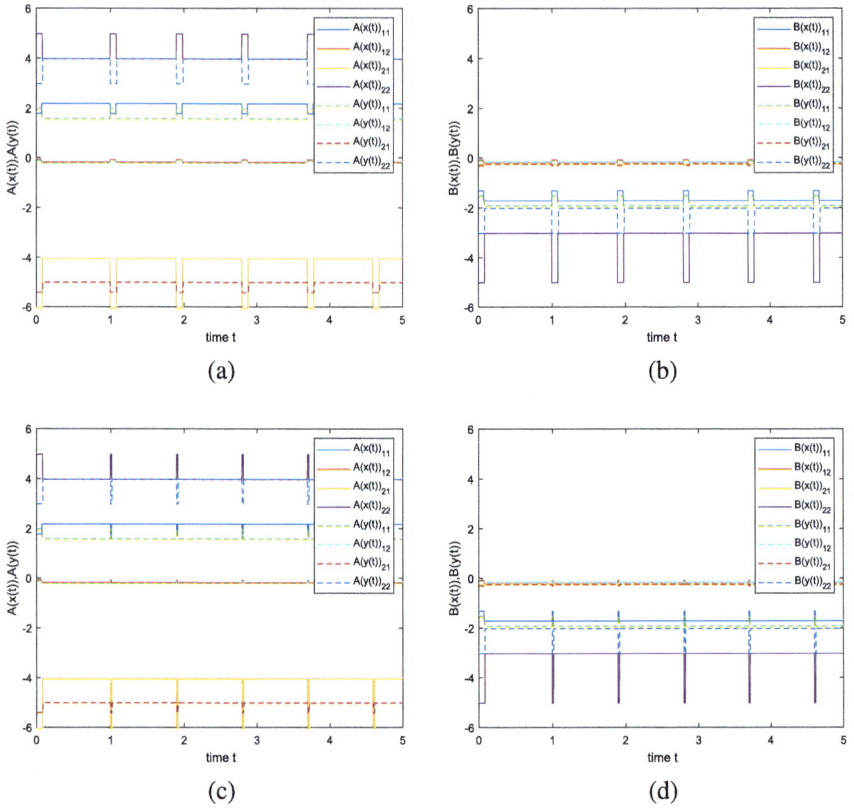

Fig. 7.14 Mismatched parameters. **a** $A(x(t))$ and $A(y(t))$ are mismatched in Theorem 2 Case 1. **b** $B(x(t))$ and $B(y(t))$ are mismatched in Theorem 2 Case 1. **c** $A(x(t))$ and $A(y(t))$ are mismatched in Theorem 2 Case 2. **d** $B(x(t))$ and $B(y(t))$ are mismatched in Theorem 2 Case 2

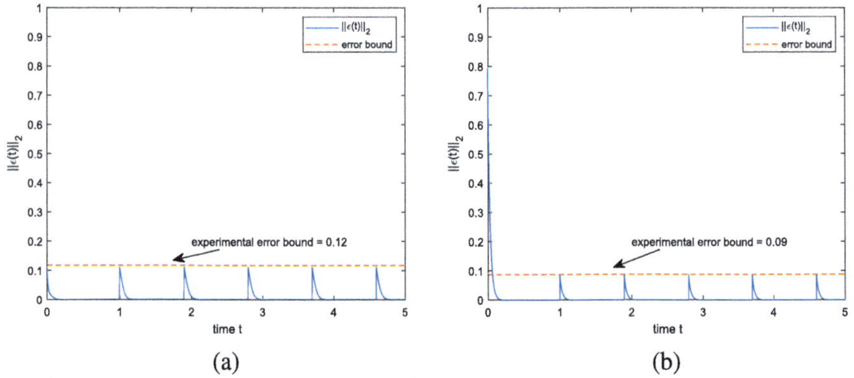

Fig. 7.15 Weak projective synchronization error. **a** WPS error of in Theorem 2 Case 1. **b** WPS error of in Theorem 2 Case 2

7.2.7 Application Example

In this section, the anti-synchronization of SMNNs, a special case of WPS, is applied to realize secure communication. In order to avoid repetition, all system parameters are the same as the parameter of Theorem 1 Case 1 in Sects. 7.2.5–7.2.6 expect $I = J = [0; 0]$, $\alpha = -1$. In this example, a image of 'S' will be communicated from the transmitter to the receiver. The procedure, which can be shown directly in Fig. 7.16, is described as follows.

First, the picture information should be expressed by binary numbers. Since the image resolution is 10×11, it can be converted into a 2640 bits binary sequence denoted as $T_s(\varsigma)$, where ς is the discrete sampling time.

Second, original signals will be encoded by states $x_i(\varsigma)$ of MS, and we choose $x_1(\varsigma)$ in this example. The arithmetic addition operation is employed for encoding. The states of $x_1(\varsigma)$ without additional signals are shown in Fig. 7.18b, and the 2640 bits binary signals are shown in Fig. 7.18a. Hence, the encoded signals $E_s(\varsigma)$ are expressed as:

$$E_s(\varsigma) = T_s(\varsigma) + (-1) \times x_1(\varsigma). \tag{7.137}$$

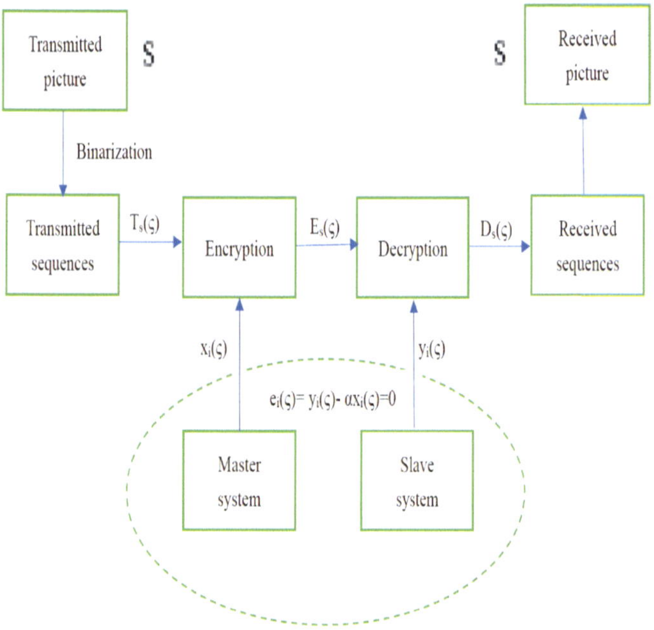

(a)

Fig. 7.16 System diagram of secure communication based on SMNNs

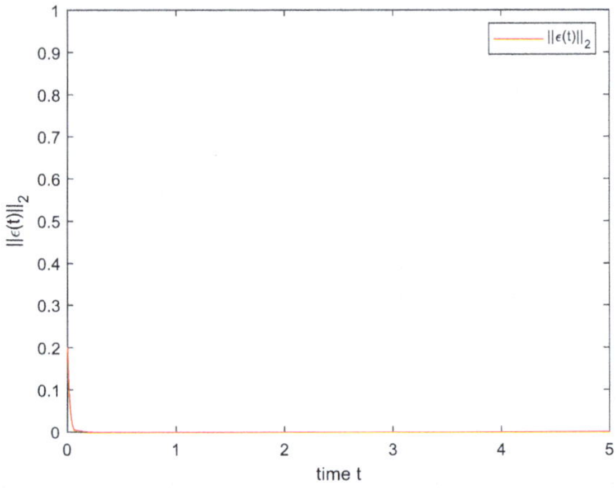

(a)

Fig. 7.17 Anti-synchronization of MS and SS

It is observed from Fig. 7.17 that the MS and SS achieve anti-synchronization after $t = 0.3$s. Then, the communication data $E_s(\varsigma)$ can be transmitted reliably after anti-synchronization. Here, we randomly select the start time of data transmission as the first second. From Fig. 7.18c it can be seen that the binary sequence is encoded and the $E_s(\varsigma)$ are transmitted in the time interval [1.0000, 1.0264], where the sampling time is 0.0001s.

Third, the received sequences $D_s(\varsigma)$ will be decoded from $E_s(\varsigma)$. The inverse operation of addition is subtraction, which means that $E_s(\varsigma)$ can be expressed as $E_s(\varsigma) - y_1(\varsigma)$. Considering that $y_1(\varsigma) = -x_1(\varsigma)$, it is derived from (7.137) that

$$D_s(\varsigma) = E_s(\varsigma) - (-x_1(\varsigma)) = E_s(\varsigma) - y_1(\varsigma) = T_s(\varsigma). \tag{7.138}$$

From Fig. 7.18a and d, it can be seen that $D_s(\varsigma)$ totally equals to $T_s(\varsigma)$. In other words, the picture of 'S' is restored and the whole communication process has been completed.

7.2.8 *Conclusion*

The WPS problem for SMNNs with mismatched parameters and impulsive control has been discussed in this section. The augmented system was established to reduce the constraint conditions and effective criteria for WPS of the master and slave systems were derived by using two types of comparison principles. In addition, the convergence rates of the ES were discussed, and the synchronization error bounds under different impulsive ranges were derived. Finally, the validity of our results was verified by numerical simulation and application example.

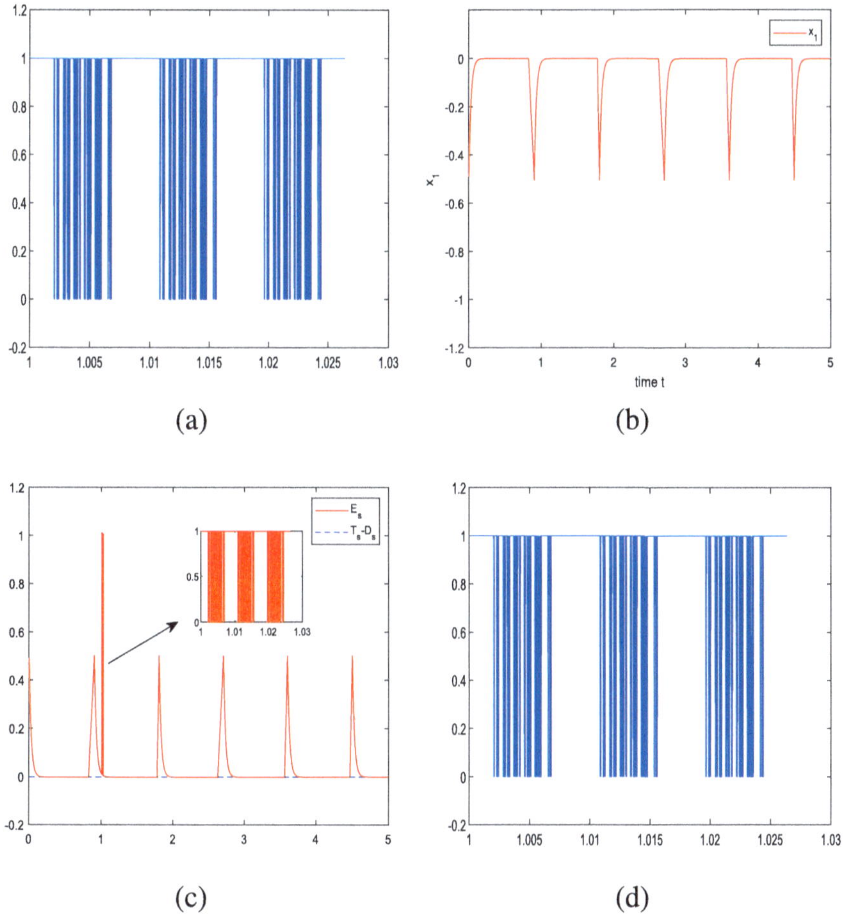

Fig. 7.18 Procedure of secure communication based on SMNNs. **a** State x_1. **b** T_s. **c** E_s and $T_s - D_s$. **d** D_s

For future works, some interesting topics are worth researching:

(1) In this section, the parameter mismatched SMNNs was discussed. However, there are several cases of dimension-mismatched MNNs. Thus, it is meaningful to study the dynamics of dimension-mismatched MNNs.

(2) In addition to the noise proposed in this section, there are many other types of external disturbances. In the future, other stochastic noises, such as multiplicative noise, can be researched.

(3) How to reduce system conservatism is an important issue. Furthermore, less conservatism methods could be utilized in the theoretical proof.

7.3 Mismatched Synchronization for Coupled Memristive Neural Networks on Time Scales

7.3.1 Introduction

In the implementation of CMNNs, the networks' performance will be reduced due to the limited switching speed of the amplifiers and the limited communication bandwidth of the devices, which means that the time delays, particularly time-varying delays, prevail throughout the signal transmission among neurons. In the vast majority of cases, time-varying delays will negatively affect the overall performance of CMNNs, such as divergence, oscillatory behavior [14, 22, 52, 109]. Therefore, time-varying delays were widely considered in the research of NNs to effectively avoid the unstable behavior of NNs caused by time-varying delays. The dynamic analysis of delayed MNNs has attracted increasing attention [110–112]. However, the existing works on the synchronization problems center on MNNs with matched delays, few concentrate on mismatched delays. Hence, the project still stays at the periphery of research, which motivates our research interest.

The CMNNs can be described in terms of a state-dependent differential equation, whose connection weights depend on the CMNNs' states. The external or internal interference can cause the system's connection weights and time-varying delays to be different at some moments. Hence, the coupled system and the isolated system may have different connection weights and time-varying delays. Under the circumstances, if only the matched weights are used to derive the criteria, the results may be somewhat inaccurate. Hence, it is necessary to investigate the synchronization between CMNNs and isolated MNNs with mismatched connection weights [113–115]. Furthermore, CMNNs own tremendously complex structures with numerous nodes, making it challenging to obtain the entire networks's information. Sometimes, an isolated system different from the existing system needs to be used to estimate the whole coupled system, that is, the coupled system and the isolated system may have mismatched dimensions and parameters. Although there were enough studies on synchronization for NNs [93, 111–120], the existing results are obtained based on whole-matched information or part-matched parameters (including matched delays, connection weights, and dimensions). Compared with existing studies, this chapter studies the synchronization problem of CMNNs with mismatched dimensions, connection weights, and time-varying delays, and most of the works published on CMNNs and MNNs can be regarded as individual cases of this chapter.

The information exchange between nodes can happen at some continuous-time intervals accompanying some discrete moments. Therefore, it is necessary and meaningful to consider both continuous-time and discrete-time in a fixed system. As a particular case of measurement chain, the framework of time scales can effectively unify the analysis of differential equations and difference equations [121–123]. Thus, the NNs based on time scales have been widely concerned by scholars [53, 124–129]. To help NNs achieve synchronization and stabilization, many control mechanisms have been reported, such as pinning control mechanism [130], sliding mode control mechanism [131], and adaptive control mechanism [132]. Compared with continuous control methods, the impulsive control mechanism can control the controlled object in some discrete moments and has the advantages of high reliability, low maintenance cost, low communication cost, and so on. Therefore, many researchers in NNs fields [50, 133, 134] have paid more attention to the impulse control mechanism. It is worth noting that few studies discussed three types of mismatched parameters among the research results on impulsive control and the time scales mentioned. Moreover, the CMNNs are often affected by instantaneous disturbances and experience abrupt changes when the networks have device failure, which may lead to the node connection failure of the neural network, and then lead to the change of the topology of the CMNNs. To ensure that the CMNNs can quickly achieve synchronization when the topology structure of CMNNs changes, the novel TPIM is proposed in this chapter. The work on the model studied with three types of mismatched parameters on time scales is challenging and significant, and the novel control mechanism presented here extends the existing control methods.

In this section, $\mathbb{R}$, $\mathbb{Z}$ and $\mathbb{N}^+$ stand for the set of real numbers, the set of integers and the set of nonnegative integers, respectively. $\mathbb{R}^n$ and $\mathbb{R}^{n\times n}$ denote the n-dimensional euclidean space and the set of $n \times n$ real matrix. $\|\cdot\|$ is the Euclidean norm in $\mathbb{R}^n$. $\mathbf{0}_{n,N}^{i,j}$ and $\mathbf{I}_{n,N}^{i,i}$ represent the $(iNn \times jNn)$-dimensional zero matrix and $(iNn \times iNn)$-dimensional identity matrix, respectively. The mathematical expectation of stochastic variable x is $\mathbb{E}\{x\}$. $\mathbb{P}\{\cdot\}$ stands for the probability of an event. $\log(\cdot)$ is the natural logarithm function. $\exp(\cdot)$ represents the exponential function. $\mathrm{diag}(()\cdots)$ represents a block-diagonal matrix. The minimum eigenvalue and maximum eigenvalue are defined as $\lambda_{\min}(\cdot)$ and $\lambda_{\max}(\cdot)$, respectively. The symmetric term in a symmetric matrix is denoted by $*$, and $\mathrm{Sy}(H) = H + H^T$. $\otimes$ denotes the Kronecker product of two matrices. $\emptyset$ is the empty set. $\mathfrak{o}_i \in \mathbb{R}^n$ stands for the column vector with the ith element being 1 and others being 0.

7.3.2 Some Preliminaries on Time Scales

To make it easy for readers to scan and read the section, the basic concepts of time scales will be presented in this subsection [121, 122].

A time scale $\mathbb{T}$ has the topology that it inherits from the real numbers with the standard topology, and it is an arbitrary nonempty closed subset of the real number set $\mathbb{R}$. For $t \in \mathbb{T}$, the forward jump operator $\sigma : \mathbb{T} \to \mathbb{T}$ and the backward jump operator $\rho : \mathbb{T} \to \mathbb{T}$ are defined as $\sigma(t) := \inf\{s \in \mathbb{T} : s > t\}$ and $\rho(t) := \sup\{s \in \mathbb{T} : s < t\}$, respectively. If $\mathbb{T}$ has a maximum $\mathscr{T}_{\max}$, then $\sigma(\mathscr{T}_{\max}) = \mathscr{T}_{\max}$ (i.e., $\inf \emptyset = \sup \mathbb{T}$); If $\mathbb{T}$ has a minimum $\mathscr{T}_{\min}$, then $\rho(\mathscr{T}_{\min}) = \mathscr{T}_{\min}$ (i.e., $\sup \emptyset = \inf \mathbb{T}$). The graininess function $\mu : \mathbb{T} \to [0, +\infty)$ is defined by $\mu(t) := \sigma(t) - t$.

If $\mathbb{T}$ has a left-scattered maximum $\mathscr{T}_{\mathrm{ls,max}}$, then $\mathbb{T}^k = \mathbb{T} - \{\mathscr{T}_{\mathrm{ls,max}}\}$, otherwise $\mathbb{T}^k = \mathbb{T}$. For the function $f^\sigma : \mathbb{T} \to \mathbb{R}$, one has $f^\sigma(t) = f(\sigma(t))$. If the function $f : \mathbb{T} \to \mathbb{R}$ is continuous at right-dense points and its left-sided limits exist at left-dense points, then the function f is called rd-continuous. The set of rd-continuous functions f will be defined by $C_{rd} = C_{rd}(\mathbb{T}) = C_{rd}(\mathbb{T}, \mathbb{R})$. $\varsigma : \mathbb{T}^k \to \mathbb{R}$ is said to be regressive, if $1 + \mu(t)\varsigma(t) \neq 0$ for any $t \in \mathbb{T}^k$. The set of all regressive and rd-continuous functions is defined as $\mathbb{R}_r$. The set $\mathbb{R}_r^+$ of all positively regressive elements of $\mathbb{R}_r$ is defined by $\mathbb{R}_r^+ = \mathbb{R}_r^+(\mathbb{T}, \mathbb{R}) = \{\varsigma \in \mathbb{R}_r : 1 + \mu(t)\varsigma(t) > 0 \text{ for all } t \in \mathbb{T}\}$.

Definition 7.5 ([121]) The function $f : \mathbb{T} \to \mathbb{R}$ is said to be $\Delta-$differentiable (when it exists and is written as f^{Δ}) at the point $t \in \mathbb{T}^k$, if for any $\epsilon > 0$, there exists a neighborhood $\mathfrak{U}$ such that the following inequality

$$|f(\sigma(t)) - f(s) - f^{\Delta}(t)(\sigma(t) - s)| \leq \varepsilon|\sigma(t) - s|, \forall s \in \mathfrak{U}$$

holds. $f : \mathbb{T} \to \mathbb{R}$ is called $\Delta-$differentiable, if it is $\Delta-$differentiable at any $t \in \mathbb{T}^k$.

Definition 7.6 ([121]) If $\varsigma \in \mathbb{R}_r$, then the exponential function on time scale $\mathbb{T}$ can be defined as follows:

$$e_{\varsigma}(t, s) = \exp(\int_s^t \pounds_{\mu(\tau)}(\varsigma(\tau))\Delta\tau), \quad s, t \in \mathbb{T}$$

where $\pounds_{\mu(\tau)}(z)$ represents the cylinder transformation and is described as: $\pounds_{\mu(\tau)}(z) = \begin{cases} \frac{\log(1+hz)}{h}, & h \neq 0 \\ z, & h = 0. \end{cases}$

Lemma 7.6 ([121]) *Assume that* $f : \mathbb{T} \to \mathbb{R}$ *and* $t \in \mathbb{T}^k$, *one has*

(1) If f *is differentiable at* t, *then* $f^{\sigma}(t) = f(t) + \mu(t)f^{\Delta}(t)$.

(2) If f_1 *and* f_2 *are both* $\Delta-$*differentiable at* t, *then* $(f_1(t)f_2(t))^{\Delta} = f_1^{\Delta}(t)f_2(t) + f_1^{\sigma}(t)f_2^{\Delta}(t)$.

Lemma 7.7 ([121]) *If* $f \in C_{rd}(\mathbb{T}, \mathbb{R})$ *and* $p \in \mathbb{R}_r^+$. *The inequality* $x^{\Delta}(t) \leq p(t)x(t)$ *implies that* $x(t) \leq x(t_0)e_p(t, t_0)$ *for all* $t \in \mathbb{T}$.

Lemma 7.8 ([135]) *For any* $\varsigma \in \mathbb{R}_r^+$ *and all* $s, t \in \mathbb{T}$, *one can obtain* $e_{\varsigma}(t, s) \leqslant e^{\varsigma(t-s)}$

Lemma 7.9 ([121]) *For all* $r, s, t \in \mathbb{T}$, *any* $\varsigma_1 \in \mathbb{R}_r$, *one has*

$$(1)\ e_{\varsigma_1}(\sigma(t), s) = (1 + \mu(t)\varsigma_1(t))e_{\varsigma_1}(t, s)$$
$$(2)\ e_{\varsigma_1}(t, t) = 1, e_0(t, s) = 1, e_{\varsigma_1}(t, r)e_{\varsigma_1}(r, s) = e_{\varsigma_1}(t, s)$$
$$(3)\ e_{\varsigma_1}(t, s) = \frac{1}{e_{\varsigma_1}(s, t)} = e_{\ominus\varsigma_1}(s, t)$$

Remark 7.20 In a continuous-time system ($\mathbb{T} = \mathbb{R}$), one can derive $\sigma(t) = t$, $\mu(t) = 0$, $f^{\Delta}(t) = \dot{f}(t)$. In a discrete-time system ($\mathbb{T} = \mathbb{Z}$), we have $\sigma(t) = t + 1$, $\mu(t) = 1$, $f^{\Delta}(t) = \Delta f(t)$. For any constant ς and all $t_0, t \in \mathbb{T}$, if $\mathbb{T} = \mathbb{R}$, then $e_{\varsigma}(t, t_0) = e^{\varsigma(t-t_0)}$; If $\mathbb{T} = \mathbb{Z}$, then $e_{\varsigma}(t, t_0) = (1 + \varsigma)^{\varsigma(t-t_0)}$. Moreover, if $\varsigma \geq 0$, then $e_{\varsigma}(t, s) \geq 1$ for all $t \geq s$ with $s, t \in \mathbb{T}$.

Remark 7.21 For any constants $a, b > 0$ and the time scale $\mathbb{T}_{a,b} = \bigcup_{k=0}^{+\infty}[k(a+b), k(a+b)+a]$, the forward jump operator can be written as:

$$\sigma(t) = \begin{cases} t, & \text{if } t \in \bigcup_{k=0}^{+\infty}[k(a+b), k(a+b)+a) \\ t+b, & \text{if } t \in \bigcup_{k=0}^{+\infty} k(a+b)+a, \end{cases}$$

and the graininess function can be derived as:

$$\mu(t) = \begin{cases} 0, & \text{if } t \in \bigcup_{k=0}^{+\infty}[k(a+b), k(a+b)+a) \\ b, & \text{if } t \in \bigcup_{k=0}^{+\infty} k(a+b)+a. \end{cases}$$

Remark 7.22 Among existing investigations, the dynamic problems of the continuous-time [46, 112, 136] and discrete-time systems [137, 138] are separated and investigated separately. However, in many systems, the information exchange between nodes can happen at some continuous-time intervals accompanying some discrete moments, for example, the rate of change of current in a circuit can be described by dynamic equations on a time scale. This is a valuable research topic to unify differential and difference equations by means of time scales [124–126]. Also, since the time variables of the dynamic equations on the measurement chain include not only the set of real numbers and the set of integers but also some other sets such as $\mathbb{T} = \{2^{n^2} : n \in \mathbb{N}^+\}$, $\mathbb{T} = \{\frac{1}{n} : n \in \mathbb{N}^+\}$, therefore, the relevant conditions obtained have more substantial theoretical significance and broader application.

7.3.3 *The Design of Novel Memristive Neural Networks*

The coupled variable-resistor model for a memristor [139] is shown in Fig. 7.19. Based on the model, one can have

$$v(t) = \left(\mathfrak{R}_{\text{on}}\frac{\omega(t)}{\mathfrak{D}} + \mathfrak{R}_{\text{off}}\left(1 - \frac{\omega(t)}{\mathfrak{D}}\right)\right) i(t) \tag{7.139}$$

$$\omega^{\Delta}(t) = \mu_v \frac{\mathfrak{R}_{\text{on}}}{\mathfrak{D}} i(t) \tag{7.140}$$

Integrating both sides of the (7.140) from t_0 to t, one has

$$\omega(t) = \mu_v \frac{\mathfrak{R}_{\text{on}}}{\mathfrak{D}} \int_{t_0}^{t} i(s)\Delta s + \omega_0 \tag{7.141}$$

where ω_0 is the initial length of doped region.

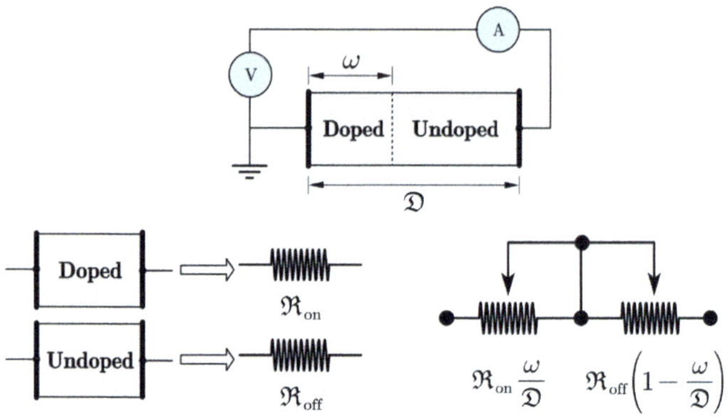

Fig. 7.19 The coupled variable-resistor model for a memristor. The TiO_{2-x} region has a lower resistance $\mathfrak{R}_{\text{on}}$, the TiO_2 region has a high resistance $\mathfrak{R}_{\text{off}}$. The full length of this memristive device is $\mathfrak{D}$, $i(t)$ and $v(t)$ represent the current and the voltage, respectively. The equal average ion mobility of ions is denoted as μ_v. The value of state $\omega(t) \in [0, \mathfrak{D}]$ depends on the physical dimensions and structure of the memristive device

From (7.139) and (7.141), one can derive

$$v(t) = (\mu_v \frac{\mathfrak{R}_{\text{on}}^2 - \mathfrak{R}_{\text{on}}\mathfrak{R}_{\text{off}}}{\mathfrak{D}^2} \int_{t_0}^{t} i(s)\Delta s + \mathfrak{R}^*) i(t) \tag{7.142}$$

where $\mathfrak{R}^* = \frac{\mathfrak{D}\mathfrak{R}_{\text{off}} + \mathfrak{R}_{\text{on}}\omega_0 - \mathfrak{R}_{\text{off}}\omega_0}{\mathfrak{D}}$.

For a current-controlled memristor whose state variable is the current flowing through the memristive device [139], one has $v(t) = \mathfrak{W}(i(t))i(t)$ and $\mathfrak{M}(i(t)) = \frac{1}{\mathfrak{W}(i(t))}$, in which $\mathfrak{W}(i(t))$ and $\mathfrak{M}(i(t))$ represent the memristor and memductance, respectively. Thus, from (7.142), one can obtain

$$\begin{cases} \mathfrak{W}^{\Delta}(i(t)) = \mu_v \dfrac{\mathfrak{R}_{\text{on}}^2 - \mathfrak{R}_{\text{on}}\mathfrak{R}_{\text{off}}}{\mathfrak{D}^2} \dfrac{v(t)}{\mathfrak{W}(i(t))} \\ \mathfrak{M}^{\Delta}(i(t)) = \mu_v \dfrac{\mathfrak{R}_{\text{on}}\mathfrak{R}_{\text{off}} - \mathfrak{R}_{\text{on}}^2}{\mathfrak{D}^2} \mathfrak{M}^3(i(t))v(t) \end{cases} \tag{7.143}$$

The above memristor model is adopted to design the isolated MNN, which the circuit diagram of MNN has been constructed in [140]. By using Kiirchhoff's current law, the isolated system can be designed as follows:

$$\begin{aligned} s_{\hat{i}}^{\Delta}(t) = & -a_{\hat{i}}^{s}(s_{\hat{i}}(t))s_{\hat{i}}(t) + \sum_{\hat{j}=1}^{m} \text{sign}_{\hat{i}\hat{j}} b_{\hat{i}\hat{j}}^{s}(s_{\hat{i}}(t)) g_{\hat{j}}(s_{\hat{j}}(t)) \\ & + \sum_{\hat{j}=1}^{m} \text{sign}_{\hat{i}\hat{j}} c_{\hat{i}\hat{j}}^{s}(s_{\hat{i}}(t)) g_{\hat{j}}(s_{\hat{j}}(t - \tau_1(t))) + \mathcal{I}_{\hat{i}}(t) \end{aligned} \tag{7.144}$$

and the $\Delta-$derivative of memristive connection weights with respect to time $t \in \mathbb{T}$ is expressed as follows:

$$\begin{aligned}(a_{\hat{i}}^s(s_{\hat{i}}(t)))^{\Delta} =& \sum_{\hat{j}=1}^{n}((b_{\hat{i}\hat{j}}^s(s_{\hat{i}}(t)))^{\Delta} + (c_{\hat{i}\hat{j}}^s(s_{\hat{i}}(t)))^{\Delta}) \\ (b_{\hat{i}\hat{j}}^s(s_{\hat{i}}(t)))^{\Delta} =& -\mu_v \mathfrak{C}_{\hat{i}}^2 \mathfrak{R}_{\hat{i}\hat{j}}(\text{sign}_{\hat{i}\hat{j}} g_{\hat{j}}(s_{\hat{j}}(t)) \\ & - s_{\hat{i}}(t))(b_{\hat{i}\hat{j}}^s(s_{\hat{i}}(t)))^3 \\ (c_{\hat{i}\hat{j}}^s(s_{\hat{i}}(t)))^{\Delta} =& -\mu_v \mathfrak{C}_{\hat{i}}^2 \mathfrak{R}_{\hat{i}\hat{j}}(\text{sign}_{\hat{i}\hat{j}} g_{\hat{j}}(s_{\hat{j}}(t-\tau_1(t))) \\ & - s_{\hat{i}}(t))(c_{\hat{i}\hat{j}}^s(s_{\hat{i}}(t)))^3\end{aligned} \tag{7.145}$$

where $\hat{i}, \hat{j} = 1, \ldots, m$, $s_{\hat{i}}(t) \in \mathbb{R}$ is the state variable of isolated system. $\tau_1(t)$ is the time-varying delay of isolated system. $g_{\hat{j}}(s_{\hat{j}}(t)) \in \mathbb{R}$ and $g_{\hat{j}}(s_{\hat{j}}(t-\tau_1(t))) \in \mathbb{R}$ stand for the activation functions with and without time-varying delays. $I_{\hat{i}}(t)$ represents the external input. $\mathfrak{R}_{\hat{i}\hat{j}} = \frac{\mathfrak{R}^2_{\text{on},\hat{i}\hat{j}} - \mathfrak{R}_{\text{on},\hat{i}\hat{j}}\mathfrak{R}_{\text{off},\hat{i}\hat{j}}}{\mathfrak{D}^2_{\hat{i}\hat{j}}}$, the memristive connection weights $a_{\hat{i}}^s(s_{\hat{i}}(t)) = \sum_{\hat{j}=1}^{n}(b_{\hat{i}\hat{j}}^s(s_{\hat{i}}(t)) + c_{\hat{i}\hat{j}}^s(s_{\hat{i}}(t))) + \frac{1}{\mathfrak{C}_{\hat{i}}\mathfrak{R}_{\hat{i}}}$, $b_{\hat{i}\hat{j}}^s(s_{\hat{i}}(t)) = \frac{\mathfrak{M}_f^{\hat{i}\hat{j}}(s_{\hat{i}}(t))}{\mathfrak{C}_{\hat{i}}}$, $c_{\hat{i}\hat{j}}^s(s_{\hat{i}}(t)) = \frac{\mathfrak{M}_{f_\tau}^{\hat{i}\hat{j}}(s_{\hat{i}}(t))}{\mathfrak{C}_{\hat{i}}}$, in which $\mathfrak{R}_{\hat{i}}$ and $\mathfrak{C}_{\hat{i}}$ are the resistor and capacitor, $\mathfrak{W}_f^{\hat{i}\hat{j}}$ and $\mathfrak{W}_{f_\tau}^{\hat{i}\hat{j}}$ represent the memristors. The function sign_{ij} satisfies $\text{sign}_{\hat{i}\hat{j}} = \begin{cases} -1, & \hat{j} = \hat{i} \\ 1, & \hat{j} \neq \hat{i} \end{cases}$.

The equivalent vector form of isolated system (7.144) can be written as follows:

$$s^{\Delta}(t) = -A^s(s(t))s(t) + B^s(s(t))g(s(t)) + C^s(s(t))g(s(t-\tau_1(t))) \tag{7.146}$$

where

$s(t) = [s_1(t), \ldots, s_m(t)]^T \in \mathbb{R}^m$,
$g(s(\cdot)) = [g_1(s(\cdot)), \ldots, g_m(s(\cdot))]^T \in \mathbb{R}^m$,
$A^s(s(t)) = \text{diag}(a_1^s(s_1(t)), \ldots, a_m^s(s_m(t))) \in \mathbb{R}^{m\times m}$,
$B^s(s(t)) = [\text{sign}_{\hat{i}\hat{j}} b_{\hat{i}\hat{j}}^s(s_{\hat{i}}(t))]_{m\times m} \in \mathbb{R}^{m\times m}$,
$C^s(s(t)) = [\text{sign}_{\hat{i}\hat{j}} c_{\hat{i}\hat{j}}^s(s_{\hat{i}}(t))]_{m\times m} \in \mathbb{R}^{m\times m}$.

Remark 7.23 In existing studies involving MNNs, there were mainly three methods to deal with the memristive connection weights, namely: state monotone models [141], state boundary models [25, 46], activation function models [111]. In fact, these mathematical assumptions have been unconsciously reached consensus for a long time. The authors generally believed that the device based on these mathematical formulas is a kind of memristor device. Recently, the reliability of these assumptions has been debated and questioned. Pershin et al. reclassified the models based on these assumptions into two types, namely "Type 1" and "Type 2", and considered that the results obtained by using these models were not relevant to the actual field of MNNs

[142]. In order to solve this controversial problem, Xiao et al. adopted the quadratic memductance function to describe the memristive characteristics [143], and Li et al. adopted the connection weight differentiable model to design MNNs [144]. Different from existing MNNs models [25, 46, 111, 141], in this novel model, the derivative of the memristive weight concerning time is a continuous function of the memristor's current and physical parameters, which implies that the memristor devices can remember past dynamics. This new idea of memristive connection weight construction has been proved to retain the memory characteristic of MNNs [144]. Therefore, the model proposed in this section can simulate the memristive characteristics well.

7.3.4 CMNNs with Multiple Mismatched Parameters and TPIM

A series of non-ideal factors (including network failures, node connection interruptions, cyber-attacks, etc.) will destroy the original structure of the CMNNs and affect the system's stability, which means that the topological structures of CMNNs will inevitably change. Therefore, CMNNs generally have some unreliable topologies under the influence of non-ideal factors. To improve the system's reliability, accuracy, and generalization, the following unreliable topological structures will be considered to design the novel CMNNs.

Defining a switching signal $\tilde{\alpha}(t) : [0, +\infty)_{\mathbb{T}} \rightarrow \{0, \ldots, N_s\}$. The unreliable network structure graph $\mathcal{G}_{\tilde{\alpha}(t)} = (\mathcal{V}, \mathcal{E}_{\tilde{\alpha}(t)})$ consists of the node set $\mathcal{V} = \{v_1, v_2, \ldots, v_N\}$ and the edge set $\mathcal{E}_{\tilde{\alpha}(t)} \subseteq \mathcal{V} \times \mathcal{V}$. For the matrix $L_{\tilde{\alpha}(t)} = [l_{\tilde{\alpha}(t),r\tilde{r}}]_{N\times N}$ in $\mathcal{G}_{\tilde{\alpha}(t)}$, if $(v_r, v_{\tilde{r}}) \in \mathcal{E}_{\tilde{\alpha}(t)}$, then $l_{\tilde{\alpha}(t),r\tilde{r}} = l_{\tilde{\alpha}(t),\tilde{r}r} > 0 (r \neq \tilde{r})$; otherwise $l_{\tilde{\alpha}(t),r\tilde{r}} = 0$. And the matrix $L_{\tilde{\alpha}(t)} = [l_{\tilde{\alpha}(t),r\tilde{r}}]_{N\times N}$ satisfies $l_{\tilde{\alpha}(t),rr} = -\sum_{\tilde{r}=1,\tilde{r}\neq r}^{N} l_{\tilde{\alpha}(t),r\tilde{r}}$. Obviously, $-L_{\tilde{\alpha}(t)}$ is the Laplacian matrix of graph $\mathcal{G}_{\tilde{\alpha}(t)}$.

Assumption 7.6 $\mathcal{G}_0$ represents the original network structure graph without the interference of non-ideal factors, $\mathcal{G}_1, \ldots, \mathcal{G}_m$ represent a series of unreliable network topological structure graphs under the influence of various non-ideal factors.

Assumption 7.7 The CMNNs model is recoverable. The changed topologies can be restored to the original topological structure if these effects of non-ideal factors disappear.

The switching signal $\alpha(t)$ is defined as $\alpha(t) : [0, +\infty)_{\mathbb{T}} \rightarrow \{1, \ldots, N_s\}$, which implies that the set $\{\tilde{\alpha}(t)\} = \{\alpha(t)\} \bigcup \{0\}$. Based on the unreliable topological structures, for any $t \in \mathbb{T}$, rth subsystem of controlled CMNNs with N identical nodes and probability switching topologies is constructed as follows:

$$
\begin{aligned}
x_r^{\Delta}(t) = & -A^x(x_r(t))x_r(t) + B^x(x_r(t))g(x_r(t)) \\
& + C^x(x_r(t))g(x_r(t-\tau_2(t))) + \beta(t)d_0\sum_{\tilde{r}=1}^{N} l_{0,r\tilde{r}}F_0^x x_{\tilde{r}}(t) \\
& + (1-\beta(t))d_{\alpha(t)}\sum_{\tilde{r}=1}^{N} l_{\alpha(t),r\tilde{r}}F_{\alpha(t)}^x x_{\tilde{r}}(t) + u_r^1(t) + u_r^2(t) + u_r^3(t)
\end{aligned}
\tag{7.147}
$$

where $r, \tilde{r} = 1, \ldots, N, x_r(t) = [x_{r,1}(t), \ldots, x_{r,n}(t)]^T \in \mathbb{R}^n$ represents the state vector of rth node at time $t \in \mathbb{T}$, $\tau_2(t)$ corresponds to the time-varying delay of system (7.147), $F_0^x = \text{diag}(f_{0,1}^x, \ldots, f_{0,n}^x)$ is the inner coupling matrix of the system under the original topology, $F_{\alpha(t)}^x = \text{diag}(f_{\alpha(t),1}^x, \ldots, f_{\alpha(t),n}^x)$ represents the inner coupling matrix of the system that has suffered from various non-ideal factors. The memristive connection weights $A^x(x_r(t)) = \text{diag}(a_1^x(x_{r,1}(t)), \ldots, a_n^x(x_{r,n}(t))) \in \mathbb{R}^{n\times n}$, $B^x(x_r(t)) = [\text{sign}_{\tilde{i}\tilde{j}} b_{\tilde{i}\tilde{j}}^x(x_{r,\tilde{i}}(t))]_{n\times n} \in \mathbb{R}^{n\times n}$, $C^x(x_r(t)) = [\text{sign}_{\tilde{i}\tilde{j}} c_{\tilde{i}\tilde{j}}^x(x_{r,\tilde{i}}(t))]_{n\times n} \in \mathbb{R}^{n\times n}$, and satisfy the form (7.145). $u_r^1(t) = [u_{r,1}^1(t), \ldots, u_{r,n}^1(t)]^T \in \mathbb{R}^n$ and $u_r^2(t) = [u_{r,1}^2(t), \ldots, u_{r,n}^2(t)]^T \in \mathbb{R}^n$ are the topology-based impulse controllers, $u_r^3(t) = [u_{r,1}^3(t), \ldots, u_{r,n}^3(t)]^T \in \mathbb{R}^n$ is the state feedback controller. $\beta(t)$ is a class of characteristic function that describes whether the CMNNs suffer from various non-ideal factors and satisfies the following form:

$$
\beta(t) = \begin{cases} 1, & \text{The topology of CMNNs is the original topology.} \\ 0, & \text{The topology of CMNNs has changed.} \end{cases}
$$

Assumption 7.8 For known constant $0 \le \beta_0 \le 1$, $\mathbb{P}\{\beta(t) = 1\} = \beta_0$, $\mathbb{P}\{\beta(t) = 0\} = 1 - \beta_0$.

Remark 7.24 $\beta(t)$ is a flexible binary stochastic variable. If CMNNs have suffered from various non-ideal factors, but the topology is not changed, that is, the topology of CMNNs is still original topology, then the value of function $\beta(t)$ is still taken as 0. If we set $\beta(t) = 0$, $u_r^1(t) = u_r^2(t) = 0$, the presented CMNNs can be simplified as existing models [145–147].

According to Assumption 7.8, it's easy to derive the following results $\mathbb{E}\{\beta(t)\} = \beta_0, \mathbb{E}\{1-\beta(t)\} = 1-\beta_0, \mathbb{E}\{\beta(t)^2\} = \beta_0, \mathbb{E}\{(1-\beta(t))^2\} = 1-\beta_0, \mathbb{E}\{\beta(t)(1-\beta(t))\} = 0$.

For the study of the synchronization problem, it is necessary to obtain the error information for state vectors $x_r(t) \in \mathbb{R}^n$ and $s(t) \in \mathbb{R}^m$ in (7.147) and (7.146). However, in this section, it is challenging to obtain relevant error information by traditional calculation methods, because there are three possible relationships between n and m, including $n = m$, $n > m$ and $n < m$. In order to solve the problem that the error information cannot be obtained due to the different dimensions of (7.147) and (7.146), a method of constructing auxiliary state variables is adopted here, such that the coupled system and the isolated system can achieve the same dimension. Next, we construct different auxiliary state variables based on the relationship between the value of m and n.

Case 1: If $n \geq m$, the auxiliary state variables of (7.146) are constructed as follows:

$$s_{m+1}(t) = \varphi_1(s_{i_1}(t)), \ldots, s_{m+\jmath}(t) = \varphi_\jmath(s_{i_\jmath}(t)), \ldots, s_n(t) = \varphi_{n-m}(s_{i_{n-m}}(t)), \tag{7.148}$$

where $\varphi_1(\cdot), \ldots, \varphi_{n-m}(\cdot)$ are known linear functions. $1 \leq \jmath \leq n-m$, $i_\jmath \in \{1, \ldots, m\}$, $s_{m+\jmath}(t) = \varphi_\jmath(s_{i_\jmath}(t))$ means that $(m+\jmath)$th auxiliary state depends on $i_\jmath$th state of the (7.146), for example, $s_{m+1}(t) = 2s_2(t)$, $s_{m+2}(t) = 4s_6(t)$.

Now, we define $\hat{s}(t) = [s_{m+1}(t), \ldots, s_n(t)]^T$, $g(\hat{s}(\cdot)) = [g_{m+1}(s_{m+1}(\cdot)), \ldots, g_n(s_n(\cdot))]^T$, the isolated system (7.146) with auxiliary state variables (7.148) can be constructed as follows:

$$\bar{s}^\Delta(t) = -A^{\bar{s}}(\bar{s}(t))\bar{s}(t) + B^{\bar{s}}(\bar{s}(t))g(\bar{s}(t)) + C^{\bar{s}}(\bar{s}(t))g(\bar{s}(t-\tau_1(t))) \tag{7.149}$$

in which $\bar{s}(t) = \left[s^T(t)\ \hat{s}^T(t)\right]^T \in \mathbb{R}^n$, $g(\bar{s}(\cdot)) = \left[g^T(s(\cdot))\ g^T(\hat{s}(\cdot))\right]^T \in \mathbb{R}^n$, $A^{\bar{s}}(\bar{s}(t)) = \mathrm{diag}(A^s(s(t)), A^1(\hat{s}(t))) \in \mathbb{R}^n$, $B^{\bar{s}}(\bar{s}(t)) = \begin{bmatrix} B^s(s(t))\ B^1(\hat{s}(t)) \\ B^2(\hat{s}(t))\ B^3(\hat{s}(t)) \end{bmatrix} \in \mathbb{R}^{n\times n}$, $C^{\bar{s}}(\bar{s}(t)) = \begin{bmatrix} C^s(s(t))\ C^1(\hat{s}(t)) \\ C^2(\hat{s}(t))\ C^3(\hat{s}(t)) \end{bmatrix} \in \mathbb{R}^{n\times n}$. The matrices $A^1(\hat{s}(t))$, $B^1(\hat{s}(t))$, $B^2(\hat{s}(t))$, $B^3(\hat{s}(t))$, $C^1(\hat{s}(t))$, $C^2(\hat{s}(t))$ and $C^3(\hat{s}(t))$ depend on auxiliary system (7.148).

In this case, the topology-based probability impulse controllers $u_r^1(t)$, $u_r^2(t)$ and feedback controller $u_r^3(t)$ are designed as follows:

$$\begin{cases} u_r^1(t) = (\beta(t)\ell_0 \sum\limits_{k=1}^{+\infty}\sum\limits_{\tilde{r}=1}^{N} l_{0,r\tilde{r}}(x_{\tilde{r}}(t) - \bar{s}(t)) \\ \qquad - (x_{\tilde{r}}(t) - \bar{s}(t)))\delta(t-t_k) \\ u_r^2(t) = ((1-\beta(t))\ell_{\alpha(t)} \sum\limits_{k=1}^{+\infty}\sum\limits_{\tilde{r}=1}^{N} l_{\alpha(t),r\tilde{r}}(x_{\tilde{r}}(t) - \bar{s}(t)) \\ \qquad - (x_{\tilde{r}}(t) - \bar{s}(t)))\delta(t-t_k) \\ u_r^3(t) = K_u(x_r(t) - \bar{s}(t)) \end{cases} \tag{7.150}$$

where ℓ_0, $\ell_{\alpha(t)} \in \mathbb{R}$ represent the strength of impulsive signal of rth node, $K_u \in \mathbb{R}^{n\times n}$ is the gain matrix. $\{t_k\}_{k=0}^{\infty} \subset \mathbb{T}$ is the impulsive sequence with $0 = t_0 < t_1 < \cdots < t_k < \cdots$, and $\lim\limits_{k\to\infty} = \infty$. $\delta(\cdot)$ represents the Dirac delta function.

Case 2: If $n < m$, the auxiliary state variables of coupled system (7.147) are constructed as follows:

$$\begin{aligned} &x_{r,n+1}(t)=\psi_1(x_{r,j_1}(t)),\ldots,x_{r,n+\iota}(t)=\psi_\iota(x_{r,j_\iota}(t)),\\ &\ldots,x_{r,m}(t)=\psi_{m-n}(x_{r,m-n}(t)), \end{aligned} \tag{7.151}$$

where $\psi_1(\cdot),\ldots,\psi_{m-n}(\cdot)$ are known linear functions. $1\le\iota\le m-n$, and for rth node, $j_\iota\in\{1,\ldots,n\}, x_{r,n+\iota}(t)=\psi_\iota(x_{r,j_\iota}(t))$ means that the $(n+\iota)$th auxiliary state depends on j_ιth state of the (7.147), for example, $x_{r,n+1}(t)=0.5x_{r,6}(t)$.

Now, let $\hat{x}_r(\cdot)=[x_{r,n+1}(\cdot),\ldots,x_{r,m}(\cdot)]^T$, $g(\hat{x}_r(\cdot))=[g_{n+1}(x_{r,n+1}(\cdot)),\ldots,g_m(x_{r,m}(\cdot))]^T$, $\hat{u}_r^1(\cdot)=[u^1_{r,n+1}(\cdot),\ldots,u^1_{r,m}(\cdot)]^T$, $\hat{u}_r^2(\cdot)=[u^2_{r,n+1}(\cdot),\ldots,u^2_{r,m}(\cdot)]^T$, $\hat{u}_r^3(\cdot)=[u^3_{r,n+1}(\cdot),\ldots,u^3_{r,m}(\cdot)]^T$. The coupled system (7.147) with auxiliary state variables (7.151) can be constructed as follows:

$$\begin{aligned} \bar{x}_r^{\Delta}(t)=&-A^{\bar{x}}(\bar{x}_r(t))\bar{x}_r(t)+B^{\bar{x}}(\bar{x}_r(t))g(\bar{x}_r(t))+C^{\bar{x}}(\bar{x}_r(t))g(\bar{x}_r(t-\tau_2(t)))\\ &+\beta(t)d_0\sum_{\tilde{r}=1}^{N}l_{0,r\tilde{r}}F_0^{\bar{x}}\bar{x}_{\tilde{r}}(t)+(1-\beta(t))d_{\alpha(t)}\sum_{\tilde{r}=1}^{N}l_{\alpha(t),r\tilde{r}}F^{\bar{x}}_{\alpha(t)}\bar{x}_{\tilde{r}}(t)\\ &+\bar{u}_r^1(t)+\bar{u}_r^2(t)+\bar{u}_r^3(t) \end{aligned} \tag{7.152}$$

where

$\bar{x}_r(t)=\left[x_r^T(t)\;\hat{x}_r^T(t)\right]^T\in\mathbb{R}^m$, $g(\bar{x}_r(\cdot))=\left[g^T(x_r(\cdot))\;g^T(\hat{x}_r(\cdot))\right]^T\in\mathbb{R}^m$,
$\bar{u}_r^1(t)=\left[(u_r^1(t))^T\;(\hat{u}_r^1(t))^T\right]^T\in\mathbb{R}^m$, $\bar{u}_r^2(t)=\left[(u_r^2(t))^T\;(\hat{u}_r^2(t))^T\right]^T\in\mathbb{R}^m$,
$\bar{u}_r^3(t)=\left[(u_r^3(t))^T\;(\hat{u}_r^3(t))^T\right]^T\in\mathbb{R}^m$, $F_0^{\bar{x}}=\mathrm{diag}(F_0^x,F_0^1)$,
$F^{\bar{x}}_{\alpha(t)}=\mathrm{diag}(F^x_{\alpha(t)},F^1_{\alpha(t)})$, $A^{\bar{x}}(\bar{x}_r(t))=\mathrm{diag}(A^x(x_r(t)),A^1(\hat{x}_r(t)))$,
$B^{\bar{x}}(\bar{x}_r(t))=\begin{bmatrix}B^x(x_r(t)) & B^1(\hat{x}_r(t))\\ B^2(\hat{x}_r(t)) & B^3(\hat{x}_r(t))\end{bmatrix}$, $C^{\bar{x}}(\bar{x}_r(t))=\begin{bmatrix}C^x(x_r(t)) & C^1(\hat{x}_r(t))\\ C^2(\hat{x}_r(t)) & C^3(\hat{x}_r(t))\end{bmatrix}$.

The matrices $A^1(\hat{x}(t))$, F_0^1, $F^1_{\alpha(t)}$, $B^1(\hat{x}(t))$, $B^2(\hat{x}(t))$, $B^3(\hat{x}(t))$, $C^1(\hat{x}(t))$, $C^2(\hat{x}(t))$, and $C^3(\hat{x}(t))$ depend on the auxiliary system (7.151).

In this case, the topology-based probability impulse controllers $\bar{u}_r^1(t)$, $\bar{u}_r^2(t)$ and feedback controller $\bar{u}_r^3(t)$ are designed as follows:

$$\begin{cases} \bar{u}_r^1(t)=(\beta(t)\ell_0\sum\limits_{k=1}^{+\infty}\sum\limits_{\tilde{r}=1}^{N}l_{0,r\tilde{r}}(\bar{x}_{\tilde{r}}(t)-s(t))\\ \qquad-(\bar{x}_{\tilde{r}}(t)-s(t)))\delta(t-t_k)\\ \bar{u}_r^2(t)=((1-\beta(t))\ell_{\alpha(t)}\sum\limits_{k=1}^{+\infty}\sum\limits_{\tilde{r}=1}^{N}l_{\alpha(t),r\tilde{r}}(\bar{x}_{\tilde{r}}(t)-s(t))\\ \qquad-(\bar{x}_{\tilde{r}}(t)-s(t)))\delta(t-t_k)\\ \bar{u}_r^3(t)=K_{\bar{u}}(\bar{x}_r(t)-s(t)) \end{cases} \tag{7.153}$$

in which $K_{\bar{u}}\in\mathbb{R}^{m\times m}$ is the gain matrix.

For the convenience of the following analysis, Case 1 and Case 2 can be unified into as a common form by means of the symbols z, y, and $\aleph$. The isolated systems (7.149) in Case 1 and (7.146) in Case 2 can be combined in the following form:

$$
\begin{aligned}
y^{\Delta}(t) = & -A^{y}(y(t))y(t) + B^{y}(y(t))g(y(t)) \\
& + C^{y}(y(t))g(y(t-\tau_1(t)))
\end{aligned}
\tag{7.154}
$$

and the coupled systems (7.147) in Case 1 and (7.152) Case 2 can be combined in the following form:

$$
\begin{aligned}
z_r^{\Delta}(t) = & -A^{z}(z_r(t))z_r(t) + B^{z}(z_r(t))g(z_r(t)) \\
& + C^{z}(z_r(t))g(z_r(t-\tau_2(t))) + \beta(t)d_0\sum_{\tilde{r}=1}^{N} l_{0,r\tilde{r}}F_0^{z}z_{\tilde{r}}(t) \\
& + (1-\beta(t))d_{\alpha(t)}\sum_{\tilde{r}=1}^{N} l_{\alpha(t),r\tilde{r}}F_{\alpha(t)}^{z}z_{\tilde{r}}(t) \\
& + \aleph_r^1(t) + \aleph_r^2(t) + \aleph_r^3(t)
\end{aligned}
\tag{7.155}
$$

and the controllers (7.150) and (7.151) can be combined in the following form:

$$
\begin{cases}
\aleph_r^1(t) = & (\beta(t)\ell_0 \sum_{k=1}^{+\infty}\sum_{\tilde{r}=1}^{N} l_{0,r\tilde{r}}(z_{\tilde{r}}(t) - y(t)) \\
& -(z_{\tilde{r}}(t) - y(t)))\delta(t-t_k) \\
\aleph_r^2(t) = & ((1-\beta(t))\ell_{\alpha(t)} \sum_{k=1}^{+\infty}\sum_{\tilde{r}=1}^{N} l_{\alpha(t),r\tilde{r}}(z_{\tilde{r}}(t) - y(t)) \\
& -(z_{\tilde{r}}(t) - y(t)))\delta(t-t_k) \\
\aleph_r^3(t) = & K_{\aleph}(z_r(t) - y(t))
\end{cases}
\tag{7.156}
$$

where $z_r(t)$, $y(t)$, $\aleph_r^1(t)$, $\aleph_r^2(t)$, $\aleph_r^3(t) \in \mathbb{R}^q$ and satisfy the following properties:

(1) If $n \geq m$, then $q = n$, and we will deal with (7.154), (7.155), (7.156) in the following two steps: (1) the symbols y, z, and $\aleph$ in the superscript of the matrices will be replaced with $\bar{s}$, x, and u, respectively; (2) the vectors $y(t)$, $z(t)$, $\aleph_r^1(t)$, $\aleph_r^2(t)$, and $\aleph_r^3(t)$ are replaced with $\bar{s}(t)$, $x(t)$, $u_r^1(t)$, $u_r^2(t)$, and $u_r^3(t)$, respectively. For example, $A^{z}(z_r(t)) = A^{x}(x_r(t))$, $A^{y}(y(t)) = A^{\bar{s}}(\bar{s}(t))$.
(2) If $n \leq m$, then $q = m$, and we will deal with (7.154), (7.155), (7.156) in the following two steps: (1) the symbols y, z, and $\aleph$ in the superscript of the matrices will be replaced with s, $\bar{x}$, and $\bar{u}$, respectively; (2) the vectors $y(t)$, $z(t)$, $\aleph_r^1(t)$, $\aleph_r^2(t)$, and $\aleph_r^3(t)$ are replaced with $s(t)$, $\bar{x}(t)$, $\bar{u}_r^1(t)$, $\bar{u}_r^2(t)$, and $\bar{u}_r^3(t)$, respectively. For example, $g(y(t-\tau_1(t))) = g(s(t-\tau_1(t)))$, $g(z_r(t-\tau_2(t))) = g(\bar{x}_r(t-\tau_2(t)))$.

Remark 7.25 In existing works on CMNNs with impulsive effects [46, 148, 149] or complex neural networks with impulsive effects [72, 150, 151], authors only considered the fixed topological structure with uncoupled impulsive effects. If the topology of CMNNs is changed, the impulsive control scheme in existing works will not apply to the new model. Although the impulsive coupling terms with the coupled gain have been researched [152–154], these models only assume that the information exchange occurs at the impulse moments, that is, the authors implicitly assumed that there was no information exchange between nodes in the non-impulse moment. Moreover, in different switching intervals, the required control signals may be different, for example, the control strength of the controller may be different. This novel control scheme considers the impulsive coupling structure and the probability coupling terms in this section, which can effectively reduce resource utilization, reduce control cost, and improve communication efficiency.

7.3.5 *Optimized CMNNs Model via Robust Analysis Approach*

For the convenience of the following descriptions, the memristive matrices $A^y(y(t))$, $B^y(y(t))$, $C^y(y(t))$, $A^z(z_r(t))$, $B^z(z_r(t))$ and $C^z(z_r(t))$ are equivalently written as: $A^y(y(t)) = \mathrm{diag}(a_1^y(y_1(t)), \ldots, a_q^y(y_q(t)))$, $B^y(y(t)) = [\mathrm{sign}_{ij} b_{ij}^y(y_i(t))]_{q\times q}$, $C^y(y(t)) = [\mathrm{sign}_{ij} c_{ij}^y(y_i(t))]_{q\times q}$. $A^z(z_r(t)) = \mathrm{diag}(a_1^z(z_{r,1}(t)), \ldots, a_q^z(z_{r,q}(t)))$, $B^z(z_r(t)) = [\mathrm{sign}_{ij} b_{ij}^z(z_{r,i}(t))]_{q\times q}$, $C^z(z_r(t)) = [\mathrm{sign}_{ij} c_{ij}^z(z_i(t))]_{q\times q}$.

In a known MNNs model, the physical parameters $\mathcal{D}_{ij}$, $\mathcal{R}_{\mathrm{on},ij}$, $\mathcal{R}_{\mathrm{off},ij}$ and μ_v are known constants. Therefore, the memristive connection weights $a_i^y(y_i(t))$, $b_{ij}^y(y_i(t))$, $c_{ij}^y(y_i(t))$, $a_i^z(z_{r,i}(t))$, $b_{ij}^z(z_{r,i}(t))$ and $c_{ij}^z(z_{r,i}(t))$ are measurable, and the lower and upper bounds can be defined as follows:
$\acute{a}_i^y = \sup_{t\in\mathbb{R}}\{a_i^y(y_i(t))\}$, $\acute{b}_{ij}^y = \sup_{t\in\mathbb{R}}\{\mathrm{sign}_{ij} b_{ij}^y(y_i(t))\}$,
$\acute{c}_{ij}^y = \sup_{t\in\mathbb{R}}\{\mathrm{sign}_{ij} c_{ij}^y(y_i(t))\}$, $\grave{a}_i^y = \inf_{t\in\mathbb{R}}\{a_i^y(y_i(t))\}$,
$\grave{b}_{ij}^y = \inf_{t\in\mathbb{R}}\{\mathrm{sign}_{ij} b_{ij}^y(y_i(t))\}$, $\grave{c}_{ij}^y = \inf_{t\in\mathbb{R}}\{\mathrm{sign}_{ij} c_{ij}^y(y_i(t))\}$,
$\acute{a}_i^z = \sup_{t\in\mathbb{R}}\{a_i^z(z_{r,i}(t))\}$, $\acute{b}_{ij}^z = \sup_{t\in\mathbb{R}}\{\mathrm{sign}_{ij} b_{ij}^z(z_{r,i}(t))\}$,
$\acute{c}_{ij}^z = \sup_{t\in\mathbb{R}}\{\mathrm{sign}_{ij} c_{ij}^z(z_{r,i}(t))\}$, $\grave{a}_i^z = \inf_{t\in\mathbb{R}}\{a_i^z(z_{r,i}(t))\}$,
$\grave{b}_{ij}^z = \inf_{t\in\mathbb{R}}\{\mathrm{sign}_{ij} b_{ij}^z(z_{r,i}(t))\}$, $\grave{c}_{ij}^z = \inf_{t\in\mathbb{R}}\{\mathrm{sign}_{ij} c_{ij}^z(z_{r,i}(t))\}$.

For $i, j = 1, \ldots, q$, one sets
$\hat{a}_i^y = \max\{\acute{a}_i^y, \grave{a}_i^y\}$, $\check{a}_i^y = \min\{\acute{a}_i^y, \grave{a}_i^y\}$, $\hat{b}_{ij}^y = \max\{\acute{b}_{ij}^y, \grave{b}_{ij}^y\}$, $\check{b}_{ij}^y = \min\{\acute{b}_{ij}^y, \grave{b}_{ij}^y\}$,
$\hat{c}_{ij}^y = \max\{\acute{c}_{ij}^y, \grave{c}_{ij}^y\}$, $\check{c}_{ij}^y = \min\{\acute{c}_{ij}^y, \grave{c}_{ij}^y\}$, $\hat{a}_i^z = \max\{\acute{a}_i^z, \grave{a}_{ij}^z\}$, $\check{a}_i^z = \min\{\acute{a}_i^z, \grave{a}_i^z\}$,
$\hat{b}_{ij}^z = \max\{\acute{b}_{ij}^z, \grave{b}_{ij}^z\}$, $\check{b}_{ij}^z = \min\{\acute{b}_{ij}^z, \grave{b}_{ij}^z\}$, $\hat{c}_{ij}^z = \max\{\acute{c}_{ij}^z, \grave{c}_{ij}^z\}$, $\check{c}_{ij}^z = \min\{\acute{c}_{ij}^z, \grave{c}_{ij}^z\}$,
$\bar{a}_i^y = \frac{1}{2}(\hat{a}_i^y + \check{a}_i^y)$, $\underline{a}_i^y = \frac{1}{2}(\hat{a}_i^y - \check{a}_i^y)$, $\bar{b}_{ij}^y = \frac{1}{2}(\hat{b}_{ij}^y + \check{b}_{ij}^y)$, $\underline{b}_{ij}^y = \frac{1}{2}(\hat{b}_{ij}^y - \check{b}_{ij}^y)$,
$\bar{c}_{ij}^y = \frac{1}{2}(\hat{c}_{ij}^y + \check{c}_{ij}^y)$, $\underline{c}_{ij}^y = \frac{1}{2}(\hat{c}_{ij}^y - \check{c}_{ij}^y)$, $\bar{a}_i^z = \frac{1}{2}(\hat{a}_{ij}^z + \check{a}_{ij}^z)$, $\underline{a}_{ij}^z = \frac{1}{2}(\hat{a}_{ij}^z - \check{a}_{ij}^z)$,
$\bar{b}_{ij}^z = \frac{1}{2}(\hat{b}_{ij}^z + \check{b}_{ij}^z)$, $\underline{b}_{ij}^z = \frac{1}{2}(\hat{b}_{ij}^z - \check{b}_{ij}^z)$, $\bar{c}_{ij}^z = \frac{1}{2}(\hat{c}_{ij}^z + \check{c}_{ij}^z)$, $\underline{c}_{ij}^z = \frac{1}{2}(\hat{c}_{ij}^z - \check{c}_{ij}^z)$.

By applying the theories of the set-valued map and differential conclusion, inspired by [155], (7.154) can be rewritten as follows:

$$\begin{aligned} y^{\Delta}(t) = &-(\bar{A}^y + \underline{A}_1^y H_1^y(t)\underline{A}_2^y)y(t) + (\bar{B}^y + \underline{B}_1^y H_2^y(t)\underline{B}_2^y) \\ &\cdot g(y(t)) + (\bar{C}^y + \underline{C}_1^y H_3^y(t)\underline{C}_2^y)g(y(t-\tau_1(t))) \end{aligned} \tag{7.157}$$

in which

$$\bar{A}^y = \text{diag}(\bar{a}_1^y, \ldots, \bar{a}_q^y),\ \bar{B}^y = [\bar{b}_{ij}^y]_{q\times q},\ \bar{C}^y = [\bar{c}_{ij}^y]_{q\times q},$$
$$\underline{A}_1^y = \underline{A}_2^y = [\sqrt{\underline{a}_1^y}\mathrm{o}_1, \ldots, \sqrt{\underline{a}_q^y}\mathrm{o}_q] \in \mathbb{R}^{q\times q},$$
$$\underline{B}_1^y = [\sqrt{\underline{b}_{11}^y}\mathrm{o}_1, \ldots, \sqrt{\underline{b}_{1q}^y}\mathrm{o}_1, \ldots, \sqrt{\underline{b}_{q1}^y}\mathrm{o}_q, \ldots, \sqrt{\underline{b}_{qq}^y}\mathrm{o}_q] \in \mathbb{R}^{q\times q^2},$$
$$\underline{B}_2^y = [\sqrt{\underline{b}_{11}^y}\mathrm{o}_1, \ldots, \sqrt{\underline{b}_{1n}^y}\mathrm{o}_q, \ldots, \sqrt{\underline{b}_{q1}^y}\mathrm{o}_1, \ldots, \sqrt{\underline{b}_{qq}^y}\mathrm{o}_q]^T \in \mathbb{R}^{q^2\times q},$$
$$\underline{C}_1^y = [\sqrt{\underline{c}_{11}^y}\mathrm{o}_1, \ldots, \sqrt{\underline{c}_{1q}^y}\mathrm{o}_1, \ldots, \sqrt{\underline{c}_{q1}^y}\mathrm{o}_q, \ldots, \sqrt{\underline{c}_{qq}^y}\mathrm{o}_q] \in \mathbb{R}^{q\times q^2},$$
$$\underline{C}_2^y = [\sqrt{\underline{c}_{11}^y}\mathrm{o}_1, \ldots, \sqrt{\underline{c}_{1q}^y}\mathrm{o}_q, \ldots, \sqrt{\underline{c}_{q1}^y}\mathrm{o}_1, \ldots, \sqrt{\underline{c}_{qq}^y}\mathrm{o}_q]^T \in \mathbb{R}^{q^2\times q},$$
$$H_1^y(t) = \text{diag}(h_1^{1y}(t), \ldots, h_q^{1y}(t)) \in \mathbb{R}^{q\times q},$$
$$H_2^y(t) = \text{diag}(h_{11}^{2y}(t), \ldots, h_{1q}^{2y}(t), \ldots, h_{q1}^{2y}(t), \ldots, h_{qq}^{2y}(t)) \in \mathbb{R}^{q^2\times q^2},$$
$$H_3^y(t) = \text{diag}(h_{11}^{3y}(t), \ldots, h_{1q}^{3y}(t), \ldots, h_{q1}^{3y}(t), \ldots, h_{qq}^{3y}(t)) \in \mathbb{R}^{q^2\times q^2},$$
$$(H_1^y(t))^T H_1^y(t) \le \mathbf{I}_{1,1}^{q,q},\ (H_2^y(t))^T H_2^y(t) \le \mathbf{I}_{q,q}^{1,1},\ (H_3^y(t))^T H_3^y(t) \le \mathbf{I}_{q,q}^{1,1}.$$

Similarly, the coupled system (7.155) can be translated into the following robust form:

$$\begin{aligned} z_r^{\Delta}(t) = &-(\bar{A}^z + \underline{A}_1^z H_1^z(t)\underline{A}_2^z)z_r(t) + (\bar{B}^z + \underline{B}_1^z H_2^z(t)\underline{B}_2^z)g(z_r(t)) \\ &+ (\bar{C}^z + \underline{C}_1^z H_3^z(t)\underline{C}_2^z)g(z_r(t-\tau_2(t))) + \beta(t)d_0\sum_{\tilde{r}=1}^{N} l_{0,r\tilde{r}}F_0^z z_{\tilde{r}}(t) \\ &+ (1-\beta(t))d_{\alpha(t)}\sum_{\tilde{r}=1}^{N} l_{\alpha(t),r\tilde{r}}F_{\alpha(t)}^z z_{\tilde{r}}(t) + \aleph_r^1(t) + \aleph_r^2(t) + \aleph_r^3(t) \end{aligned} \tag{7.158}$$

in which $\bar{A}^z = \text{diag}(\bar{a}_1^z, \ldots, \bar{a}_q^z)$, $\bar{B}^z = [\text{sign}_{ij}\bar{b}_{ij}^z]_{q\times q}$, $\bar{C}^z = [\text{sign}_{ij}\bar{c}_{ij}^z]_{q\times q}$, the matrices $\underline{A}_1^z$, $\underline{A}_2^z$, $\underline{B}_1^z$, $\underline{B}_2^z$, $\underline{C}_1^z$, $\underline{C}_2^z$, $H_1^z(t)$, $H_2^z(t)$, and $H_3^z(t)$ are defined similarly to (7.157). For example, $H_2^z(t) = \text{diag}(h_{11}^{2z}(t), \ldots, h_{1q}^{2z}(t), \ldots, h_{qq}^{2z}(t)) \in \mathbb{R}^{q^2\times q^2}$.

7.3.6 Error Model for CMNNs

The error state between coupled system (7.158) and isolated system (7.157) is defined as $e_r(t) = z_r(t) - y(t)$. Therefore, the error system can be derived as follows:

$$\begin{aligned}
e_r^{\Delta}(t) =& x_r^{\Delta}(t) - y^{\Delta}(t) \\
=& -(\bar{A}^z + \underline{A}_1^z H_1^z(t)\underline{A}_2^z)e_r(t) + (\bar{A}^y + \underline{A}_1^y H_1^y(t)\underline{A}_2^y - \bar{A}^z \\
& - \underline{A}_1^z H_1^z(t)\underline{A}_2^z)y(t) + (\bar{B}^z + \underline{B}_1^z H_2^z(t)\underline{B}_2^z)g(e_r(t)) \\
& + (\bar{C}^z + \underline{C}_1^z H_3^z(t)\underline{C}_2^z)g(e_r(t-\tau_2(t))) + (\bar{B}^z + \underline{B}_1^z H_2^z(t)\underline{B}_2^z \\
& - \bar{B}^y - \underline{B}_1^y H_1^y(t)\underline{B}_2^y)g(y(t)) + (\bar{C}^z + \underline{C}_1^z H_3^z(t)\underline{C}_2^z)g(y(t \\
& - \tau_2(t))) - (\bar{C}^y + \underline{C}_1^y H_2^y(t)\underline{C}_2^y)g(y(t-\tau_1(t))) \\
& + (1-\beta(t))d_{\alpha(t)}\sum_{\tilde{r}=1}^{N} l_{\alpha(t),r\tilde{r}}F_{\alpha(t)}^z e_{\tilde{r}}(t) \\
& + \beta(t)d_0\sum_{\tilde{r}=1}^{N} l_{0,r\tilde{r}}F_0^z e_{\tilde{r}}(t) + u_r^1(t) + u_r^2(t) + u_r^3(t)
\end{aligned} \tag{7.159}$$

where $g(e_r(\cdot)) = g(z_r(\cdot)) - g(y(\cdot))$.

Next, one defines

$e(\cdot) = [e_1^T(\cdot), \ldots, e_N^T(\cdot)]^T$, $y_N(\cdot) = \underbrace{[y^T(\cdot), \ldots, y^T(\cdot)]}_{N}{}^T$,

$g(e(\cdot)) = [g^T(e_1(\cdot)), \ldots, g^T(e_N(\cdot))]^T$, $g(y_N(\cdot)) = \underbrace{[g^T(y(\cdot)), \ldots, g^T(y(\cdot))]}_{N}{}^T$,

$\mathscr{A}^z = \mathbf{I}_{1,1}^{N,N} \otimes \bar{A}^z$, $\mathscr{A}_1^z = \mathbf{I}_{1,1}^{N,N} \otimes \underline{A}_1^z$, $\mathscr{A}_2^z = \mathbf{I}_{1,1}^{N,N} \otimes \underline{A}_2^z$, $\mathscr{B}^z = \mathbf{I}_{1,1}^{N,N} \otimes \bar{B}^z$,
$\mathscr{B}_1^z = \mathbf{I}_{1,1}^{N,N} \otimes \underline{B}_1^z$, $\mathscr{B}_2^z = \mathbf{I}_{1,1}^{N,N} \otimes \underline{B}_2^z$, $\mathscr{C}^z = \mathbf{I}_{1,1}^{N,N} \otimes \bar{C}^z$, $\mathscr{C}_1^z = \mathbf{I}_{1,1}^{N,N} \otimes \underline{C}_1^z$,
$\mathscr{C}_2^z = \mathbf{I}_{1,1}^{N,N} \otimes \underline{C}_2^z$, $\mathscr{H}_1^z(t) = \mathbf{I}_{1,1}^{N,N} \otimes H_1^z(t)$, $\mathscr{H}_2^z(t) = \mathbf{I}_{1,1}^{N,N} \otimes H_2^z(t)$,
$\mathscr{H}_3^z(t) = \mathbf{I}_{1,1}^{N,N} \otimes H_3^z(t)$, $\mathscr{A}^y = \mathbf{I}_{1,1}^{N,N} \otimes \bar{A}^y$, $\mathscr{A}_1^y = \mathbf{I}_{1,1}^{N,N} \otimes \underline{A}_1^y$,
$\mathscr{A}_2^y = \mathbf{I}_{1,1}^{N,N} \otimes \underline{A}_2^y$, $\mathscr{B}^y = \mathbf{I}_{1,1}^{N,N} \otimes \bar{B}^y$, $\mathscr{B}_1^y = \mathbf{I}_{1,1}^{N,N} \otimes \underline{B}_1^y$,
$\mathscr{B}_2^y = \mathbf{I}_{1,1}^{N,N} \otimes \underline{B}_2^y$, $\mathscr{C}^y = \mathbf{I}_{1,1}^{N,N} \otimes \bar{C}^y$, $\mathscr{C}_1^y = \mathbf{I}_{1,1}^{N,N} \otimes \underline{C}_1^y$,
$\mathscr{C}_2^y = \mathbf{I}_{1,1}^{N,N} \otimes \underline{C}_2^y$, $\mathscr{H}_1^y(t) = \mathbf{I}_{1,1}^{N,N} \otimes H_1^y(t)$, $\mathscr{H}_2^y(t) = \mathbf{I}_{1,1}^{N,N} \otimes H_2^y(t)$,
$\mathscr{H}_3^y(t) = \mathbf{I}_{1,1}^{N,N} \otimes H_3^y(t)$, $\mathscr{F}_0 = L_0 \otimes F_0^z$, $\mathscr{F}_{\alpha(t)} = L_{\alpha(t)} \otimes F_{\alpha(t)}^z$,
$\mathscr{L}_0 = L_0 \otimes \mathbf{I}_{1,1}^{N,N}$, $\mathscr{L}_{\alpha(t)} = L_{\alpha(t)} \otimes \mathbf{I}_{1,1}^{N,N}$, $\mathscr{K}_{\aleph} = \mathbf{I}_{1,1}^{N,N} \otimes K_{\aleph}$.

Note that $\mathbf{I}_{1,1}^{N,N} \otimes (\underline{B}_1^z H_2^z(t)\underline{B}_2^z) = (\mathbf{I}_{1,1}^{N,N} \otimes \underline{B}_1^z)(\mathbf{I}_{1,1}^{N,N} \otimes H_2^z(t))(\mathbf{I}_{1,1}^{N,N} \otimes \underline{B}_2^z) = \mathscr{B}_1^z \mathscr{H}_1^z(t)\mathscr{B}_2^z$, similarly, $\mathbf{I}_{1,1}^{N,N} \otimes (\underline{C}_1^z H_3^z(t)\underline{C}_2^z) = \mathscr{C}_1^z \mathscr{H}_2^z(t)\mathscr{C}_2^z$, $\mathbf{I}_{1,1}^{N,N} \otimes (\underline{B}_1^y H_1^y(t)\underline{B}_2^y) = \mathscr{B}_1^y \mathscr{H}_1^y(t)\mathscr{B}_2^y$, $\mathbf{I}_{1,1}^{N,N} \otimes (\underline{C}_1^y H_2^y(t)\underline{C}_2^y) = \mathscr{C}_1^y \mathscr{H}_2^y(t)\mathscr{C}_2^y$.

Integrating both sides of (7.159) from $t_k - \eta$ to $t_k + \eta$, and letting $\eta \to 0$, (7.159) can be further translated into the following impulsive system:

$$\begin{cases} e^{\Delta}(t) = & -(\mathscr{A}^z + \mathscr{A}_1^z \mathscr{H}_1^z(t)\mathscr{A}_2^z)e(t) + (\mathscr{A}^y \\ & + \mathscr{A}_1^y \mathscr{H}_1^y(t)\mathscr{A}_2^y - \mathscr{A}^z - \mathscr{A}_1^z \mathscr{H}_1^z(t)\mathscr{A}_2^z)y_N(t) \\ & + (\mathscr{B}^z + \mathscr{B}_1^z \mathscr{H}_2^z(t)\mathscr{B}_2^z)g(e(t)) + (\mathscr{C}^z \\ & + \mathscr{C}_1^z \mathscr{H}_3^z(t)\mathscr{C}_2^z)g(e(t-\tau_2(t))) + (\mathscr{B}^z \\ & + \mathscr{B}_1^z \mathscr{H}_2^z(t)\mathscr{B}_2^z - \mathscr{B}^y - \mathscr{B}_1^y \mathscr{H}_1^y(t)\mathscr{B}_2^y)g(y_N(t)) \\ & + (\mathscr{C}^z + \mathscr{C}_1^z \mathscr{H}_3^z(t)\mathscr{C}_2^z)g(y_N(t-\tau_2(t))) \\ & - (\mathscr{C}^y + \mathscr{C}_1^y \mathscr{H}_2^y(t)\mathscr{C}_2^y)g(y_N(t-\tau_1(t))) \\ & + (\beta(t)d_0\mathscr{F}_0 + (1-\beta(t))d_{\alpha(t)}\mathscr{F}_{\alpha(t)})e(t) \\ & + \mathscr{K}_{\aleph}e(t), \quad t \neq t_k \\ e(t_k^+) = & (\beta(t_k)\ell_0\mathscr{L}_0 + (1-\beta(t_k))\ell_{\alpha(t_k)}\hat{\mathscr{L}}_{\alpha(t_k)})e(t_k), t = t_k \end{cases} \tag{7.160}$$

where $e(t_k^+)$ and $e(t_k^-)$ are the right and left limits of $e(t)$ at t_k, and $e(t_k^-) = e(t_k)$.

Similarly, the isolated system (7.157) can be translated into the following form

$$\begin{aligned} y_N^{\Delta}(t) = & -(\mathscr{A}^y + \mathscr{A}_1^y \mathscr{H}_1^y(t)\mathscr{A}_2^y)y_N(t) + (\mathscr{B}^y + \mathscr{B}_1^y \mathscr{H}_2^y(t)\mathscr{B}_2^y)g(y_N(t)) \\ & + (\mathscr{C}^y + \mathscr{C}_1^y \mathscr{H}_3^y(t)\mathscr{C}_2^y)g(y_N(t-\tau_1(t))) \end{aligned} \tag{7.161}$$

Denoting the augmented vectors $e_y(\cdot) = [y_N^T(\cdot)\ e^T(\cdot)]^T$, $g(e_y(\cdot)) = [g^T(y_N(\cdot))g^T(e(\cdot))]^T$. Note that $y_N(t_k^+) = y_N(t_k^-) = y_N(t_k)$, then one can have

$$\begin{cases} e_y^{\Delta}(t) = & -(\mathbf{A} + \mathbf{A}_1\mathbf{H}_1(t)\mathbf{A}_2 - \beta(t)d_0\mathbf{F}_0 \\ & -(1-\beta(t))d_{\alpha(t)}\mathbf{F}_{\alpha(t)} - \mathbf{K})e_y(t) \\ & + (\mathbf{B} + \mathbf{B}_1\mathbf{H}_2(t)\mathbf{B}_2)g(e_y(t)) \\ & + (\mathbf{C} + \mathbf{C}_1\mathbf{H}_3(t)\mathbf{C}_2)g(e_y(t-\tau_2(t))) \\ & + (\hat{\mathbf{C}} + \hat{\mathbf{C}}_1\hat{\mathbf{H}}_3(t)\hat{\mathbf{C}}_2)g(e_y(t-\tau_1(t))), t \neq t_k \\ e_y(t_k^+) = & \mathbf{E}e_y(t_k) + \beta(t_k)\ell_0\mathbf{L}_0 e_y(t_k) \\ & + (1-\beta(t_k))\ell_{\alpha(t_k)}\mathbf{L}_{\alpha(t_k)}e_y(t_k), t = t_k \end{cases} \tag{7.162}$$

where

$$\mathbf{A} = \begin{bmatrix} \mathscr{A}^y & \mathbf{0}_{1,1}^{Nq,Nq} \\ \mathscr{A}^z - \mathscr{A}^y & \mathscr{A}^z \end{bmatrix}, \mathbf{A}_1 = \begin{bmatrix} \mathscr{A}_1^y & \mathbf{0}_{1,1}^{Nq,Nq} \\ -\mathscr{A}_1^y & \mathscr{A}_1^z \end{bmatrix}, \mathbf{A}_2 = \begin{bmatrix} \mathscr{A}_2^y & \mathbf{0}_{1,1}^{Nq,Nq} \\ \mathscr{A}_2^z & \mathscr{A}_2^z \end{bmatrix},$$

$$\mathbf{B} = \begin{bmatrix} \mathscr{B}^y & \mathbf{0}_{1,1}^{Nq,Nq} \\ \mathscr{B}^z - \mathscr{B}^y & \mathscr{B}^z \end{bmatrix}, \mathbf{B}_1 = \begin{bmatrix} \mathscr{B}_1^y & \mathbf{0}_{1,q}^{N,Nq} \\ -\mathscr{B}_1^y & \mathscr{B}_1^z \end{bmatrix}, \mathbf{B}_2 = \begin{bmatrix} \mathscr{B}_2^y & \mathbf{0}_{1,q}^{Nq,N} \\ \mathscr{B}_2^z & \mathscr{B}_2^z \end{bmatrix},$$

$$\mathbf{C} = \begin{bmatrix} \mathbf{0}_{1,1}^{Nq,Nq} & \mathbf{0}_{1,1}^{Nq,Nq} \\ \mathscr{C}^z & \mathscr{C}^z \end{bmatrix}, \mathbf{C}_1 = \begin{bmatrix} \mathbf{0}_{1,q}^{N,Nq} & \mathbf{0}_{1,q}^{N,Nq} \\ \mathscr{C}_1^z & \mathscr{C}_1^z \end{bmatrix}, \mathbf{C}_2 = \operatorname{diag}(\mathscr{C}_2^z, \mathscr{C}_2^z),$$

$$\hat{\mathbf{C}} = \begin{bmatrix} \mathscr{C}^y & \mathbf{0}_{1,1}^{Nq,Nq} \\ -\mathscr{C}^y & \mathbf{0}_{1,1}^{Nq,Nq} \end{bmatrix}, \hat{\mathbf{C}}_1 = \begin{bmatrix} \mathscr{C}_1^y & \mathbf{0}_{1,q}^{N,Nq} \\ -\mathscr{C}_1^y & \mathbf{0}_{1,q}^{N,Nq} \end{bmatrix}, \hat{\mathbf{C}}_2 = \operatorname{diag}(\mathscr{C}_2^y, \mathscr{C}_2^y),$$

$\mathbf{H}_1(t) = \operatorname{diag}(\mathscr{H}_1^y(t), \mathscr{H}_1^z(t))$, $\mathbf{H}_2(t) = \operatorname{diag}(\mathscr{H}_2^y(t), \mathscr{H}_2^z(t))$,
$\mathbf{H}_3(t) = \operatorname{diag}(\mathscr{H}_3^z(t), \mathscr{H}_3^z(t))$, $\hat{\mathbf{H}}_3(t) = \operatorname{diag}(\mathscr{H}_3^y(t), \mathscr{H}_3^y(t))$,

$\mathbf{F}_0 = \mathrm{diag}(\mathbf{0}_{1,1}^{Nq,Nq}, \mathscr{F}_0)$, $\mathbf{F}_{\alpha(t)} = \mathrm{diag}(\mathbf{0}_{1,1}^{Nq,Nq}, \mathscr{F}_{\alpha(t)})$,
$\mathbf{E} = \mathrm{diag}(\mathbf{I}_{1,1}^{Nq,Nq}, \mathbf{0}_{1,1}^{Nq,Nq})$, $\mathbf{L}_0 = \mathrm{diag}(\mathbf{0}_{1,1}^{Nq,Nq}, \mathscr{L}_0)$,
$\mathbf{L}_{\alpha(t_k)} = \mathrm{diag}(\mathbf{0}_{1,1}^{Nq,Nq}, \mathscr{L}_{\alpha(t_k)})$, $\mathbf{K} = \mathrm{diag}(\mathbf{0}_{1,1}^{Nq,Nq}, \mathscr{K}_{\aleph})$.

In order to obtain the main results, other definitions and lemmas will be given for utilizing in the rest of this section.

Definition 7.7 The coupled system (7.147) and isolated system (7.146) with three types of mismatched parameters are said to be multiple mismatched exponentially synchronization (MMES) in the mean square via the TPIM if for any initial values $e_y(t_0)$ and known auxiliary state variables $\varphi_1(\cdot), \ldots, \varphi_{n-m}(\cdot)$ (or $\psi_1(\cdot), \ldots, \psi_{m-n}(\cdot)$), there exist positive constants ϖ_1, ϖ_2 such that

$$\mathbb{E}\{\|e_y(t)\|\} = \mathbb{E}\{\varpi_1 \|e_y(t_0)\| e^{-\varpi_2(t-t_0)}\}$$

holds for all $t \in \mathbb{T}$. Here, ϖ_2 is the estimation of exponential convergence rate.

Definition 7.8 (*Average Impulsive Interval* [61]) The average impulsive interval T_a of the impulsive sequence $\{t_k\}_{k=0}^{\infty}$ is bounded under time scales $\mathbb{T}$, if there exists a integer $N_0 > 0$ such that:

$$\frac{T-t}{T_a} - N_0 \le N(T,t)_{\mathbb{T}} \le \frac{T-t}{T_a} + N_0, \forall t \in [0, T]_{\mathbb{T}}$$

where $N(T,t)_{\mathbb{T}}$ represents the number of impulses in the interval $(t, T)_{\mathbb{T}}$.

Definition 7.9 (*Impulse Ratio*) The impulse ratio when the CMNNs are affected by non-ideal factors is defined as $\mathfrak{J}$:

$$\mathfrak{J} = \frac{N_{\mathrm{y}}(T,t)_{\mathbb{T}}}{N(T,t)_{\mathbb{T}}} = 1 - \frac{N_{\mathrm{n}}(T,t)_{\mathbb{T}}}{N(T,t)_{\mathbb{T}}}$$

where $N(T,t)_{\mathbb{T}} = N_{\mathrm{y}}(T,t)_{\mathbb{T}} + N_{\mathrm{n}}(T,t)_{\mathbb{T}}$, $N_{\mathrm{y}}(T,t)_{\mathbb{T}}$ and $N_{\mathrm{n}}(T,t)_{\mathbb{T}}$ respectively represent the number of impulse when CMNNs are affected by non-ideal factors and not affected by non-ideal factors over the interval $(t, T)_{\mathbb{T}}$.

Assumption 7.9 For vectors $y^1(\cdot), y^2(\cdot) \in \mathbb{R}^q$, the nonlinear activation function $g(\cdot)$ satisfies the following inequality.

$$\begin{aligned}&(g(y^2(\cdot)) - g(y^1(\cdot)) - G_1(y^2(\cdot) - y^1(\cdot)))^T(g(y^2(\cdot)) - g(y^1(\cdot)) \\ &- G_2(y^2(\cdot) - y^1(\cdot))) \le 0\end{aligned}$$

where $G_1 - G_2 \le 0$, and $G_1, G_2 \in \mathbb{R}^{q \times q}$ are known matrices with appropriate dimensions.

Assumption 7.10 For known constants τ_1 and τ_2, the time-varying delays $\tau_1(t)$ and $\tau_2(t)$ are continuous and satisfy $0 \le \tau_1(t) < \tau_1, 0 \le \tau_2(t) < \tau_2$.

Lemma 7.10 ([156]) *For* $\beth$ *with* $\beth^T = \beth(\beth \in \mathbf{I}_{1,1}^{q,q})$, $\mathcal{B}_1$ *and* $\mathcal{B}_2$ *are the real matrices with appropriate dimensions, and the time-varying matrix satisfies* $\Lambda_1^T(t)\Lambda_1(t) \leq \mathbf{I}_{1,1}^{q,q}$. *Then* $\beth + \mathcal{B}_1\Lambda_1(t)\mathcal{B}_2 + \mathcal{B}_2^T\Lambda_1^T(t)\mathcal{B}_1^T < 0$ *if there exists a scalar* ν *such that the following inequality holds:*

$$\beth + \nu^{-1}\mathcal{B}_1\mathcal{B}_1^T + \nu\mathcal{B}_2^T\mathcal{B}_2 < 0.$$

7.3.7 Multiple Mismatched Synchronization Analysis

In this section, when CMNNs are affected by non-ideal factors, the topology-based probability impulsive law is adopted to derive a series of constraint conditions such that the CMNNs with multiple mismatched parameters can achieve synchronization in the framework of time scale.

For matrices $G_1, G_2 \in \mathbb{R}^{q\times q}$, any symmetric matrices P_1, P_2, Z_1, $Z_2 \in \mathbb{R}^{q\times q}$, we define $\mathbf{P} = \mathrm{diag}(()\mathbf{I}_{1,1}^{N,N} \otimes P_1, \mathbf{I}_{1,1}^{N,N} \otimes P_2)$, $\mathbf{Z} = \mathrm{diag}(()\mathbf{I}_{1,1}^{N,N} \otimes Z_1, \mathbf{I}_{1,1}^{N,N} \otimes Z_2)$, $\mathbf{G}_1 = \mathrm{diag}(()\mathbf{I}_{1,1}^{N,N} \otimes G_1, \mathbf{I}_{1,1}^{N,N} \otimes G_1)$, $\mathbf{G}_2 = \mathrm{diag}(()\mathbf{I}_{1,1}^{N,N} \otimes G_2, \mathbf{I}_{1,1}^{N,N} \otimes G_2)$.

Theorem 7.4 *Assume that Assumptions 7.6–7.10 hold, for any switching signal* $\alpha(t) = \iota$, *then the coupled system (7.147) and isolated system (7.146) can achieve MMES via TPIM with the help of auxiliary systems (7.148) or (7.151), if the following conditions hold for all* $t \in \mathbb{T}$, $k \in \mathbb{N}^+$:

(1) For known positive constants $\beta_0, d_0, d_\iota, \ell_0, \ell_\iota, \mu_{\mathrm{m}}, \varrho_2 > \varrho_1 > 1$, *and constant* $\xi \in \mathbb{R}_r^+$ *with* $\xi < 0$, *if there exist positive definite matrices* $\mathbf{P}, \mathbf{Z}$, *positive constants* ν_y, $\wp_1$, $\wp_2$ *and* $\wp_3$ *such that*

$$\begin{bmatrix} \Theta_\iota & \hat{\Theta}_1 & \nu_y\hat{\Theta}_2^T \\ * & -\nu_y\mathbf{I}_* & \mathbf{0}_* \\ * & * & -\nu_y\mathbf{I}_* \end{bmatrix} < 0 \tag{7.163}$$

$$\begin{cases} \mathbf{E}^T\mathbf{P}\mathbf{E} + \mathrm{Sy}(\ell_0\mathbf{E}^T\mathbf{P}\mathbf{L}_0) + \ell_0^2\mathbf{L}_0^T\mathbf{P}\mathbf{L}_0 \leq \varrho_1\mathbf{P} \\ \mathbf{E}^T\mathbf{P}\mathbf{E} + \mathrm{Sy}(\ell_\iota\mathbf{E}^T\mathbf{P}\mathbf{L}_\iota) + \ell_\iota^2\mathbf{L}_\iota^T\mathbf{P}\mathbf{L}_\iota \leq \varrho_2\mathbf{P} \end{cases} \tag{7.164}$$

where $\Theta_\iota = [\Theta_{\iota,ij}]_{7\times7}$, $\hat{\Theta}_1 = [\hat{\Theta}_{1,\hat{i}\hat{j}}]_{7\times4}$, $\hat{\Theta}_2 = [\hat{\Theta}_{2,\check{i}\check{j}}]_{4\times7}$, $\Theta_{\iota,11} = -\xi\mathbf{P} - \frac{1}{2}\wp_1\mathrm{Sy}(\mathbf{G}_1^T\mathbf{G}_2)$, $\Theta_{\iota,14} = \frac{1}{2}\wp_1(\mathbf{G}_1^T + \mathbf{G}_2^T)$, $\Theta_{\iota,17} = \mathbf{P} + \mathbf{Y} - \mathbf{A}^T\mathbf{Z} + \beta_0 d_0\mathbf{F}_0^T\mathbf{Z} + (1-\beta_0)d_\iota\mathbf{F}_\iota^T\mathbf{Z}$, $\Theta_{\iota,22} = -\frac{1}{2}\wp_2\mathrm{Sy}(\mathbf{G}_1^T\mathbf{G}_2)$, $\Theta_{\iota,25} = \frac{1}{2}\wp_2(\mathbf{G}_1^T + \mathbf{G}_2^T)$, $\Theta_{\iota,33} = -\frac{1}{2}\wp_3\mathrm{Sy}(\mathbf{G}_1^T\mathbf{G}_2)$, $\Theta_{\iota,36} = \frac{1}{2}\wp_3(\mathbf{G}_1^T + \mathbf{G}_2^T)$, $\Theta_{\iota,44} = -\wp_1\mathbf{I}_{1,1}^{2Np,2Np}$, $\Theta_{\iota,47} = \mathbf{B}^T\mathbf{Z}$, $\Theta_{\iota,55} = -\wp_2\mathbf{I}_{1,1}^{2Np,2Np}$, $\Theta_{\iota,57} = \hat{\mathbf{C}}^T\mathbf{Z}$, $\Theta_{\iota,66} = -\wp_3\mathbf{I}_{1,1}^{2Np,2Np}$, $\Theta_{\iota,67} = \mathbf{C}^T\mathbf{Z}$, $\Theta_{\iota,77} = \mu_m\mathbf{P} - 2\mathbf{Z}$, $\hat{\Theta}_{1,71} = -\mathbf{Z}\mathbf{A}_1$, $\hat{\Theta}_{1,72} = \mathbf{Z}\mathbf{B}_1$, $\hat{\Theta}_{1,73} = \mathbf{Z}\hat{\mathbf{C}}_1$, $\hat{\Theta}_{1,74} = \mathbf{Z}\mathbf{C}_1$, $\hat{\Theta}_{2,11} = \mathbf{A}_2$, $\hat{\Theta}_{2,24} = \mathbf{B}_2$, $\hat{\Theta}_{2,35} = \hat{\mathbf{C}}_2$, $\hat{\Theta}_{2,46} = \mathbf{C}_2$. $\mathbf{I}_*$ *and* $\mathbf{0}_*$ *represent the identity matrix and zero matrix with appropriate dimensions, respectively. The gain matrix can be designed as* $\mathbf{K} = \mathbf{Z}^{-1}\mathbf{Y}^T$.

(2) The relationship between T_a and $\mathfrak{J}$ satisfies

$$\mathfrak{J} \leq \frac{\ln \varrho_1 + \xi T_a}{\ln \varrho_1 - \ln \varrho_2}, \; T_a > -\frac{\ln \varrho_1}{\xi} \tag{7.165}$$

Proof Please see Appendix A. □

Remark 7.26 In existing investigations, there were many results on the synchronization of MNNs (or NNs) with time-varying delays [93, 103, 104, 111, 157]. However, most previous works did not consider the mismatched delays, and these results were obtained based on whole-matched information (including matched delays, connection weights, and dimensions) for MNNs. Therefore, previous research results, including theories and techniques, become ineffective in the multiple mismatched systems. Based on the analysis method in this section, the state of the coupled system can be estimated using any node in the presented novel model and can use an externally isolated system that is not affiliated with the coupled system to estimate the states of CMNNs. Furthermore, the presented novel model, analysis methods, and results extend existing works.

Remark 7.27 It is well known that MNNs are state-dependent switched nonlinear systems. For the coupled-isolated systems with different initial values, at any fixed time t_f, the uncertain parameters $h_i^{1y}(t), h_{ij}^{2y}(t)$ and $h_{ij}^{3y}(t)$ in (7.157) may not be equal to $h_i^{1z}(t)$, $h_{ij}^{2z}(t)$ and $h_{ij}^{3z}(t)$ in (7.158), that is, the weighs and uncertain parameters are mismatched [120, 155]. Moreover, the proposed analysis method can easily determine the effect of system parameter perturbations on its stability and dynamic behavior and can improve the ability of a system to resist change without adapting its initial stable configuration.

Remark 7.28 The scale function [158–160] and Jacobin matrix [161] methods have been widely adopted to study the synchronization of complex networks or chaotic systems with different dimensions. However, these methods may change the system's original structure, partly because this method relies on the scale function to compress or expand the original system. Compared with existing methods, the idea of auxiliary variables can use the original system's connection weights to establish a generalized augmented system, which retains the original system's relevant information and applies more information of the original system to the augmented system. Moreover, although some scholars have used auxiliary systems to study chaotic systems of different dimensions [162], these existing methods are unsuitable for this section's study because CMNNs contain time-varying delays, time scales, and state-dependent connection weight matrices.

If the memristive connection weights satisfy $\acute{a}_i^y = \grave{a}_i^y$, $\acute{b}_{ij}^y = \grave{b}_{ij}^y$, $\acute{c}_{ij}^y = \grave{c}_{ij}^y$, $\acute{a}_i^z = \grave{a}_i^z$, $\acute{b}_{ij}^z = \grave{b}_{ij}^z$, $\acute{c}_{ij}^z = \grave{c}_{ij}^z$, one can have $H_1^y(t) = H_2^y(t) = H_3^y(t) = 0$, $H_1^z(t) = H_2^z(t) = H_3^z(t) = 0$, then the CMNNs model and IMNN model can be reduced to CNNs and INNs, respectively.

In this case, the isolated system (7.157) can be rewritten as:

$$y^{\Delta}(t) = -\bar{A}^{y}y(t) + \bar{B}^{y}g(y(t)) + \bar{C}^{y}g(y(t-\tau_1(t))) \tag{7.166}$$

and the coupled systems (7.158) can be rewritten as:

$$\begin{aligned} z_r^{\Delta}(t) = & -\bar{A}^{z}z_r(t) + \bar{B}^{z}g(z_r(t)) + \bar{C}^{z}g(z_r(t-\tau_2(t))) \\ & + \beta(t)d_0\sum_{\tilde{r}=1}^{N} l_{0,r\tilde{r}}F_0^{z}z_{\tilde{r}}(t) \\ & + (1-\beta(t))d_{\alpha(t)}\sum_{\tilde{r}=1}^{N} l_{\alpha(t),r\tilde{r}}F_{\alpha(t)}^{z}z_{\tilde{r}}(t) \\ & + \aleph_r^1(t) + \aleph_r^2(t) + \aleph_r^3(t) \end{aligned} \tag{7.167}$$

From (7.166) and (7.167), one can get the Corollary 1.

Corollary 7.6 *Assume that Assumptions 7.6–7.10 hold, for any switching signal $\alpha(t) = \iota$, then the coupled system (7.167) and isolated system (7.166) can achieve MMES via TPIM with the help of auxiliary systems (7.148) or (7.151), if the following conditions hold for all $t \in \mathbb{T}, k \in \mathbb{N}^+$:*

(1) For known positive constants $\beta_0, d_0, d_\iota, \ell_0, \ell_\iota, \mu_m, \varrho_2 > \varrho_1 > 1$, and constant $\xi \in \mathbb{R}_r^+$ with $\xi < 0$, if there exist positive definite matrices $\mathbf{P}, \mathbf{Z}$, positive constants $\nu_y, \wp_1, \wp_2$ and $\wp_3$ such that

$$\Theta_\iota < 0$$

$$\begin{cases} \mathbf{E}^T\mathbf{P}\mathbf{E} + \mathrm{Sy}(\ell_0\mathbf{E}^T\mathbf{P}\mathbf{L}_0) + \ell_0^2\mathbf{L}_0^T\mathbf{P}\mathbf{L}_0 & \le \varrho_1\mathbf{P} \\ \mathbf{E}^T\mathbf{P}\mathbf{E} + \mathrm{Sy}(\ell_\iota\mathbf{E}^T\mathbf{P}\mathbf{L}_\iota) + \ell_\iota^2\mathbf{L}_\iota^T\mathbf{P}\mathbf{L}_\iota & \le \varrho_2\mathbf{P} \end{cases}$$

where the corresponding term $\Theta_\iota = [\Theta_{\iota,ij}]_{7\times 7}$ has been defined in Theorem 1.

(2) The relationship between T_a and $\mathfrak{J}$ satisfies

$$\mathfrak{J} \le \frac{\ln\varrho_1 + \xi T_a}{\ln\varrho_1 - \ln\varrho_2},\ T_a > -\frac{\ln\varrho_1}{\xi}.$$

7.3.8 Examples and Simulations

In this section, an example is given to corroborate the validity of the presented results.

Example 7.2 For $r = 1, \ldots, 4$ and $t \in \mathbb{T}$, the memristive device [163] is adopted to design two-dimensional CMNNs (7.147) and one-dimensional isolated system

(7.146), the parameters are taken as follows: $\mathfrak{R}_{on} = 116\Omega$, $\mathfrak{R}_{off} = 16\,\mathrm{K}\Omega$, $\mathfrak{D} = 10\,\mathrm{nm}$, $\mu_v = 10^{-14}\,\mathrm{m}^2\mathrm{V}^{-1}\mathrm{S}^{-1}$. For (7.146), we set $\acute{a}_1^s = 3.4$, $\grave{a}_1^s = 2.2$, $\acute{b}_{11}^s = 1.4$, $\grave{b}_{11}^s = -1.2$, $\acute{c}_{11}^s = 1.2$, $\grave{c}_{11}^s = -0.6$.

For (7.147), we set $\acute{a}_1^x = 4.2$, $\grave{a}_1^x = 2.4$, $\acute{a}_2^x = 3.8$, $\grave{a}_2^x = 1.8$, $\acute{b}_{11}^x = 0.8$, $\grave{b}_{11}^x = -1.2$, $\acute{b}_{12}^x = 1.1$, $\grave{b}_{12}^x = 0.9$, $\acute{b}_{21}^x = 1.2$, $\grave{b}_{21}^x = -1.4$, $\acute{b}_{22}^x = 0.8$, $\grave{b}_{22}^x = -0.6$, $\acute{c}_{11}^x = 0.4$, $\grave{c}_{11}^x = -0.2$, $\acute{c}_{12}^x = 0.5$, $\grave{c}_{12}^x = -0.5$, $\acute{c}_{21}^x = 0.6$, $\grave{c}_{21}^x = 0.4$, $\acute{c}_{22}^x = 0.8$, $\grave{c}_{22}^x = -1.6$.

The auxiliary state variable of isolated system is designed as $s_2(t) = 2s_1(t)$, that is, $s_2^{\Delta}(t) = 2s_1^{\Delta}(t)$. Then, the $\Delta-$derivative of state $s_2(t)$ can be written as $s_2^{\Delta}(t) = -2a_1^s(s_1(t))s_1(t) + 2b_{11}^s(s_1(t))g_1(s_1(t)) + 2c_{11}^s(s_1(t))g_1(s_1(t-\tau_2(t)))$, which implies that $s_2^{\Delta}(t) = -a_1^s(s_1(t))s_2(t) + 2b_{11}^s(s_1(t))g_1(s_1(t)) + 2c_{11}^s(s_1(t))$ $g_1(s_1(t-\tau_2(t)))$.

Thus, the known matrices in (7.149) can be derived as follows: $a_1^{\bar{s}}(\bar{s}_1(t)) = a_1^s(s_1(t)), a_2^{\bar{s}}(\bar{s}_2(t)) = a_1^s(s_1(t)), b_{11}^{\bar{s}}(\bar{s}_1(t)) = b_{11}^s(s_1(t)), b_{12}^{\bar{s}}(\bar{s}_1(t)) = 0, b_{21}^{\bar{s}}(\bar{s}_2(t)) = 2b_{11}^s(s_1(t))$, $b_{22}^{\bar{s}}(\bar{s}_2(t)) = 0$, $c_{11}^{\bar{s}}(\bar{s}_1(t)) = c_{11}^s(s_1(t))$, $c_{12}^{\bar{s}}(\bar{s}_1(t)) = 0$, $c_{21}^{\bar{s}}(\bar{s}_2(t)) = 2c_{11}^s(s_1(t)), c_{22}^{\bar{s}}(\bar{s}_2(t)) = 0$.

For $\alpha(t) = 1, 2$, the coupled terms are considered as follows: $d_0 = 0.8, d_1 = 0.4, d_2 = 0.7$,
$$L_0 = \begin{bmatrix} -3 & 1 & 1 & 1 \\ 1 & -3 & 1 & 1 \\ 1 & 1 & -3 & 1 \\ 1 & 1 & 1 & -3 \end{bmatrix}, \quad L_1 = \begin{bmatrix} -2 & 0 & 1 & 1 \\ 0 & -1 & 0 & 1 \\ 1 & 0 & -2 & 1 \\ 1 & 1 & 1 & -3 \end{bmatrix},$$
$$L_2 = \begin{bmatrix} -3 & 1 & 1 & 1 \\ 1 & -2 & 1 & 0 \\ 1 & 1 & -2 & 0 \\ 1 & 0 & 0 & -1 \end{bmatrix}, \Gamma_0^z = \begin{bmatrix} 0.2 & 0 \\ 0 & 0.3 \end{bmatrix}, \Gamma_1^z = \begin{bmatrix} 0.1 & 0 \\ 0 & 0.15 \end{bmatrix}, \Gamma_2^z = \begin{bmatrix} 0.12 & 0 \\ 0 & 0.4 \end{bmatrix}.$$

The time scale is assumed to be $\mathbb{T} = \bigcup_{\mathfrak{k}=0}^{+\infty}[1.0\mathfrak{k}, 1.0\mathfrak{k} + 0.9]$ $(\mathfrak{k} = 0, 1, \ldots)$. The activation functions are selected as $g(x_r(t)) = \begin{bmatrix} 0.8x_{r,1}(t) - \tanh(0.3x_{r,1}(t)) + 0.5x_{r,2}(t) \\ 0.6x_{r,2}(t) - \tanh(0.2x_{r,2}(t)) \end{bmatrix}$, $g(s(t)) = \begin{bmatrix} 0.8s_1(t) - \tanh(0.3s_1(t)) + 0.5s_2(t) \\ 0.6s_2(t) - \tanh(0.2s_2(t)) \end{bmatrix}$. The time-varying delays are taken as $\tau_1(t) = 0.8 + 0.8\sin(0.6t)$, $\tau_2(t) = 1.2 + 0.6\sin(2t)$. Obviously, we can obtain the following known parameters: $\mu(t) = \begin{cases} 0, & t \in \bigcup_{\mathfrak{k}=0}^{+\infty}[1.0\mathfrak{k}, 1.0\mathfrak{k} + 0.9) \\ 0.1, & t \in \bigcup_{\mathfrak{k}=0}^{+\infty} 1.0\mathfrak{k} + 0.9 \end{cases}$, $G_1 = \begin{bmatrix} 0.5 & 0.5 \\ 0 & 0.4 \end{bmatrix}, G_2 = \begin{bmatrix} 0.8 & 0.5 \\ 0 & 0.6 \end{bmatrix}$.

Set $\beta_0 = 0.6$, $\ell_0 = 0.3$, $\ell_1 = 0.4$, $\ell_2 = 0.45$, $\varrho_1 = e^{1.1}$, $\varrho_2 = e^{1.9}$, $\xi = -10$, by using the LMI Toolbox of MATLAB to solve the constraint conditions in Theorem 1, the gain matrix K can be represented as $K = \begin{bmatrix} -12.3496 & 4.7033 \\ 7.8787 & -4.7024 \end{bmatrix}$.

According to the condition in Theorem 1, the average impulsive interval T_a can be derived as $T_a > 0.11$. When $T_a = 0.15$, the impulse ratio can be estimated as $\mathfrak{J} \leq 0.50$. For nodes $r = 1, \ldots, 4$, the corresponding initial values are consid-

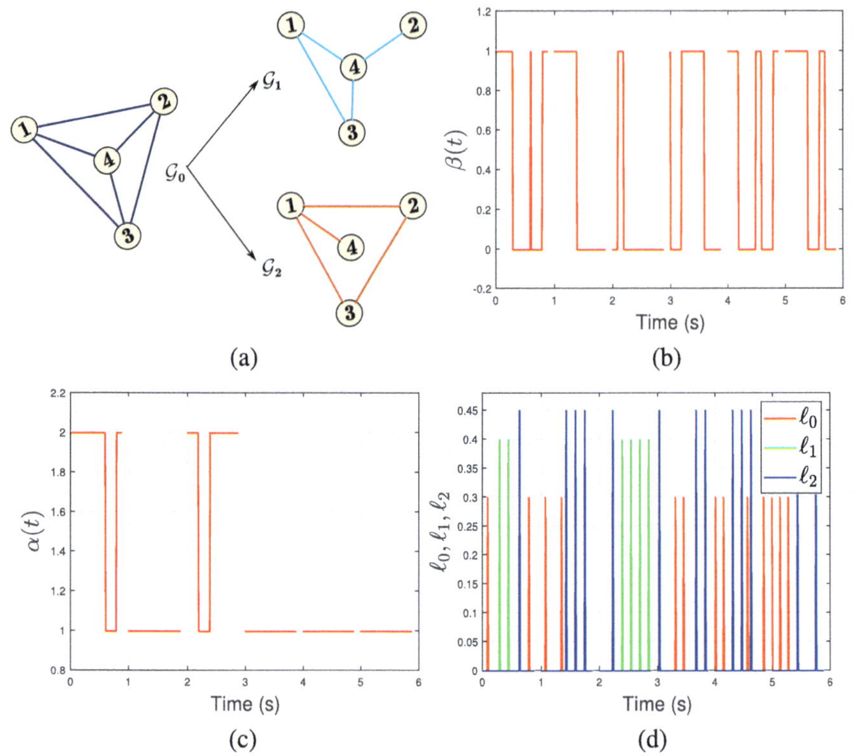

Fig. 7.20 For nodes $r = 1, \ldots, 4$, **a** the unreliable network structure graph with two types of connection structures. **b** The curve of binary stochastic variable $\beta(t)$. **c** The curve of switching signal $\alpha(t)$. **d** The impulsive instant sequence $\{t_k\}_{\mathbb{T}}$ with impulsive strength ℓ_0, ℓ_1, ℓ_2

ered as $x_1(t) = [1.3 \ \ -2.5]^T, x_2(t) = [-1.7 \ 2.8]^T, x_3(t) = [1.5 \ \ -1.6]^T, x_4(t) = [1.8 \ \ -1.1]^T$, $y(t) = [-0.3, -0.6]^T$. For the impulsive strength ℓ_0, ℓ_1 and ℓ_2, the impulsive period is defined as T_a^0, T_a^1, T_a^2, respectively. When the impulse ratio $\mathfrak{J} = 0.45$, if we set $T_a^0 = 0.14$, $T_a^1 = 0.15$ and $T_a^2 = 0.16$, that is, $T_a = 0.15$, then the simulation results are presented in Figs. 7.20, 7.21 and 7.22. When the impulse ratio $\mathfrak{J} = 0.45$, if we set $T_a^0 = 0.08$, $T_a^1 = 0.09$ and $T_a^2 = 0.07$, that is, $T_a = 0.08 < 0.11$, then the simulation results are presented in Fig. 7.23. Obviously, compared with the state responses in Figs. 7.21 and 7.22, the state responses of the error system in Fig. 7.23b are unstable because the period T_a is dissatisfied with the condition (7.165).

In the augmented system (7.162), when $t = t_k$, one can have $e_y(t_k^+) = (\mathbf{E} + \beta(t_k)\ell_0\mathbf{L}_0 + (1 - \beta(t_k))\ell_{\alpha(t_k)}\mathbf{L}_{\alpha(t_k)})e_y(t_k)$, and $\mathbf{E} + \beta(t_k)\ell_0\mathbf{L}_0 + (1 - \beta(t_k))\ell_{\alpha(t_k)}\mathbf{L}_{\alpha(t_k)} \geq \mathbf{E}$. In [61], authors have confirmed that such impulses in the augmented system (7.162) can destroy the synchronization, the larger the T_a is, the more favorable the system is to realize synchronization. When $\mathbb{T} = \mathbb{R}$, one has $\mu_m = 0$, and the (7.154) and (7.156) can be translated into continuous-time systems. By considering

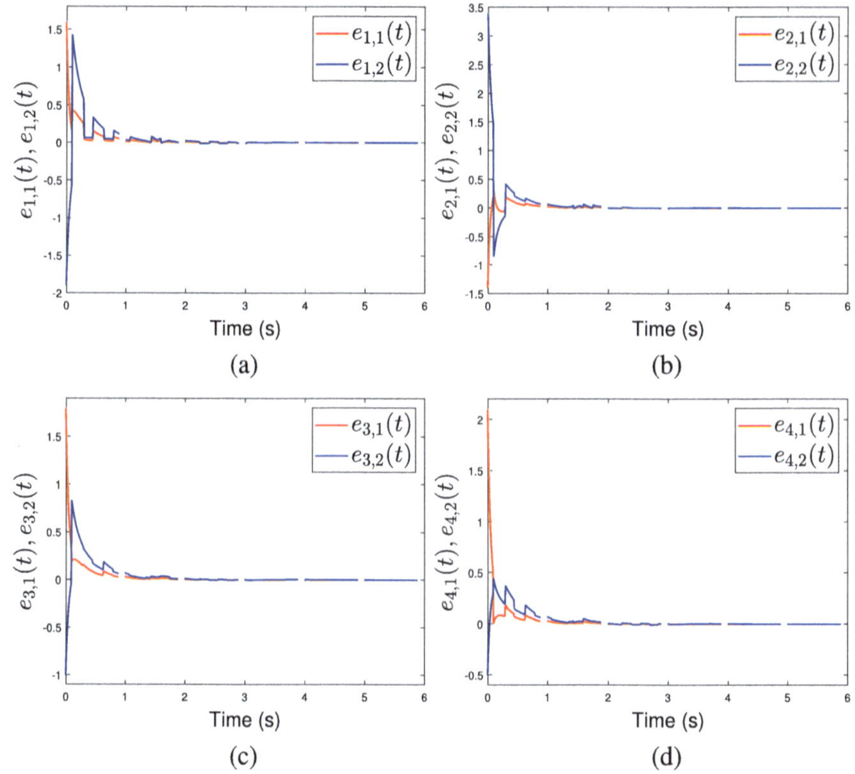

Fig. 7.21 State responses of the error system (7.160) with nodes $r = 1, 2, 3, 4$ under TPIM. **a** the trajectories of error state when $r = 1$, **b** the trajectories of error state when $r = 2$, **c** the trajectories of error state when $r = 3$, **d** the trajectories of error state when $r = 4$

the parameters shown in Fig. 7.20a–c, one can obtain $K = \begin{bmatrix} -12.6887 & 4.6192 \\ 4.2859 & -8.9766 \end{bmatrix}$ (Fig. 7.24).

Remark 7.29 The synchronization problems of neural networks with mismatch characteristics (mismatched connection weights, mismatched delays, and mismatched activation functions) have been studied extensively using the event-triggered control [113, 164], pinning control [114] and sliding mode control [165]. To the best of our knowledge, although some works have mentioned mismatch characteristics, the nonidentical dimension characteristics for MNNs have never been discussed, and the impulsive control method is rarely used to synchronize coupling-isolated systems or driven-response systems in exiting mismatched works. Moreover, by setting the parameters, our model and results can easily degenerate to the existing mismatched methods.

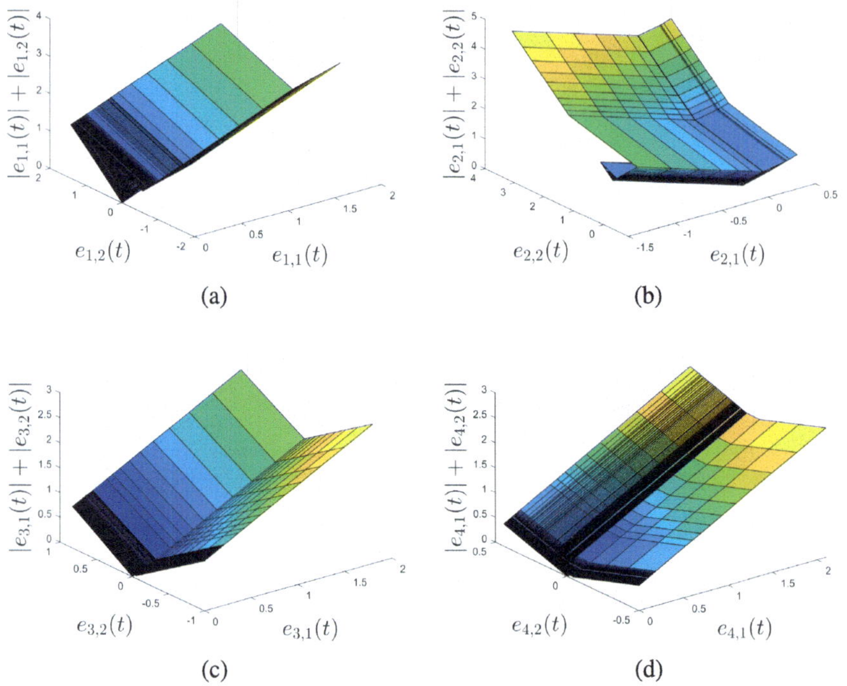

Fig. 7.22 For node $r = 1, \ldots, 4$ and time $t \in \mathbb{T}$, **a** the numerical trends of state variables $|e_{1,1}(t)| + |e_{1,2}(t)|$ on a continuous surface, **b** the numerical trend of state variables $|e_{2,1}(t)| + |e_{2,2}(t)|$ on a continuous surface, **c** the numerical trend of state variables $|e_{3,1}(t)| + |e_{3,2}(t)|$ on a continuous surface, **d** the numerical trend of state variables $|e_{4,1}(t)| + |e_{4,2}(t)|$ on a continuous surface

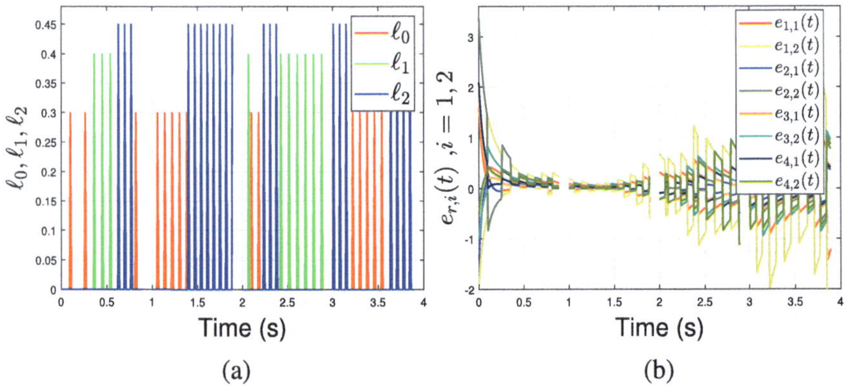

Fig. 7.23 Still consider the parameters shown in Fig. 7.20a–c, when the impulse ratio $\mathfrak{J} = 0.45$, $T_a^0 = 0.08$, $T_a^1 = 0.09$ and $T_a^2 = 0.1$, **a** the impulsive instant sequence $\{t_k\}_{\mathbb{T}}$ with impulsive strength ℓ_0, ℓ_1, ℓ_2; **b** the trajectories of error state $e_{r,1}$ and $e_{r,2}$ in (7.160) with nodes $r = 1, 2, 3, 4$

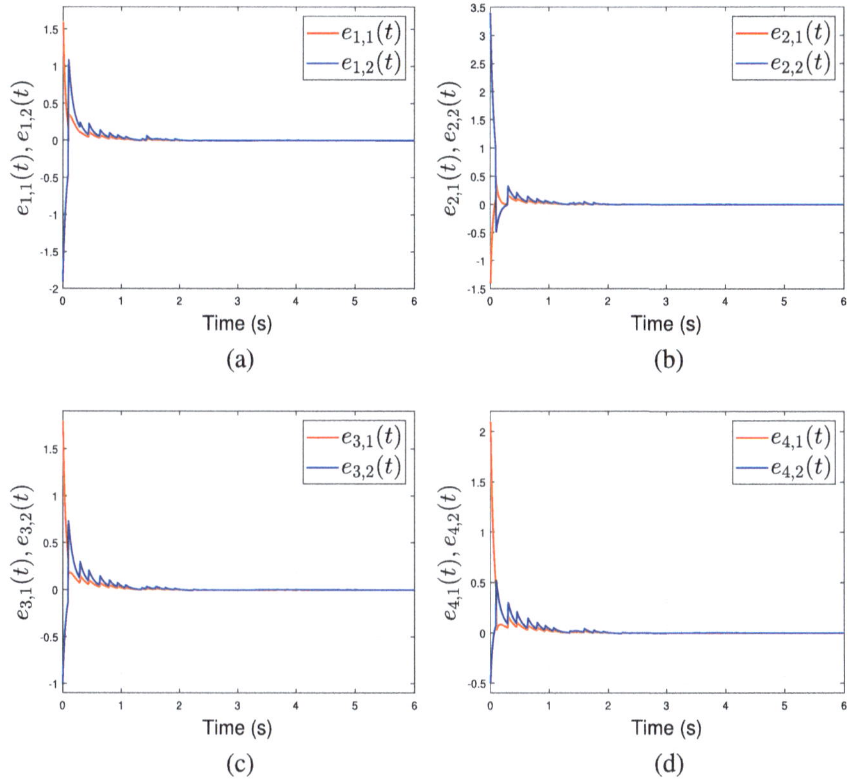

Fig. 7.24 When $\mathbb{T} = \mathbb{R}$, the state responses of error system (7.160) with nodes $r = 1, 2, 3, 4$ under impulsive control. **a** the trajectories of error state when $r = 1$, **b** the trajectories of error state when $r = 2$, **c** the trajectories of error state when $r = 3$, **d** the trajectories of error state when $r = 4$

7.3.9 Conclusion

In this section, the problem of based-topology impulsive synchronization of CMNNs with mismatched dimensions, mismatched connection weights, and mismatched time-varying delays has been studied. This section introduced the framework of time scales to unify the studies of two classes of systems (continuous-time system and discrete-time system) and adopted the method of auxiliary state variables to solve the problem of mismatch dimensions. Since a series of non-ideal factors will destroy the original structure of the CMNNs and cause the change in the topological structure of CMNNs, the novel control framework of TPIM is developed to improve the reliability and generalization of CMNNs. The results obtained can realize the synchronization of coupling-isolated systems with the same attributes and provide a solution for the information exchange between nodes with different attributes. In future work, we will study the CMNNs model with one-to-one interlayer couplings and stochastic perturbations and design different mode-dependent impulsive control schemes for each layer of CMNNs based on the state evolution method of the deterministic finite automata.

Appendix A

Proof of Theorem 1 For any switched signal $\alpha(t)$, the Lyapunov-Krasovskii functional is constructed as:

$$V(t) = e_y^T(t)\mathbf{P}e_y(t) \tag{7.168}$$

When $t \neq t_k$, the $\Delta-$derivative of $V(t)$ along the trajectory of system (7.162) can be calculated as follows:

$$\begin{aligned}
&\mathbb{E}\{V^{\Delta}(t) - \xi V(t)\}\\
=&\mathbb{E}\{(e_y^T(t))^{\Delta}\mathbf{P}e_y(t) + e_y^T(\sigma(t))\mathbf{P}e_y^{\Delta}(t) - \xi e_y^T(t)\mathbf{P}e_y(t)\}\\
\leq&\mathbb{E}\{2e_y^T(t)\mathbf{P}e_y^{\Delta}(t) + \mu_m(e_y^T(t))^{\Delta}\mathbf{P}e_y^{\Delta}(t) - \xi e_y^T(t)\mathbf{P}e_y(t)\}
\end{aligned} \tag{7.169}$$

From the (7.162), we known that there exists a positive definite matrix $\mathbf{Z}$ such that:

$$\begin{aligned}
&\mathbb{E}\{-2(\mathbf{Z}e_y^{\Delta}(t))^T(e_y^{\Delta}(t) + (\mathbf{A} + \mathbf{A}_1\mathbf{H}_1(t)\mathbf{A}_2 - \mathbf{K}\\
&- \beta(t)d_0\mathbf{F}_0 - (1 - \beta(t))d_{\alpha(t)}\mathbf{F}_{\alpha(t)})e_y(t)\\
&- (\mathbf{B} + \mathbf{B}_1\mathbf{H}_2(t)\mathbf{B}_2)g(e_y(t))\\
&- (\mathbf{C} + \mathbf{C}_1\mathbf{H}_3(t)\mathbf{C}_2)g(e_y(t - \tau_2(t)))\\
&- (\hat{\mathbf{C}} + \hat{\mathbf{C}}_1\hat{\mathbf{H}}_3(t)\hat{\mathbf{C}}_2)g(e_y(t - \tau_1(t))))\} = 0
\end{aligned} \tag{7.170}$$

Based on the Assumption 7.9, there exist positive constants $\wp_1$, $\wp_2$ and $\wp_3$ such that the following inequalities hold.

$$\begin{aligned}
&\wp_1(g(e_y(t)) - \mathbf{G}_1e_y(t))^T(g(e_y(t)) - \mathbf{G}_2e_y(t)) \leq 0\\
&\wp_2(g(e_y(t - \tau_1(t))) - \mathbf{G}_1e_y(t - \tau_1(t)))^T(g(e_y(t - \tau_1(t)))\\
&\quad - \mathbf{G}_2e_y(t - \tau_1(t))) \leq 0\\
&\wp_3(g(e_y(t - \tau_2(t))) - \mathbf{G}_1e_y(t - \tau_2(t)))^T(g(e_y(t - \tau_2(t)))\\
&\quad - \mathbf{G}_2e_y(t - \tau_2(t))) \leq 0
\end{aligned} \tag{7.171}$$

Let $\Im^T(t) = [e_y^T(t)\ \ e_y^T(t - \tau_1(t))\ \ e_y^T(t - \tau_2(t))\ \ g^T(e_y(t))\ \ g^T(e_y(t - \tau_1(t)))$ $g^T(e_y(t - \tau_2(t)))\ \ (e_y^{\Delta}(t))^T]$. From (7.168)–(7.171), one obtains

$$\begin{aligned}
&\mathbb{E}\{V^{\Delta}(t) - \xi V(t)\}\\
=&\mathbb{E}\{\Im^T(t)(\Theta_{\alpha(t)} + \hat{\Theta}_1\Delta_{\mathbf{H}}(t)\hat{\Theta}_2 + \hat{\Theta}_2^T\Delta_{\mathbf{H}}^T(t)\hat{\Theta}_1^T)\Im(t)\}
\end{aligned} \tag{7.172}$$

where $\Theta_{\alpha(t)}$, $\hat{\Theta}_1$ and $\hat{\Theta}_2$ have been defined in Theorem 1, $\Delta_{\mathbf{H}}(t) = \text{diag}(()\mathbf{H}_1(t), \mathbf{H}_2(t), \mathbf{H}_3(t), \hat{\mathbf{H}}_3(t))$. □

According to the Schur complement, Lemma 7.10 and (7.163), one can obtain

$$\Xi_{\alpha(t)} + \hat{\Theta}_1 \Delta_{\mathbf{H}}(t)\hat{\Theta}_2 + \hat{\Theta}_2^T \Delta_{\mathbf{H}}^T(t)\hat{\Theta}_1^T < 0 \tag{7.173}$$

Thus, one can derive

$$\mathbb{E}\{V^{\Delta}(t)\} < \xi \mathbb{E}\{V(t)\} \tag{7.174}$$

Note that $\xi \in \mathbb{R}_r^+$. When $t \in (t_k, t_{k+1})_{\mathbb{T}}$, by using Lemma 7.7, one can derive

$$\mathbb{E}\{V(t)\} \leq \mathbb{E}\{V(t_k^+)\}e_{\xi}(t, t_k) \tag{7.175}$$

Next, when $t = t_{k+1}$, the inequality shown below should be proved to be true for any $k \in \mathbb{N}^+$.

$$\mathbb{E}\{V(t_{k+1})\} \leq \mathbb{E}\{V(t_k^+)\}e_{\xi}(t_{k+1}, t_k) \tag{7.176}$$

If t_{k+1} is left dense, since $V(t)$ and $e_{\xi}(t, t_k)$ are both continuous-time varying functions, one obtains

$$\begin{aligned} \mathbb{E}\{V(t_{k+1})\} &= \lim_{t \to t_{k+1}^-} \mathbb{E}\{V(t)\} \\ &\leq \lim_{t \to t_{k+1}^-} \mathbb{E}\{V(t_k^+)\}e_{\xi}(t, t_k) \\ &= \lim_{t \to t_{k+1}} \mathbb{E}\{V(t_k^+)\}e_{\xi}(t, t_k) \\ &= \mathbb{E}\{V(t_k^+)\}e_{\xi}(t_{k+1}, t_k) \end{aligned} \tag{7.177}$$

By using the Lemma 7.9, one has

$$\begin{aligned} &e_{\xi}(\sigma(t), s)e_{\ominus\xi}(t, s) \\ &= e_{\xi}(\sigma(t), t) = (1 + \xi\mu(t))e_{\xi}(t, t) = 1 + \xi\mu(t) \end{aligned} \tag{7.178}$$

According to (7.178), if t_{k+1} is left-scattered (implied that $\sigma(\rho(t_{k+1})) = t_{k+1}$), one can derive

$$\begin{aligned}\mathbb{E}\{V(t_{k+1})\} &= \mathbb{E}\{V(\rho(t_{k+1})) + \mu(\rho(t_{k+1}))V^{\Delta}(\rho(t_{k+1}))\} \\ &\le \mathbb{E}\{V(\rho(t_{k+1})) + \mu(\rho(t_{k+1}))\xi V(\rho(t_{k+1}))\} \\ &= (1 + \xi\mu(\rho(t_{k+1})))\mathbb{E}\{V(\rho(t_{k+1}))\} \\ &= e_{\xi}(\sigma(\rho(t_{k+1})), \rho(t_{k+1}))\mathbb{E}\{V(\rho(t_{k+1}))\} \\ &= e_{\xi}(t_{k+1}, \rho(t_{k+1}))\mathbb{E}\{V(\rho(t_{k+1}))\} \\ &\le e_{\xi}(t_{k+1}, \rho(t_{k+1}))e_{\xi}(\rho(t_{k+1}), t_k)\mathbb{E}\{V(t_k^+)\} \\ &= \mathbb{E}\{V(t_k^+)\}e_{\xi}(t_{k+1}, t_k)\end{aligned} \tag{7.179}$$

Thus, the (7.176) holds for any $k \in \mathbb{N}^+$.

Combining (7.175) and (7.176), for any $\xi \in \mathbb{R}_r^+$, one can obtain

$$\mathbb{E}\{V(t)\} \le e_{\xi}(t, t_k)\mathbb{E}\{V(t_k^+)\}, \quad \forall t \in (t_k, t_{k+1}]_{\mathbb{T}} \tag{7.180}$$

For the function $V(t)$, when $t = t_k$, one can derive

$$\begin{aligned}\mathbb{E}\{V(t_k^+)\} &= \mathbb{E}\{e_y^T(t_k^+)\mathbf{P}e_y(t_k^+)\} \\ &= e_y^T(t_k)(\mathbf{E}^T\mathbf{PE} + \mathrm{Sy}(\beta(t_k)\ell_0\mathbf{E}^T\mathbf{PL}_0) \\ &\quad + \mathrm{Sy}((1-\beta(t_k))\ell_{\alpha(t_k)}\mathbf{E}^T\mathbf{PL}_{\alpha(t_k)}) \\ &\quad + \beta^2(t_k)\ell_0^2\mathbf{L}_0^T\mathbf{PL}_0 \\ &\quad + \mathrm{Sy}(\beta(t_k)(1-\beta(t_k))\ell_0\ell_{\alpha(t_k)}\mathbf{L}_0^T\mathbf{PL}_{\alpha(t_k)}) \\ &\quad + (1-\beta(t_k))^2\ell_{\alpha(t_k)}^2\mathbf{L}_{\alpha(t_k)}^T\mathbf{PL}_{\alpha(t_k)})e_y(t_k)\end{aligned} \tag{7.181}$$

Note that $\beta(t) = 1$ or $\beta(t) = 0$. From condition (7.164), for each $k \in \mathbb{N}^+$, there exist positive constants $1 < \varrho_1 < \varrho_2$ such that the following inequalities can be obtained.

If $\beta(t_k) = 1$, one can derive

$$\begin{aligned}&\mathbb{E}\{V(t_k^+) - \varrho_1 V(t_k)\} \\ &= e_y^T(t_k)(\mathbf{E}^T\mathbf{PE} + \mathrm{Sy}(\ell_0\mathbf{E}^T\mathbf{PL}_0) \\ &\quad + \ell_0^2\mathbf{L}_0^T\mathbf{PL}_0 - \varrho_1\mathbf{P})e_y(t_k) \le 0\end{aligned} \tag{7.182}$$

If $\beta(t_k) = 0$, one can derive

$$\begin{aligned}&\mathbb{E}\{V(t_k^+) - \varrho_2 V(t_k)\} \\ &= e_y^T(t_k)(\mathbf{E}^T\mathbf{PE} + \mathrm{Sy}(\ell_{\alpha(t_k)}\mathbf{E}^T\mathbf{PL}_{\alpha(t_k)}) \\ &\quad + \ell_{\alpha(t_k)}^2\mathbf{L}_{\alpha(t_k)}^T\mathbf{PL}_{\alpha(t_k)} - \varrho_2\mathbf{P})e_y(t_k) \le 0\end{aligned} \tag{7.183}$$

Thus, for any $\beta(t_k) \in \{0, 1\}$, $V(t_k^+)$ satisfies the following inequality:

$$\mathbb{E}\{V(t_k^+)\} \le \bar{\varrho}\mathbb{E}\{V(t_k)\}, t = t_k \tag{7.184}$$

where $\bar{\varrho} = \beta(t_k)\varrho_1 + (1 - \beta(t_k))\varrho_2$.

From (7.180) and (7.184), one can obtain

$$\begin{cases} \mathbb{E}\{V(t)\} \le e_\xi(t, t_k)\mathbb{E}\{V(t_k^+)\}, & t \in (t_k, t_{k+1}]_{\mathbb{T}} \\ \mathbb{E}\{V(t_k^+)\} \le \bar{\varrho}\mathbb{E}\{V(t_k)\}, & t = t_k \end{cases} \tag{7.185}$$

For $t \in [t_0, t_1]_{\mathbb{T}}$, one has

$$\mathbb{E}\{V(t_1^+)\} \le \bar{\varrho}\mathbb{E}\{V(t_1)\} \le \bar{\varrho}e_\xi(t_1, t_0)\mathbb{E}\{V(t_0^+)\} \tag{7.186}$$

For $t \in (t_1, t_2]_{\mathbb{T}}$, one gets

$$\mathbb{E}\{V(t)\} \le e_\xi(t, t_1)\mathbb{E}\{V(t_1^+)\} \le \bar{\varrho}e_\xi(t, t_0)\mathbb{E}\{V(t_0^+)\} \tag{7.187}$$

implies that

$$\mathbb{E}\{V(t_2^+)\} \le \bar{\varrho}\mathbb{E}\{V(t_2)\} \le \bar{\varrho}^2 e_\xi(t_2, t_0)\mathbb{E}\{V(t_0^+)\} \tag{7.188}$$

By induction method, for $t \in (t_k, t_{k+1}]_{\mathbb{T}}$, one obtains

$$\mathbb{E}\{V(t)\} \le e_\xi(t, t_k)\mathbb{E}\{V(t_k^+)\} \le \bar{\varrho}^k e_\xi(t, t_0)\mathbb{E}\{V(t_0^+)\} \tag{7.189}$$

Based on Definitions 7.8, 7.9, and Lemma 7.8, for any $t \in \mathbb{T}$, the following two cases will be discussed.

Case 1: When $\beta(t) \triangleq 1$, the CMNNs are normal operation with original topology, that is, $N_n(t, t_0)_{\mathbb{T}} = N(t, t_0)_{\mathbb{T}}$, $\bar{\varrho} = 1 * \varrho_1 + 0 * \varrho_2 = \varrho_1$. Then one can have

$$\begin{aligned} \mathbb{E}\{V(t)\} &\le \varrho_1^{N(t,t_0)_{\mathbb{T}}} e_\xi(t, t_0)\mathbb{E}\{V(t_0^+)\} \\ &\le \varrho_1^{\frac{t-t_0}{T_a}+N_0} e_\xi(t, t_0)\mathbb{E}\{V(t_0^+)\} \\ &= \varrho_1^{N_0} e^{\frac{\ln \varrho_1}{T_a}(t-t_0)} e_\xi(t, t_0)\mathbb{E}\{V(t_0^+)\} \\ &\le \varrho_1^{N_0} e^{\frac{\ln \varrho_1}{T_a}(t-t_0)} e^{\xi(t-t_0)}\mathbb{E}\{V(t_0^+)\} \\ &= \varrho_1^{N_0} e^{\left(\frac{\ln \varrho_1}{T_a}+\xi\right)(t-t_0)}\mathbb{E}\{V(t_0^+)\} \end{aligned} \tag{7.190}$$

According to condition (7.165), we know that the inequality $\frac{\ln \varrho_1}{T_a} + \xi < 0$ holds, thereby the following inequality holds.

$$\mathbb{E}\{\|e_y(t)\|^2\} \le \varrho_1^{N_0} \frac{\lambda_{\max}(\mathbf{P})}{\lambda_{\min}(\mathbf{P})} e^{\left(\frac{\ln \varrho_1}{T_a}+\xi\right)(t-t_0)}\mathbb{E}\{\|e_y(t_0)\|^2\} \tag{7.191}$$

Case 2: When $\beta(t) \in \{0, 1\}$, the CMNNs suffer from the interference of non-ideal factors in normal operation and the topology of CMNNs may be changed, that is, $N(t, t_0)_{\mathbb{T}} = N_y(t, t_0)_{\mathbb{T}} + N_n(t, t_0)_{\mathbb{T}}$. Then one can obtain

$$\mathbb{E}\{V(t)\} \leq \varrho_1^{N_n(t,t_0)_{\mathbb{T}}} \varrho_2^{N_y(t,t_0)_{\mathbb{T}}} e_\xi(t, t_0)\mathbb{E}\{V(t_0^+)\} \tag{7.192}$$

According to Definition (7.9), it is easy to obtain

$$\mathfrak{J}\left(\frac{t-t_0}{T_a} - N_0\right) \leq N_y(t, t_0)_{\mathbb{T}} \leq \mathfrak{J}\left(\frac{t-t_0}{T_a} + N_0\right)$$

$$(1-\mathfrak{J})\left(\frac{t-t_0}{T_a} - N_0\right) \leq N_n(t, t_0)_{\mathbb{T}} \leq (1-\mathfrak{J})\left(\frac{t-t_0}{T_a} + N_0\right)$$

Since $1 < \varrho_1 < \varrho_2$, one has

$$\begin{aligned}
\mathbb{E}\{V(t)\} &\leq \varrho_2^{N_y(t,t_0)_{\mathbb{T}}} \varrho_1^{N_n(t,t_0)_{\mathbb{T}}} e_\xi(t, t_0)\mathbb{E}\{V(t_0^+)\} \\
&\leq \varrho_2^{\mathfrak{J}\left(\frac{t-t_0}{T_a}+N_0\right)} \varrho_1^{(1-\mathfrak{J})\left(\frac{t-t_0}{T_a}+N_0\right)} e_\xi(t, t_0)\mathbb{E}\{V(t_0^+)\} \\
&= \varrho_{1,2}^* e^{\frac{\mathfrak{J}\ln\varrho_2}{T_a}(t-t_0)} e^{\frac{(1-\mathfrak{J})\ln\varrho_1}{T_a}(t-t_0)} e_\xi(t, t_0)\mathbb{E}\{V(t_0^+)\} \\
&\leq \varrho_{1,2}^* e^{\frac{\mathfrak{J}\ln\varrho_2}{T_a}(t-t_0)} e^{\frac{(1-\mathfrak{J})\ln\varrho_1}{T_a}(t-t_0)} e^{\xi(t-t_0)}\mathbb{E}\{V(t_0^+)\} \\
&= \varrho_{1,2}^* e^{\left(\frac{\mathfrak{J}\ln\varrho_2}{T_a}+\frac{(1-\mathfrak{J})\ln\varrho_1}{T_a}+\xi\right)(t-t_0)}\mathbb{E}\{V(t_0^+)\}
\end{aligned} \tag{7.193}$$

where $\varrho_{1,2}^* = \varrho_2^{\mathfrak{J}N_0}\varrho_1^{(1-\mathfrak{J})N_0}$.

According to condition (7.164), one has $\frac{\mathfrak{J}\ln\varrho_2}{T_a} + \frac{(1-\mathfrak{J})\ln\varrho_1}{T_a} + \xi < 0$. Thus, for any $t \in \mathbb{T}$, one can derive the following:

$$\mathbb{E}\{\|e_y(t)\|^2\} \leq \varrho_{\max}\frac{\lambda_{\max}(\mathbf{P})}{\lambda_{\min}(\mathbf{P})} e^{-\varpi_2(t-t_0)}\mathbb{E}\{\|e_y(t_0)\|^2\} \tag{7.194}$$

where $\varpi_2 = -\frac{\mathfrak{J}\ln\varrho_2}{T_a} - \frac{(1-\mathfrak{J})\ln\varrho_1}{T_a} - \xi$.

Note that the model studied under Assumption 7.7 is recoverable, so the case of $\beta(t) \triangleq 0$ does not exist.

Hence, we can assert that the coupled system (7.147) and isolated system (7.146) can achieve exponential synchronization under the action of TPIM by auxiliary systems (7.148) or (7.151).

This completes the proof.

7.4 Comments

MNNs are a new type of neural network with memory properties, and their dynamic behavior is exceptionally complex, posing significant challenges for research and applications. Compared to traditional artificial neural networks, MNNs have stronger nonvolatile memory capabilities and information processing capacities, showing broad application prospects in fields such as pattern recognition, data mining, and intelligent control. To fully exploit the unique advantages of MNNs, it is crucial to profoundly understand their dynamic mechanisms and precisely characterize their dynamic behavior.

Currently, research on MNN dynamics mainly focuses on several aspects, including time-delay activation functions, network order, number field types, dynamic analysis, and control methods. In terms of time-delay activation functions, researchers have considered the time-delay effects in neuron activation functions and conducted in-depth studies on time-varying delays, distributed delays, and other different situations, revealing the influence of time-delay factors on network dynamics. Regarding network order, the dynamic characteristics of first-order, second-order, and fractional-order MNNs have become the focus and difficulty of current research. In terms of the number of field types, MNN models based on real and complex domains have expanded research perspectives and provided possibilities for discovering new dynamic phenomena. In dynamic analysis, core issues such as synchronization, stability, and robustness involve multiple branches, including exponential synchronization, robust synchronization, fixed-time synchronization, finite-time synchronization, robust stability, exponential stability, dissipativity, and state estimation. In terms of control methods, the exploration of pinning control, sampled-data control, and impulsive control strategies contribute to the effective regulation of MNN dynamic behavior and improve network performance. The overview of research topics on memristive neural network dynamics is shown in Fig. 7.25.

The structure of neural networks directly affects the communication performance between the network and the controller. Complex neural networks with different topological structures may require different control strategies and gains. If the same control strategy is applied to structures with different neural network parameters, it will lead to resource waste. To effectively improve the real-time information exchange capability between neural networks and control strategies, and to reduce control costs, it is crucial to design new control strategies that depend on system parameters. These strategies should be able to dynamically adjust system parameters and control strategies according to the actual working environment, based on the neural network model constructed for complex environments. This is one of the urgent control design problems that need to be solved. In theoretical research, the Lyapunov functional method has become the mainstream tool for analyzing MNN dynamics due to its intuitiveness and effectiveness. However, existing research results generally have a certain degree of conservatism, restricting the practical value of the criteria. The main reasons for this conservatism lie in the unreasonable construction of Lyapunov-

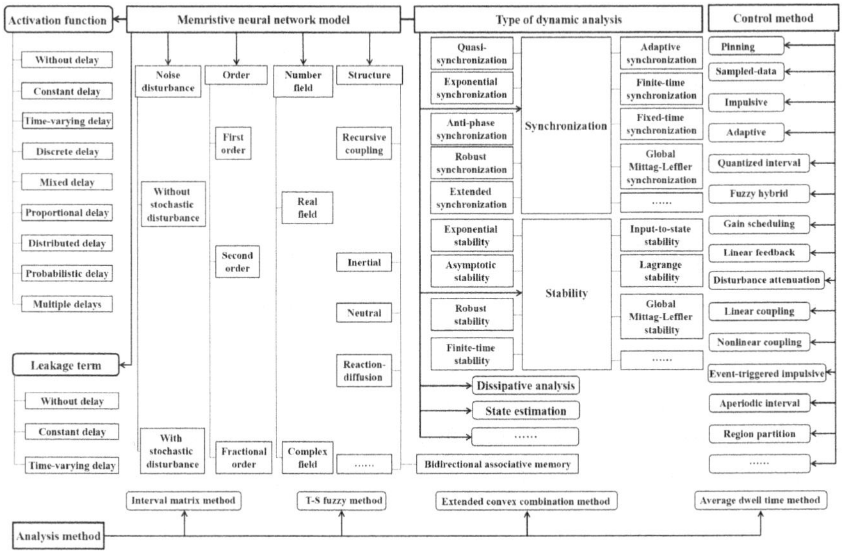

Fig. 7.25 The overview of research topics on memristive neural network dynamics

Krasovskii functionals and the estimation enlargement in the derivation process of the functionals. Therefore, constructing more refined forms of Lyapunov-Krasovskii functionals and providing more precise upper bound estimates when deriving the functionals are key to reducing the conservatism of the criteria and enhancing their practicality.

Furthermore, in the context of rapidly growing big data and complex computing demands, traditional MNNs have difficulty effectively processing complex and variable spatiotemporal information, which has given rise to the urgent need to construct complex MNNs. Complex MNNs embed MNNs into complex networks with spatial structures and establish interlayer node interconnection mechanisms, forming larger scale, more complex, and more functionally rich network models. However, various uncertain factors in complex environments, such as device parameter fluctuations, external disturbances, and communication delays, lead to significant heterogeneity among memristive neurons within complex MNNs, manifested as differences in connection weights, inconsistencies in activation functions, and delays. These heterogeneity issues seriously affect network performance, making it difficult to handle actual complex tasks. Moreover, as the network scale increases, structural parameters become difficult to accurately obtain, while interconnection and synchronization between heterogeneous nodes become more challenging. Therefore, fully considering node heterogeneity and environmental uncertainty, exploring the modeling and design of complex MNNs, and developing corresponding dynamic analysis and control methods to enhance the network's robustness, adaptability, and functionality to better support complex computing demands are of great significance for promoting the development of high-performance brain-inspired computing systems and the progress of intelligent science, and are expected to achieve new breakthroughs in the field of artificial intelligence.

References

1. Liu, P., Zeng, Z., Wang, J.: Multiple mittag-leffler stability of fractional-order recurrent neural networks. IEEE Trans. Syst. Man Cybern.: Syst. **47**(8), 2279–2288 (2017)
2. Liu, P., Zeng, Z., Wang, J.: Multistability analysis of a general class of recurrent neural networks with non-monotonic activation functions and time-varying delays. Neural Netw. **79**, 117–127 (2016)
3. Li, S., Wang, H., Rafique, M.U.: A novel recurrent neural network for manipulator control with improved noise tolerance. IEEE Trans. Neural Netw. Learn. Syst. **29**(5), 1908–1918 (2017)
4. Li, S., Wang, Z., Li, Y.: Using laplacian eigenmap as heuristic information to solve nonlinear constraints defined on a graph and its application in distributed range-free localization of wireless sensor networks. Neural Process. Lett. **37**, 411–424 (2013)
5. Li, S., Liu, B., Li, Y.: Selective positive-negative feedback produces the winner-take-all competition in recurrent neural networks. IEEE Trans. Neural Netw. Learn. Syst. **24**(2), 301–309 (2012)
6. Li, S., Lou, Y., Liu, B.: Bluetooth aided mobile phone localization: a nonlinear neural circuit approach. ACM Trans. Embedded Comput. Syst. (TECS) **13**(4), 1–15 (2014)
7. Li, S., He, J., Li, Y., Rafique, M.U.: Distributed recurrent neural networks for cooperative control of manipulators: a game-theoretic perspective. IEEE Trans. Neural Netw. Learn. Syst. **28**(2), 415–426 (2016)
8. Mohajerin, N., Waslander, S.L.: Multistep prediction of dynamic systems with recurrent neural networks. IEEE Trans. Neural Netw. Learn. Syst. **30**(11), 3370–3383 (2019)
9. Li, X., She, K., Zhong, S., Shi, K., Kang, W., Cheng, J., Yu, Y.: Extended robust global exponential stability for uncertain switched memristor-based neural networks with time-varying delays. Appl. Math. Comput. **325**, 271–290 (2018)
10. Sheng, Y., Lewis, F.L., Zeng, Z.: Exponential stabilization of fuzzy memristive neural networks with hybrid unbounded time-varying delays. IEEE Trans. Neural Netw. Learn. Syst. **30**(3), 739–750 (2018)
11. Zhang, G., Zeng, Z.: Stabilization of second-order memristive neural networks with mixed time delays via nonreduced order. IEEE Trans. Neural Netw. Learn. Syst. **31**(2), 700–706 (2019)
12. Ali, M.S., Saravanan, S.: Finite-time stability for memristor based switched neural networks with time-varying delays via average dwell time approach. Neurocomputing **275**, 1637–1649 (2018)
13. Wang, L., Shen, Y., Zhang, G.: Finite-time stabilization and adaptive control of memristor-based delayed neural networks. IEEE Trans. Neural Netw. Learn. Syst. **28**(11), 2648–2659 (2016)
14. Wu, A., Zeng, Z.: Global mittag-leffler stabilization of fractional-order memristive neural networks. IEEE Trans. Neural Netw. Learn. Syst. **28**(1), 206–217 (2015)
15. Gao, J., Zhu, P., Alsaedi, A., Alsaadi, F.E., Hayat, T.: A new switching control for finite-time synchronization of memristor-based recurrent neural networks. Neural Netw. **86**, 1–9 (2017)
16. Li, C., Lian, J., Wang, Y.: Stability of switched memristive neural networks with impulse and stochastic disturbance. Neurocomputing **275**, 2565–2573 (2018)

17. Mahfouf, M., Abbod, M.F., Linkens, D.A.: Online elicitation of mamdani-type fuzzy rules via tsk-based generalized predictive control. IEEE Trans. Syst. Man Cybern. Part B (Cybernetics) **33**(3), 465–475 (2003)
18. Mamdani, E.H.: Application of fuzzy algorithms for control of simple dynamic plant. In: Proceedings of the Institution of Electrical Engineers, vol. 121, pp. 1585–1588. IET (1974)
19. Takagi, T., Sugeno, M.: Fuzzy identification of systems and its applications to modeling and control. IEEE Trans. Syst. Man Cybern. **1**, 116–132 (1985)
20. Jang, J.S., Sun, C.T.: Neuro-fuzzy modeling and control. Proc. IEEE **83**(3), 378–406 (1995)
21. Yang, W., Yu, W., Cao, J.: Global exponential stability of impulsive fuzzy high-order bam neural networks with continuously distributed delays. IEEE Trans. Neural Netw. Learn. Syst. **29**(8), 3682–3700 (2017)
22. Shi, K., Wang, J., Tang, Y., Zhong, S.: Reliable asynchronous sampled-data filtering of t-s fuzzy uncertain delayed neural networks with stochastic switched topologies. Fuzzy Sets Syst. **381**, 1–25 (2020)
23. Wang, J.L., Wu, H.N., Huang, T., Ren, S.Y.: Analysis and Control of Output Synchronization for Complex Dynamical Networks. Springer (2019)
24. Syed Ali, M., Balasubramaniam, P., Zhu, Q.: Stability of stochastic fuzzy bam neural networks with discrete and distributed time-varying delays. Int. J. Mach. Learn. Cybern. **8**(1), 263–273 (2017)
25. Ali, M.S., Narayanan, G., Shekher, V., Alsulami, H., Saeed, T.: Dynamic stability analysis of stochastic fractional-order memristor fuzzy bam neural networks with delay and leakage terms. Appl. Math. Comput. **369**, 124896 (2020)
26. Chen, C.H., Liu, C.B.: Reinforcement learning-based differential evolution with cooperative coevolution for a compensatory neuro-fuzzy controller. IEEE Trans. Neural Netw. Learn. Syst. **29**(10), 4719–4729 (2017)
27. Mansoori, A., Eshaghnezhad, M., Effati, S.: Recurrent neural network model: a new strategy to solve fuzzy matrix games. IEEE Trans. Neural Netw. Learn. Syst. **30**(8), 2538–2547 (2019)
28. Wu, A., Zeng, Z.: Exponential passivity of memristive neural networks with time delays. Neural Netw. **49**, 11–18 (2014)
29. Jiang, P., Zeng, Z., Chen, J.: Almost periodic solutions for a memristor-based neural networks with leakage, time-varying and distributed delays. Neural Netw. **68**, 34–45 (2015)
30. Zhang, G., Zeng, Z.: Exponential stability for a class of memristive neural networks with mixed time-varying delays. Appl. Math. Comput. **321**, 544–554 (2018)
31. Wu, A., Zeng, Z.: Lagrange stability of memristive neural networks with discrete and distributed delays. IEEE Trans. Neural Netw. Learn. Syst. **25**(4), 690–703 (2013)
32. Zhang, G., Zeng, Z., Hu, J.: New results on global exponential dissipativity analysis of memristive inertial neural networks with distributed time-varying delays. Neural Netw. **97**, 183–191 (2018)
33. Zhang, J., Li, A., Lu, W.D., Sun, J.: Stabilization of mode-dependent impulsive hybrid systems driven by dfa with mixed-mode effects. IEEE Trans. Neural Netw. Learn. Syst. **31**(5), 1616–1625 (2019)
34. Wheeler, D.W., Schieve, W.: Stability and chaos in an inertial two-neuron system. Physica D **105**(4), 267–284 (1997)
35. Wan, P., Jian, J.: Passivity analysis of memristor-based impulsive inertial neural networks with time-varying delays. ISA Trans. **74**, 88–98 (2018)
36. Li, N., Zheng, W.X.: Synchronization criteria for inertial memristor-based neural networks with linear coupling. Neural Netw. **106**, 260–270 (2018)
37. Gong, S., Yang, S., Guo, Z., Huang, T.: Global exponential synchronization of inertial memristive neural networks with time-varying delay via nonlinear controller. Neural Netw. **102**, 138–148 (2018)
38. Xiao, Q., Huang, T., Zeng, Z.: Global exponential stability and synchronization for discrete-time inertial neural networks with time delays: a timescale approach. IEEE Trans. Neural Netw. Learn. Syst. **30**(6), 1854–1866 (2018)

39. Yang, S., Guo, Z., Wang, J.: Robust synchronization of multiple memristive neural networks with uncertain parameters via nonlinear coupling. IEEE Trans. Syst. Man Cybernet.: Syst. **45**(7), 1077–1086 (2015)
40. Xiao, J., Zhong, S., Li, Y.: New passivity criteria for memristive uncertain neural networks with leakage and time-varying delays. ISA Trans. **59**, 133–148 (2015)
41. Liu, Y., Li, C., Huang, T., Wang, X.: Robust adaptive lag synchronization of uncertain fuzzy memristive neural networks with time-varying delays. Neurocomputing **190**, 188–196 (2016)
42. Gu, Y., Yu, Y., Wang, H.: Synchronization for fractional-order time-delayed memristor-based neural networks with parameter uncertainty. J. Franklin Inst. **353**(15), 3657–3684 (2016)
43. Chen, H., Kang, W., Zhong, S.: A new global robust stability condition for uncertain neural networks with discrete and distributed delays. Int. J. Mach. Learn. Cybern. **10**, 1025–1035 (2019)
44. Zhu, W., Wang, D., Liu, L., Feng, G.: Event-based impulsive control of continuous-time dynamic systems and its application to synchronization of memristive neural networks. IEEE Trans. Neural Netw. Learn. Syst. **29**(8), 3599–3609 (2017)
45. Duan, S., Wang, H., Wang, L., Huang, T., Li, C.: Impulsive effects and stability analysis on memristive neural networks with variable delays. IEEE Trans. Neural Netw. Learn. Syst. **28**(2), 476–481 (2016)
46. Bao, H., Park, J.H., Cao, J.: Exponential synchronization of coupled stochastic memristor-based neural networks with time-varying probabilistic delay coupling and impulsive delay. IEEE Trans. Neural Netw. Learn. Syst. **27**(1), 190–201 (2015)
47. Yang, X., Lu, J., Ho, D.W., Song, Q.: Synchronization of uncertain hybrid switching and impulsive complex networks. Appl. Math. Model. **59**, 379–392 (2018)
48. Zhang, W., Tang, Y., Miao, Q., Du, W.: Exponential synchronization of coupled switched neural networks with mode-dependent impulsive effects. IEEE Trans. Neural Netw. Learn. Syst. **24**(8), 1316–1326 (2013)
49. Zhang, H., Zhang, W., Miao, Q., Cui, Y.: Synchronization of switched coupled neural networks with distributed impulsive effects: an impulsive strength dependent approach. Neural Process. Lett. **50**, 515–529 (2019)
50. Yang, X., Li, X., Lu, J., Cheng, Z.: Synchronization of time-delayed complex networks with switching topology via hybrid actuator fault and impulsive effects control. IEEE Trans. Cybern. **50**(9), 4043–4052 (2019)
51. Guo, Z., Wang, J., Yan, Z.: Global exponential dissipativity and stabilization of memristor-based recurrent neural networks with time-varying delays. Neural Netw. **48**, 158–172 (2013)
52. Xin, Y., Li, Y., Cheng, Z., Huang, X.: Global exponential stability for switched memristive neural networks with time-varying delays. Neural Netw. **80**, 34–42 (2016)
53. Xiao, Q., Huang, T., Zeng, Z.: Passivity and passification of fuzzy memristive inertial neural networks on time scales. IEEE Trans. Fuzzy Syst. **26**(6), 3342–3355 (2018)
54. Feng, G.: Analysis and Synthesis of Fuzzy Control Systems: A Model-Based Approach. CRC Press (2018)
55. Li, C., Feng, G., Huang, T.: On hybrid impulsive and switching neural networks. IEEE Trans. Syst. Man Cybern. Part B (Cybernetics) **38**(6), 1549–1560 (2008)
56. Liu, X., Zhong, S., Ding, X.: Robust exponential stability of nonlinear impulsive switched systems with time-varying delays. Nonlinear Anal.: Modell. Control **17**(2), 210–222 (2012)
57. He, G., Fang, J.a., Li, Z.: Synchronization of hybrid impulsive and switching dynamical networks with delayed impulses. Nonlinear Dyn. **83**, 187–199 (2016)
58. Zhang, J., Li, C., Huang, T.: Finite-time synchronization of coupled switched impulsive neural networks. In: 2017 13th IEEE International Conference on Control & Automation (ICCA), pp. 577–581. IEEE (2017)
59. Ren, W., Xiong, J.: Exponential stability of stochastic impulsive switched delayed systems based on vector lyapunov functions. In: 2017 11th Asian Control Conference (ASCC), pp. 1888–1893. IEEE (2017)
60. Zha, L., Fang, J.a., Li, X., Liu, J.: Event-triggered output feedback h_∞ control for networked markovian jump systems with quantizations. Nonlinear Anal.: Hybrid Syst. **24**, 146–158 (2017)

61. Lu, J., Ho, D.W., Cao, J.: A unified synchronization criterion for impulsive dynamical networks. Automatica **46**(7), 1215–1221 (2010)
62. Chen, W.H., Zheng, W.X.: Global exponential stability of impulsive neural networks with variable delay: An lmi approach. IEEE Trans. Circuits Syst. I Regul. Pap. **56**(6), 1248–1259 (2008)
63. Wu, A., Zeng, Z.: Exponential stabilization of memristive neural networks with time delays. IEEE Trans. Neural Netw. Learn. Syst. **23**(12), 1919–1929 (2012)
64. Wang, H., Dong, R., Xue, A., Peng, Y.: Event-triggered $l_2 - l_\infty$ state estimation for discrete-time neural networks with sensor saturations and data quantization. J. Franklin Inst. **356**(17), 10216–10240 (2019)
65. Li, N., Cao, J.: New synchronization criteria for memristor-based networks: adaptive control and feedback control schemes. Neural Netw. **61**, 1–9 (2015)
66. Wang, L., Shen, Y., Yin, Q., Zhang, G.: Adaptive synchronization of memristor-based neural networks with time-varying delays. IEEE Trans. Neural Netw. Learn. Syst. **26**(9), 2033–2042 (2014)
67. Zhang, W., Qi, J., He, X.: Input-to-state stability of impulsive inertial memristive neural networks with time-varying delayed. J. Franklin Inst. **355**(17), 8971–8988 (2018)
68. Zhang, W., Huang, T., He, X., Li, C.: Global exponential stability of inertial memristor-based neural networks with time-varying delays and impulses. Neural Netw. **95**, 102–109 (2017)
69. Wang, L., Zeng, Z., Ge, M.F., Hu, J.: Global stabilization analysis of inertial memristive recurrent neural networks with discrete and distributed delays. Neural Netw. **105**, 65–74 (2018)
70. Huang, Z., Cao, J., Raffoul, Y.N.: Hilger-type impulsive differential inequality and its application to impulsive synchronization of delayed complex networks on time scales. Science China Inf. Sci. **61**(7), 78201 (2018)
71. Wang, H., Duan, S., Li, C., Wang, L., Huang, T.: Exponential stability analysis of delayed memristor-based recurrent neural networks with impulse effects. Neural Comput. Appl. **28**, 669–678 (2017)
72. He, W., Qian, F., Cao, J.: Pinning-controlled synchronization of delayed neural networks with distributed-delay coupling via impulsive control. Neural Netw. **85**, 1–9 (2017)
73. Qi, J., Li, C., Huang, T.: Stability of inertial bam neural network with time-varying delay via impulsive control. Neurocomputing **161**, 162–167 (2015)
74. Ali, M.S., Zhu, Q., Pavithra, S., Gunasekaran, N.: A study on $\langle(\mathcal{Q}, \mathcal{S}, \mathcal{R}) - \gamma\rangle$-dissipative synchronisation of coupled reaction-diffusion neural networks with time-varying delays. Int. J. Syst. Sci. **49**(4), 755–765 (2018)
75. Ali, M.S., Gunasekaran, N., Agalya, R., Joo, Y.H.: Non-fragile synchronisation of mixed delayed neural networks with randomly occurring controller gain fluctuations. Int. J. Syst. Sci. **49**(16), 3354–3364 (2018)
76. Gunasekaran, N., Saravanakumar, R., Joo, Y.H., Kim, H.S.: Finite-time synchronization of sampled-data T - S fuzzy complex dynamical networks subject to average dwell-time approach. Fuzzy Sets Syst. **374**, 40–59 (2019)
77. Gunasekaran, N., Zhai, G., Yu, Q.: Sampled-data synchronization of delayed multi-agent networks and its application to coupled circuit. Neurocomputing **413**, 499–511 (2020)
78. Ali, M.S., Palanisamy, L., Gunasekaran, N., Alsaedi, A., Ahmad, B.: Finite-time exponential synchronization of reaction-diffusion delayed complex-dynamical networks. Discrete & Contin. Dyn. Syst. **14**(4), 1465–1477 (2021)
79. Zhu, Q., Cao, J.: Mean-square exponential input-to-state stability of stochastic delayed neural networks. Neurocomputing **131**, 157–163 (2014)
80. Shi, K., Wang, J., Zhong, S., Tang, Y., Cheng, J.: Hybrid-driven finite-time H∞ sampling synchronization control for coupling memory complex networks with stochastic cyber attacks. Neurocomputing **387**, 241–254 (2020)
81. Lv, X., Li, X., Cao, J., Duan, P.: Exponential synchronization of neural networks via feedback control in complex environment. Complexity, pp. 1–13 (2018)

82. Hu, J., Sui, G., Lv, X., Li, X.: Fixed-time control of delayed neural networks with impulsive perturbations. Nonlinear Analy.: Modell. Control **23**(6), 904–920 (2018)
83. Karthick, S., Sakthivel, R., Alzahrani, F., Leelamani, A.: Synchronization of semi-markov coupled neural networks with impulse effects and leakage delay. Neurocomputing **386**, 221–231 (2020)
84. Karthick, S., Sakthivel, R., Wang, C., Ma, Y.K.: Synchronization of coupled memristive neural networks with actuator saturation and switching topology. Neurocomputing **383**, 138–150 (2020)
85. Mahmoud, G.M., Farghaly, A.A., Abed-Elhameed, T.M., Darwish, M.M.: Adaptive dual synchronization of chaotic (hyperchaotic) complex systems with uncertain parameters and its application in image encryption. Acta Phys. Pol., B **49**(11), 1923–1939 (2018)
86. Hassan, M.F., Hammuda, M.: A new approach for constrained chaos synchronization with application to secure data communication. J. Franklin Inst. **356**(12), 6697–6723 (2019)
87. Pyragienė, T., Pyragas, K.: Anticipating synchronization in a chain of chaotic oscillators with switching parameters. Phys. Lett. A **379**(47), 3084–3088 (2015)
88. Wang, Q., Wang, J., Huang, Y., Ren, S.: Generalized lag synchronization of multiple weighted complex networks with and without time delay. J. Franklin Inst. **355**(14), 6597–6616 (2018)
89. Bao, H., Cao, J.: Projective synchronization of fractional-order memristor-based neural networks. Neural Netw. **63**, 1–9 (2015)
90. Wang, Z., Wu, H.: Projective synchronization in fixed time for complex dynamical networks with nonidentical nodes via second-order sliding mode control strategy. J. Franklin Inst. **355**(15), 7306–7334 (2018)
91. Chen, C., Ding, Z.: Projective synchronization of nonidentical fractional-order memristive neural networks. Discrete Dynamics in Nature and Society (2019)
92. Zhang, H., Wang, X.: Complex projective synchronization of complex-valued neural network with structure identification. J. Franklin Inst. **354**(12), 5011–5025 (2017)
93. Ding, S., Wang, Z., Zhang, H.: Quasi-synchronization of delayed memristive neural networks via region-partitioning-dependent intermittent control. IEEE Trans. Cybern. **49**(12), 4066–4077 (2019)
94. Shi, K., Tang, Y., Zhong, S., Yin, C., Huang, X., Wang, W.: Nonfragile asynchronous control for uncertain chaotic lurie network systems with bernoulli stochastic process. Int. J. Robust Nonlinear Control **28**(5), 1693–1714 (2018)
95. Yang, N., Yu, Y., Zhong, S., Wang, X., Shi, K., Cai, J.: Exponential stability of markovian jumping memristor-based neural networks via event-triggered impulsive control scheme. IEEE Access **8**, 32564–32574 (2020)
96. Li, X., O'Regan, D., Akca, H.: Global exponential stabilization of impulsive neural networks with unbounded continuously distributed delays. IMA J. Appl. Math. **80**(1), 80–99 (2015)
97. Liu, Y., Guo, B., Park, J.H., Lee, S.: Nonfragile exponential synchronization of delayed complex dynamical networks with memory sampled-data control. IEEE Trans. Neural Netw. Learn. Syst. **29**(1), 118–128 (2018)
98. Shi, K., Wang, J., Zhong, S., Tang, Y., Cheng, J.: Non-fragile memory filtering of T–S fuzzy delayed neural networks based on switched fuzzy sampled-data control. Fuzzy Sets Syst. **394**, 40–64 (2020)
99. Cai, X., Wang, J., Zhong, S., Shi, K., Tang, Y.: Fuzzy quantized sampled-data control for extended dissipative analysis of $T-S$ fuzzy system and its application to wpgss. J. Franklin Inst. **358**(2), 1350–1375 (2021)
100. Luo, J., Li, M., Liu, X., Tian, W., Zhong, S., Shi, K.: Stabilization analysis for fuzzy systems with a switched sampled-data control. J. Franklin Inst. **357**, 39–58 (2020)
101. Wang, H., Duan, S., Huang, T., Tan, J.: Synchronization of memristive delayed neural networks via hybrid impulsive control. Neurocomputing **267**, 615–623 (2017)
102. Tang, H., Wang, J., Wang, L., Hu, X., Zhou, Y., Duan, S.: Impulsive control for passivity and exponential synchronization of coupled neural networks with multiple weights. J. Franklin Inst. **356**(10), 5434–5463 (2019)

103. Tang, Z., Park, J.H., Feng, J.: Impulsive effects on quasi-synchronization of neural networks with parameter mismatches and time-varying delay. IEEE Trans. Neural Netw. Learn. Syst. **29**(4), 908–919 (2018)
104. Kumar, R., Sarkar, S., Das, S., Cao, J.: Projective synchronization of delayed neural networks with mismatched parameters and impulsive effects. IEEE Trans. Neural Netw. Learn. Syst. 1–11 (2019)
105. Wen, S., Zeng, Z., Huang, T.: Dynamic behaviors of memristor-based delayed recurrent networks. Neural Comput. Appl. **23**, 815–821 (2013)
106. Guan, Z., Liu, Z., Feng, G., Wang, Y.: Synchronization of complex dynamical networks with time-varying delays via impulsive distributed control. IEEE Trans. Circuits Syst. I Regul. Pap. **57**(8), 2182–2195 (2010)
107. Li, M., Li, X., Han, X., Qiu, J.: Leader-following synchronization of coupled time-delay neural networks via delayed impulsive control. Neurocomputing **357**, 101–107 (2019)
108. Qin, X., Wang, C., Li, L., Peng, H., Yang, Y., Ye, L.: Finite-time projective synchronization of memristor-based neural networks with leakage and time-varying delays. Physica A **531**, 121788 (2019)
109. Ali, M.S.: Stability analysis of markovian jumping stochastic cohen-grossberg neural networks with discrete and distributed time varying delays. Chin. Phys. B **23**(6), 060702 (2014)
110. Zhang, G., Hu, J., Zeng, Z.: New criteria on global stabilization of delayed memristive neural networks with inertial item. IEEE Trans. Cybern. **50**(6), 2770–2780 (2019)
111. Guo, Z., Gong, S., Wen, S., Huang, T.: Event-based synchronization control for memristive neural networks with time-varying delay. IEEE Trans. Cybern. **49**(9), 3268–3277 (2018)
112. Gong, S., Guo, Z., Wen, S., Huang, T.: Finite-time and fixed-time synchronization of coupled memristive neural networks with time delay. IEEE Trans. Cybern. **51**(6), 2944–2955 (2019)
113. Cao, Y., Wang, S., Guo, Z., Huang, T., Wen, S.: Synchronization of memristive neural networks with leakage delay and parameters mismatch via event-triggered control. Neural Netw. **119**, 178–189 (2019)
114. Li, Y., Luo, B., Liu, D., Yang, Z.: Robust synchronization of memristive neural networks with strong mismatch characteristics via pinning control. Neurocomputing **289**, 144–154 (2018)
115. Xin, Y., Li, Y., Huang, X., Cheng, Z.: Quasi-synchronization of delayed chaotic memristive neural networks. IEEE Trans. Cybern. **49**(2), 712–718 (2017)
116. Ali, M.S.: Novel delay-dependent stability analysis of takagi-sugeno fuzzy uncertain neural networks with time varying delays. Chin. Phys. B **21**(7), 070207 (2012)
117. Ali, M.S., Saravanakumar, R.: Improved delay-dependent robust h_∞ control of an uncertain stochastic system with interval time-varying and distributed delays. Chin. Phys. B **23**(12), 120201 (2014)
118. Abdurahman, A., Jiang, H., Teng, Z.: Exponential lag synchronization for memristor-based neural networks with mixed time delays via hybrid switching control. J. Franklin Inst. **353**(13), 2859–2880 (2016)
119. Ali, M.S., Gunasekaran, N., Rani, M.E.: Robust stability of hopfield delayed neural networks via an augmented lk functional. Neurocomputing **234**, 198–204 (2017)
120. Fan, Y., Huang, X., Li, Y., Xia, J., Chen, G.: Aperiodically intermittent control for quasi-synchronization of delayed memristive neural networks: an interval matrix and matrix measure combined method. IEEE Trans. Syst. Man Cybern.: Syst. **49**(11), 2254–2265 (2018)
121. Bohner, M., Peterson, A.: Dynamic Equations on Time Scales: An Introduction with Applications. Springer Science & Business Media (2001)
122. Agarwal, R., O'Regan, D., Saker, S., et al.: Dynamic Inequalities on Time Scales, vol. 2014. Springer (2014)
123. Bohner, M., Georgiev, S.G., et al.: Multivariable Dynamic Calculus on Time Scales. Springer (2016)
124. Liu, X., Zhang, K.: Synchronization of linear dynamical networks on time scales: pinning control via delayed impulses. Automatica **72**, 147–152 (2016)
125. Xiao, Q., Zeng, Z.: Scale-limited lagrange stability and finite-time synchronization for memristive recurrent neural networks on time scales. IEEE Trans. Cybern. **47**(10), 2984–2994 (2017)

126. Lu, X., Zhang, X., Liu, Q.: Finite-time synchronization of nonlinear complex dynamical networks on time scales via pinning impulsive control. Neurocomputing **275**, 2104–2110 (2018)
127. Xiao, Q., Huang, T., Zeng, Z.: Stabilization of nonautonomous recurrent neural networks with bounded and unbounded delays on time scales. IEEE Trans. Cybern. **50**(10), 4307–4317 (2019)
128. Taousser, F.Z., Defoort, M., Djemai, M., Djouadi, S.M., Tomsovic, K.: Stability analysis of a class of switched nonlinear systems using the time scale theory. Nonlinear Anal. Hybrid Syst **33**, 195–210 (2019)
129. Huang, Z., Cao, J., Li, J., Bin, H.: Quasi-synchronization of neural networks with parameter mismatches and delayed impulsive controller on time scales. Nonlinear Anal. Hybrid Syst **33**, 104–115 (2019)
130. Liu, X., Ho, D.W., Song, Q., Xu, W.: Finite/fixed-time pinning synchronization of complex networks with stochastic disturbances. IEEE Trans. Cybern. **49**(6), 2398–2403 (2018)
131. Yueying, W., Shen, H., Reza, K., Dengping, D.: Dissipativity-based fuzzy integral sliding mode control of continuous-time ts fuzzy system. IEEE Trans. Fuzzy Syst. (2017)
132. Li, Y., Tong, S.: Adaptive fuzzy output-feedback stabilization control for a class of switched nonstrict-feedback nonlinear systems. IEEE Trans. Cybern. **47**(4), 1007–1016 (2016)
133. Liu, X., Zhang, K., Xie, W.C.: Pinning impulsive synchronization of reaction-diffusion neural networks with time-varying delays. IEEE Trans. Neural Netw. Learn. Syst. **28**(5), 1055–1067 (2016)
134. Zhou, Y., Zeng, Z.: Event-triggered impulsive control on quasi-synchronization of memristive neural networks with time-varying delays. Neural Netw. **110**, 55–65 (2019)
135. Wang, C.: Almost periodic solutions of impulsive bam neural networks with variable delays on time scales. Commun. Nonlinear Sci. Numer. Simul. **19**(8), 2828–2842 (2014)
136. Chen, Y., Wang, Z., Yuan, Y., Date, P.: Distributed h_∞ filtering for switched stochastic delayed systems over sensor networks with fading measurements. IEEE Trans. Cybern. **50**(1), 2–14 (2018)
137. Li, Q., Shen, B., Wang, Z., Huang, T., Luo, J.: Synchronization control for a class of discrete time-delay complex dynamical networks: A dynamic event-triggered approach. IEEE Trans. Cybern. **49**(5), 1979–1986 (2018)
138. Yan, H., Zhang, H., Yang, F., Zhan, X., Peng, C.: Event-triggered asynchronous guaranteed cost control for markov jump discrete-time neural networks with distributed delay and channel fading. IEEE Trans. Neural Netw. Learn. Syst. **29**(8), 3588–3598 (2017)
139. Strukov, D.B., Snider, G.S., Stewart, D.R., Williams, R.S.: The missing memristor found. Nature **453**(7191), 80–83 (2008)
140. Wen, S., Bao, G., Zeng, Z., Chen, Y., Huang, T.: Global exponential synchronization of memristor-based recurrent neural networks with time-varying delays. Neural Netw. **48**, 195–203 (2013)
141. Guo, Z., Yang, S., Wang, J.: Global exponential synchronization of multiple memristive neural networks with time delay via nonlinear coupling. IEEE Trans. Neural Netw. Learn. Syst. **26**(6), 1300–1311 (2014)
142. Pershin, Y.V., Di Ventra, M.: On the validity of memristor modeling in the neural network literature. Neural Netw. **121**, 52–56 (2020)
143. Xiao, M., Zheng, W.X., Jiang, G., Cao, J.: Qualitative analysis and bifurcation in a neuron system with memristor characteristics and time delay. IEEE Trans. Neural Netw. Learn. Syst. **32**(5), 1974–1988 (2020)
144. Li, L., Xu, R., Gan, Q., Lin, J.: Synchronization of a novel model for memristive neural networks via sliding mode control. ISA Trans. **106**, 31–39 (2020)
145. Zhang, W., Li, C., Huang, T., Huang, J.: Stability and synchronization of memristor-based coupling neural networks with time-varying delays via intermittent control. Neurocomputing **173**, 1066–1072 (2016)
146. Chen, W., Huang, Y., Ren, S.: Passivity of coupled memristive delayed neural networks with fixed and adaptive coupling weights. Neurocomputing **313**, 346–363 (2018)

147. Li, J., Jiang, H., Hu, C., Alsaedi, A.: Finite/fixed-time synchronization control of coupled memristive neural networks. J. Franklin Inst. **356**(16), 9928–9952 (2019)
148. Yang, X., Cao, J., Qiu, J.: Pth moment exponential stochastic synchronization of coupled memristor-based neural networks with mixed delays via delayed impulsive control. Neural Netw. **65**, 80–91 (2015)
149. Zhang, W., Li, C., Huang, T., He, X.: Synchronization of memristor-based coupling recurrent neural networks with time-varying delays and impulses. IEEE Trans. Neural Netw. Learn. Syst. **26**(12), 3308–3313 (2015)
150. Lv, X., Cao, J., Li, X., Abdel-Aty, M., Al-Juboori, U.A.: Synchronization analysis for complex dynamical networks with coupling delay via event-triggered delayed impulsive control. IEEE Trans. Cybern. **51**(11), 5269–5278 (2020)
151. Tang, Z., Park, J.H., Wang, Y., Feng, J.: Impulsive synchronization of derivative coupled neural networks with cluster-tree topology. IEEE Trans. Netw. Sci. Eng. **7**(3), 1788–1798 (2019)
152. Yang, S., Guo, Z., Wang, J.: Global synchronization of multiple recurrent neural networks with time delays via impulsive interactions. IEEE Trans. Neural Netw. Learn. Syst. **28**(7), 1657–1667 (2016)
153. Lv, X., Li, X., Cao, J., Perc, M.: Dynamical and static multisynchronization of coupled multistable neural networks via impulsive control. IEEE Trans. Neural Netw. Learn. Syst. **29**(12), 6062–6072 (2018)
154. Chen, J., Chen, B., Zeng, Z., Jiang, P.: Effects of subsystem and coupling on synchronization of multiple neural networks with delays via impulsive coupling. IEEE Trans. Neural Netw. Learn. Syst. **30**(12), 3748–3758 (2019)
155. Yang, X., Ho, D.W.: Synchronization of delayed memristive neural networks: robust analysis approach. IEEE Trans. Cybern. **46**(12), 3377–3387 (2015)
156. Dong, H., Wang, Z., Ho, D.W., Gao, H.: Robust h_∞ fuzzy output-feedback control with multiple probabilistic delays and multiple missing measurements. IEEE Trans. Fuzzy Syst. **18**(4), 712–725 (2010)
157. Liu, D., Zhu, S., Sun, K.: Global anti-synchronization of complex-valued memristive neural networks with time delays. IEEE Trans. Cybern. **49**(5), 1735–1747 (2018)
158. Dai, H., Si, G., Zhang, Y.: Adaptive generalized function matrix projective lag synchronization of uncertain complex dynamical networks with different dimensions. Nonlinear Dyn. **74**, 629–648 (2013)
159. Fan, A., Li, J.: Adaptive neural network prescribed performance matrix projection synchronization for unknown complex dynamical networks with different dimensions. Neurocomputing **281**, 55–66 (2018)
160. Zhang, C., Wang, X., Ye, X., Zhou, S., Feng, L.: Robust modified function projective lag synchronization between two nonlinear complex networks with different-dimensional nodes and disturbances. ISA Trans. **101**, 42–49 (2020)
161. Zhang, D., Mei, J., Miao, P.: Global finite-time synchronization of different dimensional chaotic systems. Appl. Math. Model. **48**, 303–315 (2017)
162. Zheng, S.: Adaptive modified function projective synchronization of unknown chaotic systems with different order. Appl. Math. Comput. **218**(10), 5891–5899 (2012)
163. Hu, X., Feng, G., Duan, S., Liu, L.: A memristive multilayer cellular neural network with applications to image processing. IEEE Trans. Neural Netw. Learn. Syst. **28**(8), 1889–1901 (2016)
164. Wang, Q., Wang, J.L.: Finite-time output synchronization of undirected and directed coupled neural networks with output coupling. IEEE Trans. Neural Netw. Learn. Syst. **32**(5), 2117–2128 (2020)
165. Xiong, J.J., Zhang, G.B., Wang, J.X., Yan, T.H.: Improved sliding mode control for finite-time synchronization of nonidentical delayed recurrent neural networks. IEEE Trans. Neural Netw. Learn. Syst. **31**(6), 2209–2216 (2019)

Chapter 8
Memristor-Based Neural Network and Its Application

Memristive neural networks have emerged as a promising paradigm for implementing intelligent systems, leveraging the unique properties of memristors to emulate synaptic behavior and enable efficient neuromorphic computing. This chapter explores the application of memristive neural networks in four advanced tasks, showcasing their potential to revolutionize various domains, including image processing, computer vision, and information retrieval.

Section 8.1 focuses on the application of memristor crossbar-based multilayer cellular neural networks (MCM-CNNs) in edge detection. By incorporating a memristor crossbar composed of a pair of memristors, an MCM-CNN is designed to exploit the programmability and compactness of memristors for efficient edge detection in both binary and color images. The figure of merit (FOM) is used to evaluate the performance of the proposed MCM-CNN structure against traditional edge detection operators, demonstrating a threefold improvement in edge detection accuracy.

In Sect. 8.2, the focus shifts to the challenge of excessive writing errors in single memristor-based crossbar arrays, which hinder the effective generation of image captions. To address this issue, a novel approach is proposed, involving the construction of a crossbar array of memristor modules, linearly mapping neural network weights to memristance values, and designing a memristive neural network based on the memristor module. The resulting neural image caption generator, based on VGG-16-LSTM and the memristor module crossbar array, achieves remarkable performance, reaching up to 99.38%.

Section 8.3 presents a novel framework that integrates a Hopfield neural network crawler with information entropy and a memristor model to optimize the crawling strategy for hypertext links. By employing a single-layer Hopfield neural network, adopting a breadth-best-first crawling strategy, designing a memristor-based activation function, and leveraging information entropy and a cubic nonlinearity memristor model, the proposed approach efficiently calculates hyperlink weights. Experimental results on Tibet development topics demonstrate the superior performance of the

Y. Yu et al., *Memristive Computing*,
https://doi.org/10.1007/978-981-96-0332-9_8

proposed crawlers in terms of topic-relevant similarity and precision compared to other crawling techniques.

The diverse applications presented in this chapter highlight the immense potential of memristive neural networks in tackling complex tasks across various domains. By harnessing the unique characteristics of memristors, such as their synaptic-like behavior and nanoscale dimensions, these applications showcase the enhanced performance, efficiency, and adaptability of memristive neural networks compared to traditional approaches. This chapter demonstrates the promising future of memristor-based neuromorphic computing and lays the foundation for further research in this rapidly evolving field.

8.1 Memristor Crossbar Array-Based Multilayer Cellular Neural Network

8.1.1 Introduction

Cellular neural networks (CNNs) are the most extensively used in image processing including noise reduction, edge detection, and character recognition [1–3]. However, most CNN models can only handle fundamental tasks such as binary image and grayscale image because they are based on single-layer CNNs [4]. Looking for suitable CNN models which can be used in more complex and multitask areas is needed.

The multilayer CNN (MCNN) is a more effective model than the single-layer CNN, which has been proved to be able to deal with many complicated tasks [5]. MCNN is actually extended from a signal-layer CNN with characteristics of real-time and high-speed parallel computing, and it is used in color image processing and image segmentation in [6, 7]. Yet, on the one hand, the lack of programmability in connecting components between cells of CNN causes design complexity. On the other hand, the circuit scale is always limited in large-scale integration due to the size and high-power consumption of traditional components. Exploring large-scale implementable, low-power consumption, high-speed MCNN models based on better components is presented in this chapter.

The crossbar structure formed by two memristors is gradually focused due to its high density and low-power consumption, and it has become the main form of ANN construction [8]. In this chapter, we continue to use this simple structure, for its strong scalability, which is a necessary condition for MCNN. Simultaneously, compared to the MCNN formed by traditional components, its circuit structure is more straightforward, the power consumption is lower, and it is easy to integrate.

8.1.2 Basics of Memristor

Memristors are similar to neurons because their resistance R can change continuously between R_{on} and R_{off}, as shown in Fig. 8.1, and the resistance can be remembered after power off [9]. The electrical characteristics of the memristor theoretically proposed by Chua illustrated the relationship between the charge and the magnetic flux [10]. Subsequently, Hewlett-Packard (HP) Labs published an article in Nature to announce that a memristor had been manufactured which gave the mathematical model and the circuit simulation curve [11]. According to achievements of HP Labs, a novel memristor model proposed in [11] is considered in this chapter and the mathematical expression is written as follows:

$$M = \frac{w(t)}{D} R_{on} + \left(1 - \frac{w(t)}{D}\right) R_{off} \tag{8.1}$$

where $w(t)$ is the width of the doped region and changes when the current is passed through the memristor; M is the memristor resistance; D is the total length of the thin film; R_{on} is the minimum resistance when $w(t) = D; R_{off}$ is the maximum resistance when $w(t) = 0$. The change rate of $w(t)$ with times is expressed as:

$$\frac{\mathrm{d}w(t)}{\mathrm{d}t} = \mu_v \frac{R_{on}}{D} i(t) \tag{8.2}$$

where μ_v denotes the dopant mobility and $i(t)$ represents the current. To avoid the second-order boundary effects at thin film edges causes the influence of nonlinearities to most fabricated devices [12], window function $f(x)$ is expressed as:

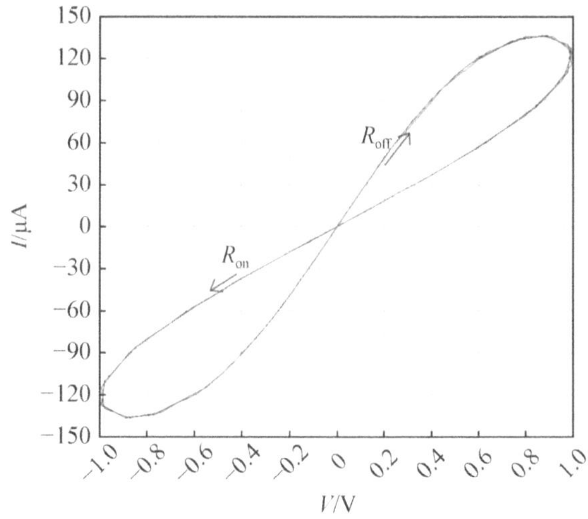

Fig. 8.1 I–V curve result of the memristor based on the proposed model simulated in LTspice

$$f(x) = 1 - (2x - 1)^{2p} \tag{8.3}$$

where p is a positive integer used to eliminate the nonlinearity. We set the appropriate parameters: $R_{\text{on}} = 1\,\text{k}\Omega$, $R_{\text{off}} = 100\,\text{k}\Omega$, $D = 10\,\text{nm}$, $\mu_v = 20\,\text{m}^2 \cdot \text{v}^{-1} \cdot \text{s}^{-1}$, $p = 10$, and the input sinusoidal voltage $y = sin(2\pi t)$. The voltage-current curve of the memristor based on the proposed model is simulated in LTspice Simulator as shown in Fig. 8.1.

8.1.3 Design of MCM-CNN

8.1.4 Crossbar Architecture of Memristor

Figure 8.2 shows an equivalent circuit diagram of a cell neuron including capacitance, resistance, control power, and independent power [13]. E, C, R_x, R_y, and I represent independent voltage source, capacitance, two resistors, and the multiple control current sources respectively.

In this chapter, replacing the weight module in the traditional CNN with a memristor crossbar can easily adjust the weight and improve the integration. Figure 8.3 shows a schematic diagram of the structure of replacing a voltage-controlled current source with a memristor crossbar. In Fig. 8.3, V_{in}, V_{out}, and G represent the input voltage, the output voltage, and the memristor conductance (reciprocal of memristor resistance) respectively. For the $M \times N$ dimensional cellular neural network, the cell in the ith row and the jth column is denoted as $C(i, j)$, and its neighbor cells are $C(k, l)$, and k and l represent the row of $C(k, l)$ with columns. V_{xij} is the state voltage of $C(i, j)$. V_{uij} is the input voltage of $C(i, j)$, V_{yij} is the output voltage of $C(i, j)$, $I_{xu(i,j;k,l)}$,$I_{xu(i,j;k,l)}$ are linear voltage-controlled current sources, and $I_{yx(i,j;k,l)}$ is a current source with piecewise-linear voltage control.

The crossbar architecture is frequently used to connect nano-devices such as memristors to reduce the number of cross-wires and achieve high density [14]. Figure 8.4 shows the circuit structure of the jth neuron in the hidden layer, which is

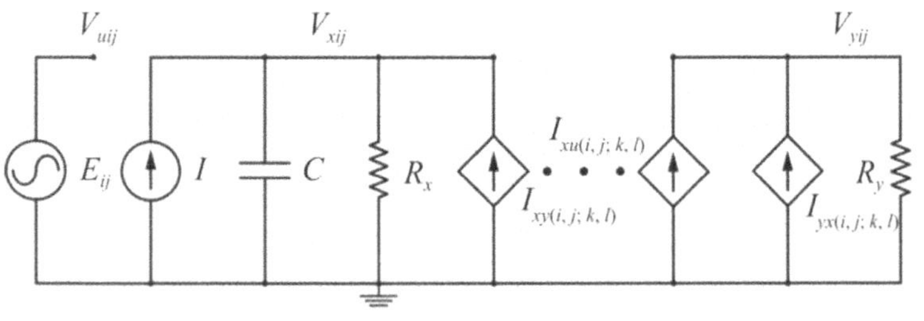

Fig. 8.2 Equivalent circuit diagram of the CNN

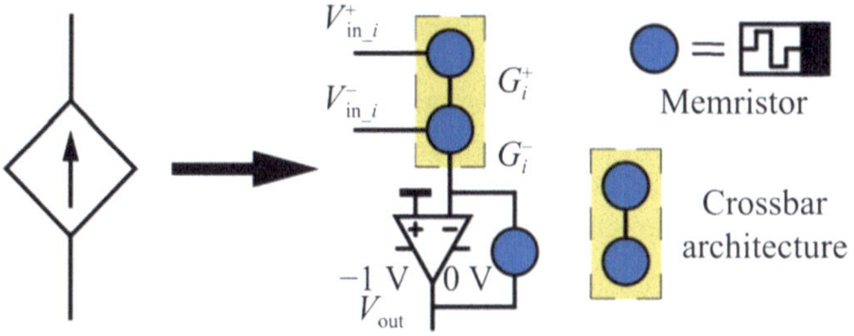

Fig. 8.3 Replacing traditional components with memristor crossbar

Fig. 8.4 The jth neuron circuit

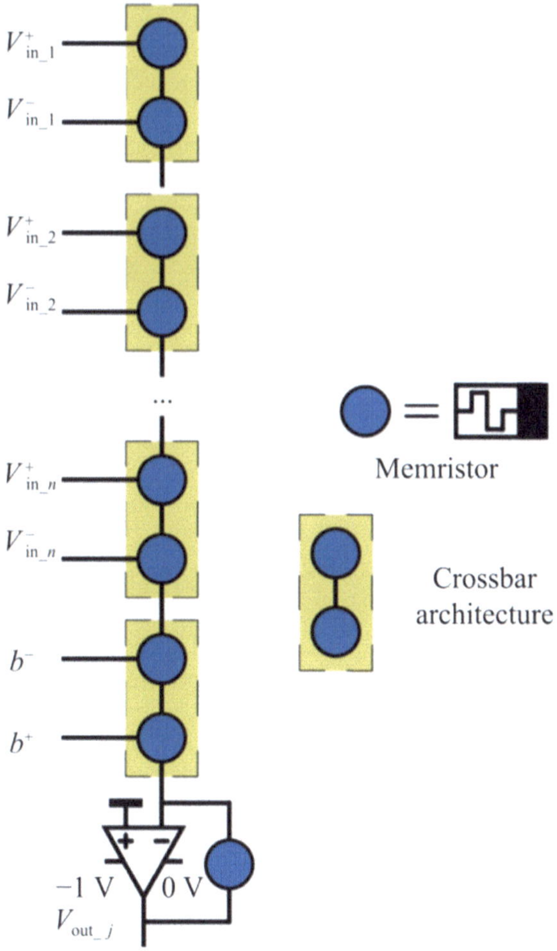

also the jth column of the memristor crossbar. For a single neuron, it accepts many input signals for the dot product and yields only one output signal.

The memristor conductive characteristic is adopted to represent the weight in the dot product operation. Considering the value of the memristor resistance is only positive, we use two memristors to store one weight together obeying some rules to satisfy the design requirement of the neural network [15]. That is, for one neuron with N inputs, there are $2N + 2$ memristors, and the first N memristors store positive values, the next N memristors store negative values, and the last two memristors are used to store positive and negative offsets, respectively. However, the traditional circuit uses two operational amplifiers, and the waveform of the output result after operation is opposite [16]. In this chapter, the optimization of the circuit structure is that the first N memristors are used to store negative values and the second N memristors are used to store positive values. Only one operational amplifier is needed to obtain positive results which can reduce the power consumption and area of the circuit.

In this section, off-chip training is adopted as a coding method for the memristor crossbar circuit. The matrix is trained by software and it is coded into the circuit according to the following formulas:

$$\begin{cases} \sigma_w^+ = \frac{\Delta\sigma}{|W|} W^+ \oplus \sigma_{\min} \\ \sigma_w^- = \frac{\Delta\sigma}{|W|} W^- \oplus \sigma_{\min} \\ \sigma_b^+ = \frac{\Delta\sigma}{|b|} b^+ \oplus \sigma_{\min} \\ \sigma_b^- = \frac{\Delta\sigma}{|b|} b^- \oplus \sigma_{\min} \\ \Delta\sigma = \sigma_{\max} - \sigma_{\min} \\ W = W^+ - W^- \\ b = b^+ - b^- \\ \sigma_{\max} = \frac{1}{R_{\text{on}}} \\ \sigma_{\min} = \frac{1}{R_{\text{off}}} \end{cases} \tag{8.4}$$

where σ_w and σ_b are the matrices of conductance values encoded by the memristor for storing weights and offsets, respectively. W^+ and W^- denote the negative and positive values of the weight matrix after training, respectively. $\oplus$ represents adding a constant to each element in the matrix. Then, the result of the dot product operation is sent to the operational amplifier to implement the activation function, and the output value of the activation function is the final result of the neuron.

8.1.5 Design of Activation Function

A piecewise-linear function shown in the following is adopted to replace the traditional sigmoid function in our proposed design:

$$y = \begin{cases} 0, \ x < -\frac{t}{2} \\ \frac{1}{t}x + \frac{1}{2}, \ |x| \leqslant \frac{t}{2} \\ 1, \ x > \frac{t}{2} \end{cases} \tag{8.5}$$

where $1/t$ is the slope and y is the output value of the activate function.

The output of the piecewise-linear function is monotonic and continuous between 0 and 1 which is similar to a sigmoid function and closest to biological neurons at the physical level. Furthermore, this piecewise-linear function can be realized with the circuit structure given in this chapter without additional circuit components, which is different from the traditional sigmoid function. Thus, the proposed design effectively simplifies the complexity of the circuit and reduces power consumption.

In the ideal case, let $t = 8$, the curve of the activate function in the circuit is shown in Fig. 8.5.

In order to simplify the calculation, the sensitive resistance value R_m of the operational amplifier is set to:

$$R_m = \frac{\max(|W|)}{t\Delta\sigma}. \tag{8.6}$$

The final output signal value V_{out_i} of the jth neuron circuit can be obtained by:

$$V_{\text{out}\_j} = \frac{1}{t}\left(\sum_{i=1}^{n} V_{\text{in}}\sigma_j + b_j\right). \tag{8.7}$$

CNN is a large-scale nonlinear analog system that can process real-time signals. Each basic circuit unit of CNN is defined as a cellular neuron, which is formed by the

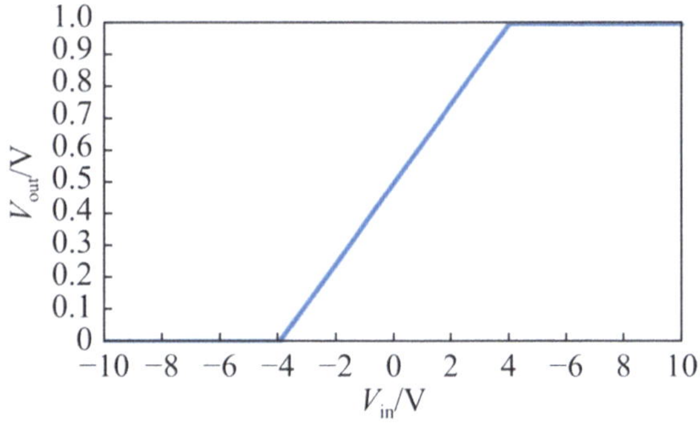

Fig. 8.5 $V_{in} - V_{out}$ curve of the piecewise-linear function

same cell neurons connected regularly in space. These cellular neurons interact only with interfacing cell neurons, each of which has states associated with input, output, and kinetic rules. However, non-adjacent cell neurons can also have an indirect effect since CNNs have continuous dynamic behavior with time.

8.1.6 Two-Dimensional CNN Structure

In this chapter, a two-dimensional CNN structure of 25 cells formed by five rows and five columns is taken as an example, as shown in Fig. 8.6a. The middlemost cell located in row i and column j is denoted as $C(i, j)$. It is associated with eight adjacent cells, with a black solid wire to indicate this connection.

MCNN is an extension of CNN, and Fig. 8.6a also shows the structure of a three-layer CNN. The cells in each layer also interact with cells in interface layers as shown in Fig. 8.6b. For CNN with M rows and N columns, its equation of state is expressed as follows:

$$C\frac{\mathrm{d}x_{ij}(t)}{\mathrm{d}t} = -\frac{1}{R}x_{ij}(t) + \sum_{C(k,l)\in N_r(i,j)} A(i, j; k, l)y_{kl}(t) + \sum_{C(k,l)\in N_r(i,j)} B(i, j; k, l)u_{kl}(t) + I \tag{8.8}$$

where

$$\boldsymbol{A} = \begin{pmatrix} A_{11} & \cdots & A_{1i} & \cdots & A_{1n} \\ \vdots & \ddots & \vdots & & \vdots \\ A_{i1} & \cdots & A_{ii} & \cdots & A_{in} \\ \vdots & & \vdots & \ddots & \vdots \\ A_{n1} & \cdots & A_{ni} & \cdots & A_{nn} \end{pmatrix}, \boldsymbol{B} = \begin{pmatrix} B_{11} & \cdots & B_{1i} & \cdots & B_{1n} \\ \vdots & \ddots & \vdots & & \vdots \\ B_{i1} & & B_{ii} & \cdots & B_{in} \\ \vdots & & \vdots & \ddots & \vdots \\ B_{n1} & \cdots & B_{ni} & \cdots & B_{nn} \end{pmatrix}, \boldsymbol{C} = \begin{bmatrix} C_1 & & & \\ & \ddots & & \\ & & C_2 & \\ & & & \ddots \\ & & & & C_n \end{bmatrix},$$

$$\boldsymbol{R} = \begin{pmatrix} R_1 & & & & \\ & \ddots & & & \\ & & R_i & & \\ & & & \ddots & \\ & & & & R_n \end{pmatrix}, \boldsymbol{x}_{ij} = \begin{pmatrix} x_{1ij} \\ x_{2ij} \\ \vdots \\ x_{nij} \end{pmatrix}, \boldsymbol{y}_{ij} = \begin{pmatrix} y_{1ij} \\ y_{2ij} \\ \vdots \\ y_{nij} \end{pmatrix}, \boldsymbol{u}_{ql} = \begin{pmatrix} x_{1ql} \\ x_{2ql} \\ \vdots \\ x_{nql} \end{pmatrix}, \boldsymbol{I} = \begin{pmatrix} I_1 \\ I_2 \\ \vdots \\ I_n \end{pmatrix}.$$

Specifically, feedback template N indicates the effect of neighborhood cell output on central cell $C(i, j)$, control template B represents the effect of domain cell input on central cell $C(i, j)$, and I represents the threshold.

Fig. 8.6 CNN structure

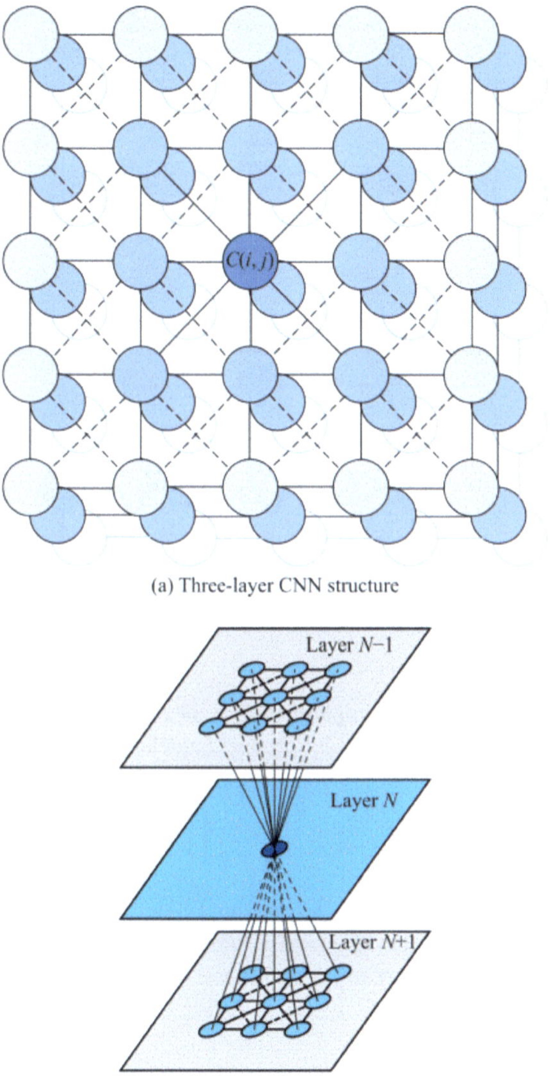

(a) Three-layer CNN structure

(b) Interaction between each layer of cells

The output equation of CNN is given by:

$$y_{ij}(t) = \frac{1}{2}\left(\left|\boldsymbol{x}_{ij}(t) + 1\right| - \left|\boldsymbol{x}_{ij}(t) - 1\right|\right),$$
$$1 \leqslant i \leqslant M;\ 1 \leqslant j \leqslant N. \tag{8.9}$$

Figure 8.7 shows a crossbar structure consisting of three rows (input signals) and three columns (output signals) without bias units. The memristor representing a

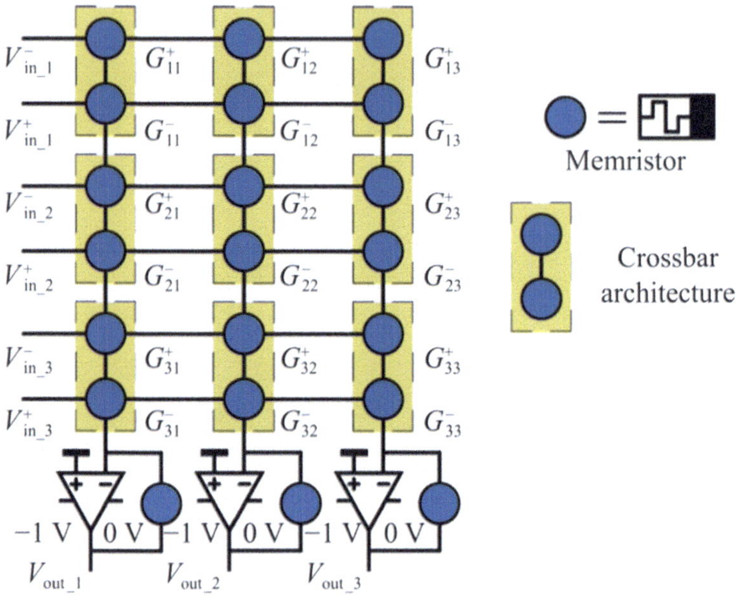

Fig. 8.7 The 3 × 3 memristor crossbar

synapse weight is implemented in the cross point of every two wires. It can perform many parallel analog dot product calculations as

$$\begin{cases} I_c = V_{\text{in}} G \\ V_{\text{out}} = I_c R_m \end{cases} \tag{8.10}$$

where V_{in}, V_{out}, and I_c are the input signal, the output signal, and the current, respectively. G denotes the memristor conductance value, and R_m is the sensitive resistance value of the operational amplifier.

8.1.7 Edge Detection Based on MCM-CNN

In this section, the experiment mainly uses the weight changeability of CNN and adjusts the appropriate feedback operator A and control operator B to apply to the edge detection of different images. The algorithm flow diagram of the experiment is shown in Fig. 8.8.

Table 8.1 lists the edge detection algorithms used in this section and their main advantages or disadvantages. Furthermore, the model proposed in this chapter is compared with existing edge detection algorithms in the following experiments to illustrate the advantages of the model.

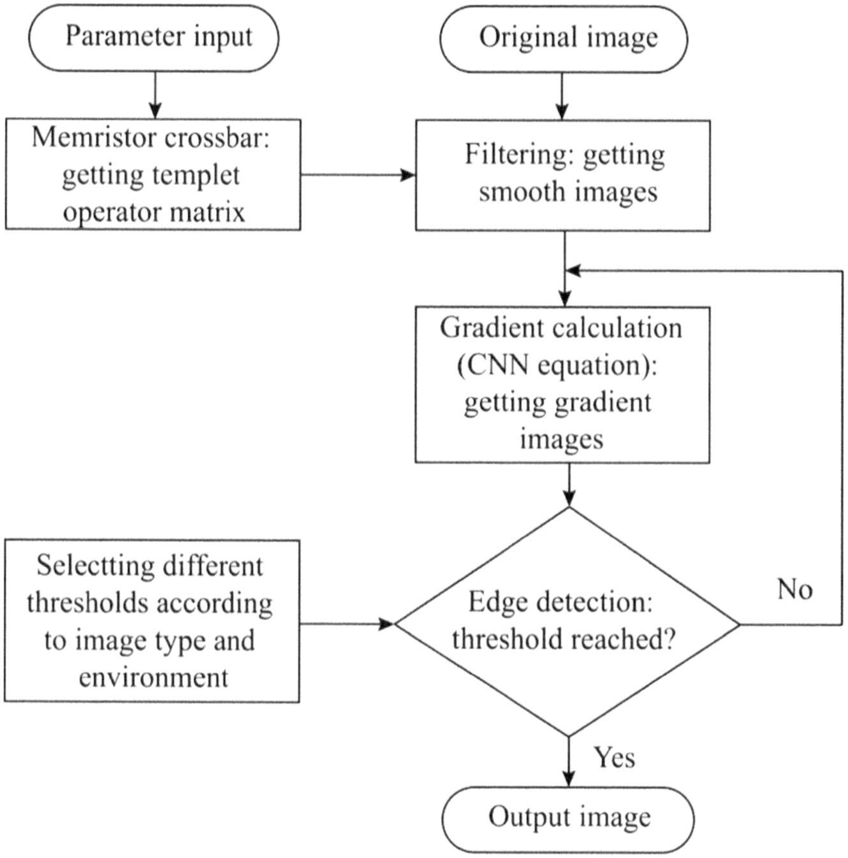

Fig. 8.8 Operational diagram of MCM-CNN

Table 8.1 Advantages and disadvantages of several edge detection algorithms

Algorithm	Main advantages or disadvantages
Canny	Anti noise interference/edges are marked only once
Gaussian	Accurate edge positioning/noise sensitivity
Log	Noise sensitivity
Prewitt	Anti noise interference
Roberts	Low positioning accuracy
Sobel	Anti noise interference/low precision
MCM-CNN	Variable template/hardware implementation

According to the mathematical expression of CNN, MATLAB modeling and simulation are adopted to demonstrate the advantages of the three-layer MCM-CNN in edge detection. The feedback template A and the control template B are selected the same as [17], as shown in the following to process a binary image Camel:

$$\begin{cases} A = \begin{bmatrix} 0\ 0\ 0 \\ 0\ 2\ 0 \\ 0\ 0\ 0 \end{bmatrix} \\ B = \begin{bmatrix} -1 & -1 & -1 \\ -1 & 8 & -1 \\ -1 & -1 & -1 \end{bmatrix}. \\ I = -1 \end{cases} \tag{8.11}$$

The results are shown in Fig. 8.9. On the one hand, we can see that Fig. 8.9b–d are much more effective than the three other traditional operators, and the detected noise is significantly less than other operators. On the other hand, in order to quantitatively evaluate the edge detection results, this chapter quantitatively compares the performance of each contour detection operator by using the commonly used performance index, FOM. The definition of FOM is as follows:

$$\mathrm{FOM} = \frac{1}{\max(N_i, N_t)} \sum_{i=1}^{N_t} \frac{1}{1 + \alpha d_i^2} \tag{8.12}$$

where N_i and N_t are the number of the ideal and detected contour pixels, α is the compensation coefficient (usually 1/9), and d_i is the distance between the detected edge contour point and its nearest contour point. The value of FOM is between 0 and 1, and the larger the value, the better the detection effect.

In this chapter, the results obtained by the MCM-CNN and several other traditional operators are calculated to receive the respective FOM values which are shown in Fig. 8.10.

It can be seen from Fig. 8.10 that the results of several traditional operators are similar. The results of the first layer MCM-CNN algorithm are not good and the best result is the third-layer MCM-CNN. Its results are also three times more than the results of traditional operators.

Afterward, we change the values of templates A, B, and threshold I as shown in the following to process the classic Lena image:

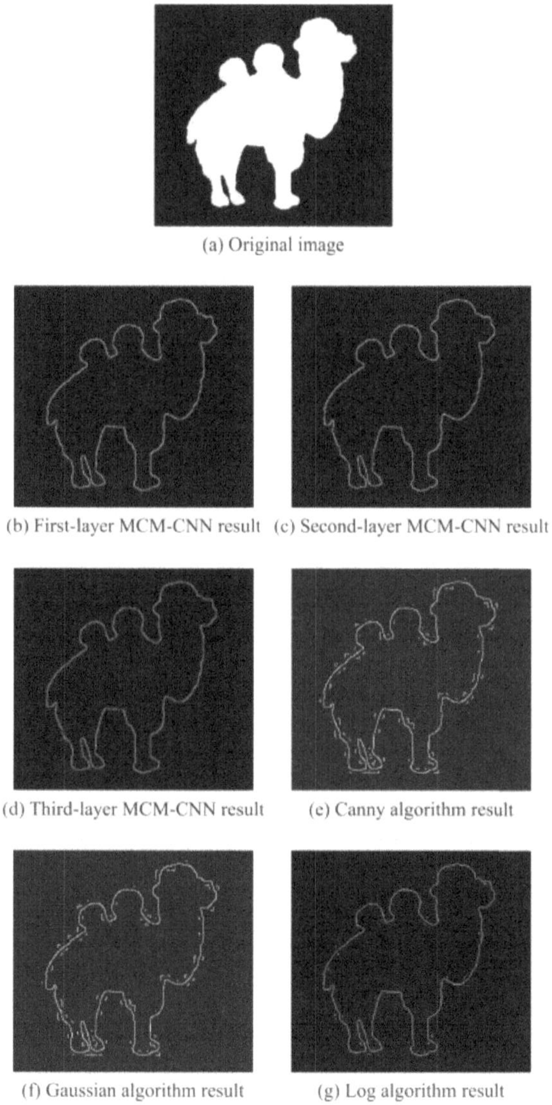

Fig. 8.9 Comparison of Edge detection algorithms on a camel image

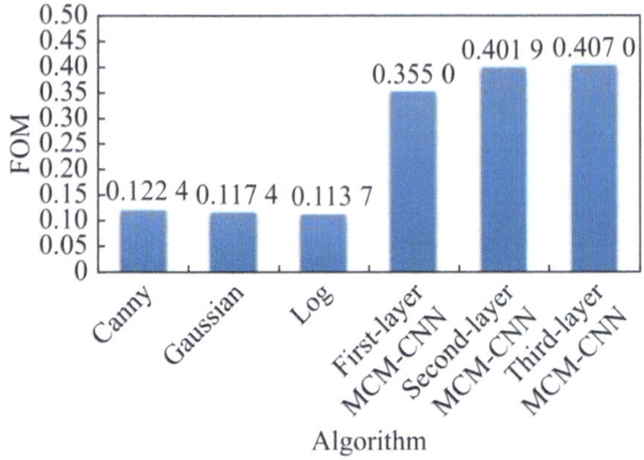

Fig. 8.10 FOM values for different algorithms of camel image

$$\begin{cases} \boldsymbol{A} = \begin{bmatrix} 0 & 0 & 0 \\ 0 & 1.6 & 0 \\ 0 & 0 & 0 \end{bmatrix} \\ \boldsymbol{B} = \begin{bmatrix} -1.004 & -1.004 & -1.004 \\ -1.004 & 10 & -1.004 \\ -1.004 & -1.004 & -1.004 \end{bmatrix}. \\ I = -0.5 \end{cases} \tag{8.13}$$

At the same time, this chapter also compares the effects of other traditional edge detection operators on the Lena image as shown in Fig. 8.11.

Figure 8.12 summarizes the FOM values of several edge detection operators to compare with our proposed MCM-CNN. The results demonstrate that the FOM of our proposed third-layer MCN-CNN is three times more than that of other conventional operators.

By changing the template operator matrix, we can perform edge detection on other images. It can be seen from Figs. 8.10 and 8.12 that it is better than other commonly traditional operators. Compared with several traditional edge detection algorithms, MCM-CNN is used for edge detection with faster processing and higher stability in this chapter.

8.1.8 Conclusions

A memristor crossbar is a high-density and low-power synaptic structure that can be used to construct large-scale artificial neural networks. In this chapter, we propose an MCM-CNN image edge detection model based on the memristor crossbar method. We design an MCM-CNN circuit based on the memristive crossbar's synaptic structure and optimize the activation function. In this novel model, we can adjust the

Fig. 8.11 Lena image edge detection effect comparison

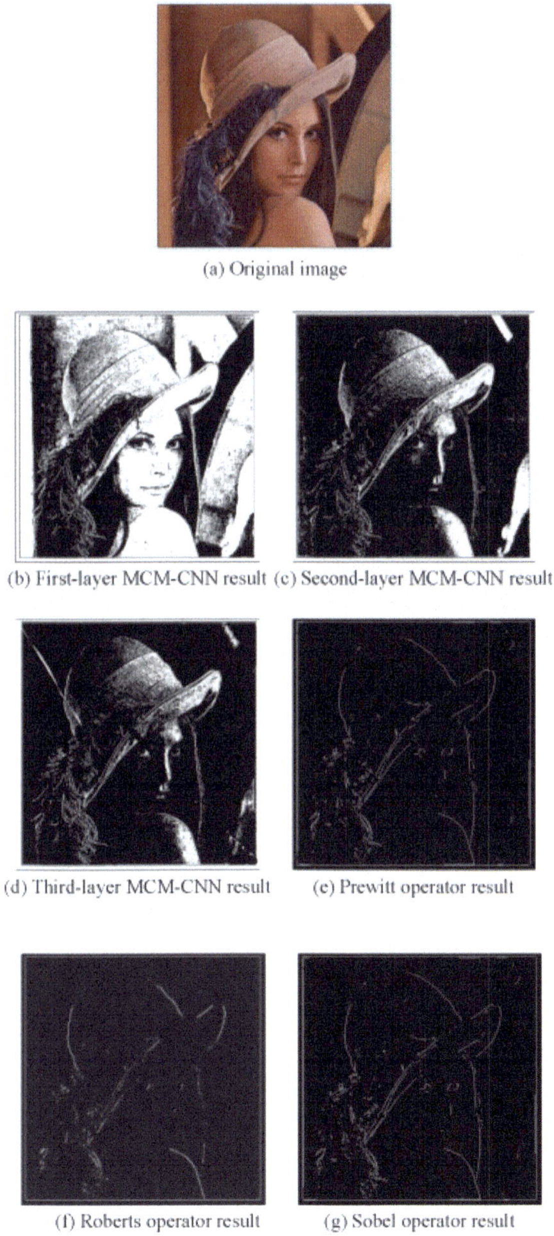

(a) Original image

(b) First-layer MCM-CNN result (c) Second-layer MCM-CNN result

(d) Third-layer MCM-CNN result (e) Prewitt operator result

(f) Roberts operator result (g) Sobel operator result

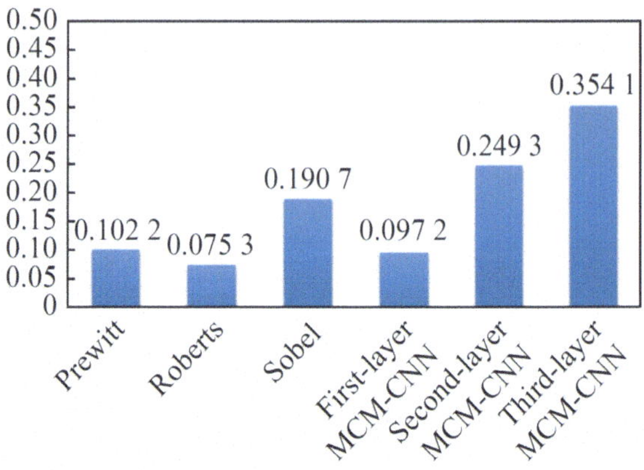

Fig. 8.12 FOM values for different algorithms of lena image

value of the feedback template and control template in MCM-CNN by changing the memristor crossbar switch's input voltage to realize various applications of MCM-CNN. The image edge detection applications are implemented by using the designed MCM-CNN. Simulation results verify the efficiency and flexibility of MCM-CNN. We believe that the proposed MCM-CNN will have great application value in the field of image processing.

8.2 Memristor Crossbar Array-Based Neural Image Caption Generator

8.2.1 Introduction

Generating a caption for an arbitrary image is a demanding task. Problems combining image and language understanding such as image caption generating is a complex and challenging task because the models must be powerful enough to overcome the computer vision challenges of correctly discriminating objects in an image and expressing their relationships in a natural language simultaneously. To obtain high performance, many previous works focused on fine-grained visual processing or multiple steps of reasoning [18–22] have proposed an encoder-decoder structure in an attempt to accomplish the task, in which a Convolutional Neural Network (CNN) has been used as an encoder, and a Long-Short Term Memory (LSTM) network has been used as a decoder.

Several studies on memristor-based neuromorphic computing have proposed feasible ways to construct CNN [23] or LSTM [24] on built memristor arrays. The most recent one has already constructed a fully hardware memristor-based CNN network to accomplish the MNIST handwriting digits recognition [25] task. However, recognizing handwriting digits, which only uses a five-layer CNN, is one of the simplest jobs humans require neural networks to do. When it comes to a more complicated task like image caption generating, memristive neural networks can hardly achieve the performance of the software results due to their basic design of memristor crossbar, which uses multi-level memconductance states to store the weights of neural networks.

In this chapter, a novel implementation of a memristor crossbar that uses memristance values which linearly map the weights of neural networks has been proposed to solve the inaccuracy of writing errors brought by the previous design of memristor. Such memristor crossbar design assures a theoretically high accuracy to finish the image caption-generating task requiring a huge VGG-16-LSTM network. In this system, the memristance values are predetermined by ex-situ trained networks, and the values can be easily changed when the ex-situ training results change. A feedback weight programming algorithm has been proposed to ensure that each memristor in the crossbar can be tuned to correct fast and precise values. Furthermore, a pipeline has been introduced to compress the calculating time.

8.2.2 Crossbar Design Based on Memristor Modules

8.2.2.1 Weight Mapping and the Memristor Module

To avoid the errors of the weight written on the memristor, the linear mapping, and the memristor module can realize the operation proposed in this section. The memristor model published by Strukov [11] describes the memristor as a device with current-controlled doped and undoped areas. When the ratio of the width of the doped and undoped area is not larger than 1, it will change linearly with DC inputs, as shown in Fig. 8.13. This change is described by following (8.14) to (8.16), where r represents the ratio of doped and undoped layers in the memristor, $i(t)$ represents the time-varying current, u_v presents electron mobility, D represents the width of the memristor, and R_{on} represents the doped layer resistance. $f(r)$ defines a window function, and p is a positive integer, which is 20 [26] in this chapter. k represents the scaling factor between the boundary drift rate and the current.

$$dr = kf(r)i(t)dt \tag{8.14}$$

$$f(r) = 1 - (2r - 1)^{2p} \tag{8.15}$$

$$k = \frac{\mu_y R_{on}}{D^2} \tag{8.16}$$

The memristance of a memristor can be defined as (8.17), in which R_{off} represents the undoped layer resistance.

$$R_{mem} = R_{on} \cdot r + R_{off} \cdot (1 - r) \tag{8.17}$$

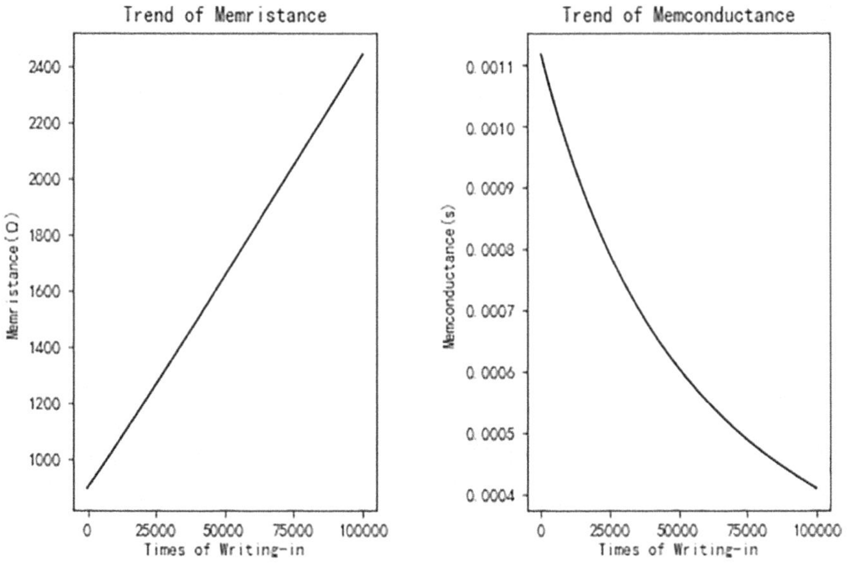

Fig. 8.13 The change of memristance and memconductance in 10,000 times of pulses written-in

With the linear change of r, weights w can be easily mapped with r. By following the rules, larger weights will receive larger output; the W is directly mapped with $1 - r$ instead of r because the memristance will get larger when $1 - r$ increases, as shown in (8.17). So, the mapping can be defined as (8.18).

$$w = 1 - r \tag{8.18}$$

As shown in Fig. 8.14, to realize such mapping, a memristor module has been proposed to replace a single memristor in the crossbar structure. As memristance has been used to store the weights, current must be used as the input to accomplish the task of multiplications due to Ohm's law. To sum up the result in voltage form, an analog $v - i$ converter [27] has been used to change the voltage result into the current to finish addition easily in the circuit. Such a module maximally utilizes the analog characteristic of the memristor, which offers a continuous storage space of weights in theory. This ensures the accuracy of the VMM and offers possibilities for huge neural networks' realization on memristive hardware.

8.2.2.2 Crossbar Structure of Memristor Modules

To accomplish the task of VMM fast and with low energy consumption, Crossbars of Memristor Modules (CMM) should be used in the network. Memristor modules connect with each other by their four pins to build CMM. Since memristance has non-negative values, double memristor modules are used to store both positive and

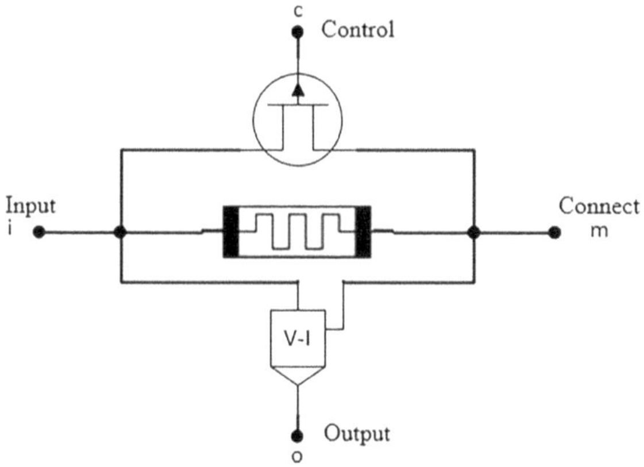

Fig. 8.14 The structure of a memristor module. It contains an input port to receive the input currents, a control port to control the write-in process, a connect port to connect with other modules, and an output port to output the results

negative matrix values to obtain real VMM [23]. The detailed structure of CMM has been introduced separately with the introduction of VGG-16 and LSTM network realization on CMM.

Figure 8.15 shows the write-in circuit of the CMM that allows writing predetermined weights or training updates into the memristor modules. To save write-in time and assure accuracy, the ex-situ software calculates the heaviest current it can use to assure 70% of the write-in process within 20% inaccuracy, and then the write-in circuit uses predetermined low current to finish the tuning. As (8.14), (8.15) suggest, the maximum can be defined by (8.19). To finish 70% of the write-in process, the maximum time can be confirmed as (8.20) shows.

$$I \cdot dt < 20\% \cdot |w - (1 - r)| \frac{D^2}{\mu_\varphi R_{\varphi n}} \tag{8.19}$$

$$t < \frac{70\% \cdot \mu_u \cdot R_{\sigma n}}{20\% \cdot I \cdot D^2} \tag{8.20}$$

Such a circuit allows fast writing-in with heavy-current at first to save write-in time, and precise tuning with low current, at last, to assure high accuracy. The control circuit will help choose which column of the memristor module should be written-in by using the control port of the module. It works like an $N - 2N$ decoder that offers low at the written-in column to deactivate the transistor at the control port of the memristor module and offers high at other columns to ensure that other memristor modules are short-circuited.

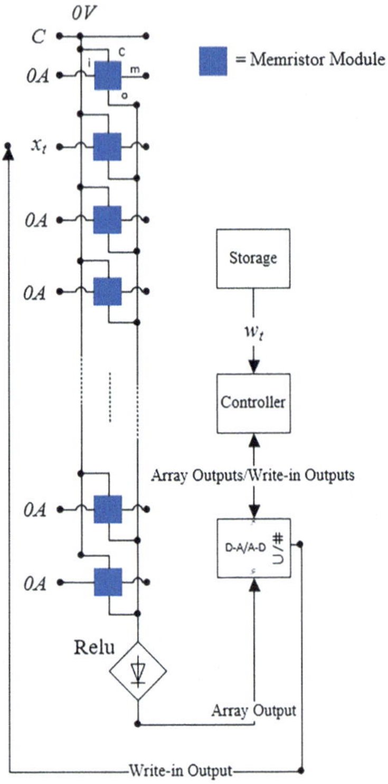

Fig. 8.15 The write-in circuit of CMM. The controller will calculate and output the current pulses, then refer to the output of the column to the row that is written-in

8.2.3 Image Caption Generator Based on CMM

The previous section discusses how CMM works, and this section introduces how VGG-16 and LSTM networks are built on CMM and how they perform the task of Image caption generation.

Remark 8.1 As for software implementation, the transformer-based model has significant performance compared with the LSTM on the image caption tasks. However, the Transformer-based model needs the integrated text without segmentation as the training data, which means the complex memristor circuit with the immense size is required. Furthermore, the number of parameters of the Transformer-based model is much more than the LSTM, which makes the hardware implementation more difficult. Considering the aforementioned facts, the LSTM-based model is adopted to achieve hardware implementation in this chapter.

8.2.3.1 VGG-16 Realization on CMM

VGG-16 network is a deep convolutional neural network, which has a broad range of applications in image recognition [28–30]. It allows input-kernel VMMs to finish the recognition of the image. VGG-16 network has 13 convolutional layers, 5 pooling layers, and 3 fully connected layers, which in total contains 138,144,128 weights [31].

Since each layer in VGG-16 has high similarity on the CMM structure (the fully connected layer actually does the same operation as the convolutional layer), here we use the first layer of VGG-16, which contains two convolutional layers and a pooling layer, to show how VGG-16 is implemented on CMM. As shown in Fig. 8.16, a column of CMM stores a kernel of the convolutional layer. $2N + 1$ memristor modules are used to store N weights in a kernel and a bias that is used in CNN operations. The $2N$ part is divided into two arrays that contain N parameters to store possible N positive or negative values of the weights in the kernel. The inputs and the kernels should be both stretched into a vector vertically to obtain VMM. This process can be illustrated in detail in Fig. 8.17 and (8.21), (8.22).

$$f\begin{bmatrix} x_{1,1} & \cdots & x_{1,n} \\ x_{2,1} & \cdots & x_{2,n} \\ \vdots & \ddots & \vdots \\ x_{m,1} & \cdots & x_{m,n} \end{bmatrix} = \left[x_{1,1}, \ldots, x_{1,n}, x_{2,1}, \ldots, x_{2,n}, x_{m,1}, \ldots, x_{m,n}\right]^T \tag{8.21}$$

$$y_j = \sum_{i=0}^{N-1} \left[x_i^+ w_{ij}^+ + x_i^- w_{ij}^-\right] + w_b \tag{8.22}$$

The function f flattens the matrix.

Rectified Linear Unit (ReLU) [31] is applied after each convolutional operation as an activation function, which can be easily realized through a rectifier circuit with diodes by filtering the negative current.

These single columns of CMM can build a single layer of the convolutional layer to deal with a 2-dimensional convolution operation. The columns connect with each other in series and each row receives the same current (input) from the previous layer. The output of each column will be stored in Static Random Access Memories (SRAMs) by Analog-to-Digital Converters (ADCs) for the next layer's operation. As shown in Fig. 8.17, the convolutional layers of the VGG-16 network on CMMs are designed in 3-dimension to reduce the physics distance with each other and reduce the attenuation of the current when the results of the layers are summed up. The number of kernels of the previous convolutional layer will become the number of layers of the next convolutional layer. As shown in Fig. 8.18, the average pooling layer can be constructed by 4 memristor modules as the VGG-16 network instructed. The pooling operation of the VGG-16 network can be regarded as a 2×2 local matrix multiplied with a 2×2 kernel [23]. The memristance of each memristor module is demonstrated by (8.23).

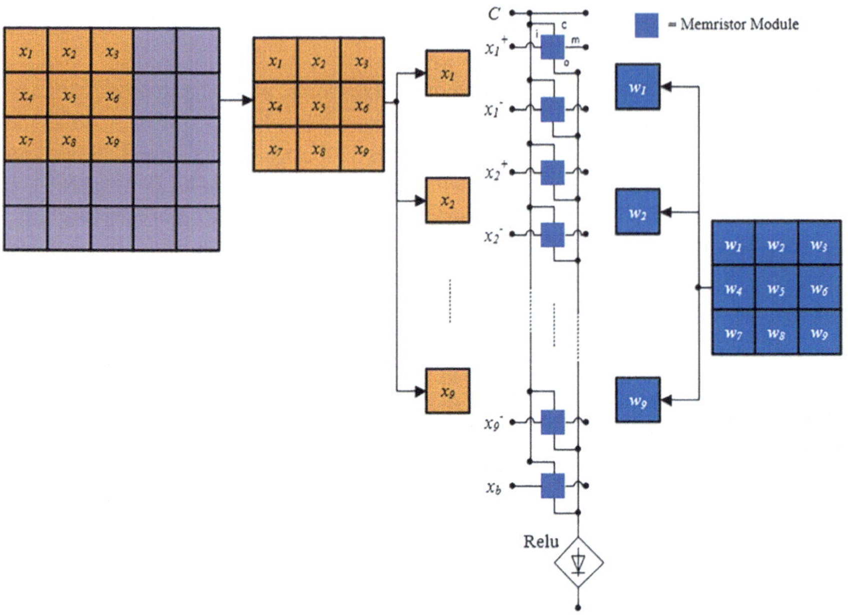

Fig. 8.16 A kernel of the VGG-16 network and the data flow of a single convolution operation. 18 memristor modules are used for both possible positive and negative values in kernels, and one is used for bias. Both the input and the kernel values are stretched to vectors before the convolution operation

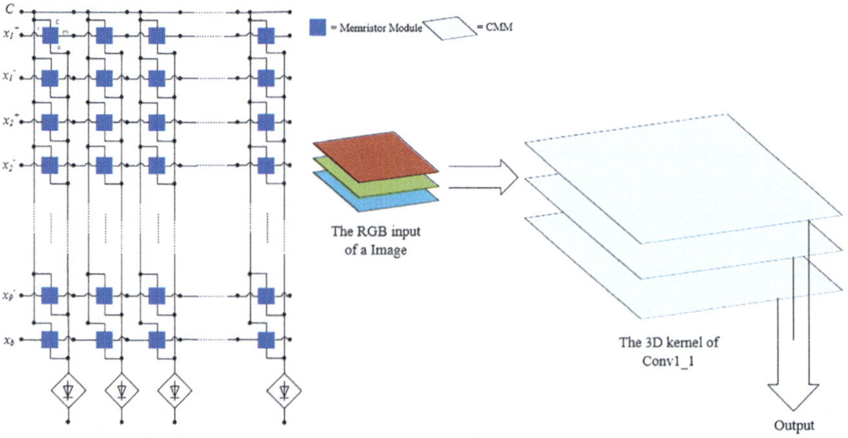

Fig. 8.17 The CMM and the 3-Dimensional structure of the first convolutional layer in VGG-16. To receive the RGB 3-channel inputs in parallel, a 3-layer-CMM is needed, and the outputs are the mean output of each layer

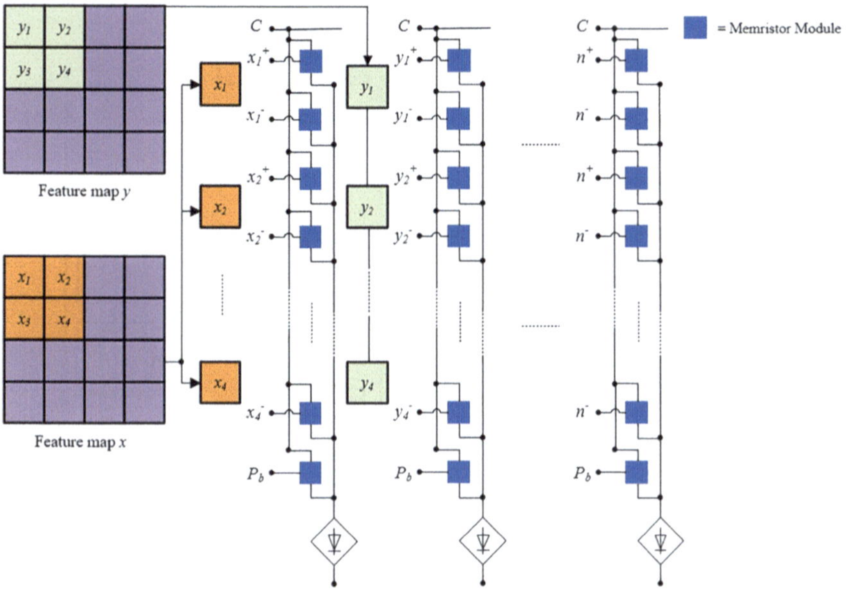

Fig. 8.18 The pooling layer on CMM. Each column receives different inputs from different feature maps

$$R_{mem} = R_{on} \cdot \frac{3}{4} + R_{off} \cdot \frac{1}{4} \tag{8.23}$$

The whole layout of one layer of the VGG-16 network is illustrated in Fig. 8.19. The data are stored in SRAMs and are converted into current pulses by Digital-to-Analog Converters (DACs). The DACs and the ADCs used in this chapter should at least be 11-bit to ensure the accuracy of the write-in process of the CMM. The CMM is integrated with a DAC as input and an ADC as output. SRAMs are used as stacks to store intermediate data processed in each layer. A pipeline is introduced in this layout since the next layer does not need to wait for the previous layer to finish all its work. As calculated, the second convolutional layer in the layout should wait for $2K + 3$ cycles to connect the data it needs, in which K^2 refers to the input size of the previous layer. Similarly, the pooling layer should wait for $K + 2$ cycles instead of K^2 cycles to get the complete data from the previous layer. The final fully connected layer is abandoned and the data is sent into the LSTM network demonstrated in the next section.

8.2.3.2 LSTM Network Realization on CMM

The LSTM network is an advanced Recurrent Neural Network (RNN) [32, 33], which has great applications in the field of data prediction, natural language processing, and translation [34–38]. As shown in Fig. 8.20, several gates, including input gate,

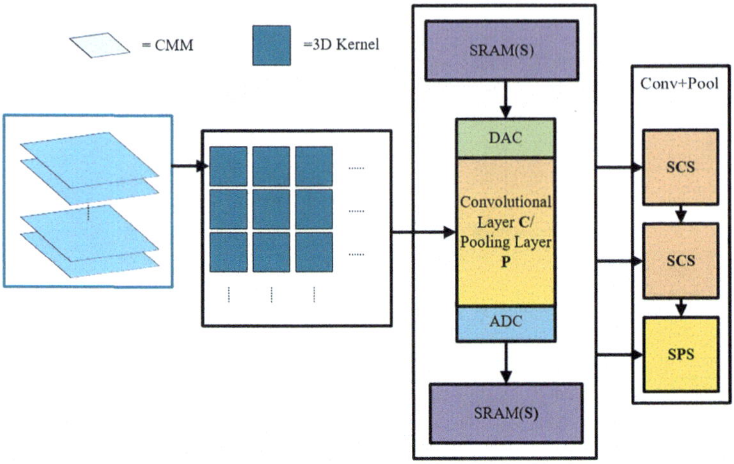

Fig. 8.19 Layout of one layer of the VGG-16

Fig. 8.20 Structure of the LSTM unit

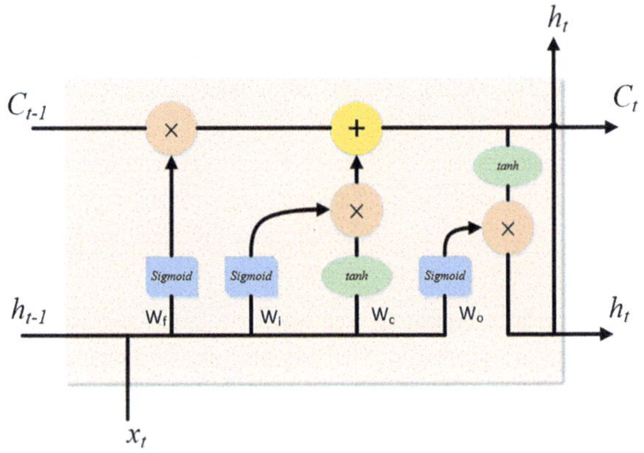

forget gate, updating gate, and output gate [33], are introduced into the network to reinforce the short memory effects. These gate operations can be defined by (8.24) to (8.28).

$$i_t = \sigma\left([x_t, h_{t-1}] \times [w_{ix}, w_{ih}] + b_i\right) \tag{8.24}$$

$$o_t = \sigma\left([x_t, h_{t-1}] \times [w_{ox}, w_{oh}] + b_o\right) \tag{8.25}$$

$$f_t = \sigma\left([x_t, h_{t-1}] \times [w_{fx}, w_{fh}] + b_f\right) \tag{8.26}$$

$$c_t = f_t \odot c_{t-1} + i_t \odot \tanh\left([x_t, h_{t-1}] \times [w_{cx}, w_{ch}] + b_c\right) \tag{8.27}$$

$$h_t = o_t \odot \tanh(c_t) \tag{8.28}$$

where i, o, f and c, indicate the input gate, output gate, forget gate, and the status of the LSTM network cell, respectively. n refers to the current status, and $n-1$ refers to the last status. σ refers to a sigmoid function, and tanh refers to a hyperbolic tangent function. h refers to the output of the cell, and b refers to the bias. $\odot$ represents the production of the corresponding elements of each matrix. To ensure the accuracy of the LSTM network (that is, to reduce the use of the nonlinear function that is realized by hardware), Eqs. (8.24) to (8.26) is realized by CMM to accelerate the process and Eqs. (8.27), (8.28) is operated by software because they contain hyperbolic tangent functions that can be hardly imitated by specific hardware [24]. The sigmoid function is defined in (8.29) as a linear function to reduce the hardware complexity [23, 33].

The data flow illustration and the topography of the whole LSTM network on CMM are shown in Fig. 8.21. When using CMM to perform the VMM tasks, there are some differences between using the LSTM network and the VGG-16 network since the LSTM network uses the true VMM definition in mathematics, but the VMM in VGG-16 just multiples the corresponding elements in the matrix. The rows in the LSTM network's CMM are independent of each other, and each row receives an input current pulse of a column value from the input vector and stores the weights of the corresponding row in the LSTM unit. The multiplication is accomplished by Ohm's Law in each memristor module, and KCL achieves the accumulation to perform the VMM task. The fully connected layer used in the LSTM network is just the same as the one the VGG-16 network uses, which can be demonstrated by convolutional layers introduced in the previous section, containing 512×5000 parameters.

Unlike the VGG-16 network, the parameters of the LSTM network could not be predetermined since the word embedding matrix and the output of the LSTM are completely mapped together. As a result, some tiny inaccuracy of the write-in

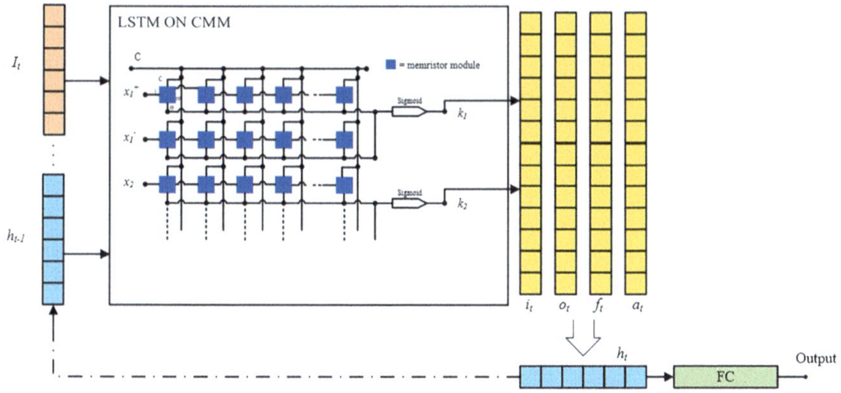

Fig. 8.21 The structure and the data flow of LSTM on CMM. The i_t, f_t, o_t and the second half part of the c_t, $[x_t, h_{t-1}] \times [w_{cx}, w_{ch}] + b_c$, defined as a_t, are calculated on CMM. Double rows are used for possible positive and negative values in the LSTM unit, and the output of the memristor modules in each row are connected together to execute the VMM operation

process of the LSTM units' weight could ruin the whole network, inducing the error output. The training of the LSTM network is introduced in the next section.

8.2.3.3 Model

When dealing with the image caption-generating task, the VGG-16 and LSTM networks should work together organically [19]. As shown in Fig. 8.22, after 15 layers of CNN image feature extraction, the result will be fed to the LSTM as the first input. Before feeding the result of the fc7 of the VGG-16, which is in the shape of 1×4096, the output result is transformed into the shape of 512×8 to fit the memory size of the LSTM network. Then the 512×8 matrix will be compassed to 1-dimension as the first input of the LSTM network. Also, all words contended will be embedded into a vector with a size of 512 to fit the input of each LSTM unit. The word output of the last LSTM network unit will be transformed into the vector by word embedding and fed as the input into the next LSTM unit. The whole process can be described by (8.29) to (8.31) below:

$$x_{-1} = VGG_{16}\,(\text{Image}) \tag{8.29}$$

$$x_n = W_e T_n, n \in \{0, \ldots, N-1\} \tag{8.30}$$

$$p_{n+1} = LSTM(X_n) \tag{8.31}$$

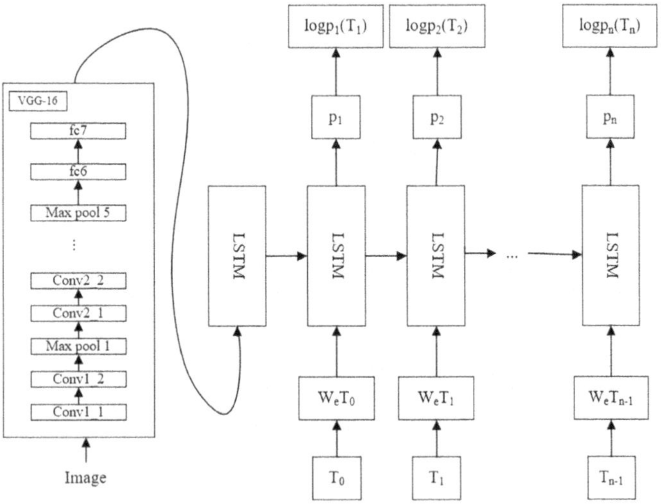

Fig. 8.22 The structure of the image caption generator. The output of the fc7 of the VGG-16 is fed as the first input to the LSTM network. The output word of each LSTM unit is fed as the input to the next LSTM unit

where x is the input of each LSTM unit, W_e is the word embedding matrix, T_n is the word in the ground truth sentence of the input image and p refers to the probability matrix of the output one-hot vector of the true word. Training is performed to optimize the loss function that calculates the sum of the $log(p)$ at each step as (8.32) suggests:

$$L\,(\text{Image}, T) = -\sum_{n=1}^{N} \log p_n\,(T_n) \tag{8.32}$$

To train the LSTM network in a memristive way, the update of the weights should use the write-in circuit mentioned in Sect. 8.2.2. Adam algorithm [39] has been used as the optimizer in the chapter, and the update value of each time can be described below [24, 38]:

$$\Delta W = -\alpha \times \frac{\hat{m}_n}{\sqrt{\hat{v}_n} + \varepsilon} \tag{8.33}$$

where α is the learning rate, $\hat{m}_n$ and $\hat{v}_n$ are momentum matrices in the Adam algorithm. The update use circuit in Sect. 8.2.2 can be described by (8.34):

$$\Delta W = Round\left(\frac{\Delta W}{dr}\right) * dr \tag{8.34}$$

where Δ is the true value of the update, $Round$ is the integer-valued function, and dr is the change of the ratio of the doped and the undoped area of each write-in pulses applied.

8.2.4 Experimental Results

8.2.4.1 Training Details

The dataset used in this chapter is MSCOCO, which contains 82,783 images for training [40], 40,504 images for validation, and 40,775 images for testing. The VGG-16 network is pre-trained on ImageNet, and its weight is written into the network and remain unchanged. For the long and short-term memory network (LSTM) trained on the CNN, the unit memory size is set to 512 dimensions, the maximum caption length is set to 20, the decoding layer is set to 1, the initial weight is randomly set, and the number of training iterations is 5. The initial learning rate α is 0.0001, the two exponential decay rates are 0.9 for β_1 and 0.999 for β_2, and the zero ϵ is 1e-6 in the chapter.

8.2.4.2 Write-in Accuracy Analysis

For the writing inaccuracy of the CMM, the maximum inaccuracy value can be calculated theoretically. According to (8.14) to (8.16), the ratio change of the doped layer to the undoped layer in each write is given by (8.35):

$$\Delta r = \frac{\mu_v R_{on}}{D^2}(1-(2r-1)^{20})i \cdot dt \tag{8.35}$$

Given $i \cdot dt$, the current and the current pulse length (or the time interval for the write-in circuit to read the output of the CMM), the inaccuracy is given by (8.36):

$$E = Cell\left(\frac{w_{t\arg et} - w_n}{\Delta r}\right) * \Delta r - (w_{t\arg et} - w_n) \tag{8.36}$$

where w_{target} is the target write-in weight value, w_n is the current weight value, and $Cell$ is the round-up function. According to (8.36) and the value of each parameter in this chapter, the maximum error for a given $i \cdot dt$ can be calculated by (8.37):

$$Max(E) = \frac{\mu_v R_{on}}{D^2} \cdot i \cdot dt = 1 \times 10^{-4} \cdot i \cdot dt \tag{8.37}$$

where, μ_v is $10^{-14} \cdot m^2 \cdot s^{-1} \cdot V^{-1}$, R_{on} is 100 Ω, and D is 10^{-8} m.

If $i \cdot dt$ takes 10^{-8} (it-8 for short), when writing to the first convolution layer (Conv1_1) of the VGG-16 network, the absolute distribution of the inaccuracy is shown in Fig. 8.23. Due to the previously-mentioned reasons, half of the memristor modules (1750) have the value of 0, and the error written to the value of 0 is the same and very small. Ignoring these half 0 values, the error distribution is almost random. The maximum value of the writing error is about $9.99742e^{-05}$, which is consistent with the theoretically calculated maximum error, the average value is about $2.33629e^{-05}$, and the standard deviation is about $3.47368e^{-05}$. For other $i \cdot dt$ values and the traditional memristor crossbar, the writing inaccuracy of the first

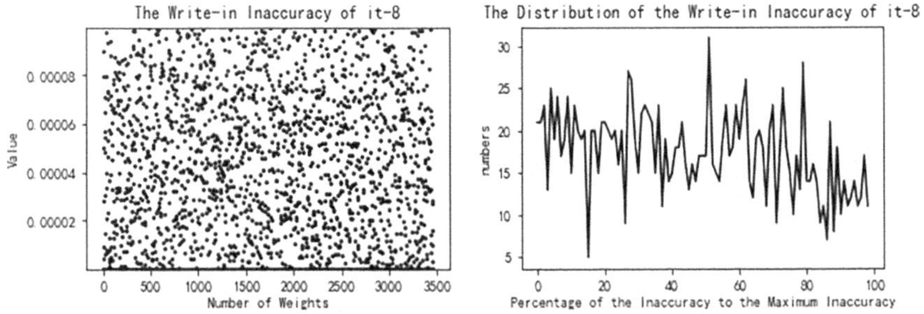

Fig. 8.23 The distribution of the write-in inaccuracy of the first layer of VGG-16. The distribution is relatively random

Table 8.2 The average, max, and standard deviation value of the write-in inaccuracy in different write-in conditions. The Mem32 [41–43] the traditional design of the memristor crossbar array using a single 32-bit memristor as storage of the weights

i·dt	Average value	Max value	Standard deviation
$1e^{-10}$	$2.35119e^{-07}$	$9.98953e^{-07}$	$3.10589e^{-07}$
$1e^{-9}$	$2.28469e^{-06}$	$9.98714e^{-06}$	$3.04855e^{-06}$
$1e^{-8}$	$2.33629e^{-05}$	$9.99269e^{-05}$	$3.06684e^{-05}$
$1e^{-7}$	$2.31049e^{-04}$	$9.99155e^{-04}$	$3.07779e^{-04}$
$1e^{-6}$	$2.37653e^{-03}$	$9.99086e^{-03}$	$3.10242e^{-03}$
Mem32	$4.11224e^{-3}$	$1.61257e^{-2}$	$5.24595e^{-3}$

convolutional layer (Conv1_1) is shown in Table 8.2. The conventional design of the memristor crossbar array using a single 32-bit memristor as storage of the weights (Mem32) is also shown in the table for comparison. It can be seen that the values of the mean, maximum, and standard deviation of the error do not change significantly with the value of $i \cdot dt$, but their magnitude changes with the magnitude of $i \cdot dt$.

8.2.4.3 Generation Result Comparison

To perform the simulation of the proposed model, a neural network environment (Ref. for short in the table) is built on a general-purpose computer. The pre-trained parameters [44] of ImageNet are written into the VGG-16 network and the LSTM network is trained. A CMM simulation written and trained by the methods of it-10, it-9, it-8, it-7, and it-6 is performed to simulate the VGG-16 network and LSTM (an additional 300% noise has been added due to the real circuit situation). Establishing a traditional memristor crossbar array simulation (referred to as Mem32) written and trained by using the 32-bit memristor weight value for the VGG-16 network and LSTM for comparison. The evaluation results of BLEU [45], ROUGE [46], METEOR [47] are shown in Table 8.3. Table 8.4 compares the scores of some classical models and the proposed model, including Soft-Attention [48], Up-Down [49], SGAE [49], and GCN-LSTM [18]. From Table 8.4, it can be seen that the performances of these models [18, 48, 49, 49] are almost similar to our reference model on the evaluation metrics of BLEU2, BLEU3, and BLEU4. However, the main purpose of this chapter is to explore the possibility of tackling image captioning tasks with memristive neural networks, which is why we propose a way to design a novel crossbar array based on the memristor module and make a breakthrough in memristive neural network solution for the cross-modal image caption generation task. Ablation experiments have been performed to determine if the inaccuracy remains stable when changing the whole network slightly. Two main experiments are performed and are called "Abs-1" for Ablation Study 1 and "Abs-2" for Ablation Study 2. In Abs-1, we reduced the layers of the VGG-16-LSTM architecture, such that fewer LSTM units

Table 8.3 The scores in different write-in conditions

Conditions	BLEU1	BLEU2	BLEU3	BLEU4	ROUGE	METEOR
Ref.	0.646	0.460	0.314	0.214	0.454	0.190
it-10	0.642	0.451	0.305	0.206	–	–
it-9	0.634	0.443	0.296	0.197	–	–
it-8	0.629	0.438	0.293	0.195	0.326	0.105
it-7	0.509	0.281	0.145	0.078	0.326	0.105
it-6	0.480	0.211	0.087	0.033	0.320	0.103
Mem32	0.382	0.192	0.082	0.040	–	–
Abs-1	0.498	0.270	0.113	0.043	0.360	0.140
Abs-2	0.412	0.211	0.079	0.042	0.332	0.097

Table 8.4 The scores of each models

Conditions	BLEU1	BLEU2	BLEU3	BLEU4	ROUGE	METEOR
Ref.	0.646	0.460	0.314	0.214	0.454	0.190
it-10	0.642	0.451	0.305	0.206	–	–
GCN-LSTM [1]	0.809	–	–	0.383	0.585	0.286
Soft-Attention [6]	0.707	0.492	0.344	0.243	–	0.239
Up-Down [7]	0.798	–	–	0.363	0.569	0.277
SGAE [8]	0.810	–	–	0.390	0.589	0.284

(256 units as opposed to 512 units in the original architecture) and lower dimensions (initial LSTM layer and the attention layer have dimensions of 256 units) are used to train the model. In Abs-2, we added a few layers of the original architecture to note and compare results. We added more weight layers to the original model (more convolutional layers) to ensure that the architecture supports 19 layers. The results of these experiments have been updated in Table 8.3. It can be seen that the results for both Abs-1 and Abs-2 are significantly less than the reference scores, which is why it is not advisable to add or reduce layers of the VGG-16-LSTM architecture. Hence, due to lower scores in write-in conditions, experimental results from Abs-1 and Abs-2 conclude that the inaccuracy does not remain stable when changing the whole network slightly.

It can be seen from Table 8.3 that smaller writing errors will make the memristive neural network perform better, but when $i \cdot dt$ is less than $10e^{-8}$, further reducing of $i \cdot dt$ will not improve the performance of the network. Similarly, when $i \cdot dt$ is $10e^{-8}$, the actual hardware requirements are not too demanding like it-9 or it-10, which requires a micro-current constant current source. Therefore, it is recommended to use it-8 as the circuit implementation solution in the actual circuit. Additionally, the manual evaluation results of the output of inputting 9 randomly given pictures to the reference network, it-10, it-8, it-6, and Mem32 are shown in Fig. 8.24. In the

Fig. 8.24 The outputs in different input conditions of 9 randomly given images

figure, green represents good, yellow represents medium, and red represents poor. It can be seen that the output of it-10 and it-8 (and theoretically it-9) are highly similar to the output of the reference network, though having slight differences in outputs of the same picture. However, in the selected 9 pictures, none of them can be described well by it-6 and Mem32, and some off the output for each picture is the same. This output can be attributed to the fact that the VGG-16 network fails to extract image features, and the weight value of the LSTM network may also converge to an optimal local value.

8.2.5 Conclusion

The chapter was motivated by the traditional memristor crossbar array structure to propose and design a memristive neural network based on a memristor module that solves the discontinuous weight values writing, large errors, and nonlinearities problems in traditional ways way memristive array. The network is implemented to perform the cross-modal image caption generation task successfully. Its output is highly correlated with the output of the reference network, which breaks the bottleneck of the traditional memristor crossbar array structure that cannot output the correct title. At the same time, this chapter discusses and experiments with the writing parameters of the memristive neural network based on the CMM and gives the optimal parameter it-8 scheme in hardware implementation. Its BLEU-1 score is 0.629, which is only 0.017 smaller than that of the reference network, and the performance is 97.37% of the reference network. Experimental results show that the network has great noise immunity. This chapter provides a breakthrough in a memristive neural network solution for the cross-modal image caption generation task, and at the same time, provides a reference for the memristive implementation of other large neural networks. It also provides hardware implementation of future high-speed. However, this chapter only discusses the implementation achievability of the basic model of image caption generation task on the memristor, the possibility of the application of the memristor-based Transformer network will be studied in future works.

8.3 Memristor and Information Entropy-Based Neural Network Crawler

8.3.1 Introduction

With the rapid growth of network resources, vertical search engine [50] can meet specific user requirements compared to the general-purpose search engine [51]. A special-purpose web crawler, known as a focused crawler, is a good solution to the problem of low precision [52]. Most focused crawlers utilize anchor text to analyze hyperlinks and calculate topic-relevant similarity. Compared to traditional focused crawlers based on content or link analysis, such as BFS spider [53], PageRank spider [54], Shark-Search [55], and smart-crawler [56], the dominant focused crawlers based on artificial intelligence (AI) model improve the performance.

To design a more efficient focused crawler algorithm, a combination of the link-based and content-based analysis is proposed in [57]. However, this simple structure did not integrate the merits of the two methods efficiently. Thus, how to combine the two methods effectively is the focus of this chapter. A neural network model that has the characteristics of neurons and link weight catches our attention. A neural network spider based on a spreading activation algorithm is proposed in [58], which is used to model the Web as a weighted neural network. This spider performs well for specific requirements, but can lead to insufficient precision. How to design the algorithm of topic-relevant similarity of link and node in a neural network to improve the precision is the goal of work.

Researches [59–65] show that memristor can be used for neuronal synaptic design. Similarly, we use the memristor model to design the link weight of the Hopfield net spider. A cubic nonlinearity memristor mathematical model is proposed in [66], which replaces the Chua diode with a flux-controlled memristor. Therefore, we apply the memristor model to the design of the internal structure of the neural network algorithm. Through the analysis of focused crawlers literatures above, we design a Neural Network Crawler with Information Entropy (NNCIE) and Neural Network Crawler with Memristor Model (NNCMM) in this chapter.

8.3.2 *Framework and Workflow*

The overview of the Hopfield neural network is introduced in subsection A of this section. In the following part, the system framework and workflow of our neural network crawlers which is based on the Hopfield net spider will be introduced. The differences between our crawlers and the conventional spider can be summarized as follows.

- Chau's research uses a best-first crawling strategy in the framework of spider, which leads to a problem of local optimum. In our research, we compare several common crawling strategies and combine the Breadth-First strategy with the Best-First strategy for our crawler framework to solve this problem.
- The activation function of Chau's research adopts a modified sigmoidal function for normalizing the node weight. In our research, we introduce a memristor and adopt a monotone-increasing and piecewise-linear memristor mathematical model as an activation function for high performance.
- Chau's research presents a Hopfield net spider, which does not have a precise method to calculate the relevance of link weight. Our research provides an information entropy algorithm and a memristor mathematical model to improve accuracy.

8.3.2.1 Hopfield Neural Network

Hopfield net is a recurrent neural network that connects the output to input by feedback. Since the output of the feedback neural network has feedback to its input, the Hopfield network generates continuous state changes under the activation of the input. If the Hopfield network is a stable network with convergence, the changes made by this feedback and iterative calculation are getting smaller. Once the stable equilibrium state is reached, the Hopfield network will output a stable constant value.

8.3.2.2 System Framework

To efficiently crawl topic-relevant web sources, we convert the Hopfield neural network model to a forward-propagating activated neural network. The process of spreading activation is shown in Fig. 8.25. It can be seen that the output in the current state (iteration 0) of the Hopfield net continues to be activated as input in the next state (iteration 1) until the neural network becomes stable. The system of neural net crawler starts with a set of seed URLs, then the neural net crawler activates their neighbor URLs, combines link weights, and calculates the weights of newly discovered nodes. In this spreading activation process, the nodes with low weights are discarded. A Hopfield neural network framework is designed in Fig. 8.26. It can be seen that the core of this neural network framework is spreading activation. The two-part design is the most crucial unit in the design of spreading activation, which is the neural network synapse and activation function. The synapse is the link weight in our neural network crawler. Therefore, we propose the design of an information entropy algorithm and cubic nonlinearity memristor model for calculating link weight, which is topic-relevant similarity. For the activation function, we apply the monotone-increasing and piecewise-linear memristor model instead of the sigmoid function.

8.3.2.3 System Workflow

System workflow is shown in Fig. 8.27. Firstly, the crawler system needs to be initialized. Seed URLs are chosen by the manager, which is a set of relevant web pages with the target domain. The crawler fetches each seed URL as a node with a weight of 1 and puts them into the neural network.

The crawler adds the seed URLs to the module of the URL manager. The module of Crawler Manager visits URLs, which are fetched from the URL manager and stores all structures of web pages as a DOM tree. The Parse Center parses the DOM tree and gets the textual content of the web page and a set of new URLs. Textual content is stored in the MongoDB database. The link weights of new URLs are calculated by the Topic-relevant similarity module, which is designed by the information entropy or memristor model. Next, node weights are calculated by an activation function designed by the memristor model in Parse Center. Next, the stop condition is checked. If it does not match the stop condition, filter the URLs and add them to the URL database. The stop condition is that the crawler has already crawled the predetermined number of web pages or the average weight of all web nodes in an iteration is smaller

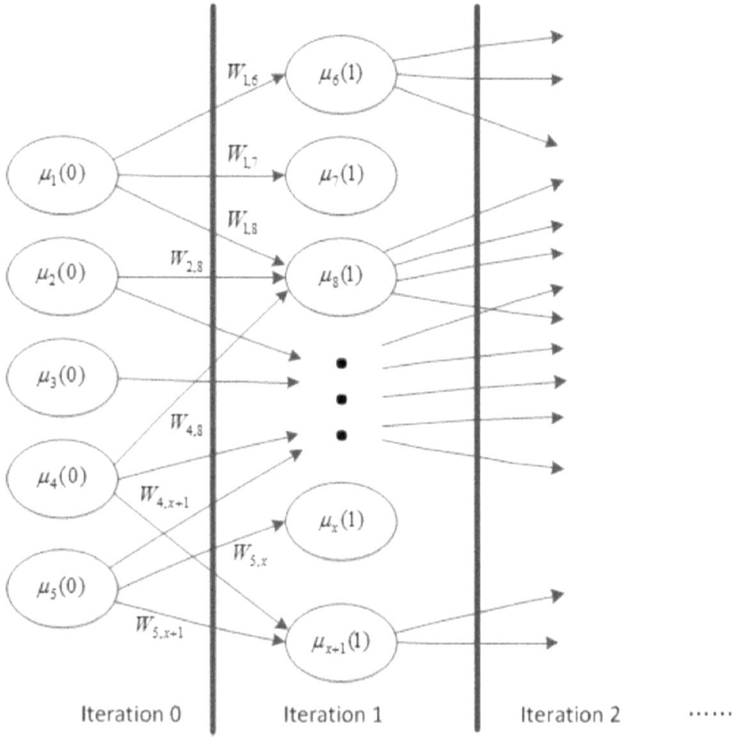

Fig. 8.25 Process of spreading activation

than a maximum allowable error (a small number). If the stop condition is met, the crawling job is completed and exits the system. Otherwise, the next layer is iterated.

For a well-designed focused crawler, it is essential to adopt an efficient crawling strategy algorithm. Breadth-First and Best-First search algorithms are the common choices in focused crawlers. However, the Best-First Search strategy is easy to fall into the local optimum, and the Breadth-First strategy can visit and collect plenty of useless web pages, which leads to a decrease in precise rate. To Combine the advantages of Breadth-First Search and Best-First Search, the Breadth-Best-First strategy is adopted in our framework. The detailed design will be introduced in Sect. 8.3.3.

8.3.3 Analysis

This section presents a theoretical analysis of the main aspects of our research in this chapter. First, we analyze the Breadth-Best-First Search and then propose an

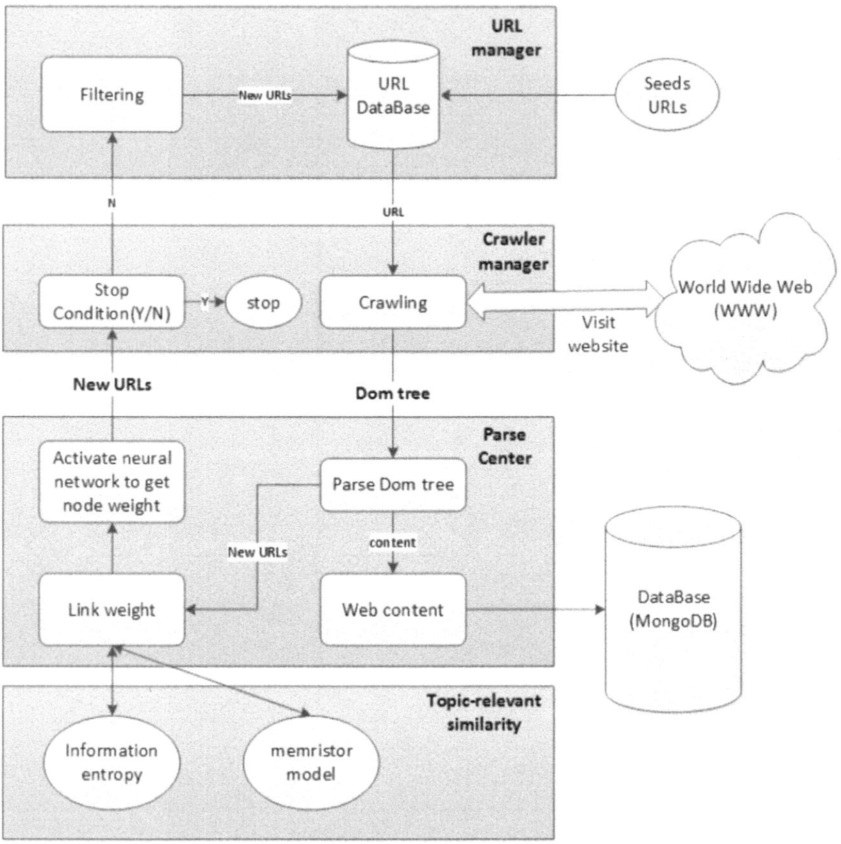

Fig. 8.26 System framework of the focused crawler

activation function in a neural network crawler and two algorithms for calculating link weight, which are information entropy and memristor model.

8.3.3.1 Breadth-Best-First

An improved crawling strategy, best-breadth-first, is proposed to improve crawling efficiency and coverage.

Best-Breadth-First is the combination of Best-First and Breadth-First, which takes advantage of Best-First crawling strategy and avoids the local optimum of Best-First by applying Breadth-First. Figure 8.28 illustrates the process of this crawling strategy. It can be seen that web page 1.1 and web page 1.2 are linked from web page 1. Similarly, web page 2 links to web page 2.1 and web page 2.2, and web page 3 links to web page 3.1 and web page 3.2. In an iteration, firstly, the crawler puts these web

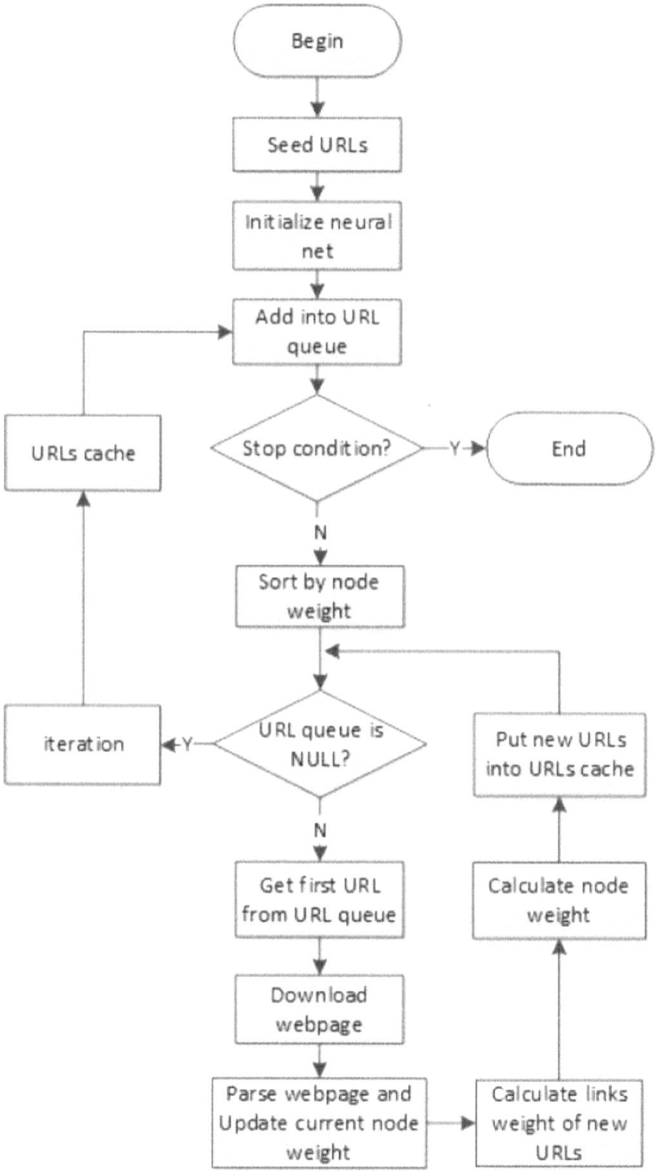

Fig. 8.27 Flowchart of focused crawler

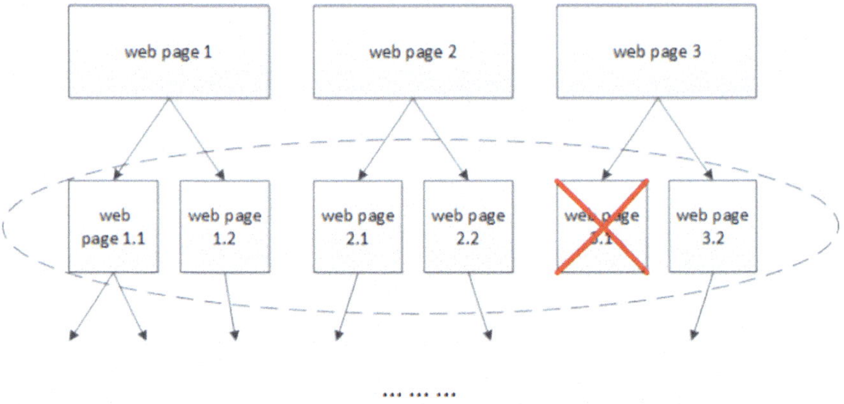

Fig. 8.28 Breadth-Best-First algorithm

pages, which are in the dotted circle, into the URL queue. Then, the weight of each node in the URL queue is calculated. Next, these URLs of web pages with weight are sorted and the URLs of low weight, such as web page 3.1, are filtered. Finally, when all of the high-weight URLs, such as web page 1.1, are downloaded, the new iteration will start. In other words, we must complete the crawling and analysis of all URLs in the URL queue before the next iteration.

8.3.3.2 Calculation of Node Weight

In order to calculate the node weight in our neural network crawlers, we design a memristor mathematical model function as the activation function. The node weight function is defined as follows:

$$\mu_i(t+1) = f_m\left(\sum_{every,known,parent,bo,f,i} \omega_{b,i}\mu_b(t)\right) \tag{8.38}$$

where $\mu_i(t+1)$ denotes the weight of node i (web page) at an iteration $t+1$, and $\omega_{b,i}$ is the weight of the link between two nodes, which is designed in Sect. 8.3.3.4. The activation function f_m is the memristive function, and we illustrate it in Sect. 8.3.3.3.

After an iteration, we update the weight of nodes through web contents. The function is presented as follows:

$$\mu_i(t+1) = f_m\left[\mu_i(t+1) \times P_i\right] \tag{8.39}$$

where P_i is a weight of relevance for web textual content, and it shows the real relevance of web pages. The larger P_i is, the more relevant. The function P_i is defined as follows:

$$P_i = \text{number of keywords in web content} \tag{8.40}$$

8.3.3.3 Activation Function

The HP Company presents a nanometer-size solid-state two-terminal device called the memristor in [11], and it uses a nonlinear constitutive relation, which is between the device terminal voltage and terminal current, to describe the memristor:

$$v = M(q)\,i \tag{8.41}$$

where the nonlinear function called the memristance is defined by:

$$M(q) \triangleq \frac{d\phi(q)}{dq} \tag{8.42}$$

The activation function in the neural net crawler is designed by the function of the memristor, which is characterized by the monotone-increasing nonlinearity [67].

In research [59], the memristor is defined by

$$\phi(q) = bq + 0.5(a - b)(|q + 1| - |q - 1|) \tag{8.43}$$

where $a, b > 0$. The memristance $M(q)$ is defined by:

$$M(q) = \frac{d\phi(q)}{dq} = \begin{cases} a, & |q| < 1 \\ b, & |q| > 1 \end{cases} \tag{8.44}$$

Equation (8.43) is monotone-increasing and piecewise linear, and the slope of (8.43) is the value of memristor. Therefore, we adopt a memristive function to define our activation function in neural net crawlers. It is defined by:

$$f_m(x) = bx + 0.5(a - b)(|x + 1| - |x - 1|) \tag{8.45}$$

$$M(q) = \frac{df_m(x)}{dx} = \begin{cases} a, & [;\, 0 \le |x| < 1 \\ b, & [;\, |x| \ge 1 \end{cases} \tag{8.46}$$

where a is a positive number, which is less than 1, and b is a small positive number.

8.3.3.4 Information Entropy Algorithm

In this section, we illustrate how to design a judgment of the hyperlink relevance with an information entropy algorithm from a theoretical perspective for a Neural Net Crawler with Information Entropy (NNCIE).

The link weight is for evaluating whether the URL, which is between two web pages, is relevant to the target domain.

Information entropy is an abstract concept in mathematics. From the view of information dissemination, information entropy presents the value of information. The greater the value of information entropy is, the more information it contains. Let D indicate a set of target domains, there are n keywords in the target domain, and it is defined as:

$$D = \{key_1, key_2, \ldots, key_n\} \tag{8.47}$$

where $key_i, i = 1, 2, \ldots, n$, indicate each key word in target domain. Let $E(D)$ indicate the information entropy of target domain D, we adopt information entropy function as follows:

$$ED(D) = -\sum_{k=1}^{n} p_k \log_2 pk \tag{8.48}$$

where n is the number of elements in the target domain D, and is the frequency of keyword p_k in the target domain appearing in the current anchor text. When we put anchor text of hyperlink between node b and node i into (8.48), we can get the value of information entropy $E(D)$, and the weight of link $\omega_{b,i}$ between node and node is defined as follows:

$$\omega_{b,i} = f_l(E(D)) \tag{8.49}$$

where f_l is the corresponding function of information entropy and link weight. It is defined as follows:

$$f_l = x^2 \tag{8.50}$$

Here, we use information entropy function (8.48) to maintain the amount of information of keywords in current anchor text. Equation (8.50) makes the value of link weight bigger if there is more relevant information in the anchor text. In contrast, the value of link weight is smaller if there is less relevant information in anchor text.

The function is defined as follows:

$$pk = \frac{number\ of\ keyword\ in\ anchor\ text}{total\ number\ of\ words\ in\ abchor\ text} \tag{8.51}$$

where number denotes the number of keywords k in the current anchor text, and total number denotes the number of words in anchor text. Considering the number of words in Chinese sentences that need to be counted, the jieba segmentation algorithm is used to cut the sentences and count the number.

8.3.3.5 Memristor Model Algorithm

We present the design of the link weight of NNCMM with a memristor mathematical model. The memductance function $W(\phi)$ is defined by:

$$W\left(\phi\right) \triangleq \frac{dq\left(\phi\right)}{d\phi} \tag{8.52}$$

The relationship between the voltage across $v(t)$ and the current through $i(t)$ the memristor is thus given by:

$$i\left(t\right) \triangleq \frac{dq}{d}\frac{d\phi}{dt} = W\left(\phi\left(t\right)\right)v\left(t\right) \tag{8.53}$$

The cubic nonlinearity memristor model is defined by:

$$q\left(\phi\right) = \alpha\phi + \beta\phi^3 \tag{8.54}$$

Therefore, we can obtain the memductance function $W(\phi)$ as given below:

$$W\left(\phi\right) = \frac{dq\left(\phi\right)}{d\phi} = \alpha + 3\beta\phi^2 \tag{8.55}$$

The (8.54) is cubic and nonlinear, and (8.55) is the memductance value. Therefore, we adopt this memristor mathematical model to design the link weight function of memristive neural net crawler. It is defined by:

$$f_{mem} = \alpha x + \beta x^3 \tag{8.56}$$

where α is a small positive number, and β is a positive number, which is more than 1. This is the sum of the frequency of the keyword in (8.47) that appeared in anchor text.

Similarly, for designing the link weight of the memristive neural net crawler, we also adopt jieba segmentation algorithm to cut the sentences, then use (8.51) to calculate the frequency of every keyword that appeared in anchor text. The sum of every frequency is as the variable of (8.56). The result of the function is the value of the link weight.

8.3.4 Experiments

In this section, in order to systematically evaluate the performance of our focused crawlers, NNCIE and NNCMM, an experiment to compare with the Breadth-First Search (BFS) spider, Breadth-Best-First search (BBFS) spider, and Hopfield net spider is designed, which is based on Tibet big data project. The crawling platform is implemented by using Python language and deployed on the Windows 10 operating system.

8.3.4.1 Experiment Design

Table 8.5 shows the keyword lists and seed URLs in this experiment. Firstly, we need to set up a set of seed URLs, which are highly relevant web pages with target domains, and a standard that if one web page can be classified as the relevant web page. Therefore, we build two keyword lists as the target domain for calculating the relevance of web nodes and hyperlinks. In this experiment, the main topic is the development and construction achievement in Tibet. Therefore, the words in the keyword lists are about places of Tibet and development. For each crawler, we use the same seed URLs and target domain, then we run their processes until they have crawled 10,000 web pages.

To compare the performance of these five crawlers in our experiment, we run their processes to analyze their relevance of web pages and time cost. We define the standard of relevance, which is if each keyword list has one word that could be found three times in web content, then this web page is relevant. Therefore, we define the precision rates of the crawlers as follows:

$$precision\,rate = \frac{number\ of\ relevant\ pages\ visited}{number\ of\ all\ pages\ visited}$$

Because we set up a set of seed URLs from some relevant websites and the crawler must visit some websites and crawl the unknown web pages, there are plenty of web resources on the website, but we couldn't accurately count the total number of relevant pages. The keyword lists and seed URLs are shown in Fig. 8.29. Therefore, we focus on precision rate. We also compare efficiency by the time each crawler spends.

Table 8.5 The comparison of performance for five crawlers

Crawler	Content			
	Number of pages visited	Number of relevant pages visited	Precision (%)	Time (s)
BFS	10000	1702	17.03	5737.30
BBFS	10000	3571	35.71	4233.89
Hopfield net spider	10000	4177	41.77	3876.79
Our NNCIE	10000	5178	51.78	3546.33
Our NNCMM	10000	5369	53.68	3698.43

Keyword List 1	Keyword List 2	Seed URLs
1.西藏(Tibet)	1.发展(Development)	1.URL: http://www.tibet.cn
2.拉萨(Lhasa)	2.建设(Construction)	2.URL: http://www.xizang.gov.cn
3.青藏(Qinghai-Tibet)	3.成果(Payoff)	3.URL: http://www.chinatibetnews.com
4.昌都(Qamdo)	4.成绩(Grade)	4.URL: http://www.tibetcn.com
5.日喀则(Xikaze)	5.伟业(Great cause)	5.URL: http://www.tibetmagazine.net/xzsh
6.林芝(Nyingchi)	6.创举(Pioneering work)	6.URL: http://www.tibetagri.gov.cn/sites /MainSite
7.山南(Shannan)	7.事业(Cause)	
8.那曲(Nagqu)	8.业绩(Performance)	
9.阿里(Ngari)	9.成就(Achievement)	
	10.果实(Fruit)	
	11.进展(Progress)	
	12.经济(Economy)	
	13.增收(Increase revenue)	
	14.科技(Technology)	
	15.致富(Increase income)	

Fig. 8.29 The keyword lists and seed urls

8.3.4.2 Experiment Result and Discussion

Table 8.5 illustrates the efficiency of BFS, BBFS, Hopfield net spider, NNCIE, and NNCMM for fetching 10000 pages from a pre-setting target domain. The result shows that BFS has the lowest precision and takes the longest time. NNCIE and NNCMM crawled over 1000 web pages more than the Hopfield net spider. The precision of NNCIE and NNCMM is improved by over 10% corresponding to BFS, BBFS, and Hopfield net spider. In the meantime, the time spent by NNCIE and NNCMM has also dropped.

Figures 8.30 and 8.31 show the amount and percentage of relevant pages changing with the number of web pages crawled respectively, which means that the topic-relevant similarity is increased efficiently. They illustrate the effectiveness of proposed algorithms in terms of relevant web pages obtained by BFS, BBFS, Hopfield net spider, NNCIE, and NNCMM. With the proposed algorithms, NNCIE and NNCMM

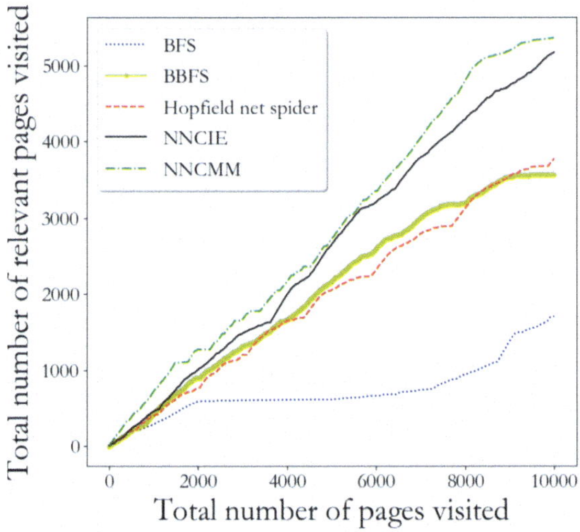

Fig. 8.30 Amount of relevant pages changing with number of pages visited

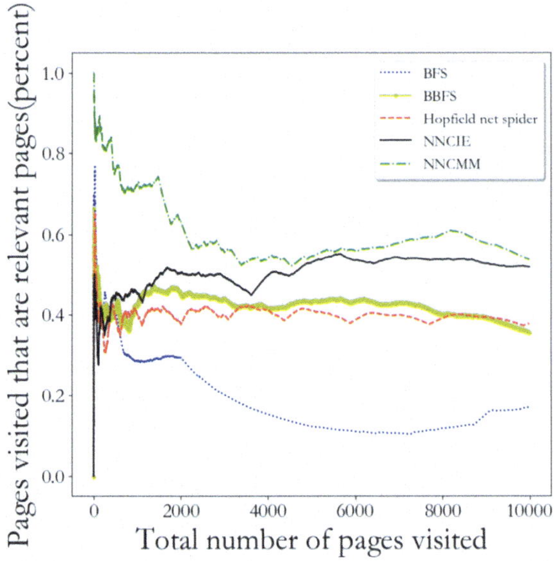

Fig. 8.31 Percentage of relevant pages changing with the crawling pages

can avoid spending too much time crawling unrelevant sites. At the same time, NNCIE and NNCMM can visit more relevant web pages. At the 1% level, NNCIE and NNCMM have significantly better precision than the other three algorithms. In this aspect, the purpose of our improvement has been achieved.

The BFS spider obtains web pages by crawling the URLs close to the starting URLs, which tend to link relevant pages. It does not consider link weights. Therefore, the BFS spider crawls less relevant web pages than the others. The BBFS combines the breadth-first algorithm with the best-first algorithm. It uses link weights to evaluate the relevance of hyperlinks. The Hopfield net spider based on a neural network combines link analysis with content analysis, which uses content analysis to correct the node weights. Therefore, BBFS should be more precise than BFS, and Hopfield net spider performs better than BFS and BBFS. The detailed data analysis also shows that Hopfield net spider crawls more than 606 relevant pages than BBFS, and BBFS crawls more than 1,869 relevant pages than BFS while crawling 10,000 web pages.

Based on the Hopfield net spider, we propose information entropy and cubic nonlinearity memristor model to evaluate the relevance of hyperlinks and adopt monotone-increasing and piecewise-linear memristor model as an activation function to make the node weights of low relevant web pages become smaller, and the node weights of high relevant web pages become bigger. Therefore, NNCIE and NNCMM should be more precise and have a higher precision rate. The detailed experimental data proves that NNCIE and NNCMM have a higher precision rate than the other three algorithms.

In addition, we take into account time costs. Figure 8.32, shows the crawling time cost changing with the number of pages visited. We can see that our crawlers take less time than Hopfield Net Spider, BBFS, and BFS. According to Table III, it shows that the cost of NNCIE and NNCMM is less than the other three algorithms. In fact, the BFS spider spends the longest time, because it needs to visit and analyze many low-quality web pages.

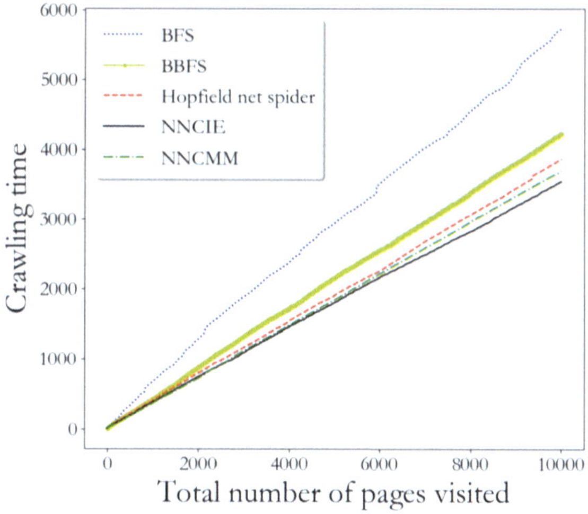

Fig. 8.32 Crawling time cost with the number of pages visited

8.3.5 Conclusion

In this section, a framework of the Hopfield neural network crawler with information entropy and memristor model is presented. We propose a new activation function of the piecewise-linear memristor model and design two novel topic-relevant similarity algorithms with information entropy and cubic nonlinearity memristor model in the framework of neural network. Two novel neural network crawlers are designed for Tibet big data project, which are proven to efficiently increase the topic-relevant similarity and achieve higher precision than other crawlers.

8.4 Comments

This chapter explores the applications of memristive neural networks in various domains, including edge detection, image caption generation, and focused web crawling. By leveraging the unique properties of memristors, these applications achieved significant advancements in their respective fields, surpassing the performance of traditional methods. These diverse applications clearly demonstrate the adaptability and effectiveness of memristive neural networks in addressing a wide range of computational challenges. The research findings highlight the immense potential of memristors in the field of neuromorphic computing. Despite the remarkable progress made, there are still numerous research directions to be explored in the field of memristive neural networks and their applications:

Developing more efficient training algorithms for memristor arrays. Existing works have shown that the scale of memristor crossbar arrays has a significant impact on system performance. Therefore, it is essential to research and develop efficient training algorithms suitable for large-scale memristor arrays. This may involve improving existing techniques such as matrix decomposition, gradient quantization, and sparse coding to fully exploit the parallel computing potential of memristor arrays, enabling fast and accurate weight updates.

Constructing memristor-based heterogeneous computing platforms. Although memristors have advantages in terms of energy efficiency and integration density, they still face challenges in implementing certain computational functions. Therefore, exploring the combination of memristors with traditional devices such as CMOS and FPGA to build heterogeneous computing platforms is a promising research direction. This not only leverages the complementary strengths of different devices to achieve higher performance but also accelerates the industrialization process of memristors.

Continuous learning architectures for online learning. Most of the memristor neural networks discussed in this chapter are based on offline training, with little focus on online continuous learning. However, the dynamic evolution characteristics of memristors are highly similar to the plasticity of biological synapses, making them well-suited for implementing continuous learning systems. Future research can focus on developing memristor-based continuous learning architectures that enable systems to continuously optimize and expand based on new input data streams without requiring offline retraining. This is of great significance for the environmental adaptation and behavioral optimization of autonomous intelligent agents.

Exploring the biological mechanisms of memristor neural networks. Understanding the working mechanisms and dynamical behaviors of memristor neural networks from a biological perspective is crucial for revealing the intrinsic mechanisms of memory and learning. By establishing the connection between memristor synaptic plasticity and biological synaptic plasticity, we can deepen our understanding of biological neural circuits. Moreover, drawing inspiration from the structural and functional organization principles of biological neural systems can provide important insights for building more efficient and robust memristor neural networks. The research works covered in this chapter demonstrate the tremendous potential of memristors in realizing efficient intelligent computing. Continued in-depth research in theoretical foundations, methodological innovations, and architectural designs are expected to push artificial intelligence to new heights.

References

1. Mazumder, P., Li, S.R., Ebong, I.E.: Tunneling-based cellular nonlinear network architectures for image processing. IEEE Trans. Very Large Scale Integ. (VLSI) Syst. **17**(4), 487–495 (2009). DOI https://doi.org/10.1109/TVLSI.2009.2014771
2. Lee, W.H., Mazumder, P.: Motion detection by quantum-dots-based velocity-tuned filter. IEEE Trans. Nanotechnol. **7**(3), 355–362 (2008). https://doi.org/10.1109/TNANO.2007.915019
3. Cruz, J., Chua, L.: A 16× 16 cellular neural network universal chip: the first complete single-chip dynamic computer array with distributed memory and with gray-scale input-output. Analog Integr. Circ. Sig. Process **15**, 227–237 (1998)
4. Domınguez-Castro, R., Espejo, S., Rodrıguez-Vázquez, A., Carmona, A., Foldesy, P., Zarándy, A., Szolgay, P., Szirányi, T., Roska, T.: A 0.8 mm cmos two-dimensional programmable mixed-signal focal-plane array processor with on-chip binary imaging and instructions storage. IEEE J. Solid-State Circ. **32**, 1013–1026 (1997)
5. Chua, L., Roska, T.: The cnn paradigm. IEEE Trans. Circuits Syst. I: Fundam. Theory Appl. **40**(3), 147–156 (1993). https://doi.org/10.1109/81.222795
6. Duan, S., Hu, X., Dong, Z., Wang, L., Mazumder, P.: Memristor-based cellular nonlinear/neural network: design, analysis, and applications. IEEE Trans. Neural Netw. Learn. Syst. **26**(6), 1202–1213 (2015). https://doi.org/10.1109/TNNLS.2014.2334701
7. Sziranyi, T., Zerubia, J., Geldreich, D., Kato, Z.: Cellular neural network for markov random field image segmentation. In: 1996 Fourth IEEE International Workshop on Cellular Neural Networks and their Applications Proceedings (CNNA-96), pp. 139–144 (1996). https://doi.org/10.1109/CNNA.1996.566509
8. Ekstrom, M.: Digital image processing. IEEE Trans. Acoust. Speech Signal Process. **28**(4), 484–486 (1980). https://doi.org/10.1109/TASSP.1980.1163437
9. Mokhtar, S.M.A.B., Abdullah, W.F.H.: Re-model fabricated memristor behavior in lt-spice and applied in logic circuit. In: 2014 IEEE Symposium on Computer Applications and Industrial Electronics (ISCAIE), pp. 106–110 (2014). https://doi.org/10.1109/ISCAIE.2014.7010219
10. Chua, L.: Memristor-the missing circuit element. IEEE Trans. Circuit Theory **18**(5), 507–519 (1971). https://doi.org/10.1109/TCT.1971.1083337
11. Strukov, D.B., Snider, G.S., Stewart, D.R., Williams, R.S.: The missing memristor found. Nature **453**(7191), 80–83 (2008)

12. Joglekar, Y.N., Wolf, S.J.: The elusive memristor: properties of basic electrical circuits. Eur. J. Phys. **30**(4), 661 (2009)
13. Chua, L., Yang, L.: Cellular neural networks: theory. IEEE Trans. Circuits Syst. **35**(10), 1257–1272 (1988). https://doi.org/10.1109/31.7600
14. Strukov, D.B., Likharev, K.K.: Cmol fpga: a reconfigurable architecture for hybrid digital circuits with two-terminal nanodevices. Nanotechnology **16**(6), 888 (2005)
15. Yakopcic, C., Hasan, R., Taha, T., McLean, M., Palmer, D.: Memristor-based neuron circuit and method for applying learning algorithm in spice. Electron. Lett. **50**(7), 492–494 (2014)
16. Fei, H., Zhiqiang, Y., Peng, L., et al.: Circuit design of convolutional neural network based on memristor crossbar arrays. J. Comput. Res. Dev. **55**(5), 1097–1107 (2018)
17. Hu, X., Feng, G., Duan, S., Liu, L.: A memristive multilayer cellular neural network with applications to image processing. IEEE Trans. Neural Netw. Learn. Syst. **28**(8), 1889–1901 (2017). https://doi.org/10.1109/TNNLS.2016.2552640
18. Yao, T., Pan, Y., Li, Y., Mei, T.: Exploring visual relationship for image captioning. In: Proceedings of the European Conference on Computer Vision (ECCV), pp. 684–699 (2018)
19. Vinyals, O., Toshev, A., Bengio, S., Erhan, D.: Show and tell: a neural image caption generator. In: Proceedings of the IEEE Conference on Computer Vision and Pattern Recognition, pp. 3156–3164 (2015)
20. Kiros, R., Salakhutdinov, R., Zemel, R.: Multimodal neural language models. In: International conference on machine learning, pp. 595–603. PMLR (2014)
21. Vendrov, I., Kiros, R., Fidler, S., Urtasun, R.: Order-embeddings of images and language. In: Y. Bengio, Y. LeCun (eds.) 4th International Conference on Learning Representations, ICLR 2016, San Juan, Puerto Rico, May 2-4, 2016, Conference Track Proceedings (2016). http://arxiv.org/abs/1511.06361
22. Yu, N., Hu, X., Song, B., Yang, J., Zhang, J.: Topic-oriented image captioning based on order-embedding. IEEE Trans. Image Process. **28**(6), 2743–2754 (2018)
23. Yakopcic, C., Alom, M.Z., Taha, T.M.: Memristor crossbar deep network implementation based on a convolutional neural network. In: 2016 International joint conference on neural networks (IJCNN), pp. 963–970. IEEE (2016)
24. Li, C., Wang, Z., Rao, M., Belkin, D., Song, W., Jiang, H., Yan, P., Li, Y., Lin, P., Hu, M., et al.: Long short-term memory networks in memristor crossbar arrays. Nat. Mach. Intell. **1**(1), 49–57 (2019)
25. Cai, F., Correll, J.M., Lee, S.H., Lim, Y., Bothra, V., Zhang, Z., Flynn, M.P., Lu, W.D.: A fully integrated reprogrammable memristor-cmos system for efficient multiply-accumulate operations. Nat. Electronics **2**(7), 290–299 (2019)
26. Zaplatilek, K.: Memristor modeling in matlab® & simulink®. In: Proceedings of the 5th European Conference on European Computing Conference, pp. 62–67 (2011)
27. Hung, C.C., Ismail, M., Halonen, K., Porra, V.: A low-voltage rail-to-rail cmos vi converter. IEEE Trans. Circuits Syst. II: Analog Digital Signal Proc. **46**(6), 816–820 (1999)
28. Cireşan, D., Meier, U.: Multi-column deep neural networks for offline handwritten chinese character classification. In: 2015 international joint conference on neural networks (IJCNN), pp. 1–6. IEEE (2015)
29. Ciregan, D., Meier, U., Schmidhuber, J.: Multi-column deep neural networks for image classification. In: 2012 IEEE Conference on Computer Vision and Pattern Recognition, pp. 3642–3649. IEEE (2012)
30. Zeiler, M.D., Fergus, R.: Visualizing and understanding convolutional networks. In: Computer Vision–ECCV 2014: 13th European Conference, Zurich, Switzerland, September 6-12, 2014, Proceedings, Part I 13, pp. 818–833. Springer (2014)
31. Simonyan, K., Zisserman, A.: Very deep convolutional networks for large-scale image recognition (2014). arXiv:1409.1556
32. Mikolov, T., Karafiát, M., Burget, L., Cernockỳ, J., Khudanpur, S.: Recurrent neural network based language model. In: Interspeech, vol. 2, pp. 1045–1048. Makuhari (2010)
33. Hochreiter, S., Schmidhuber, J.: Long short-term memory. Neural Comput. **9**(8), 1735–1780 (1997)

34. Schmidhuber, J., Wierstra, D., Gomez, F.J.: Evolino: Hybrid neuroevolution/optimal linear search for sequence prediction. In: Proceedings of the 19th International Joint Conferenceon Artificial Intelligence (IJCAI) (2005)
35. Bao, W., Yue, J., Rao, Y.: A deep learning framework for financial time series using stacked autoencoders and long-short term memory. PLoS ONE **12**(7), e0180944 (2017)
36. Jia, R., Liang, P.: Data recombination for neural semantic parsing (2016). arXiv:1606.03622
37. Wu, Y., Schuster, M., Chen, Z., Le, Q.V., Norouzi, M., Macherey, W., Krikun, M., Cao, Y., Gao, Q., Macherey, K., et al.: Google's neural machine translation system: Bridging the gap between human and machine translation (2016). arXiv:1609.08144
38. Xiong, W., Wu, L., Alleva, F., Droppo, J., Huang, X., Stolcke, A.: The microsoft 2017 conversational speech recognition system. In: 2018 IEEE International Conference on Acoustics, Speech and Signal Processing (ICASSP), pp. 5934–5938. IEEE (2018)
39. Kingma, D.P., Ba, J.: Adam: A method for stochastic optimization (2014). arXiv:1412.6980
40. Lin, T.Y., Maire, M., Belongie, S., Hays, J., Perona, P., Ramanan, D., Dollár, P., Zitnick, C.L.: Microsoft coco: Common objects in context. In: Computer Vision–ECCV 2014: 13th European Conference, Zurich, Switzerland, September 6-12, 2014, Proceedings, Part V 13, pp. 740–755. Springer (2014)
41. Yakopcic, C., Hasan, R., Taha, T.M.: Memristor based neuromorphic circuit for ex-situ training of multi-layer neural network algorithms. In: 2015 International Joint Conference on Neural Networks (IJCNN), pp. 1–7. IEEE (2015)
42. Alibart, F., Gao, L., Hoskins, B.D., Strukov, D.B.: High precision tuning of state for memristive devices by adaptable variation-tolerant algorithm. Nanotechnology **23**(7), 075201 (2012)
43. Oblea, A.S., Timilsina, A., Moore, D., Campbell, K.A.: Silver chalcogenide based memristor devices. In: The 2010 International Joint Conference on Neural Networks (IJCNN), pp. 1–3. IEEE (2010)
44. Deng, J., Dong, W., Socher, R., Li, L.J., Li, K., Fei-Fei, L.: Imagenet: A large-scale hierarchical image database. In: 2009 IEEE Conference on Computer Vision and Pattern Recognition, pp. 248–255. Ieee (2009)
45. Papineni, K., Roukos, S., Ward, T., Zhu, W.J.: Bleu: a method for automatic evaluation of machine translation. In: Proceedings of the 40th annual meeting of the Association for Computational Linguistics, pp. 311–318 (2002)
46. Anderson, P., Fernando, B., Johnson, M., Gould, S.: Spice: Semantic propositional image caption evaluation. In: Computer Vision–ECCV 2016: 14th European Conference, Amsterdam, The Netherlands, October 11-14, 2016, Proceedings, Part V 14, pp. 382–398. Springer (2016)
47. Denkowski, M., Lavie, A.: Meteor universal: Language specific translation evaluation for any target language. In: Proceedings of the Ninth Workshop on Statistical Machine Translation, pp. 376–380 (2014)
48. Xu, K., Ba, J., Kiros, R., Cho, K., Courville, A., Salakhudinov, R., Zemel, R., Bengio, Y.: Show, attend and tell: Neural image caption generation with visual attention. In: International Conference on Machine Learning, pp. 2048–2057. PMLR (2015)
49. Anderson, P., He, X., Buehler, C., Teney, D., Johnson, M., Gould, S., Zhang, L.: Bottom-up and top-down attention for image captioning and visual question answering. In: Proceedings of the IEEE Conference on Computer Vision and Pattern Recognition, pp. 6077–6086 (2018)
50. Curran, K., Mc Glinchey, J.: Vertical search engines. ITB J **8**(2), 4 (2007)
51. Jin, Y., Ishizuka, M., Matsuo, Y.: Extracting inter-firm networks from the world wide web using a general-purpose search engine. Online Inf. Rev. **32**(2), 196–210 (2008)
52. Yu, Y.B., Huang, S.L., Tashi, N., Zhang, H., Lei, F., Wu, L.Y.: A survey about algorithms utilized by focused web crawler. J. Electron. Sci. Technol. **16**(2), 129–138 (2018)
53. Chau, M., Chen, H.: Comparison of three vertical search spiders. Computer **36**(5), 56–62 (2003). https://doi.org/10.1109/MC.2003.1198237
54. Haveliwala, T., et al.: Efficient computation of pagerank. Technical Report, Stanford University (1999)
55. Hersovici, M., Jacovi, M., Maarek, Y.S., Pelleg, D., Shtalhaim, M., Ur, S.: The shark-search algorithm. An application: tailored web site mapping. Comput. Netw. ISDN Syst. **30**(1–7), 317–326 (1998)

56. Khare, A., Dalvi, A., Kazi, F.: Smart crawler for harvesting deep web with multi-classification. In: 2020 11th International Conference on Computing, Communication and Networking Technologies (ICCCNT), pp. 1–5 (2020). https://doi.org/10.1109/ICCCNT49239.2020.9225369
57. Pal, A., Tomar, D.S., Shrivastava, S.: Effective focused crawling based on content and link structure analysis (2009). arXiv:0906.5034
58. Chau, M., Chen, H.: Personalized and focused web spiders. In: Web Intelligence, pp. 197–217. Springer (2003)
59. Adhikari, S.P., Yang, C., Kim, H., Chua, L.O.: Memristor bridge synapse-based neural network and its learning. IEEE Trans. Neural Netw. Learn. Syst. **23**(9), 1426–1435 (2012). https://doi.org/10.1109/TNNLS.2012.2204770
60. Meng, Z., Xiang, Z.: Stability analysis of stochastic memristor-based recurrent neural networks with mixed time-varying delays. Neural Comput. Appl. **28**, 1787–1799 (2017)
61. Duan, S., Dong, Z., Hu, X., Wang, L., Li, H.: Small-world hopfield neural networks with weight salience priority and memristor synapses for digit recognition. Neural Comput. Appl. **27**, 837–844 (2016)
62. Li, B., Wang, Y., Wang, Y., Chen, Y., Yang, H.: Training itself: Mixed-signal training acceleration for memristor-based neural network. In: 2014 19th Asia and South Pacific Design Automation Conference (ASP-DAC), pp. 361–366 (2014). https://doi.org/10.1109/ASPDAC.2014.6742916
63. Hu, M., Chen, Y., Yang, J.J., Wang, Y., Li, H.H.: A compact memristor-based dynamic synapse for spiking neural networks. IEEE Trans. Comput. Aided Des. Integr. Circuits Syst. **36**(8), 1353–1366 (2017). https://doi.org/10.1109/TCAD.2016.2618866
64. Yu, Y., Wang, X., Zhong, S., Yang, N., Tashi, N.: Extended robust exponential stability of fuzzy switched memristive inertial neural networks with time-varying delays on mode-dependent destabilizing impulsive control protocol. IEEE Trans. Neural Netw. Learn. Syst. **32**(1), 308–321 (2021). https://doi.org/10.1109/TNNLS.2020.2978542
65. Yongbin, Y., Nijing, Y., Chenyu, Y., Tashi, N.: Memristor bridge-based low pass filter for image processing. J. Syst. Eng. Electron. **30**(3), 448–455 (2019). https://doi.org/10.21629/JSEE.2019.03.02
66. Muthuswamy, B.: Implementing memristor based chaotic circuits. Int. J. Bifurc. Chaos **20**(05), 1335–1350 (2010)
67. Itoh, M., Chua, L.O.: Memristor oscillators. Int. J. Bifurc. Chaos **18**(11), 3183–3206 (2008)

The manufacturer's authorised representative in the EU is Springer Nature Customer Service Centre GmbH, Europaplatz 3, 69115 Heidelberg, Germany. If you have any concerns regarding our products, please contact ProductSafety@springernature.com

Printed and bound by CPI Group (UK) Ltd, Croydon, CR0 4YY
01/10/2026
02225937-0014